TIME, LIFE, AND MAN

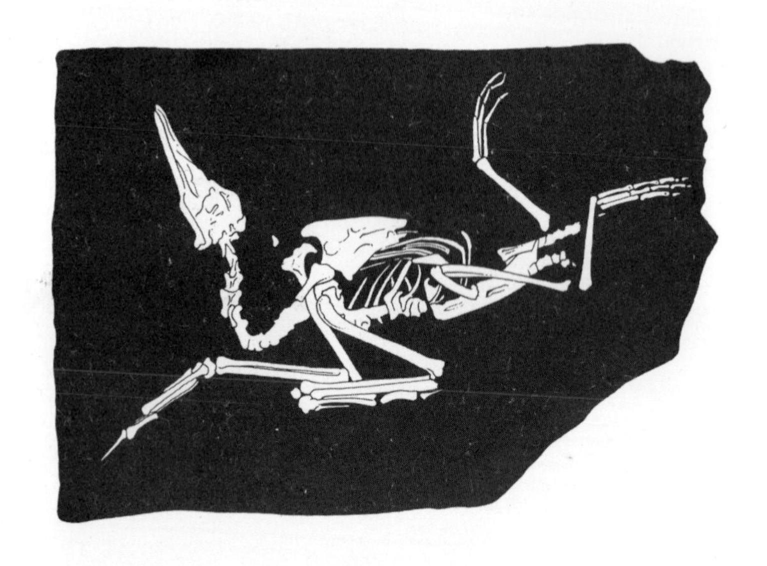

NEW YORK · JOHN WILEY & SONS, INC.

London · Chapman & Hall, Limited

TIME, LIFE, AND MAN

THE FOSSIL RECORD

R. A. STIRTON

Professor of Paleontology

Director, Museum of Paleontology

University of California

Berkeley

Library of Congress Catalog Card Number: 59-5876

Printed in the United States of America

for Lillian

Preface

THIS BOOK is designed primarily as a text for a course in introductory paleontology, though most of it is intended also for the interest of general readers. It is based on the subject matter, covering both plants and animals, which I have presented in Paleontology I at the University of California for a number of years. It is written to arrest the attention of students with little or no training in the biological and the earth sciences, yet for more advanced students brings into focus the over-all picture of paleontology.

The first five chapters outline the objectives and general principles to which attention is directed throughout the book. There is also a chapter on the history of paleontology which traces the changes in concepts and the evolution of thought on the subject as more fossil evidence was uncovered and as society gradually became more tolerant of studies in the antiquity and evolution of life. Throughout the book attention is directed to individuals who have contributed to our terminologies, to our knowledge of fossils, and to the sequence of life.

A simplified classification is presented to give an idea of the plants and animals, both living and extinct, which are included under each hierarchy, their known geologic ranges, and some of the more important characters used in outlining the scope of the phyla, classes, orders, etc. Scientific and common names, in their various usages for fossils, are utilized to acquaint the student with the terms he may encounter in his outside or assigned readings.

Most of the book deals with a chronologic presentation of the sequence of life from the Pre-Cambrian to the Pleistocene. In many respects this is much like historical geology, but the fossil record is emphasized more than the physical events in earth's history.

The correlations, including allocation of boundaries between Eras,

Periods, and Epochs, are those that seem to have been most widely adopted by paleontologists and stratigraphers today and do not necessarily represent my own views. Sections in the bibliography which refer to the more recent publications on charts and citations to bibliographies, particularly in North America, are included.

The last nine chapters are based on selected lectures. Many others could have been chosen, but space did not permit. In these lectures the student may realize the application of the knowledge he has acquired in the preceding chapters. There is some repetition. This, however, is not only unavoidable, it is also one of the most effective methods in teaching. Some chapters are more technical than others, but it is felt that some of the more technical aspects should be introduced in a beginning course.

I am most grateful to a number of my colleagues at the University of California for their helpful suggestions, proofreading, and criticisms. These are Z. M. Arnold, D. I. Axelrod, C. L. Camp, W. A. Clemens, G. H. Curtis, J. W. Durham, W. L. Fry, G. T. James, F. H. Kilmer, R. M. Kleinpell, J. M. Langenheim, R. L. Langenheim, Jr., E. G. Linsley, M. C. McKenna, L. H. Miller, F. E. Peabody, J. H. Peck, Jr., D. E. Savage, R. H. Tedford, R. L. Usinger, D. E. Weaver, S. P. Welles, and G. D. Woodard. Professor Kleinpell not only read and criticized the chapter *Foraminifera and Oil* but contributed many facts not otherwise available to me. Donald E. Weaver kindly made available his observations on the type of the Miocene. The fossils for Figures 92, 99, 103, 107, 113, 116, and 124 were selected by R. L. Langenheim, Jr., those for Figures 132, 142, and 146 by J. H. Peck, Jr., and those for Figures 158 and 159 by J. W. Durham.

Outside the University most useful comments and suggestions have been received from Sir Douglas Mawson, University of Adelaide; Professor Bernhard Peyer, University of Zurich; Professor G. H. R. von Koenigswald, University of Utrecht; Dr. H. B. S. Cooke, University of the Witwatersrand; Dr. Brian Daily, South Australian Museum; Professor Samuel P. Ellison, Jr., University of Texas; Dr. Wann Langston, Jr., National Museum of Canada; and Dr. Theodore Downs, Los Angeles County Museum.

All but twenty-one of the illustrations were prepared by Owen J. Poe, staff artist, Museum of Paleontology, University of California, and Figures 89, 91, 186, 206, and 208 were drawn by H. E. Hamman. I, myself, prepared the sketches for Figures 15, 19, 21, 22, 23, 24, 49, 68, 69, 83, 84, 85, 90, 102, 173, and 184. The manuscript was typed by Lillian Stirton.

R. A. Stirton

Berkeley, California
November 1958

Contents

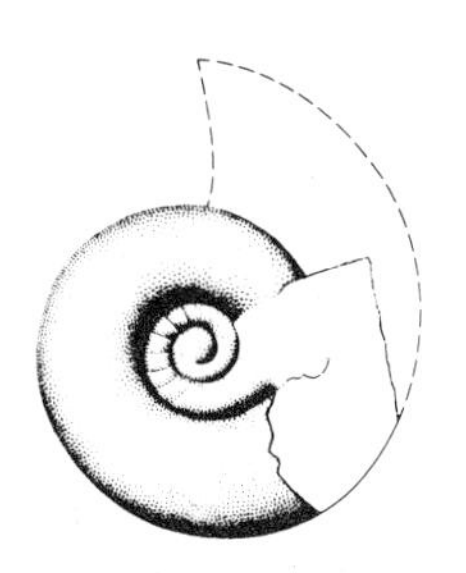

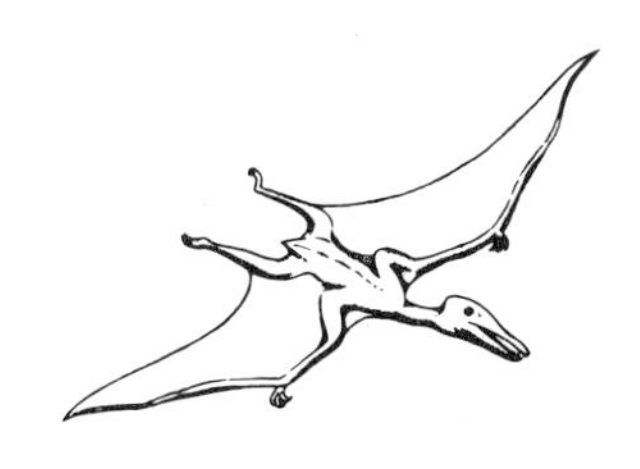

1 Paleontology: Objectives, Definition, and Scope

PALEONTOLOGY IS THE STUDY of prehistoric life. Evidence of such life is preserved in rocks of the earth's crust in the form of petrified remains and other indications of animals or plants. Since ancient times man has attempted to learn all he can about it. He has been greatly concerned with the origin and succession of life, and justifiably so, since the patterns of evolution and extinction displayed by fossils will, in principle, undoubtedly be followed in the future; but in man they may be guided by his knowledge, purpose, and ethics.

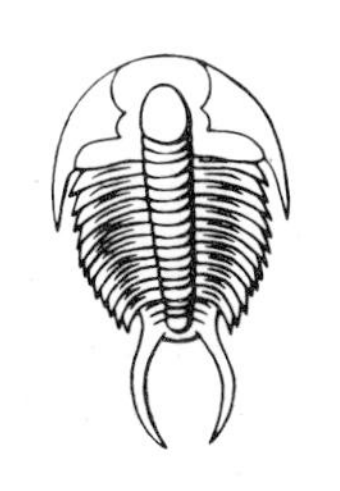

Application

A question frequently asked by the uninitiated is, "why study paleontology"? The answer is a simple one. Man is an inquisitive creature and has probed into everything imaginable to satisfy his curiosity. The satisfaction he gets from increasing his knowledge has

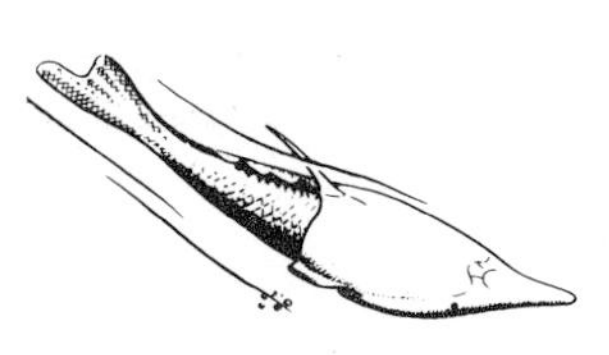

carried him to the peak in the domination of all other forms of life. Intellectual satisfaction is an educational result and is not necessarily remunerative, though frequently it is both.

Education and Relaxation. A paleontologist finds as much pleasure and relaxation in tramping over hills and through badlands in search of evidence of some rare creature that no man has ever seen as a fisherman experiences from landing the largest trout in a stream or a businessman derives from the biggest deal of his career. Furthermore, unlike the unfortunate trout, the fossil is not consumed, nor is it stashed away in a private bank account, but it is made available to society through the media of scientific descriptions and popular discourse.

Public interest in paleontology is well demonstrated by the thousands of people who visit museums throughout the world to see something of the famous dinosaurs, the three-toed horse, or some other form of life entirely unknown to them. Others come to educational institutions with fossils they have found, possibly in or near their own back yards, to learn what nature of beast may have existed there in past ages. Radio and television programs as well as popular accounts in magazines have brought new information to millions. Many of these people — particularly the youngsters — wish to see the fossil evidence for themselves.

The greatest value of paleontology, then, is in the personal satisfaction and pleasure that it affords the individual. Many youth organizations, such as the Boy Scouts, go on fossil-hunting trips in the healthy atmosphere of the wide open spaces, ever conscious of the possibility that one of the group may make some amazing discovery. Here, then, is a way of directing the attention of our youth to beneficial and pleasurable activities. Most of the renowned paleontologists became interested in the subject in their early teens.

No earth science course is complete without some instruction in the history of life on earth. The demand is constantly increasing for people with training at almost every level of instruction in the earth sciences and in the biological sciences. One of our greatest problems today is that of supplying our schools with inspired teachers.

Economic Application. Economic application of paleontology has increased tremendously during the past 25 years; this is especially true in the petroleum industry. It has been estimated that more than 75 per cent of the world's reservoir of petroleum has not been located. Search for gas and oil is being conducted on a greater scale than ever before, and this activity should be increased during the next 30 years, if not longer. These products grease and propel the wheels of industry, and their importance in everyday life should be realized by everyone.

Paleontologists and geologists, with their knowledge of the life of the past and their information on the structure of the earth's crust, hold the keys that unlock the vaults of these vast sources of lubricants and energy. All kinds of information derived from basic research, though seemingly of no practical importance at the moment, may become tremendously significant in the future. The unknown is not available to us until we find it. In the fields of foraminiferology, invertebrate paleontology, and stratigraphy alone there has been a demand for at least five times our present output in trained personnel for industry and for teaching.

Knowledge of fossils as a guide to the sequence of rocks in the earth's crust has been used in locating gold and other ore deposits. For example, gold occurs in auriferous conglomerates at the base of the Cambrian over vast areas in Australia. Proximity to these gold-bearing formations can be determined by the age of fossils in the overlying sedimentary rocks. Coal beds have been located by similar methods in the United States and elsewhere.

When polished, many limestones and fine-grained sandstones with well-preserved fossils make some of the most beautiful interior wall surfaces. An example is the Salem oölitic limestones of the Mississippian Period near Bedford, Indiana. There the limestone is made up primarily of small fossil shells and rounded, calcareous oölites. This stone has been quarried extensively for building purposes. Elsewhere limestones with chain corals and other fossils have been similarly used. Everyone who has visited the Grand Canyon has seen the beautiful polished table tops made of the silicified logs from the Petrified Forest of Arizona. Recently it has been found that some rich concentrations of uranium occur in fossil wood. Even dinosaur bones and bones of fossil mammals contain uranium.

So we see that there are many reasons why we should study paleontology. To those who understand it, paleontology is an interesting and fascinating subject.

Objectives

In dealing with fossil remains, once he has the specimens at hand, a paleontologist may have in mind any combination of six objectives: *identification, form and function, associations of plants and animals, evolution in the different groups of organisms, dispersal and distribution of plants and animals through time and space, and correlation.*

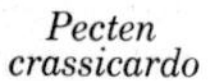

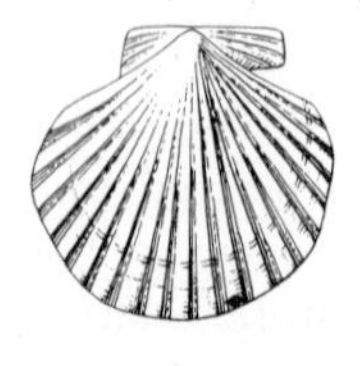

Pecten crassicardo

Pecten raymondi

Fig. 1. Identifiable features in two species of Tertiary pectens.

Identification. Foremost of these objectives is identification (Fig. 1). Any investigator must know as precisely as possible what he is dealing with; inaccurate information is worse than no information, because it is likely to lead to erroneous conclusions. Frequently the fossil is too incomplete or the group to which it belongs is not known well enough for species identification. Nevertheless, the class, order, and family can usually be recognized by anyone conversant with his field of study. Careful comparison must then be made with the known genera and species. If the species can be recognized, the specimen or specimens will be much more useful than if only their generic status can be established. The student must not be influenced by preconceived ideas of what the fossil may be. It is important to know whether a fossil shell in the Los Angeles marine embayment in the late Miocene is the same species as one in rocks of the same age in the San Joaquin Valley area. We must not assume that they are the same because the formations in which they occur are the same age. On the other hand, an early Pliocene horse from Nevada is not necessarily different from another in the Great Plains region of Nebraska and South Dakota because they were widely separated geographically. The identification must be correct if the fossil is to serve a useful purpose.

Form and Function. By studying the structure of a fossil, and drawing a parallel with living organisms, the habits of extinct animals or plants can be realized. The form of the fossil, then, offers clues to how it functioned in the environment in which it lived.

A good example is the skeleton of the saber-toothed cat, *Smilodon californicus*. A comparison with the mountain lion, *Felis concolor*, reveals a marked difference in these big cats. We know the mountain lion is a fast runner and that it springs upon deer and crushes their necks with its powerful jaws. The limb bones in the saber-toothed cat are not so slender as those of the mountain lion, and the skeleton as a whole is stockier. The cranium and mandibles (Fig. 2) show even more conspicuous differences. Long saberlike canines in the upper jaw of the saber-

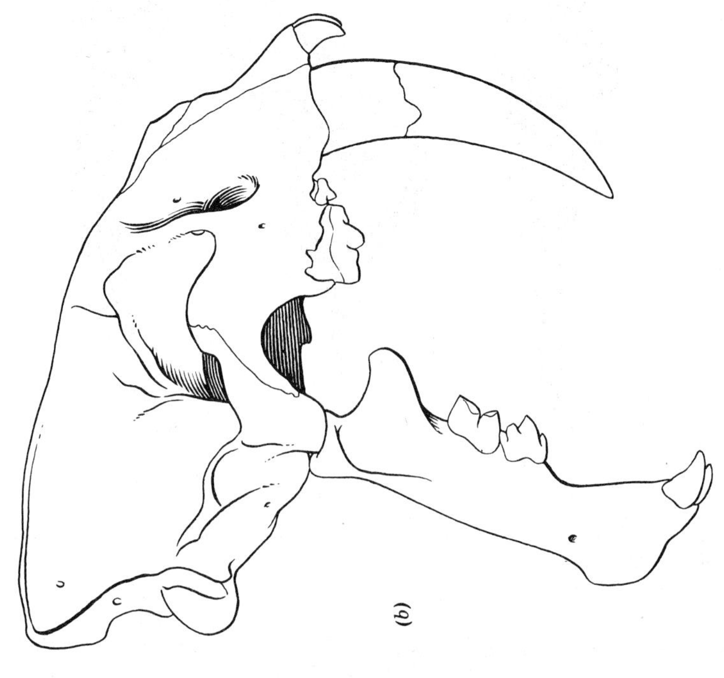

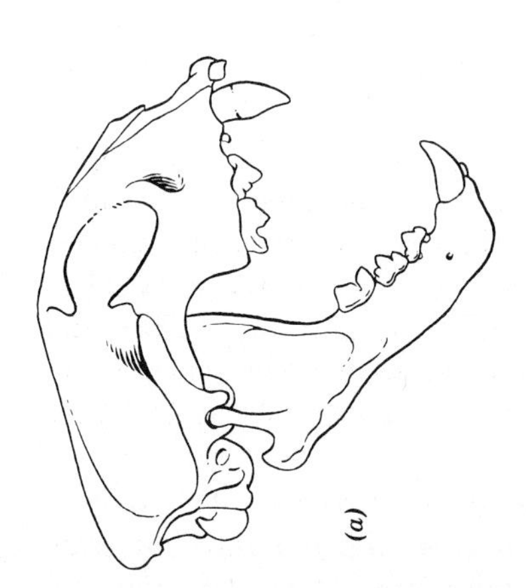

Fig. 2. Skulls of (a) *mountain lion,* Felis concolor, *and* (b) *saber-toothed cat,* Smilodon californicus, $\times \frac{1}{4}$.

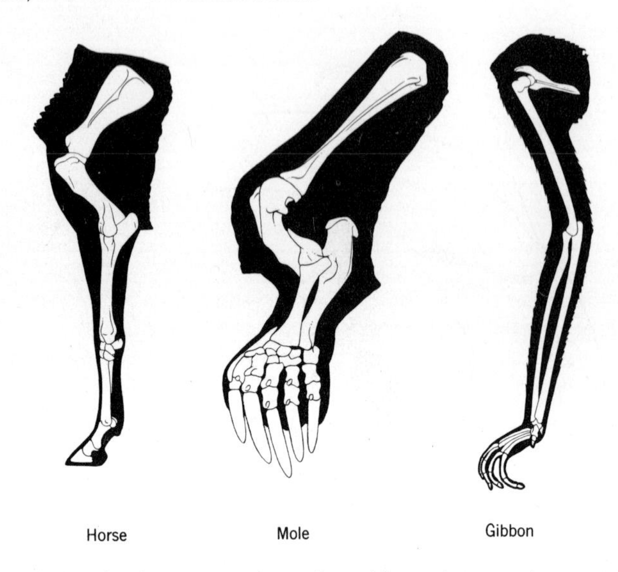

Fig. 3. Forelimbs of a horse, a mole, and a gibbon, showing their specializations. Not drawn to scale.

toothed cat are remarkably different from those of the mountain lion; also, its wide, gaping lower jaw, because of its construction, could open to a position at right angles to the upper row of teeth. These and other differences lead us to infer that *Smilodon californicus* was a stabbing cat that fed upon slower-moving animals or those that were otherwise handicapped. Certainly its saberlike canines made it a more formidable adversary than any of the true cats, such as the mountain lion.

The structure and texture of fossil leaves, when compared with the leaves on trees growing in subtropical areas of heavy rainfall and with those in temperate climates of moderate rainfall, offer excellent indications of climates of the past. In the Eocene of Oregon the predominantly broad, thick-textured leaves indicate an area of relatively heavy rainfall in which the trees were green throughout the year, and freezing temperatures seldom if ever prevailed. In Miocene rocks of the same area, formed many millions of years later in geologic time, the smaller and finer textured leaves give evidence of alterations in their functions to meet the requirements of a colder and dryer climate. Today that area is almost a desert.

Parallels of form and function can be drawn between the forelimbs of the horse, the mole, and the gibbon (Fig. 3), and those of fossil limbs that show similar adaptive modifications. An adaptation for rapid running over open grassland is seen in the horse. In the foot all of the side toes are lost, with only the central digit remaining. Though the limb bones are angulate fore and aft, they are not well constructed for lateral movement. The gibbon, on the other hand, has lost no digits, and the entire limb is adapted for rotation and movement in any direction. This permits him to swing through the trees, reach high above his head to pick fruit, and also to manipulate his hands for many purposes. The forelimb of the mole is greatly shortened, and the broad front foot is modified for pushing dirt aside in the construction of his underground tunnels. On the surface of the ground he can make little headway in trying to escape, but in his subterranean runways he moves quickly.

These examples show that certain functions can be attributed to the parts we see preserved in fossils, whether they are conspicuous or subdued.

Associations of Plants and Animals. Assemblages of plants and animals and the nature of the sediments in which they are entombed offer evidence of local environmental conditions of the past. It must be realized that fossils are not dead objects in the mind's eye of a paleontologist because he visualizes them as they existed in their environments.

The famous Black Hawk Ranch fossil quarry on the lower slopes of Mount Diablo in Contra Costa County, California, and the surrounding area has offered insight into an environment that existed some 10 million years ago in the early Pliocene. The Sierra Nevada was a relatively low mountain chain at that time, and the San Francisco Bay area, including the present site of Mount Diablo, was a low, alluvial plain with numerous shallow lakes fed by meandering streams from the adjacent hills. There must have been wide stretches of grassland dotted here and there with clumps of brush or groves of trees. Willows, sycamores, poplars, and elms outlined the stream courses, and willows extended into the marshy flats with the heavier grasses and reeds. On higher ground and in the low hills was the woodland-chaparral association, with its oaks, sumac, mountain mahogany, and probably many other plants not yet found as fossils. The landscape varied, but there was not the diversity over the same area that there is today.

Rabbits, *Hypolagus*, about the size of a cottontail must have ventured from sheltered brush patches during the cool of the day, always ready to scurry to cover if the ever-watchful fox, *Vulpes vafer*, were to make his appearance. A large ground squirrel, referable to the liv-

Fig. 4. A restoration of an early Pliocene environment and landscape in the San Joaquin Valley of middle California. Horse, Pliohippus, *and mastodont,* Gomphotherium.

ing genus, *Citellus*, was also present but probably inhabited the dryer and higher ground away from the stream. He, too, had numerous enemies to keep his numbers reduced. These included a small gray fox, *Urocyon*, a ringtailed cat, *Bassariscus parvus*, and a large mustelid, probably with habits much like the wolverine. There was a raccoonlike animal probably distantly related to the coatis, *Nasua narica,* now living in Central and South America. Lizards basked in the warm sunshine but moved quickly from sight when danger seemed to threaten.

Elephantlike creatures known as mastodonts, *Gomphotherium simpsoni,* ambled about in herds much as modern elephants (Fig. 4). They differed from the elephant in that their heads were flatter and their trunks were probably shorter; in addition they had two pairs of tusks and molars 6 inches long with a thick coat of enamel. Unless they were trapped in a mudhole near a stream, where they loved to bathe, these four-tuskers probably were not disturbed by the larger carnivores. Many mastodonts must have been stuck in quagmires along the old stream courses, much as cows and horses bog down today. Such death traps were common, and this seems to be the most likely manner in which so many of these huge pachyderms met their death. Once they sank in to their knees they were never able to extricate themselves and within a few days would die of starvation and fatigue. Some of their bones remained in the mud and others deteriorated, but some got into the stream channels and in times of heavy rainfall washed downstream where they sank into deeper holes or lodged and were buried with the remains of other animals and plants.

Horses were abundant, particularly a three-toed species known as *Hipparion forcei.* These little horses must have ranged over the grassland in great herds where they were associated with camels and small hornless antilocaprids, *Capromeryx. Hipparion forcei* was heavier but not taller than the mule deer of the Sierras. On each side of his foot were side toes with small hoofs much like the dew claws of the cow and the pig. A slightly larger horse, *Pliohippus leardi,* was less numerous; it was much more like the modern horse, as it too had lost its side toes.

At least three kinds of camels inhabited the San Francisco Bay area during the early Pliocene. One, *Paracamelus,* was larger than the camels living today in Asia and Africa. Others were smaller and less numerous, but the genera and species cannot be identified from the fragmentary fossils available at this time.

The little hornless antilocaprid, *Capromeryx,* was not as tall as a sheep and was more delicately constructed, yet it must have been as fast as a greyhound. These little ruminants came to the water holes in droves, for the quarry is replete with their jaws, teeth, and bones.

The peccaries, *Prosthennops,* probably restricted their activities to heavily wooded areas near streams where they fed on acorns and other fruits much like their living relatives in tropical America. It was not a simple matter for a pair of carnivores to move in on a herd of peccaries and effect a kill.

There were large cats, *Pseudaelurus thinobates,* lurking in the brush near the water hole ready to pounce upon the unsuspecting herbivores. These big felines, related to and much like the mountain lion, probably stalked the horses and camels, springing on their prey and crushing the neck vertebrae with their powerful jaws. Some saber-toothed cats were present, but little is known about them.

The marshland around the lakes and streams, with its reeds and tallgrass, was thickly populated with small beaver, *Eucastor lecontei.* These rodents were the size of muskrats and probably had habits much like other marsh-dwelling rodents. At times they must have died in large numbers, for their teeth and jaws are abundant in the quarry. Additional evidence of a marshland was the presence of a crane, *Grus conferta.* There were other bird bones too fragmentary for identification.

Perhaps the most peculiar animals in the fauna were the hyenoid dogs. Though related to true dogs, they were adapted as scavengers, much like the modern hyenas. There were two kinds, *Aelurodon aphobus,* the larger and rarer one, and *Osteoborus diabloensis,* the smaller and commoner species; in some respects these carrion eaters probably behaved like the condor and the vulture of today. *Osteo-*

borus, possibly due to its greater numbers and smaller size, was more likely to find dead animals, but when *Aelurodon* appeared on the scene the smaller dogs retreated to a respectful distance until the king had had his fill.

Evolution in the Different Groups of Organisms. It has been shown that any type of animal or plant developed from an earlier and simpler ancestral form by transmission and change of hereditary characters. Changes were rapid in some groups of organisms and slow in others, or they may have been greatly accelerated at certain times.

The classic example of a relatively rapid rate is the evolution of the horse. It changed from a four-toed, flat-footed animal the size of a fox terrier to our modern horse in a time interval of 60 million years. Progressive changes occurred especially in the skull, teeth, brain, and feet. These changes were of such magnitude that relatively short periods of time can be established by the evidence from horse remains in the rocks; even part of a tooth offers clues for the recognition of the different parts of a Tertiary Epoch. Evolution in these structures is illustrated in a later chapter.

On the whole, plants and invertebrates evolved much more slowly than vertebrate animals, but there are exceptions to this generalization. Some groups of marsupials — pouched animals like the opossum — and coelacanth fish with lunglike structures have changed very little in the last 100 million years or more. On the other hand, most mammals evolved rapidly into diverse forms during the Cenozoic. Within 600,000 years man evolved rapidly, particularly the brain. From the late Devonian to Pennsylvanian time amphibians evolved rapidly into diverse forms, and many groups of dinosaurs went through similar changes in the Mesozoic.

Paleontologists are continually contributing to our knowledge on the progressive changes in the different groups of organisms. In many orders and families the record is excellent and in others it is sketchy, but the basic patterns are essentially the same. Such factual information is not only basic but necessary in tracing the history of life on earth.

Dispersal and Distribution of Plants and Animals in Time and Space. Much of the evidence for dispersal and distribution is based on the evolutionary progression of animals and plants in the orders, families, and genera and the occurrence of their fossils in rock sequences in the earth's crust. A good example is the origin of proboscideans (mammals with trunks—mastodonts and elephants) as primitive amphibious creatures in the Eocene of Africa and their subsequent evolution and

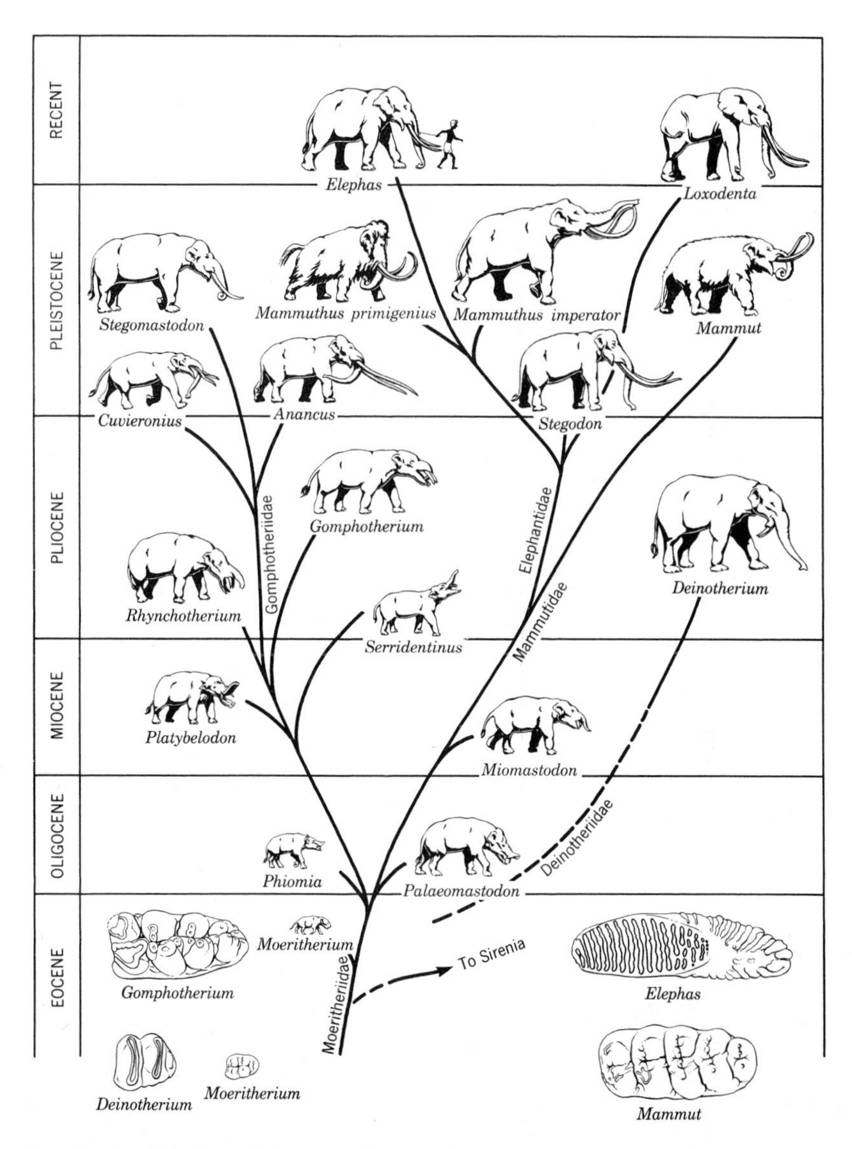

Fig. 5. A phylogeny of the mastodonts and elephants. Restorations after Osborn, $\times \frac{1}{200}$; lower molar teeth drawn to scale. By permission of the American Museum of Natural History.

dispersal to distant regions of the earth (Fig. 5). (*Note:* the beginning student needs to refer frequently to the geologic table until he has mastered the terminology.)

Somewhere, probably in Africa, between late Cretaceous and early Eocene times, a land-living species that tended to frequent the water gave rise to the sirenians (sea cows and dugongs) and the moeritheres, *Moeritherium.* Skeletal remains of the moeritheres have been found in the late Eocene and early Oligocene of Egypt. Though these animals were related to primitive sirenians, they had more in common with the earliest mastodonts. They were about the size of a Newfoundland dog but with a somewhat superficial resemblance to a hippopotamus. The external openings for the ears and the eyes were situated high on the head. They had no trunks. These and other features showed that they were close to but not in the direct ancestry of the mastodonts.

Moeritheres were still living in northeastern Africa when the first primitive mastodonts appeared there in the early Oligocene. These first mastodonts, *Palaeomastodon* and *Phiomia,* though larger and of different build than the moeritheres, had four tusks, small mastodont molars, and trunks like those of other proboscideans. The crowns of the molars in these earliest mastodonts were constructed in two or three transverse lophs separated by intermediate valleys. In *Palaeomastodon* the valleys remained open, but in *Phiomia* they were partly obstructed by extensions from the lophs. According to Henry Fairfield Osborn (see Chapter 6), who wrote two enormous volumes on the Proboscidea, these two African Oligocene genera were the forerunners of the two major families of later mastodonts.

Next, we find at least five genera of mastodonts in the early Miocene of Eurasia, and there is no evidence of their having been there previously. From these evolved other groups as the mastodonts became more abundant in mammalian faunas. There are records of both mastodont families reaching North America in the middle Miocene; these, like the mastodonts in the Old World, evolved into diverse kinds and spread all over the continent during the late Miocene and the Pliocene.

In the early Pleistocene, when the water barrier between North and South America was elevated above sea level, at least three kinds of mastodonts spread into the southern continent. Meanwhile, the elephants and mammoths evolved from the stegodonts, probably in southern Asia. The stegodonts were strictly Old World forms which had descended from one of the earlier Pliocene groups of mastodonts.

Mammoths dispersed in North America by way of an Alaskan-Siberian land connection in the middle Pleistocene. As with the earlier mastodonts, the mammoths spread rapidly all over the continent, but their pathway into South America, where they surely would have gone, was blocked by the development of a rain forest across the area of northwestern Colombia and Panama.

All proboscideans died out in the Americas in late Pleistocene time, but there is evidence that the first men to reach the New World saw mammoths and American mastodons. The only living representatives of this great order of mammals are the Indian and African elephants.

Another peculiar group of proboscideans were the deinotheres. They differed from all other proboscideans in their peculiar lower jaws that were so conspicuously bent downward. Their molars with only two simple transverse lophs, were also quite diagnostic. The deinotheres showed up first in the early Miocene of Eurasia, and the last of them, insofar as we know, died out in the Pleistocene of Africa.

Correlation. Much of a paleontologist's time is spent in efforts to recognize synchrony of biologic and geologic events demonstrated by evidence in the rock units of the earth's crust and in determining whether these events occurred in different continents or in adjacent mountains and valleys.

The biological events demonstrated by the Black Hawk Ranch fauna may be fairly clear in his mind, but he will want to know what animals and plants were living in Nevada, in Europe, or even in Australia at that time. A correlation between the areas of California and Nevada is not too difficult. Because of the proximity of the two areas they have numerous genera and some species in common. It is much more difficult to establish a correlation between Europe and California; the distance is greater, and in all probability no one species would be found in both places. But the evolution and dispersal between the two hemispheres of groups like the mastodonts, horses, dogs, cats, and beavers offer evidence of close approximations in time. A correlation with Australia is extremely difficult, since there were no land connections to that island continent in the Pliocene; hence terrestrial vertebrate animals could not get there except by unusual methods of dispersal. In such cases the paleontologist must rely on tie-ins with marine faunas that can be dated, sometimes with interpretations of climatic conditions, or on the stage of evolution of the organisms.

Geologic events pertain to the times of elevation of mountains or the times of deposition of certain rock formations. Here, too, the paleontologist may want to know the positions of oceans and land masses at any given time. Frequently he can determine when and where the continents were connected by land or disconnected by oceanic waters. This, of course, is paleogeography.

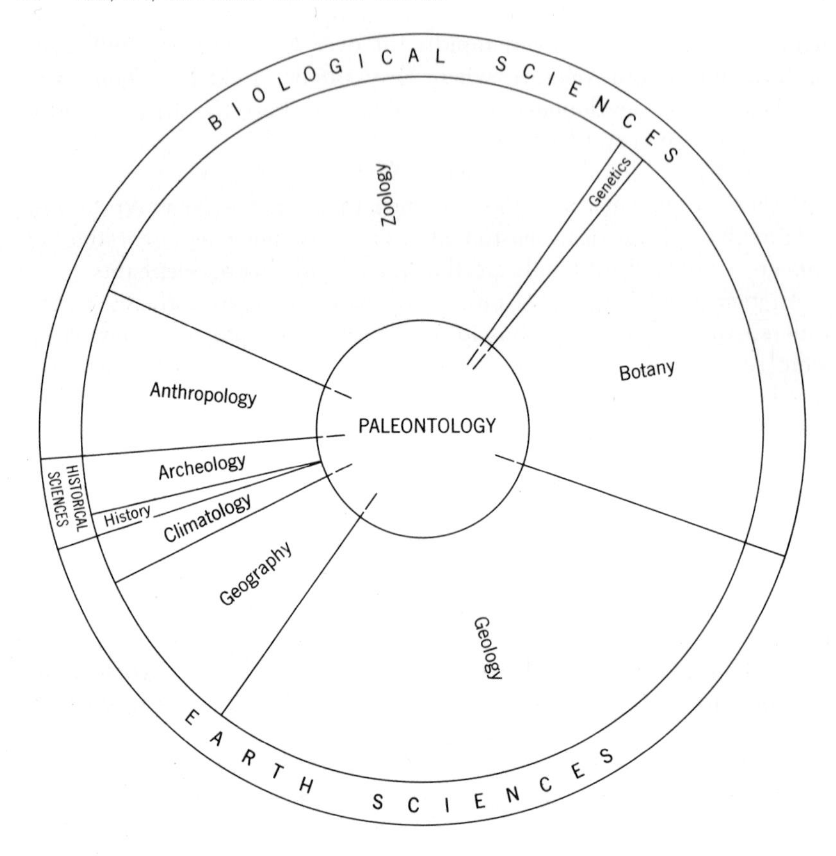

Fig. 6. Paleontology and related fields of science.

Relationship of Paleontology to Other Fields of Science

BIOLOGICAL SCIENCES

Nearly all fields of the biological sciences have a bearing on paleontology. Without our knowledge of living organisms some of the structures in fossils would be difficult to interpret. This was demonstrated at the close of the eighteenth and the beginning of the nineteenth centuries when the famous French paleontologist Baron Georges Cuvier (see Chapter 6) compared all his fossils from the Paris Basin with the large collection of Recent mammal skeletons in the Musée Nationale de l'Histoire naturelle and Jean Baptiste de Lamarck compared his fossil shells with those living in the sea.

Zoology. Knowledge of many fields in zoology is indispensible to paleontologists. These include *protozoology*—the study of protozoans (microscopic foraminifers and radiolarians); *conchology*—the study of shells or mollusks; *entomology*—the study of insects; *ichthyology*—the study of fish; *herpetology*—the study of amphibians and reptiles; *ornithology*—the study of birds; *mammalogy*—the study of mammals; *comparative anatomy*—applicable to the structure of all animals; *embryology, serology,* and many others.

Botany. An equivalent number of fields could be listed in the botanical sciences. Some are *ecology*—the interrelationships of animals and plants in their environments; *limnology*—the study of fresh waters, including their biological, physical, and chemical conditions; *oceanography*—the study of oceans and their phenomena, which includes both zoology and botany.

Genetics. In its contributions to the study of evolution genetics is also applicable to zoological and botanical phenomena.

Anthropology. The fossil record of man and perhaps all the primates, in which everyone is interested, is also very closely related to paleontology.

EARTH SCIENCES

Geology. As one of the physical sciences, *geology* is directly associated with every problem in paleontology, for the fossils come from the rocks in the earth's crust. Paleontologists work side by side with geologists in their search for petroleum. Many fields, such as *stratigraphy, structural geology,* and *petrography* (the description and systematic classification of rocks), are constantly employed by the paleontologist.

Geography. Geography plays an important role in constructing paleogeographic maps of the positions of continents and oceans of the past.

Climatology. Climates of the past can be interpreted by the nature of the sediments and the plants and animals in the rocks. These can be more readily understood by one who is familiar with the climates and their causes that prevail today.

HISTORICAL SCIENCES

History. *Recorded history, archeology,* and *paleontology* cover the history of life on earth. It is difficult to define where one stops and the other begins. From the written history of man's surroundings and activities we go back into archeology where these conditions must be interpreted from the artifacts, relics, and monuments that he left behind. Paleontology continues this history back through geologic time.

Paleontology, then, in all its ramifications is related to all these fields of science. It encompasses a much broader scope than is usually recognized; in fact, it is broader than any of the fields of science with which it is related. Its closest relationships are with geology and biology, but it is no closer to one than to the other, irrespective of the strong arguments that have been voiced on both sides of the question.

2 Fossils and Their Preservation

A skeleton in clay stone

UNACCOUNTABLE TRILLIONS of animals and plants have lived and died during the billion or more years that life has been on earth. A relatively small number with hard parts suitable for preservation got into a position which resulted in their fossilization. By far the most abundant were pollens and spores and the shells of microscopic organisms that settled to the bottoms of prehistoric seas. Such shellfish as oysters and lamp shells were covered along strand lines or in water offshore. Countless numbers were abraded and broken into small pieces by shifting sands and the action of tides. Larger vertebrate animals, such as whales, huge marine reptiles, and fish, met a similar fate.

More spectacular were the four-footed land animals and the forests of later geologic time, but they were much less likely to be preserved than the shells in the sea. These animals and plants frequently died on the open prairies or in humid forests where even their hardest parts deteriorated by weathering and decay, with no chance of being buried. But floods sometimes covered their remains with shifting sediments in stream courses, on vast flood plains, or in deltaic deposits like those built up at the mouths of rivers. Others were covered by shifting sands or by layers of volcanic ash blown from the craters of erupting volcanos. In these and many other ways the record of past life was preserved for us in the rocks of the earth's crust. Thus, irrespective of the medium of preservation, whether sandstone, tar, amber, or ice, *if organic materials or any traces of them are to be preserved as fossils, they must be buried quickly or protected in some manner so that bacterial action or weathering cannot destroy the specimens.* In essence, fossils are preserved in much the same manner as canned fruit or vegetables, but fossils are much more permanent. Fossil bones, teeth, shells, or wood, showing the effect of surface weathering before they were buried, are commoner than perfectly preserved specimens. They usually show evidence of fracture, abrasion, or exfoliation.

Movements in the earth's crust elevated the rock formations into hills and mountains, or streams cut through them, leaving erosional escarpments in which fossils have been exposed. Earth-moving operations also have uncovered many fossils. Even so, an extremely meager representation of the relatively few that were buried in the first place has come to our attention. Nevertheless, it is a fascinating record and we have learned a great deal from it.

Fossils

Fossils are the remains or traces of any recognizable organic structures preserved since prehistoric time. They may be the imprints of leaves, shells of mollusks, parts or complete skeletons of vertebrate animals, or even the trackways of animals in old mud flats. The word fossil was introduced by the Romans for any peculiar object dug out of the earth. This, of course, included many oddly shaped stones and other things not organic in origin. But toward the close of the eighteenth century fossils were recognized and referred to by the leading scientists as organic relics characteristic of successive geologic Periods. Fossils, as such, do not necessarily represent extinct species. Some species of rodents in Europe and Australia were first found as fossils and later were discovered living on those continents. All fossils are not petrified; they may be preserved in various ways.

Kinds of Preservation

Petrifaction. Petrifaction, widely used in reference to fossils, literally means "turned into stone," but in fact it seldom, if ever, happens completely. Common petrifying materials are silica, collophane, calcium carbonate, iron oxide, and iron sulfide. These may occur in fossils in any combination.

Silica. The best replacement agent is *silica*, though it has been questioned whether the replacement is actually complete. The fossil logs in the Petrified Forest of Arizona are one of the best examples. Logs and parts of logs from coniferous trees that were 120 feet or more in height and several feet in diameter have been found there in considerable numbers. Buried deeply in the old Triassic flood plains where the sedimentary rocks are loaded with silica from volcanic activity, these logs were in an excellent position to be replaced. Much of the original woody material was dissolved and replaced, molecule by molecule, by silica. When sectioned and polished, the cell structure of the fossil wood can be observed under the microscope. Most fossil wood is preserved in this way. The process is sometimes referred to as *histometabasis*.

The siliceous shells of microscopic foraminifers and other shells occur abundantly in the form of chalcedony or quartz in the fossil record. These shells were probably calcareous, siliceous, or opaline in the first place. On the other hand, siliceous sponge spicules are frequently replaced by calcite. Silica offers us one of the most beautiful methods of preservation. Some of the most decorative table and desk tops are made of polished petrified wood.

Collophane. Nearly all fossil bone tissue is petrified by collophane or calcite. Little is silicified. Collophane is an amorphous (uncrystallized), hydrous, calcium-carbonate phosphate mineral residue of organic bone tissue and phosphorite rock. After the animal matter in the bone disappears percolating solutions in the sediments deposit collophane. This method is frequently combined with others in preserving fossils. Collophane is commonly replaced by calcite. Other minerals, sometimes associated with collophane as cavity fillings in fossil bones, are quartz, chalcedony, opal, dolomite, aragonite, barite, pyrite, dahllite, and wavellite.

Calcite (Calcium carbonate). Another closely related agent is calcium carbonate, which occurs in abundance in nearly all stratified rocks. It is carried in solution in underground water and may be added to the original porous structure of bone or shell. The bones or shells, of course, are heavy and appear to be turned to stone. Once hard parts are well *permineralized* in this way they will stand considerable surface weathering. Without chemical analysis it is practically impossible to differentiate between collophane and calcium-carbonate methods of preservation. If buried bones are subjected to underground waters with considerable calcium carbonate or collophane in solution, permineralization can sometimes take place fairly rapidly in terms of geologic time. Certainly a very hard bone is not necessarily older than one less well preserved.

Iron Oxide. Sometimes fossils are infiltrated by certain iron compounds, such as hematite, limonite, or vivianite. This is usually detected by black, brown, rusty, or blue coloration in specimens that have been exposed to the climate on the surface of the ground.

Iron Sulfide. Calcareous bones in some specimens are replaced with an iron sulfide called pyrite—a heavy, pale, brass-yellow granular mineral. It is probably formed from the interaction of iron in the sediments and sulfur derived from decaying organic matter in saline or brackish water and possibly even in some fresh waters. Unfortunately, it is not a satisfactory method of preservation, since oxidation of the pyrite causes the specimens to disintegrate. This is commonly called "pyrite's disease." There are seventeen fine skeletons of duck-billed dinosaurs, *Iguanodon*, in l'Institut Royal des Sciences naturelles of Brussels, which were excavated from a coal mine in Belgium and are slowly disintegrating because of this reaction of pyrite.

Distillation. When animals or plants are buried quickly with other decaying organic matter in fine-grained sediments the volatile organic materials are distilled out, leaving a residue of carbon. This carbon film shows an outline of the soft parts that are usually lost in fossils. Some unique fossils are preserved in this way.

The outlines of the dolphinlike body forms of marine reptiles called ichthyosaurs were perfected by the carbon-film method in the dark Holzmaden shales of Bavaria. Otherwise the shape of the fleshy parts of the tail and the dorsal fins would never have been detected.

Much of the structure of the leaves and bark of the vegetation of the Pennsylvanian coal swamps is clearly preserved for us in black shales. These range from delicate fern leaves to the characteristic scars on the bark of the towering scale trees. Even in the early part of our geologic record, high in the Selkirk Mountains of British Columbia, Cambrian black shales have yielded the outlines of the soft anatomy of a multitude of small marine creatures that otherwise would have escaped our attention.

Compressions. These are best exemplified in fossil leaves that were buried in the mud or sand of lakes, lagoons, and streams. Air and water in the tissues of the leaves was forced out by the pressure of overlying sediments, and as the mud changed to shale and the sand became sandstone the leaves were compressed into lignite or coal. The specimens in shales are much more compressed than those in sandstones, though they always retain some thickness. Other compressed leaves are found in volcanic ash. Some invertebrate fossils such as bryozoans (moss animals) are preserved by compression.

Impressions. When a fossil shell or bone is buried in fine-grained sediments, the surrounding silt settles tightly around it. If the fossil is removed, it can be seen that the sediments have made a perfect mold around its external surface. This is known as an *external mold.* If the fossil is a bivalve shell, like a clam, fine sediments usually fill the inside cavity once occupied by the soft parts of the mollusk. This fine silt not only fills the inner space, but on its vaulted surface it gives an outline in reverse of the inner surfaces of the two shells. This is an *internal mold.* External and internal molds, then, are negatives of the original. Many molds or parts of molds are seen in exposures in which fossil shells occur in sandstones, shales, and limestones.

Molds of bones in the rocks are frequently very useful when parts of the bone have been eroded. The intervening mold can be filled with plaster of Paris, and in this way the total length and one surface of the missing part can be duplicated.

A most unusual mold is one of the body of a rhinoceros from the state of Washington. Evidently the bloated body of the animal was floating at the edge of a small body of water. A lava flow spread into the lake surrounding the carcass, and the water cooled the lava quickly before it could destroy the body. The cavity in the lava rock was discovered through an opening in the cliffs at Blue Lake. A plaster of Paris duplicate of the inner surface was made, and when the different pieces were put together a recognizable replica of the carcass was formed. Further identification was possible from parts of the lower jaw and foot bones found in the cavity.

Hard parts of animals or plants are made up of substances that may eventually dissolve and be carried away in solution in underground water. The space previously occupied by the specimen may then be filled with some infiltrating materials which make a perfect cast of the original. A *cast* is a duplicate of the original made by filling the space between the molds with some mineral or sediment. In museums plaster of Paris casts are made of specimens and sent to other institutions for comparison with other fossils.

Other *impressions* are *without thickness*. The impressions of leaves in thin, bedded shales or in volcanic ash are among the best examples. The venation, the outline, and other diagnostic features are clearly preserved.

Many impressions of the soft parts of animals were found in the Solnhofen shales also of Bavaria. These shales were worked for years for etchings and engravings in lithographic printings. Among the numerous fossils found there were two skeletons of a Jurassic bird *Archaeopterix*. The skeletons of these oldest birds were so reptile-like that they surely would have been called reptiles if their feathers had not been faithfully outlined as impressions in the shales.

Tracks and *trackways* (more than one track) of animals appear as imprints or compressions. These fossils are frequently clearly displayed on old mud flats or on moist ground where animals happened to pass. In the prison yard at Carson City, Nevada, are the footprints of mammoths, horses, camels, ground sloths, and other animals of the Pleistocene Epoch that lived just prior to Recent time. The tracks of the ground sloth showed so much resemblance to gigantic human tracks that there was much speculation about their identity. Dinosaur trackways have long been recognized, though the first ones observed in the Connecticut valley were thought to have been made by birds because they were three-toed.

Many marine or fresh-water shales and sandstones show the trailways of mollusks over and through ripple marks. Frequently the identity of the long-dead wanderer can be determined by comparison with similar marks made by its living relatives.

Careful study of the abundant trackways in the Triassic Moenkopi formation of northern Arizona has revealed much about the habits and about the anatomy of the feet of an abundant reptile and amphibian fauna that otherwise left meager evidence of their existence. Some

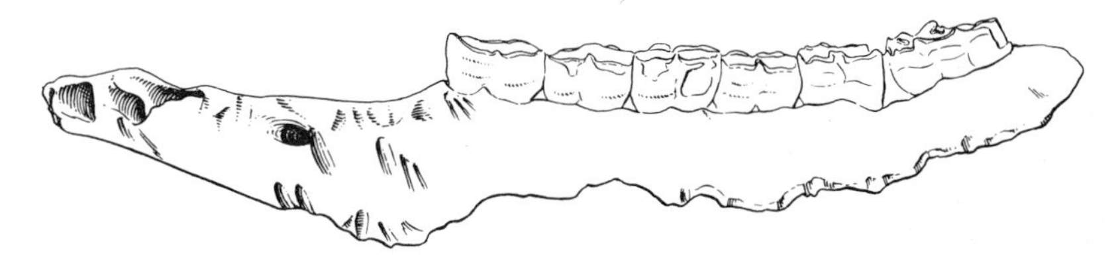

Fig. 7. Rodent tooth marks on the lower jaw of a horse from Pliocene rocks in South Dakota, $\times \frac{1}{2}$.

frequented the stream courses, others spread out over the fluvial mud flats; some walked nearly upright on their hind legs, and others lumbered about on all four feet. Furthermore, their trackways indicate whether they were running or moving slowly.

Many *pseudomorphs* (false replicas) simulate the single tracks of animals. If part of a trackway is present, this is assurance that the observer has found the fossil tracks of an animal. Pseudomorphs, objects and shapes that resemble tracks, can be quite deceiving.

Tooth Marks. There are other indications of the presence or habits of animals of the past. Rodents with a deficiency of calcium or other minerals frequently gnaw on fresh bones before they are buried. Later these bones, clearly showing the tooth marks, are preserved as fossils. A fragmentary lower jaw of a small, three-toed horse from the Pliocene of South Dakota, (Fig. 7), shows incisor tooth marks 2 millimeters wide at numerous places along the edge of the broken jawbone. The incisor marks across the broken surfaces show that the bone was gnawed after it had been fractured. Many fossil bones from cave deposits show similar tooth marks.

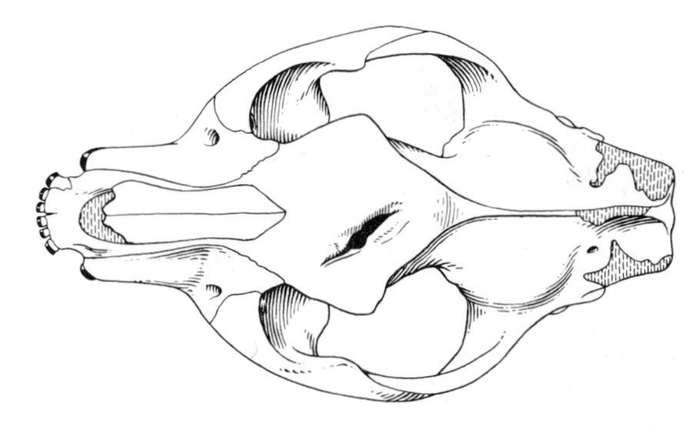

Fig. 8. A skull of a cat, Nimravus bumpensis, *showing puncture through the frontal bones,* $\times \frac{1}{2}$.

Carnivorous animals have also left their tooth marks on bones. Many bones of this kind have been found at Rancho La Brea in Hancock Park in the heart of Los Angeles. In the fossil collection of the South Dakota School of Mines is the skull of a cat, *Nimravus bumpensis,* with a puncture in its frontal bone nearly an inch long, (Fig. 8). Measurements show that this may have been inflicted by one of its Oligocene contemporaries, a saber-toothed cat of the genus *Hoplophoneus.* The callous, irregular surface, indicative of partial healing along the margin of the wound, offers evidence that it was inflicted while the animal was still alive. Partly healed lacerations on the skulls of the large phytosaur reptiles, *Machaeroprosopus,* from the Petrified Forest area of Arizona, demonstrate that these creatures must have fought much as crocodiles do today.

Coprolites. Fossil dung or excreta known as coprolites are abundant in some formations. This is particularly true of certain fish and reptiles. A crocodilian coprolite from the Miocene of Colombia, South America, contains the tooth of a rodent; others have fish scales. A coprolite closely associated with the skeleton of a fossil dog from the Ricardo Pliocene in California contained rabbit and pocket-mouse teeth, together with bones of these and other small mammals.

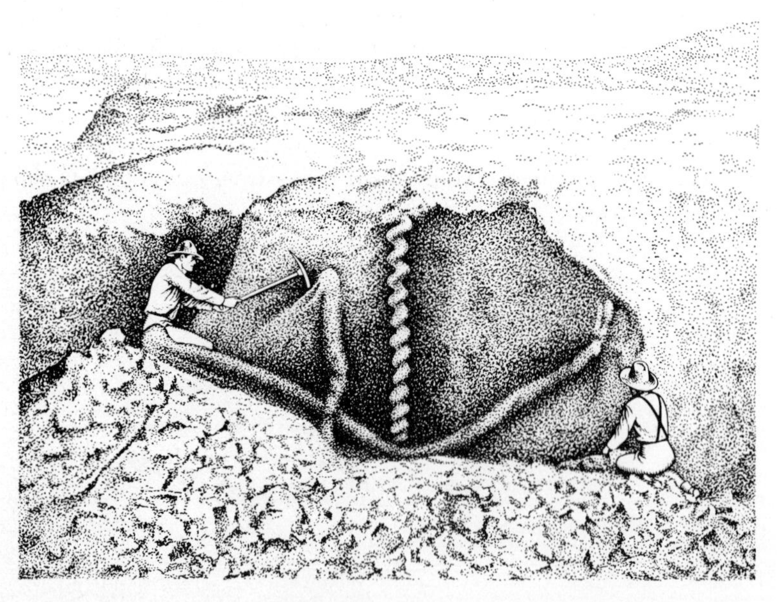

Fig. 9. It has been suggested that the spiraled structures of Daemonelix *in the early Miocene Harrison formation of western Nebraska are fossil burrows of an early beaver.*

Fossil dung of a ground sloth found in Gypsum Cave in Nevada was made up of fibers of Joshua trees and other desert plants. Even the stomach content of the woolly mammoth in the frozen tundras of Siberia revealed bits of pine, fir, willow, buttercups, and other kinds of plants now found in the lower Arctic Circle. Another source of information is regurgitated owl pellets. These are replete with the bones of small birds, mammals, and reptiles.

Gastroliths. Highly polished stomach stones from the gizzards of birds or the stomachs of reptiles are also fossils. Such stones have been found within or near the rib baskets of dinosaurs, and no other stones like them were found in the formations where the skeletons were preserved. These too are considered fossils.

Burrows. Sometimes worm burrows or the burrows of such extinct animals as the bivalve mollusk offer evidence of life in the past. One of the most famous is the *Daemonelix*, spiraled structures of the Harrison formation of western Nebraska, (Fig. 9). It is still argued that these were the burrows of an extinct beaver, *Palaeocastor*, or that the cavities were left after plant roots had deteriorated and were later occupied by the rodents. Certain structures in some of the oldest rocks are thought to be worm burrows.

SPECIAL KINDS OF PRESERVATION

Incrustation. Caves and fissures have given us some of our most interesting fossil remains. Caverns in western Europe, in South Africa, and in China have yielded not only the bones and skulls of early man, but also the bones of many of the animals which he may have killed for food. Elsewhere, cave lions, hyenas, and other flesh-eating mammals dragged parts of carcasses back into their cavernous lairs. Other animals met their death by falling into fissures. Most of the smaller mammals and birds were carried into these places by owls and by rodents with the habits of pack rats.

Caves or caverns like the Carlsbad Caverns in New Mexico or the Mammoth Caves in Kentucky were formed by the gypsum or limestone in underlying rocks being carried away in solution. Water, with calcium carbonate in solution, dripping from the ceilings in such caverns formed stalactites much like huge icicles. On the floor of the caves the rock debris, bones, teeth, and wood were coated by the drippings and were incrusted in stalagmites. This is the *incrustation* method of preservation. Usually there is little or no replacement or permineralization of the bones. The extremely hard surface coating and the delicate nature of the bone render them difficult to prepare.

Similar incrustations take place in hot springs where the growth of blue-green and green algae have covered objects with calcium carbonate or silica. Most of the examples known to us are referable to the Pleistocene Epoch, but older deposits are also recorded.

Preservation in Peat. The Irish deer, *Megaloceros*, was a spectacular mammal. No deerlike creature has exceeded this animal's tremendous spread of antlers, which measured as much as 10 feet from tip to tip in some specimens. Irrespective of the size of their wide antlers, the body was no larger than that of the moose of the Canadian woods. Irish deer or giant fallow deer, as they are sometimes called, lived in western Europe during the Pleistocene. They also ranged into England and Ireland across land connections with the mainland.

In Ireland and elsewhere there were extensive swampy places in which the fibrous substances of bog mosses and other vegetation accumulated in thick layers. Skulls and parts of skeletons of *Megaloceros* have been discovered deep in the peat deposits near or on the surface of the underlying marl. Once the bones sank into the peat bogs a chemical reaction prevented decay, and though the bones were perfectly preserved there was little or no permineralization or replacement. This kind of preservation has been found in peat beds in many parts of the world.

Preservation in Tar. One of the most perfect materials for preservation is asphalt or tar, and there is no better example than the renowned tar pits at Rancho La Brea (Fig. 10). The drama of life and death about the old tar seeps could easily challenge the best in screen and television today. Nearly all of the animals and many of the plants of that area in the later Pleistocene were in the cast. From the immense mammoths, fierce saber-toothed cats, and giant condors to small mice, tiny birds, and insects, they found their way to the open pools of unctuous bitumen. By one means or another they became entangled in this death trap. A quail may have alighted on the surface at dusk, and an owl may have come in after the quail; or a bison may have reached too far for a tuft of green grass. Flesh eaters became too courageous in their efforts to get at a trapped pronghorn. And so one after another they sank deeper as their flesh, skin, and feathers deteriorated. Upwelling oil and gas causing movement in the plastic matrix, together with the trampling of large mammals in their efforts to pull themselves out of the tar, separated and mingled the bones in black entombment. There was no chance for replacement or permineralization by carbonates or other petrifying minerals, but the asphalt penetrated to the most remote recesses of the

Fig. 10. A saber-toothed cat, Smilodon, *and a dead bison in a tar seep at Rancho La Brea.*

bones. They were literally pickled in tar. In time the active, upwelling petroleum shifted to adjacent sites, leaving behind a record of life to be uncovered by man.

Other tar pits with bones occur in California, Trinidad, and Ecuador. There are records in Europe of early Cenozoic tar seeps in which animals were evidently caught and preserved much like the ones at Rancho La Brea, but these beds have been solidified into rock by heat and pressure.

Preservation in Amber. "Insects in amber" has become a common expression. It has a long history. Even before the days of Pliny, fossil rosin in which insects were eventually observed was utilized by man. Amber amulets shaped into human idols and one of a young wild horse were fashioned by men of the late Stone Age. When the first insect was recognized in amber is not known. Certainly it was long before 1260 A.D., when there was a profitable industry in making rosaries. One of the greatest accumulations of amber occurs in a bluish-gray marine glauconitic clay of the early Oligocene Age under the Baltic Sea off the coast of Samland near Danzig, Germany. Heavy storms dislodged the amber by wave action, and, since rosin is light, it was carried shoreward and deposited along the beach. Many pieces contained flies, spiders, blossoms, and even bits of feathers, all indicative of life in the past. Even the scales on the wings of moths and the finest hairs on mosquitos are faithfully preserved. All of these, except the thick chitinous exoskeletons of beetles, tend to disintegrate into dust when the amber nodule is opened.

After the evidence from all sources was evaluated, it was recognized that a subtropical forest with its varied animal inhabitants had occu-

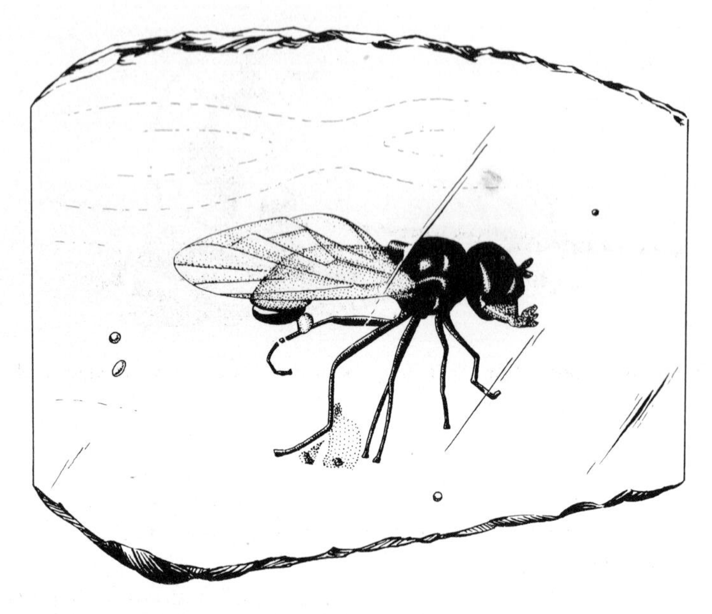

Fig. 11. A fly in Baltic amber.

pied that area. The amber was derived from tree resin that dropped to the ground, where it solidified into rosin and was later washed into marine waters. Numerous insects and other arthropods became entangled in the resin and were finally enveloped in the sticky substance. This could have happened in the trees or on the ground because earthworms and land snails were also found in the amber. All of the major groups of insects now living were represented in the Baltic forest, though some of the kinds found there are now in distant parts of the world. The largest insects that must have been present avoided or pulled out of the resin. There were ants, termites, earwigs, cockroaches, moths, cicadas, praying mantises, flies (Fig. 11), mosquitoes, antlions, anteating bugs, beetles, pseudoscorpions, and many others. A bee with pollen on its legs never got back to its hive. Birds in their attempt to snap up the trapped insects sometimes left parts of their feathers in the resin. A woodpecker, a titmouse, and a motmot have been identified. Mammal hair, mammal fleas, and even the footprint of a small mammal were recognized. A small lizard and reptile scales were also discovered.

An idea of the forest was built up from the pollens, blossoms, and small leaves that stuck in the resin. From this we can visualize a subtropical forest with pines, oaks, beeches, chestnuts, maples, holly, junipers, date palms, magnolias, cinnamons, laurels, sandalwood, palmetto pines, cycadlike plants, olives, silky camellias, mistletoe, and geraniums. All of this was revealed by small pieces of amber — a fascinating record. Promising recent discoveries of insects in amber have been made in the Tertiary of the Chiapas in Mexico and in the Cretaceous of Alaska. These should open new chapters in our knowledge of fossil insects.

Mummification. On February 25, 1928, Richard S. Lull of Yale University received a letter telling of the discovery of a remarkable animal. The unusual specimen was found in a funnel-shaped pit in the Aden-crater, an extinct volcano near Deming, New Mexico. The specimen was one of the small ground sloths, *Nothrotherium shastense.* It was located in a cavern on the floor of an old fumarole almost completely covered with dry, loose bat guano. This was some 100 feet below the floor of the crater. Evidently the animal lost its footing at the edge of the pit, and as it slid down the sloping wall it partly broke its fall by clinging as best it could by its sharp claws. There it was doomed to die of starvation. In that extremely dry climate parts of the skin, tendons, and claw sheaths were mummified. Something gnawed most of the hide from the skeleton, leaving its tooth marks still plainly visible. The skeleton was complete, and, interestingly, the bones were held in articulation by the original tendons and ligaments. Even the stomach content, still well preserved as a food ball, gave information on the food habits of the animal and some notion of the plant life and environment in the area at the time the *Nothrotherium* slid into the pit. Bits of twigs, roots, and seed coats suggest that the animal fed on short, hairy, desert scrub. Evidently the climatic conditions were little different from today's.

An example of similar preservation was found in Gypsum Cave near Las Vegas, Nevada, where parts of another *Nothrotherium* were discovered. Dry caves or caverns offer the best conditions. Eberhard Cavern at Last Hope Inlet in Patagonia has yielded skin with the hair and the horny sheaths of claws of a South American ground sloth, *Glossotherium.* The skull of an extinct camel, *Camelops hesternus,* with dried tendons still attached, was found in Utah. Even evidence of a dinosaur mummy, *Anatosaurus annectens,* is known to have existed. When found by the veteran fossil collector Charles H. Sternberg, delicate impressions of the skin in the form of molds and casts showed that the carcass had remained in a dehydrated condition for some time before it was covered by fine river sand mixed with clay.

Refrigeration. In the Arctic tundra of Siberia and Alaska many animals have been found where they have been under refrigeration for thousands of years. Wolf packs and sledge dogs have eaten their meat.

Fig. 12. A woolly mammoth, Mammuthus primigenius, *breaking through the ice.*

Parts of frozen specimens have been brought back to museums for attractive displays. Some of these animals broke through thin ice and were quickly carried under thicker layers to an icy grave (Fig. 12). Others tumbled into deep crevasses or bogged down in soggy terrain and were soon covered and frozen; thus they became fossils by refrigeration.

A most realistic account of a woolly mammoth that had been under refrigeration in the permafrost earth of Siberia has come to us from Benkendorf, a young Russian surveyor. This unusual experience occurred when he was exploring the Indigirka River during the abnormally heavy rainy season of May 1846. A new stream course cut by the floods was undercutting the river bank, carrying away masses of peat and loamlike chaff. There was a sudden gurgling and movement in the water under the bank. As one of the men shouted and pointed, a peculiar, shapeless mass was rising and sinking at the edge of the swirling stream. Those in the boat pulled up close and waited excitedly for the mysterious object to show itself again. Soon the huge black mass bobbed up like a fisherman's cork. There was the monster, a woolly mammoth, hardly 12 feet away, with the white of its half-open eyes showing, its tusks gleaming, and its trunk twisting about in the water. A rope was cast over its head and chains were fastened to its tusks, but when the creature was anchored to a stake far back on the bank it was realized that the hind legs were still embedded in the frozen earth. Evidently it had mired down in the soft peat where it soon froze as if standing in a normal position. Twenty-four hours later, when the currents had completely freed the hind legs, it took 50 men and help from the native Yakuts' horses to drag the beast 12 feet back from the water's edge. It was fat and nearly full grown, an excellent specimen, but decay progressed rapidly. The tusks were cut off, and an effort was made to sever the huge head; and, though the stench was dreadful, the stomach content was examined. The men were engrossed in their work, and time slipped by quickly. An alarm from the Yakuts was the first warning that the ground was sinking underfoot. Some of the party scampered into the boat, others retreated to safety farther ashore as the carcass of the mammoth was again claimed by the turbulent river, into which it sank, never to appear again.

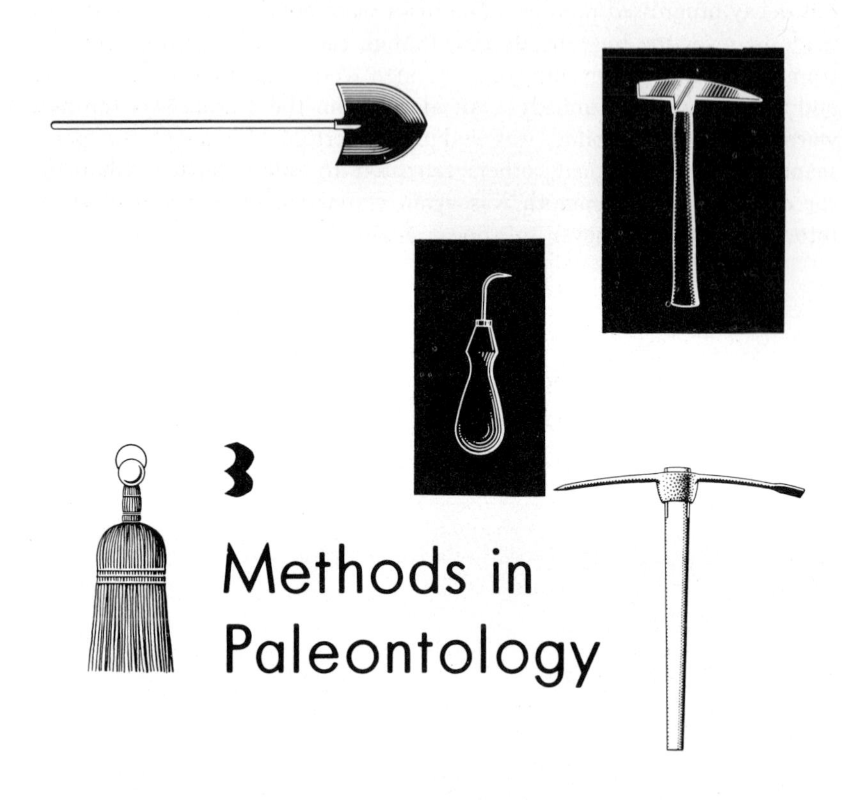

3 Methods in Paleontology

Where Fossils Are Found

Fossils are not necessarily far away; they may turn up in your own back yard. This may occur when any excavation is done. Usually, though, they are discovered as a result of diastrophic movements or deformation in the earth's crust or lithosphere that elevate the landscape into hills or mountains. Erosion then cuts through the formations to expose the fossils. In some areas only a slight uplift is sufficient for streams to cut long erosional escarpments along their courses, as may be seen in the Great Plains of Kansas, Nebraska, and Oklahoma. Great badlands may develop, as in South Dakota or in the Siwalik Hills of India. Though usually dry and hot in the summer months, areas like these are the fossil hunter's paradise. In eastern United States rocks deposited in the ancient inland seaways that date back hundreds of millions of years have an abundance of marine fossils.

If fossils are in the rocks, they may appear in any number of places. When prospecting for them search should be made in any exposure of sedimentary rocks. Extensive badlands where little or no vegetation can gain a foothold is, of course, ideal, but the sloping escarpments well back from a stream may also yield clues to important fossil beds. Fresh cuts in stream banks, sea cliffs, or roads may offer the thrill of a lifetime. In hydraulic operations or when sluicing for gold in mountainous areas many a curiosity in the form of a fossil has struck the imagination of the miner. In Alaska these sluicing operations have exposed the carcasses of mammoths, horses, caribou, and many other Pleistocene mammals. It is said that some were uncovered in the positions in which they sank into the bog and appeared to be standing on all four feet. At times, while digging their dens, coyotes, badgers, or ground squirrels throw fossils out on the surface. This is sometimes called squirrel-hole geology and paleontology. Strange as it may seem, big red ants have brought together in their mounds not only sand grains but, among other objects, some of the rarest of tiny mammal teeth.

After years of experience a paleontologist may recognize a likely fossil-producing exposure from a considerable distance. In South Australia a Pliocene exposure was seen nearly 4 miles away, where later prospecting resulted in the discovery of an excellent quarry site on an escarpment along the west side of Lake Palankarinna.

Area Selected. The area or region selected for fossil hunting usually depends on the interest of the paleontologist. He will, therefore, go where he is most likely to find evidence of the problem with which he is concerned. Almost invariably, though, something else turns up of even greater interest.

A paleontologist frequently learns of the possibilities in an area from the previous reports of geologists or other paleontologists made while reconnaissance work was being done. In the early exploration of the Western Territories of the United States numerous fossil sites were noted. All of these have still not been adequately prospected. Geologists occupied with regional geological mapping in the search for petroleum sometimes send in a few specimens that lead to important discoveries. Evidence of this kind led to the discovery of a large fauna of extinct mammals and reptiles in the upper Magdalena River basin of Colombia, South America.

Another important source of information is the average layman. An Australian farmer located bones of fossil kangaroos and those of other pouched mammals on a hillside pasture when he went out to bring in the cows. A cowboy riding the range in Wyoming located the first skull of a large horned dinosaur on the face of a cliff and pulled a horn off with his lariat. Road crews cutting a new highway in Mexico uncovered an abundance of fossil leaves in a small hill. Gert Terblanch, a schoolboy, stumbled upon the skull of a primitive man in an erosional remnant of an old cave breccia on the Kromdraii farm in South Africa; and a boy scout found the bone of a flamingo in the hills east of Oakland, California. These and many other similar finds are reported by thoughtful discoverers so that specimens may be collected and prepared without endangering the fossil. Furthermore, the reports of laymen frequently lead to more important finds in the same area.

Areas may be selected in which the discovery of fossils will offer outstanding contributions to our knowledge of little or unknown animals and plants. The American Museum of Natural History sent expeditions into the heart of Mongolia because it was thought that the early ancestors of man or the predecessors of the earliest-known horse might be found there. Instead, many other fossils of equal or of greater importance, including the only known Cretaceous mammal skulls, dinosaurs and their eggs, and a tremendous rhinoceroslike beast, *Baluchitherium,* the largest terrestrial land mammal now known, were discovered. The University of California and the South Australian Museum explored the center of Australia for the early ancestors of the kangaroo, koala, duckbilled platypus, and related forms. These attempts to fill the tremendous gaps in our knowledge have met with partial success, but much more needs to be done. Locally, a paleontologist may spend weeks or months searching for fossils in some formation which might yield important information on the age of the rocks.

Collection and Preparation of Fossils

Different methods are used in collecting and preparing fossils. Well-preserved fossils of vertebrate animals are much more difficult to find than most invertebrates. Many vertebrate remains are large and require more time than smaller organisms, though fossil flowers and fruits of plants, fossil insects, and certain other invertebrates seldom encountered also demand special attention. Consequently, every precaution must be exercised to insure the protection of these rare fossils both in the field and in the laboratory (Fig. 13).

Discovery and Outlining the Block. Several years ago a paleontologist was carefully prospecting in the badlands east of the old Ricardo post office in California. As he rounded a small knoll of Pliocene flood-plain deposits, his trained eyes spotted fragments of small bones

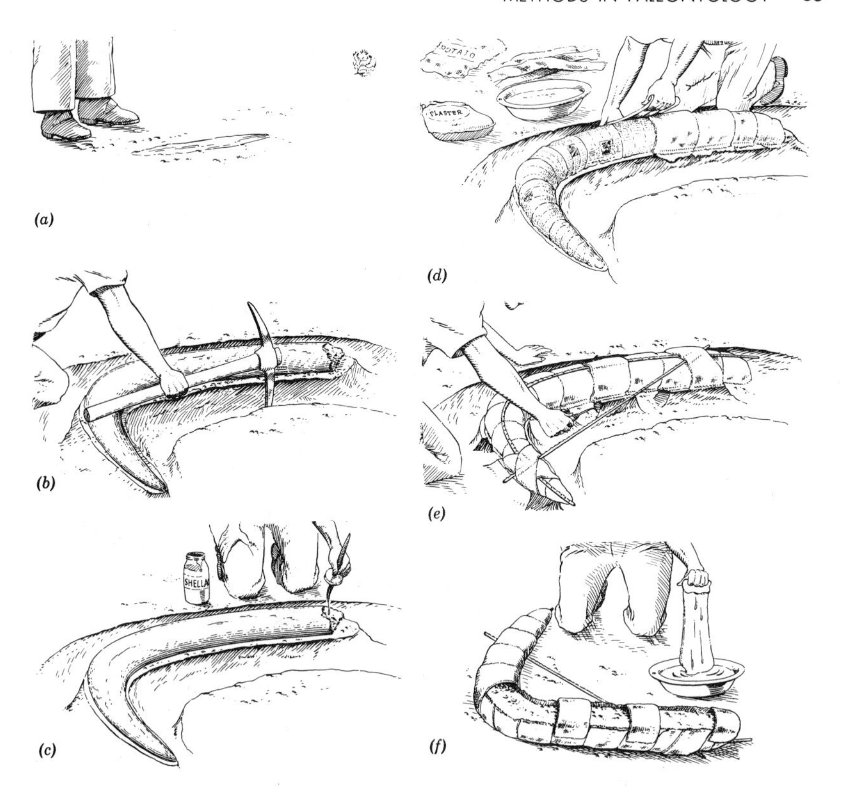

Fig. 13. Uncovering a mammoth tusk and preparation of a plaster block in the field: (a) *discovery;* (b) *outlining the block;* (c) *application of shellac;* (d) *first plaster strips over wet newspaper;* (e) *re-enforcing block;* (f) *plaster strips covering under side of block.*

on the surface. *Experience cautioned him to stop in his tracks. For a few minutes he carefully analyzed the situation to determine his procedure.* After bringing up his tools and other necessary equipment, he proceeded to the bottom of the small wash some 20 feet below the spot where he had first seen the bits of bone. He then progressed slowly up the wash picking up every fragment. Later the silt in the wash was screened for other bits of bone. At last all the surface materials were carefully wrapped, and his field number for the specimen was written on each package. Thus the danger of stepping on or losing part of the specimen was eliminated. Two or three permineralized toe bones and a claw bone indicated that he had found part of a fossil dog skeleton. No bone was seen sticking out of the soft, weathered surface, but the position of the fragment farthest up the slope was marked.

The slope was carefully brushed with a whiskbroom, and two or three additional pieces of bone were recovered. Near the position of the uppermost fragment greater care was exercised. At this stage a soft paint brush was employed instead of the whiskbroom. At last the end of a bone was located in place in the underlying fine-grained sandstone. Thus the level in the sedimentary rocks where the remainder of the skeleton lay was located.

The exposed bone was covered with paper, and soft dirt was placed on top to hold the paper in place. Pick and shovel were then used to remove nearly three feet of overlying sediments. When the overburden had been cut down to a level 10 inches above the bone the heavier tools were cast aside. A small pick and scratch awls were used to work on down until bone was again located. The outline of the skeleton was then determined. This was done carefully by exposing as little of the bone as possible.

When bone was exposed, but still in the matrix, it was permitted to dry for a while. Then Glyptal Lacquer cement No. 1276 (a General Electric product) was applied with the tip of a sash brush. Since the glyptal had been thinned with solvent thinner 1511-M, it soaked into the bone readily. Wherever the surface of the bone was cracked a piece of thin Kleenex was stippled on with the glyptal and the terminal hairs of the sash brush, thus holding the shattered pieces of bone permanently in place. Later when this application had dried the exposed bones were covered with moist sand. Slightly different methods could have been employed at this stage, depending on the preservation of the bone, the area of the bone surface exposed, and the amount of moisture in the rock matrix. Sometimes glyptal cannot be used if the fossil is too wet.

The outline of the dog skeleton determined the shape of the block to be removed. A trench wide enough for elbow room was then dug around the block and down to a level several inches below the skeleton. All edges were undercut, particularly at both ends of the rather elongated block.

Making the Plaster Block. The paleontologist, thoroughly thrilled by his find, by this time had visualized every conceivable kind of dog it might be, but he would not jeopardize the specimen by uncovering more bone to satisfy his curiosity. He was then ready to proceed with another phase of his most fascinating job, for this one specimen had made the entire trip more than worthwhile. Soft, flexible burlap was cut into strips to fit over the block and into the undercuts; also, two long strips were cut to fit around the edges of the block. These were first soaked in water and wrung out so that they would not absorb too much water from the mixture of plaster of Paris (called plaster by nearly all paleontologists) which was soon to be applied. A wash basin was brought out in which water and plaster were mixed to the consistency of thick cream. Then the damp strips of burlap were soaked one at a time in the plaster and laid over the block, overlapping an inch or two, much as the roof of a house is shingled. Each strip with its plaster was kneaded tightly to the surface of the block by the paleontologist with his fingertips and the palms of his hands. The plaster was mixed in quantities so small that it could be applied before it started to set. Three pans of plaster were mixed before the final clinching strips were put around the edges of the block and the last handful of plaster from the wash basin was smoothed out over the top. The block was then under control and would have time to partly dry out before the next day.

Early next morning a wide tunnel was cut under the middle of the block, and two long, wide strips of burlap soaked in plaster were shoved through with the ends overlapping on the top of the block. There the block stood on two sandstone pedestals. Eventually, when the last application of plaster had set, the block was turned over. Once it was firmly placed upside down in soft sand the ends of the plastered strips that were kneaded into the undercuts and the parts of the strips that were wrapped around through the median tunnel were cut away with a sharp knife. The matrix on the underside of the block was then reduced to within an inch or so of the bone. This side was also covered with burlap strips and plaster. Just to make sure that there would be no torsion or twisting during transportation, two sturdy boughs from a nearby bush were cut and plastered to the block, again with burlap strips. Soon it would be ready for transportation to the museum.

Larger blocks frequently require several coats of burlap and plaster and greater reinforcement with timbers or steel. Each specimen removed in plaster blocks offers an engineering problem of its own. Enormous skeletons like those of dinosaurs or mammoths are necessarily removed in numerous blocks. Smaller, well-preserved specimens such as horse teeth and foot bones are tightly wrapped in paper. All of these specimens must bear the field numbers of the collector.

Recording Information in the Field. Before leaving the locality the exact geographical location illustrated by small sketch maps is recorded in field notes. The location is also marked on aerial photographs on hand for that purpose. Further, stratigraphic positions with the strikes and dips (see page 72) and other observations on the sedimentary rocks

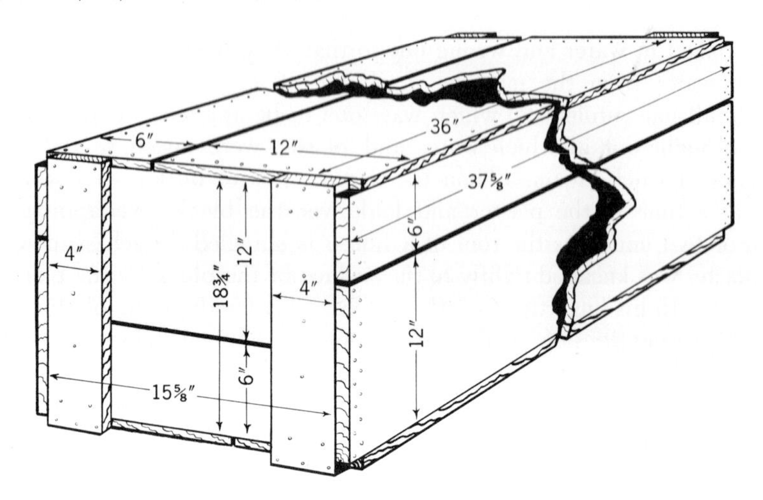

Fig. 14. A box recommended for packing specimens for shipment. Adopted from Camp.

are carefully recorded so that anyone can locate the site in the future. Without accurate information specimens are worse than useless. They may also be misleading.

Transportation of Specimens in the Field. Small blocks and paper packages are easily carried back to base camps, but getting the heavier blocks out of the mountains or desert areas sometimes poses a difficult problem. In Australia two or more men carried blocks of the giant marsupial, *Diprotodon,* lashed between two-by-four timbers, out of the old moundspring bog at Lake Callabonna. A large crocodilian skull far back in the badlands of Colombia was tied between two bamboo poles and strapped in shaft fashion to the sides of two unsuspecting pack mules. In eastern El Salvador blocks of the huge ground sloth, *Megatherium,* with dry banana leaves for packing, were transported back to the hotel in San Miguel in ox carts. About 50 years ago blocks containing dinosaur bones were hauled to the railway by team and wagon. Nowadays the jeep and landrover get into almost impossible places to bring out their precious cargos.

CRATES AND PACKING BOXES. Some specimens require crating with strong lumber or scantlings in the quarries. These are then ready to ship by freight or to be hauled in trucks back to the museum. In the past well-made boxes could be obtained from local stores, but today because of the extensive use of pasteboard cartons this source is not so dependable as it was 25 years ago. It is better to construct packing boxes of good pine lumber to the size desired. Specifications in inches for boxes made for many years by many museums are as follows: top and bottom pieces, two 1 x 6 x 36; two 1 x 12 x 36. Sides, two 1 x 6 x 37⅝; two 1 x 12 x 37⅝. Ends, two 1 x 6 x 15⅝; two 1 x 12 x 15⅝; four uprights 1 x 4 x 18¾. The length, width, or depth may be altered to accommodate blocks of different shapes. As can be seen, these boxes are strongly reinforced by the way the boards fit together. They may be reinforced additionally by banding iron or by other boards (Fig. 14).

While the blocks and packages are being packed in excelsior, straw, banana leaves, or whatever may be available, cleat boards can be nailed inside the shipping boxes to prevent the heavier blocks from shifting and thus endangering the other specimens. Everything must be tightly packed. Freight is handled roughly.

Preparation in the Museum. Preparation of the specimens in the museum is a painstaking job. Impatient persons should never attempt to prepare fossils. The old slogan "anything worth doing is worth doing right" certainly holds in the laboratory. The blocks must be cut open and the bones hardened with glyptal, cellulose cement, or other materials as the job proceeds. Care must be exercised to keep the field numbers always with the specimen until the locality and specimen numbers are inked on the bone. Additional preparation and other methods are employed when skeletons are selected for displays. All of this requires much more explanation than can be included in an elementary text book.

Collecting and Preparing Larger Invertebrates. The fossils of larger invertebrates (Porifera, Coelenterata, Brachiopoda, Echinodermata, Mollusca, or Arthropoda — see classification) frequently occur in abundance in rocks that were laid down as sediments along strand lines of the old seaways. Whether they lie in soft sands and shales or in hard rocks they are usually collected in bulk (i.e., with considerable stone matrix). In the museum the best specimens are prepared carefully with hand tools, or, when the numbers of individuals are great, the stone is broken along the bedding planes until good examples are secured. Some of these rocks will break down in water, but there is also danger of the fossil disintegrating. For example, certain limestones will weather away if rain water highly charged with carbon dioxide is allowed to wash over them. The calcium carbonate in the limestone is then carried away in solution. Another popular method these days is leaching the fossils out of the rock with acids. Although invertebrates are usually preserved as some form of calcium carbonate, many are permineralized with other minerals. In this method the limestone or sandstone is dissolved by the acids, leaving the fossils behind as a residue. Some of the most delicate structures are freed from the matrix in this way. In other instances, several sections of the stone and embedded fossils are made to disclose the characters of the specimen. Sand blasts or heat directed at certain points are sometimes used in removing matrix.

When rare specimens are recognized in the field they can be hardened with glyptal or thin cellulose cement and carefully removed. Even plaster of Paris is employed to give them a firm base. Molds of fossils which have weathered away can be heated and filled with plaster of Paris to give the collector a perfect cast of the original.

Collecting and Preparing Microfossils. Foraminifers which have become so important in petroleum exploration are usually the most numerous microfossils in museum collections. Other microfossils include pollens, diatoms, radiolarians, ostracods, conodonts, and the like.

Larger microfossils are sometimes examined in the field with a hand lens, but many rock samples are brought to the laboratory to test them for microscopic organisms. In addition to recording exacting field data, great care should be exercised to prevent contamination. This is much more dangerous than many collectors realize. The specimens are so small that a bit of dust will sometimes carry a dozen or more foraminifers. The field sample should be taken well down in the rocks and the pick cleaned thoroughly before another is taken. The fresh sample is placed in a double paper bag, then tightly wrapped and sealed with gummed paper. Packages should be tightly packed for transportation to prevent holes being worn through the bags. Paper bags are preferable to cloth bags because foraminifers lodge in the fabric, and if the bag is used again there is the possibility of mixing fossils from different formations. Nevertheless, cloth sample bags are used in the tropics because paper bags tend to mildew and deteriorate, but even the most expensive cloth bags should not be used again for samples of foraminifers.

After part of the sample is taken from the paper bags in the laboratory the remainder of the matrix is stored in the original bags for future reference, if needed.

In the laboratory the rock sample is crushed and placed in a cylindrical water beaker. Crushing does break some fossils, but they are so small and numerous that such breakage is negligible. The beaker containing water and the sample is agitated, usually on a mechanical roller, to free the foraminifers from the crushed matrix. The fine materials are then washed through a 20-mesh screen (400 mesh to the square inch) where the coarse detritus is eliminated; only the fine sand and foramini-

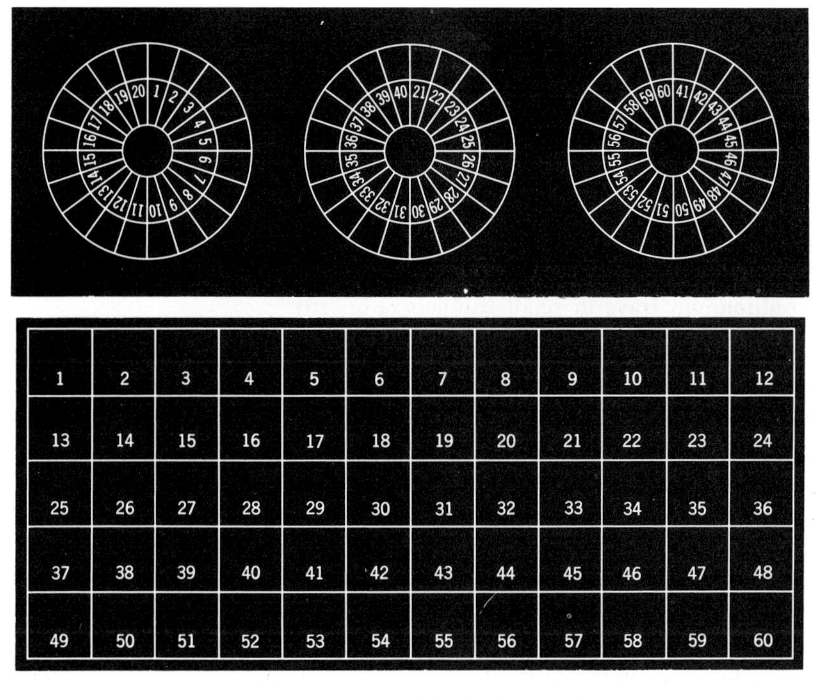

Fig. 15. Microscope slides for foraminifers.

fers sift through to a 115-mesh screen (13,225 mesh to the square inch) below. This residue is permitted to dry. Then under microscopic projection foraminifers are picked out with the moist tip of a camel's-hair brush.

Representative assemblages from the collection sites are usually placed in slides with three compartments. The slide is made up of a rectangular piece of black paper, a thin cardboard with three holes in it, and a glass microscope cover slide. In classifying the materials a work slide with 60 numbered squares on black paper is sometimes used to keep the different genera or species apart after they are recognized and segregated. The foraminifers adhere to a gum tragacanth solution (a white or reddish gum substance) that is spread over the black paper. Specimens can be freed from the gummed surface by application of moisture. This, then, is called a work slide. Type specimens rate the exclusive distinction of being mounted in the center of a large compartment on one slide. This makes them readily accessible for comparison. Another kind of slide is even more satisfactory for comparative purposes. This has a large round compartment arranged like a wheel. The spokes in the wheel separate it into as many as 60 subdivisions. In each subdivision are stuck numerous individual foraminifers of each species of a genus. The gum tragacanth holds them in place. In the hub, directly in the center of the field, is the spot for the identification of foraminifers. Thus the specimen can be readily compared with all the species on the slide with a minimum effort. In the classroom twenty-compartment slides are used to give better representation on one slide. The research collection of slides with specimens are numbered, labeled, and stored on aluminum trays in small cabinets (Fig. 15).

Collecting and Preparing Fossil Plants. The same caution against contamination is required in collecting and curating fossil pollens. Airborne pollen will sometimes get mixed with the fossil samples. Freeing fossil pollens from their rock matrices is a much more complicated procedure than it is with foraminifers. The rocks are digested by hydrofluoric acid. Part of the sample is then placed in a centrifuge and later decanted (liquid drained off). This is repeated many times by applying other acids and adding distilled water and alcohol for cleaning the pollen.

When fine-grained sandstones, shales, or volcanic ash are parted along the bedding plane compressed fossil leaves as lignitic materials are sometimes exposed. These leaves are nearly, but never completely, flat. Most identifications of fossil leaves have been made from specimens exposed in this way.

Techniques known as the *Peel and Transfer Method,* developed by John Walton, University of Glasgow, and by the late William Henry Lang, University of Manchester, have made it possible to examine both sides of the fossil leaf. The stone is cut to a flat surface above, but close to, the fossil leaf. The side with the exposed fossil is coated with a cellulose acetate solution and then fused to a glass slide with hot Canada balsam. After moistening the slide and rock containing the specimen, both are coated with hot paraffin wax. Later when the wax is hard it is cut away to expose the rock. The siliceous rock is then dissolved with hydrofluoric acid, leaving the delicate fossil attached to the balsam on the slide. The newly exposed surface is almost perfect. Finally, the balsam may be dissolved and the specimen can be mounted on a cellulose acetate film. These are excellent for detailed study.

Fossil wood is best studied by means of thin sections. These sections are mounted on glass slides for observation under the microscope. Nearly every museum has a fossil log or part of one on display.

Curating Fossils

Care of specimens in the collections of any museum is one of its most important functions. Modern museums are not old stuffy buildings, replete with cobwebs and guarded by bearded custodians, but active research organizations. Some use much of their space and the time of their trained staff for arranging public displays, whereas others concentrate on teaching and research. Men dedicate their lives to these activities. It is a genuine lifelong interest, not a whim or fancy.

A curator must have the utmost respect for the care and preservation of specimens at all times. Not only the specimens but all facts and records pertaining to the fossils must be guarded for present and future workers. This may be called a *museum conscience.* As new techniques in preparation and research become known, collections made early in the history of an institution's activities may assume new values. We now find that most fossils show traces of radioactivity. How important this may be is not yet known.

It is sometimes difficult to convince individuals against "private-hoarder" tendencies. No individual should lay personal claim to museum specimens. Many people have the mistaken notion that fossils have great monetary value because a few individuals who could afford it paid liberally for certain specimens or because some museums sacrificed their precious funds to purchase private collections. Occasionally people selfishly attach monetary value to fossil specimens, without recognizing that their value to society is through scientific study.

Several years ago a fossil horse skull was unearthed by a road crew in Nevada. The man who first laid hands on it claimed it as his own. John C. Merriam of the University of California learned of the discovery. Later Merriam called at the man's home but could not persuade him to part with his find. Two years later Merriam happened by and again inquired about the fossil horse skull. The man's wife was not sure but she thought it was in the chicken coop. Though they searched the place the fossil could not be found. About that time her son arrived home from school to tell them it had been thrown out in the back lot. There they found it badly shattered. Merriam gathered up the pieces he could find. All that remains of that beautiful skull is now in the Museum of Paleontology at the University of California.

Catalogue Cards and Labels. As soon as specimens reach the museum an *accession card* should be filled out, preferably by the collector or by the curator. This card bears all kinds of pertinent data. It may refer to a single tooth sent in by a local farmer or to many cases of fossils

ACCESSION CATALOGUE — UNIVERSITY OF CALIFORNIA MUSEUM OF PALEONTOLOGY

Date Rec'd 15 January 1954 Acc'n No. 1493
Nature of Material Eight boxes of fossil mammalian specimens from Australia
Rec'd from Paul F. Lawson How Obtained South Aust. Mus. and Univ. of Calif. Expedition
Address South Australian Museum, North Terrace, Adelaide, S. Australia
Correspondence Lawson - Stirton: 4 Nov. '53 Stirton - Lawson: 27 Jan. '54 Field Notes R. H. Tedford (1953. R. A. Stirton; 1953.
Collector N. B. Tindale; P. F. Lawson; R. D. Woodard; R. H. Tedford; H. E. Reynolds; R. A. Stirton When Collected March - August, 1953
Locality and Loc. No. Lake Menindee, V5371; Lake Callabonna, V5374; Lake Palankarinna, V5367.
Catalogue Nos. 44380-44401; 44463-44540; 44542-44560; 44567-44582; 44587-44644; 45343.
Remarks Joint expedition with South Australian Museum
Date of Entry 15 March 1954 Entered by

LOCALITY CATALOGUE — UNIVERSITY OF CALIFORNIA Museum of Paleontology

Loc. No.	1st Acc'n No.	Description of Locality
V5367	1493	Lake Palankarinna

Escarpment on west side of Lake Palaukarinna, east of Lake Eyre; 18 mi. S. 75° W. of Etadunna Station homestead. Military grid reference 656431, ordinance sheet Marree, South Australia, H54/1.2.5.6, zones 5 and 6, first ed. 1942, scale 1:506880. Fossils from interbedded sandy stream channel about 35' above conglomerate in greenish and reddish gypsiferous clays.

Numerical Catalogue—VERTEBRATES

Acc'n Nos. 1493 Date of Entry 9 June 1954 Entered by
General Locality Lake Palankarinna

Mus. No.	Name	Date	Collector	Orig. No.	Loc. No.
44381	Fig. Prionotemnus palankarinnicus Stirton Holotype, mand.	4 Aug, 1953	R.A. Stirton	RAS 4759	V5367
44382	Fig. " " Paratype, max.	"	"	RAS 4760	"
44383	Fig. " " Mt. IV, phalanges	8 Aug, 1953	"	RAS 4794	"
44384	" " " max.	6 Aug, 1953	R. H. Tedford	RHT 262	"
44385	" " " mand.	30 July, 1953	R. A. Stirton	RAS 4736	"
44386	" " " "	6 Aug, 1953	"	RAS 4770	"
44387	" " " "	7 Aug., 1953	"	RAS 4779	"
44388	" " " "	"	R.H.T		
44389	" " " "	4 Aug, 1953			
44390	" " "	7 Aug., 1953			

Systematic Catalogue—VERTEBRATES — UNIVERSITY OF CALIFORNIA MUSEUM OF PALEONTOLOGY

Case No. 470
Name Prionotemnus palankarinnicus Drawer No. 5

Mus. No.	Loc. No.	Place	Acc'n No.	Material
44381	V5367	Lake Palankarinna	1493	mandible (Holotype), figured
44382	"	" "	"	maxillary (Paratype), "
44383	"	" "	"	Mt. IV, phalanges (Paratype)
44384	"	" "	"	" " "
44385	"	" "	"	maxillary "
44386	"	" "	"	mandible "
44387	"	" "	"	" "
44388	"	" "	"	" "
44389	"	" "	"	" "
44390	"	" "	"	" "

Fig. 16. Museum catalogue cards. Adopted from Camp.

from an expedition to South Africa. Next, *locality cards* are written for all separate sites. This should be done by the collector, since the source of the information is associated with his field numbers. After the specimens are prepared the different individuals are entered on *numerical cards*, which refer to the date, the collector, his field number, and the museum locality number. Once the specimens are adequately identified the names and numbers of the individuals are entered on *systematic cards* (Fig. 16). This makes all specimens referable to a genus or to a species more readily accessible. All such records, including specimen labels, are written in Higgins eternal writing ink because most inks in time will fade. Field notes with the collector's field numbers should also be written in eternal writing ink, since they are the final source of information for all specimens.

Museum Interpreted as a Book. A well-curated museum of paleontology could be interpreted as a story of prehistoric life on earth. The chapters could be the different fields of study, i.e., paleobotany, foraminiferology, paleomammalogy, paleoherpetology, paleornithology, and any others represented in the museum collections. Major divisions of the chapters could include the assemblages of plants and animals from different Eras, Periods, and Epochs of geologic time. The leaves in the books could be the trays of specimens in the cases. And, finally, the illustrations and text on the pages could be represented by the individual specimens, labeled and arranged in a systematic manner. A most fascinating book it would be to those who took the time to read it carefully.

Illustrations and Photography. One of the most important functions in the museum is the work of the scientific illustrators who make drawings to accompany the specimens being described or discussed in technical and semipopular articles by members of the museum staff. An accurate illustration of a specimen is frequently more important than the writer's description of it. Our opinions or interpretations of a fossil may change, but a drawing accurately executed is of permanent value.

This is of paramount importance with microfossils. Thousands of individual organisms are represented in a single sample. This is especially true in the foraminifers among which many kinds are still unknown. Until the fauna of a given area and stratum becomes exceedingly well known, individual variation is so marked that, unless the species are well illustrated, species lists and check lists will remain inadequate sources of reference. Too often such lists have become sources of unfortunate taxonomic misconceptions.

Methods. Larger fossils are measured as accurately as possible with proportional dividers, using a right angle for a control in taking the measurements. Microscopic organisms, grass seeds, or tiny teeth may be done with the aid of a reticule in the eyepiece of a microscope to facilitate such delicate work. Some illustrators prefer to use a camera lucida from which the reflection of the fossil is traced on the paper.

Different methods are used in the preparation of black-and-white illustrations. In all drawings the main features are first outlined lightly in pencil. Light is always directed from the upper left-hand corner. Both *wash* and *lead pencil* drawings are done on Strathmore board which has a kid finish. When the drawing is completed a fixative is applied to the surface to prevent smears. These methods are employed in illustrating foraminifers and other microscopic organisms, though wash drawings are frequently used for bones and teeth of larger specimens. Another similar method is with *carbon pencil*, using blending stubs on Ross board, a type of scratch board with a chalk-coated finish (Fig. 17).

The greatest number of megascopic scientific illustrations are in *pen* and *ink*. Both *line* and *stipple* methods are widely adopted. These are done on plate-finished Strathmore board with Higgins India drawing ink. Shadows and highlights are accentuated by the weight of the lines or

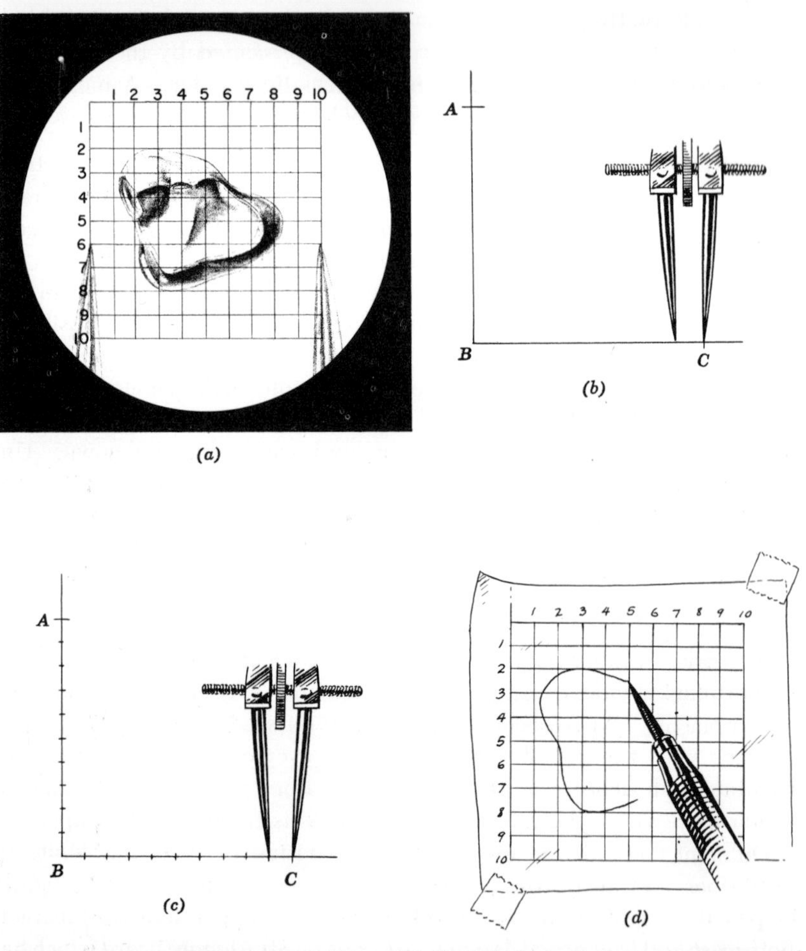

Fig. 17. A microscopic specimen drawn with the aid of a reticule in the eyepiece of the microscope. (a) Specimen as seen through the eyepiece of the microscope with the reticule superimposed and the bow-dividers measuring, in focus, the width of grid. (b) Legs of right angle A-B, B-C equal the number of magnifications desired or stepped off by the bow-dividers. (c) Right angle is then divided into ten equal parts to correspond to the reticule grid. (d) Grid completed on paper and tracing paper in place. The drawing by reticulation is now begun.

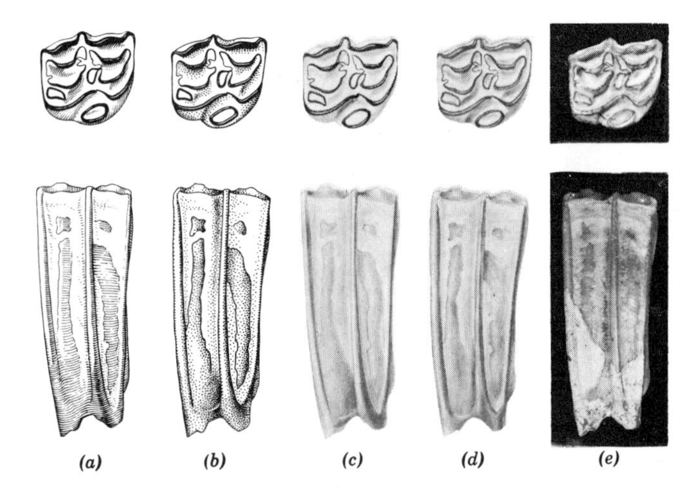

Fig. 18. Five methods of illustrating a fossil horse tooth: (a) *line drawing;* (b) *stipple;* (c) *wash;* (d) *carbon pencil on Ross board;* (e) *photograph,* $\times\ ^{2}/_{3}$.

stipples. Delicate work is done with a fine pen. The Esterbrook 352 pen is flexible and lends itself to shading. For larger specimens coarse pens, such as the Gillott 170 and 404 or the Esterbrook 356 points, are excellent. Pen and ink drawings are much cheaper to reproduce than other kinds of illustrations, since they can be done by zinc etchings on the text paper, whereas halftone productions require copper cuts printed on coated paper, a more expensive process (Fig. 18).

Museum illustrators frequently make maps, charts, and labels. Zip-A-Tone is now used in replacing the older method of crosshatching to show aerial patterns on both maps and charts. It is also used to indicate vacuities in skulls, bones, or other fossils. This material can be purchased in numerous patterns from art-supply houses. Nowadays most lettering is done with a Leroy lettering set, but when the illustrator does the hand lettering he uses a speedball pen.

Another widely adopted method of illustration is *photography*. Photographs are usually easy to obtain but costly to reproduce. Nevertheless, excellent results can be had from modern methods. Unfortunately, the perspective renders the periphery in most photographs indistinct, whereas drawings can be done in orthographic projection. Some fossils, because of their preservation, will not photograph well without a treatment of ammonium hydroxide and hydrochloric acid fumes to give them a white coating of ammonium chloride.

Many microscopic structures are made available by photomicrography by means of attaching a camera to a microscope.

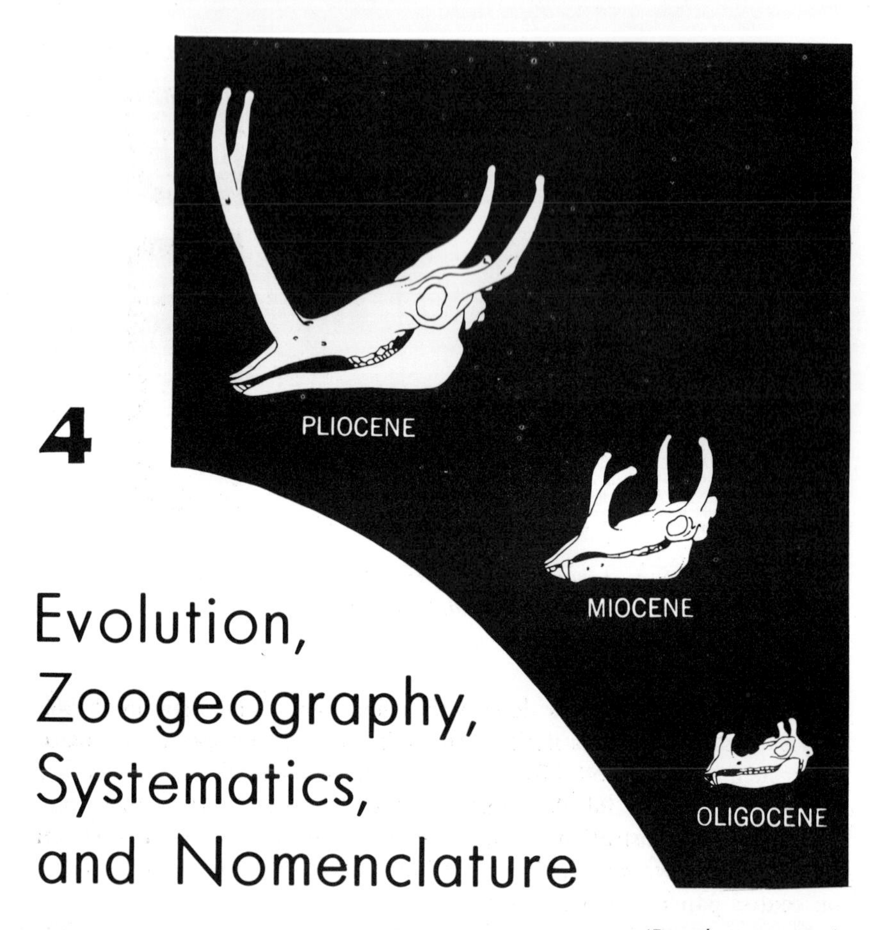

4 Evolution, Zoogeography, Systematics, and Nomenclature

(From bottom to top):
Protoceras, Syndyoceras, Synthetoceras

No subject has caused more argument and misunderstanding than that of evolution, most of which has arisen because of an inadequate knowledge of the facts. From the irrefutable evidence of the long history of life on earth it is now clearly apparent that profound changes in organisms have transpired.

Organic Evolution

The fossil record continues to show that smaller and simpler animals and plants occur in the oldest formations and that their descendents became progressively more complex in their structures and functions in the succeeding Eras and Periods of geologic time. Organisms have evolved into almost every conceivable possibility in their adaptive radiations. In some families, for instance, that of the horse (Chapter 31), one line of descent (there were many different lines) continued for 60 million years, from eohippus, *Hyracotherium*, an animal no larger than a fox terrier which had four functional toes on nearly flat front feet, up to our modern horse, *Equus*, with its single-toed hoofs. Such changes can be traced through almost imperceptible intergradations from one species to the next and eventually from an advanced species of one genus into the first species of the next genus, and so on. Though the rate of evolution was not always the same, and in some features it was even somewhat reversed, there were divergent phyletic lines leading to different adaptations for life. Some groups faded and died along the way, but others survived and propagated their species.

This is the kind of evidence available on the organic evolution and basic relationships of plant and animal life. It is supported further in the comparative anatomy of the organisms, in their embryonic development, and in their physiology. This evolutionary pattern has been compared to the growth of a tree from the time its cotyledons broke the ground surface to its different stages of growth and finally to the periphery of foliage and bark of a large elm as representing the end products.

Organic evolution opposes the ideas that life was *spontaneously generated* (disproved by Francisco Redi, Italian naturalist and court physician (1626–1698), that it arose by *special creation* (as thought by Linnaeus, Cuvier, Owen, and others), or, as the *cosmozoic theory* would explain, that it reached the earth as living matter from some other source in the universe. It supports the conclusion that any given plant or animal, past or present, descended from an earlier and somewhat different ancestral form by transmission and change in hereditary characters.

Mutability of Species: Darwin. One of the most stimulating contributions to the biological sciences was Charles Robert Darwin's *On the Origin of Species by Means of Natural Selection, or the Preservation of Favoured Races in the Struggle for Life* (1859). The idea of evolution was conceived by scientists prior to that date. Even the great Greek zoologist and philosopher Aristotle thought there was some mysterious tendency for organisms to change in time from simple to complex forms. But the abundance of evidence to support organic evolution had not been assembled, and nothing convincing had been proposed as to how it took place. The distinguished French anatomist and taxonomist Jean Baptiste Pierre Antoine de Monet, Chevalier de Lamarck suggested his theory of "inheritance of acquired characteristics" as early as 1801. He felt that the development or loss of certain organs or structures in the animal were effected through use and disuse. He thought that these characteristics were inherited by the individuals and passed on to the next generation, suggesting an inheritable progression in evolution. But Darwin would have none of "Lamarck's nonsense," as he called it. Even so, Lamarck was the first recognized scientist to outline a modern concept of evolution, even though there is no evidence that body cells through use and disuse effect germ cells, as he had suggested.

In 1831, when Darwin was a young man, a British ship the H.M.S. *Beagle* was scheduled to circumnavigate the globe (1831–1835). It was recommended to the captain that Darwin join the venture as a naturalist. Unfortunately, he suffered severely from seasickness throughout the voyage, which not only inhibited his work but impaired his health for the remainder of his life. Irrespective of this, Darwin was a keen observer of all of the manifestations of life about him. On the trip around the world he gathered data concerning animals and geology. In Argentina he saw and collected fossil bones of some of the peculiar mammals that had inhabited that area during the Pleistocene. In South America and elsewhere he observed the peculiar existence of many closely related species within relatively restricted areas, yet not living together. Because of the influence of his early belief in special creation it was difficult for this young naturalist to understand why it had been necessary to create all of these species with such slight differences in narrowly restricted and adjacent areas. The question of the mutability of species (mutations, p. 52) was foremost in his mind from that time on.

When he returned to England his bad health greatly curtailed his activities, and in 1842 he and his self-sacrificing wife settled in the small town of Down in Kent. There he reviewed all his data from the field and continued his observations on animals as best he could. He recognized the variability in animals, even down to the individuals in a litter or in a family. No two, except in the case of identical twins, were alike. As early as 1838 he had read Thomas Robert Malthus' *An Essay on the Principle of Population* (1798) in which it was shown that if all individuals produced were to live, within a few hundred years there would not be enough space for them on earth. This struggle for existence gave Darwin his idea of *natural selection*. It is interesting to note that Alfred Russell Wallace, a British naturalist and close friend of Darwin, who had been collecting and studying the floras and faunas of Malaya and the East Indies, had come to the same conclusion independently. Nevertheless, it was Darwin's clear exposé, based on indisputable facts and Thomas Henry Huxley's brilliant lectures, that convinced the scientific world and many others that evolution had taken place.

It is now granted that Darwin was correct in his idea that evolution took place by natural selection, but his explanations as to how and why it came about were not entirely correct. Much discussion and criticism centered around these points in his theory of evolution during the next 70 years. There were two schools of thought widely known as *neodarwinian* and *neolamarckian.* These newer versions, as the names imply, did not fully subscribe to either Darwin's or Lamarck's theories but incorporated into them new ideas, which in some ways were no nearer to the truth.

The science of genetics began to play an important role in the study of evolution at the beginning of the twentieth century. Geneticists were able to show that germ plasm alone was responsible for the hereditary transmission of characters. This was initiated as changes in the *genes,* which are supposed to be chemical units in the *chromosomes* of the nuclei in the germ cells. Genes are not observable, though it has been demonstrated that they have definite positions in the chromosomes.

Synthetic Theory of Evolution. During the past 25 years paleontologists, geneticists, morphologists, and systematic neontologists have pooled their efforts and thus derived a *synthetic theory of evolution.* This is based on the study of interbreeding populations. The paleontologist has been able to show what has happened in different populations in various lines of descent through millions of years. Observations made during the lifetime of an individual or even within a few hundred years are far too limited to demonstrate the evolution from one species to the next. Studies in paleontology have shown that evolution did take place and what the patterns of evolution were like, and geneticists and biologists have given us clues as to how the changes came about in individuals and in populations. Ecologists have shown that existence is both a competitive and cooperative process. Numerous factors enter into evolutionary changes, many of which we do not fully understand, but this in no way invalidates the factual evidence now available on the subject. All in all, evolution is an extremely intricate process which cannot be fully discussed here.

New characters and characteristics in individuals of local populations usually but not always arise as minute differences called *mutations.* Thus offspring are never totally like their parents, though they bear marked resemblances to one or both. Nor are the young exactly alike except in identical twins developed from the same egg. This is *variation.* Mutations may arise by *random fixation* of new chromosomal combinations in sexual reproduction, as *gene mutations* which apparently involve a change in the composition of the chemical unit that makes up the gene, or by a loss of one quarter of the parental genes, as in small populations in genetic drift. Even so, it has not been discovered what causes these changes. These mutations, however, are slight in comparison with the more persistent characters in the race that are inherited from the parents and their immediate ancestry. In the continuous appearance of new combinations of gene and chromosome arrangements in sexual reproduction, characters and characteristics appear upon which natural selection can operate. As Dobzhansky states, "The human species, and all other species of organisms which live in the world at present, were not preformed in the primeval amoebae or in the primordial viruses. They have evolved gradually in the history of the earth under the control of natural selection."

Gene mutations, new chromosomal combinations in sexual reproduction, and loss of one fourth of parental genes in small populations, then, are thought to produce the variability we see in interbreeding populations or in the individuals of families, litters, etc. In every population some individuals have more offspring than others. This is differential reproduction. Greater fecundity (number of offspring produced), sexual activity, and other factors in succeeding populations will result in greater numbers with these qualities. As a result, *natural selection* progressing toward adaptive features will favor those that contribute most to the gene pool of the interbreeding population. Thus the fittest are not necessarily the ones that appear to be the most physically fit. They may possess other qualities that make them more suited to survive in all extremes of their environment. This differs from the Darwinian concept. While the environment changes, as it does almost continously to a lesser or greater degree, those individuals with the hereditary potential to cope best with these changes will tend to survive in greater numbers and to invade and sustain themselves in other ecologic niches, for various reasons not available to their predecessors. In this way the environment is the directive force in evolution, though the environment does not directly change the genetics of the organisms.

Zoogeographic Regions and Provinces. It has long been recognized that different parts of the land surface and the oceanic waters of the earth have their characteristic animals and plants. The largest units are called *zoogeographic regions* (sometimes called realms). Their areas vary somewhat when based on the different groups of living organisms (mollusks or foraminifers, mammals or spiders), but for the most part the regions and their subdivisions known as *faunal provinces* can be synthesized into a useful terminology. These patterns of distribution have developed and fluctuated throughout geologic time (Fig. 19).

CONTINENTAL ZOOGEOGRAPHIC REGIONS. The continental zoogeographic regions are currently called *Palaearctic*, *Nearctic*, *Neotropical*, *Ethio-*

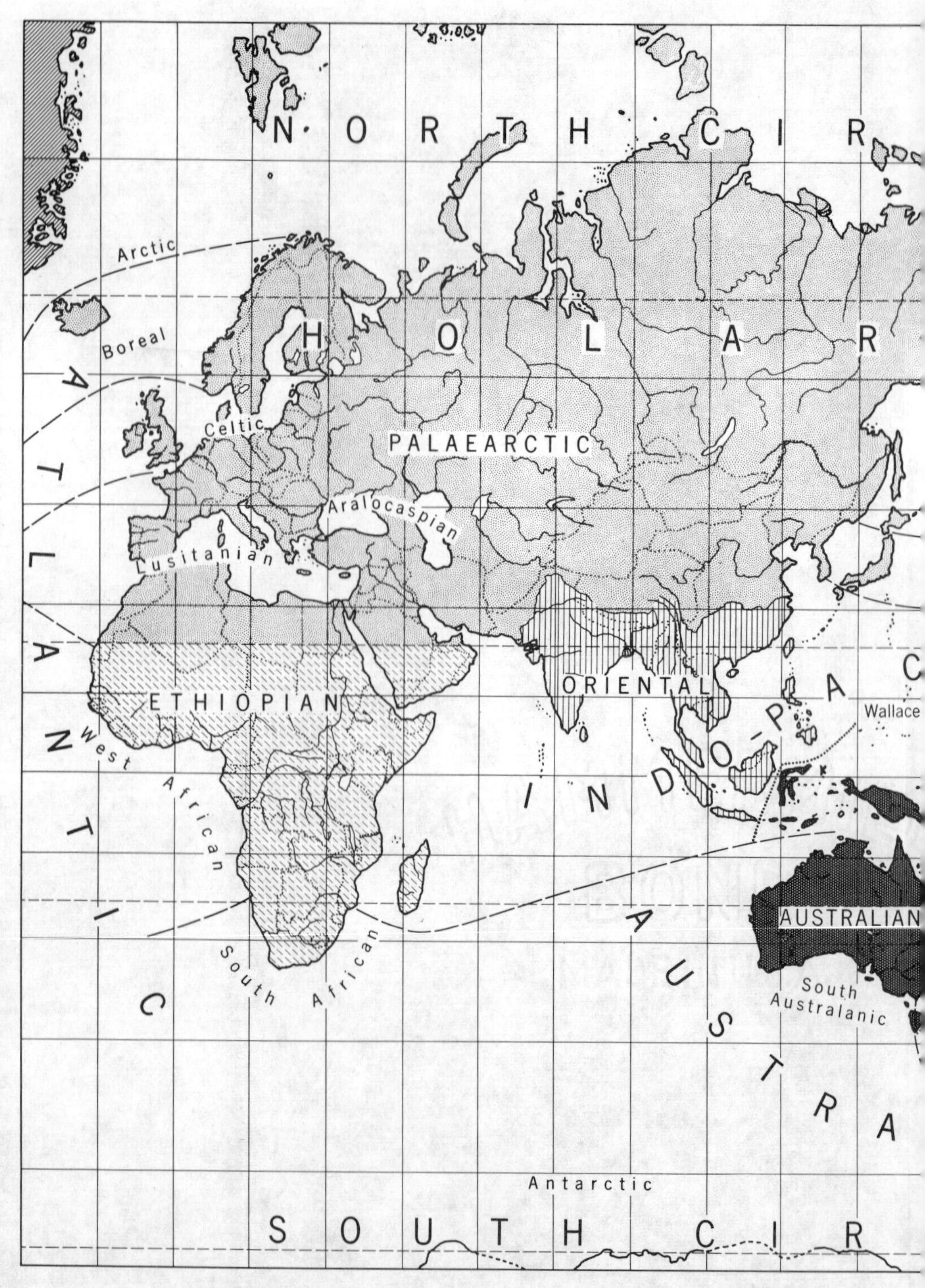

Fig. 19. Zoogeographic regions and provinces after Davies and others.

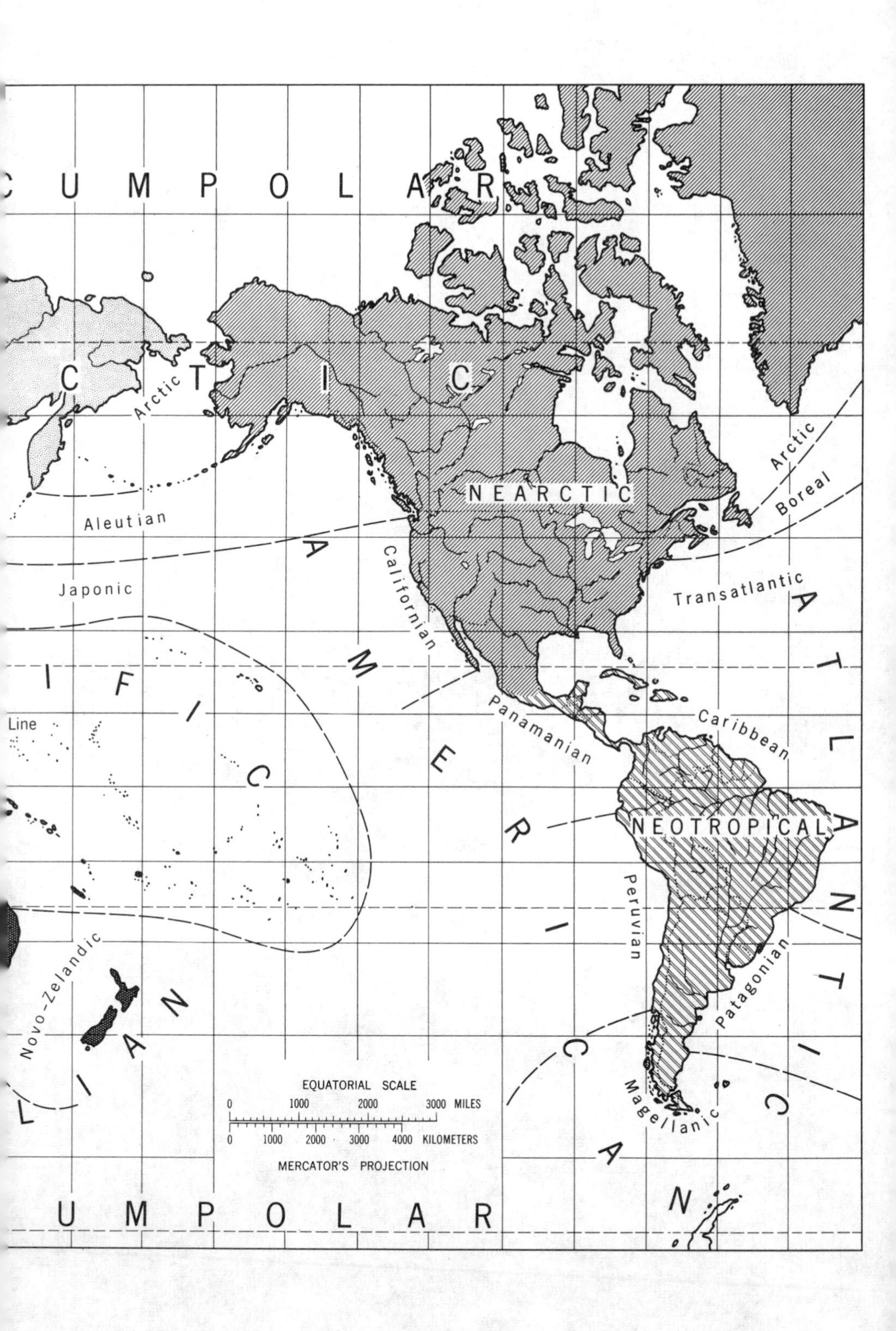

CUMPOLAR
ARCTIC
NEARCTIC
Arctic
Aleutian
Japonic
Californian
Arctic
Boreal
Transatlantic
AMERICAN
PACIFIC
Line
Panamanian
Caribbean
NEOTROPICAL
Peruvian
Patagonian
Novo-Zelandic
Magellanic
ATLANTIC
EQUATORIAL SCALE
0 1000 2000 3000 MILES
0 1000 2000 3000 4000 KILOMETERS
MERCATOR'S PROJECTION
UMPOLAR

pian, Oriental, and *Australian.* The evidence from mammals, birds, some reptiles (snakes and lizards), insects, and flowering plants for the most part are used in outlining these geographic distributional units. The Australian region is more distinct faunistically than the other regions. This is because it has been isolated from the other continents at least since Mesozoic time (the Era of dinosaurs), and as a result its fauna includes the egg-laying platypus, the kangaroo, the emu, the Australian lungfish, and many other kinds of animals not found elsewhere in the world. On the other hand, there has been intermittent mingling of certain groups of terrestrial animals between the Palaearctic and Nearctic regions (together they are frequently called *Holarctic*) since the Devonian, when life on land started expanding. Nevertheless, the faunas as a whole from each region are distinctive. Though there are many mammalian families common to the two regions, deer, bears, dogs, horses, rhinoceroses, and the like, few genera are found common to both regions. Some families not found in both regions are Palaearctic—hyenas, giraffes, cattle, pigs, etc.; Nearctic—raccoons, pronghorns, oreodonts, pocket gophers, etc. There is much greater distinction in the genera and species. Similar differences are seen in the Nearctic and Neotropical and in the Palaearctic, Ethiopian, and Oriental. Though these regions were based on the distribution of living organisms, the same principle can be applied in the geologic Epochs and Periods of the past.

MARINE ZOOGEOGRAPHIC REGIONS. Marine life also lends itself to regional and provincial designations. Six regions based on living benthonic (bottom-dwelling) mollusks have been designated by Philip Lutley Sclater, Alfred Russel Wallace and Samuel Pickworth Woodward: *Australian, American, Atlantic, Indo-Pacific, North Circumpolar,* and *South Circumpolar.* As with life on land, such groups of marine organisms as foraminifers, corals, and marine arthropods do not fit into these zoogeographic units of distribution, though some of them may do so. The Australian region has been divided on the basis of the mollusks into the *Novo-Zelandic* and *South Australian* provinces; the American, into the *Aleutian, Californian, Magellanic, Panamanian,* and *Peruvian;* the Atlantic, into the *Aralocaspian, Boreal, Caribbean, Celtic, Lusitanian, Transatlantic, Patagonian, West African,* and *South African;* the Indo-Pacific, into the *Indo-Pacific* proper and the *Japonic;* the North Circumpolar with the *Arctic;* and the South Circumpolar with the *Antarctic.* The Aralocaspian is based on invertebrate faunas, many of which are middle and late Tertiary, more or less intermediate between those living in marine and fresh-water environments.

Smaller zoogeographic units both on land and in the sea are called *communities* or *stations.*

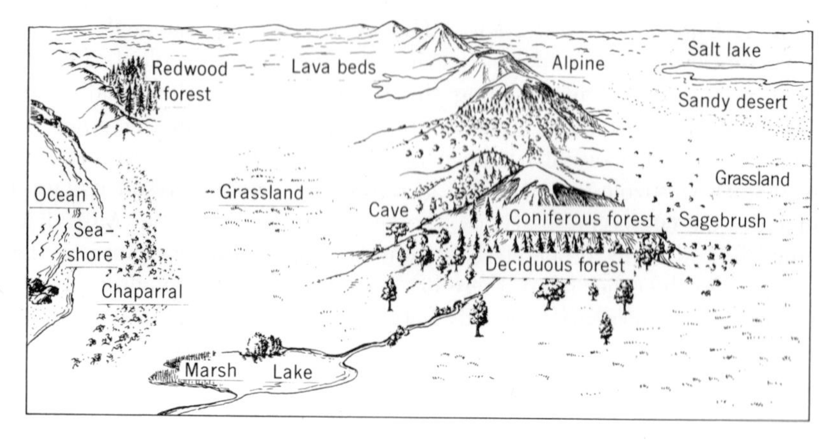

Fig. 20. An example of land environments in the middle latitudes of North America.

Environment. The environment is the physical and biological surroundings in which an organism lives.

ON LAND. Phenomena affecting land environments include solar radiation, atmospheric conditions, altitude, precipitation, soil, water, temperature, winds, and other kinds of life both plant and animal. These determine the *habitats* or the places in which animals reproduce, grow, live, and die. There are many kinds of environments, such as lakes, swamps, streams, islands, lake and stream borders, chaparral, deciduous forests, savannahs, grasslands, conifer forests, eucalyptus forests, rain forests, sagebrush, saltbrush, gibber plains, sandy deserts, mountains above snow line (alpine), caves, and tundras. Each of these has its own kind of life, though some species may range from one into the other. For example, the North American pronghorns range from the grasslands into sagebrush deserts, but for the most part the environmental tolerances of plants and animals show marked restrictions (Fig. 20).

Diastrophic movements in the earth's crust resulting in changes in topography, atmospheric conditions, and other factors have caused an almost continuous alteration in continental environments. All the while the *animals and plants either remain in the same area and evolve, disperse elsewhere and evolve, or die out.*

IN THE SEA. Marine environments have been recognized on the basis of open water or bottom conditions, the depth of the water, the shore and tidal areas, proximity to the shore on the continental shelf, the continental slope, and in the ocean deeps. These environments may be affected by temperature, winds, typhoons, light, water pressure, local currents, sediments derived from rivers, or nature of the life near or on the ocean surface or on the bottom (benthos).

The *Benthic* environments mostly on the floor of the ocean include the Supratidal, Neritic, and Oceanic. The *Supratidal* encompasses the shore areas of storm tides and sprays as well as environments along shores affected by saline water (Fig. 21).

The *Neritic* is the floor of the continental shelf between the high-tide mark and the edge of the shelf at 50 to 100 fathoms in depth (fathom = 6 feet). This is divided into the Intertidal, Middle, and Lower Neritic environments. The *Intertidal* (Upper Neritic) is the floor surface between the highest and lowest points of the tide; whereas the *Middle Neritic* is between the lowermost point when the tide is out and a depth of 25 fathoms; and the *Lower Neritic* is the bottom on the shelf between the point at 25 fathoms and the edge of the shelf somewhere between 50 and 100 fathoms in depth.

In past geologic Periods and Epochs the areas included in the Neritic fluctuated considerably with the subsidence and elevation of continental platforms and the presence of epeiric seas and marine troughs. Embayments spread over and retreated from many of the areas that are now land. Nearly all of the marine fossil record has come from these areas.

The Oceanic benthos environments are the Bathyal, the Abyssal, and the Hadal. The *Bathyal* is the steep surface of the continental slope. It is a somewhat varied environment, dropping from the edge of the continental shelf down to 1000 or 2000 fathoms in different places. There the light diminishes as the depth increases.

The *Abyssal*, by far the largest ecological unit in the world, covers the dark depths of the ocean floor below the Bathyal. Some of the deepest trenches in the abyss are more than 6 miles below the surface waters. This is called *Hadal*.

Organisms greatly restricted in their ranges within certain depths are referred to as *stenotopic*. Others with greater latitude in their depth range tolerance are called *eurytopic*.

Pelagic environments are in the open water above the benthos. Organisms with two kinds of habits occupy these waters. *Planktonic* forms have no means of locomotion of their own but float or are moved about by ocean currents, winds, or waves. Though some adult animals may be benthonic, the larvae are frequently planktonic. *Nektonic* organisms, on the other hand, swim about by their own efforts.

In the Neritic area over the continental shelf or near the coast the Pelagic environment is called *Neritopelagic*. In the open ocean (Oceanic) there are three environments. The *Epipelagic* is in the upper oceanic waters; in different latitudes it ranges between 40 and 100 fathoms. Be-

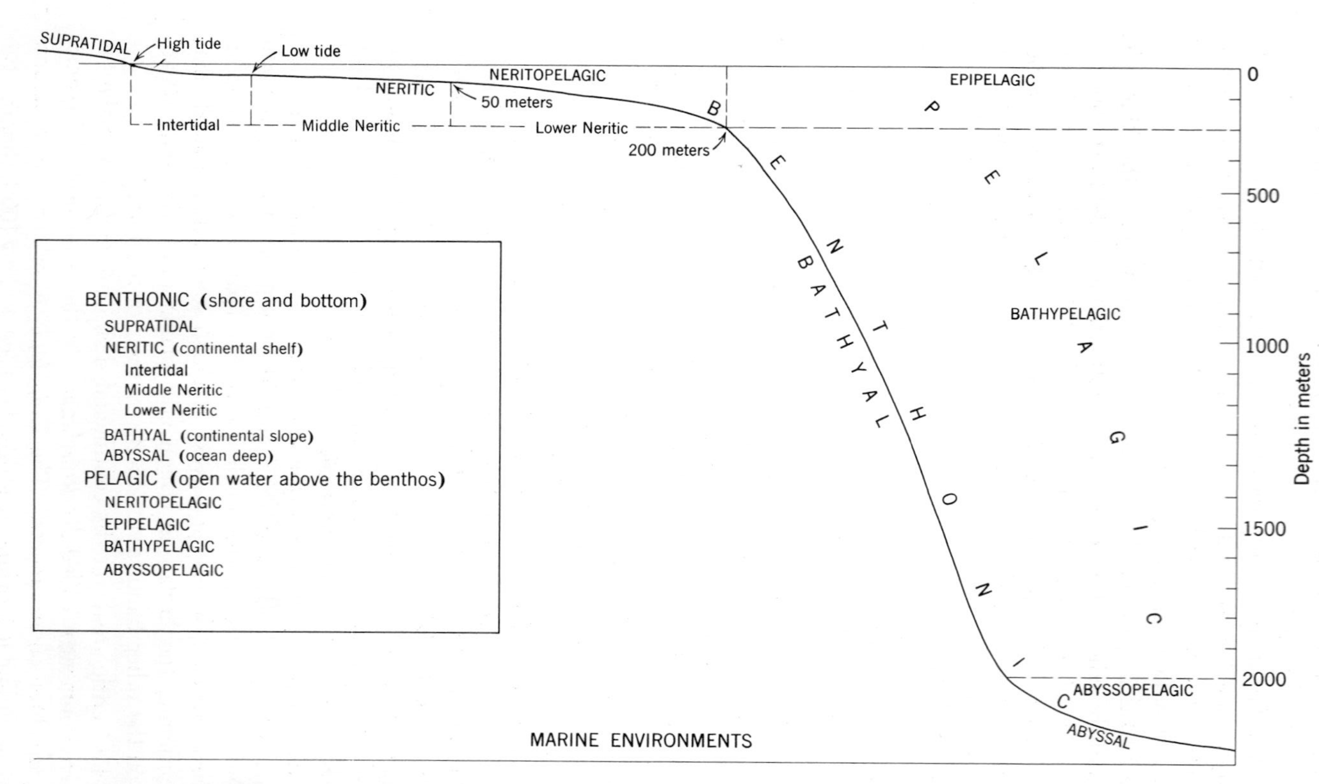

Fig. 21. A classification of marine environments, modified from a "Report of the Committee on a Treatise on Marine Ecology and Paleontology, 1948–1949."

low this where little light penetrates is the *Bathypelagic* in which strange deep-sea fishes and many highly modified invertebrates are found. The lower limit is between 1000 and 2000 fathoms. The *Abyssopelagic* is the environment below the Bathypelagic into the ocean deeps.

Distribution. Dispersal, evolution, and speciation are intricately related processes. As their numbers increase, species and subspecies spread to the extreme limits of the environments suitable to their existence. This has been accomplished between such distant areas as Alaska and Argentina by the puma (mountain lion) in a relatively short time. Thus the same species occurs in both places, with its subspecies interspersed geographically between. If their original center of dispersal had been the western United States, as it may well have been, those populations dispersing to the south became tolerant of the most adverse conditions in warm, moist, southern latitudes, and those spreading northward became acclimated to winter extremes and other conditions in northern areas. It should be constantly borne in mind that physiographic features of the earth's crust change continuously while animals and plants are mutating, and new environments are constantly appearing.

It is reasonable to assume that distributions like that of the puma are the result of a normal spread of populations into new and suitable areas of occupation. This is accomplished mostly by small evolutionary steps in which the succeeding generations of offspring through their mutations are able to exist during the extremes of certain environments that their ancestors could not have survived.

Occasional distribution may result from the seasonal migrations of birds, mammals, or fish, from the spawning migrations, not necessarily seasonal, of some fish, and from the one-way emigrations of man, squirrels, lemmings, grasshoppers, or crickets. In this way certain individuals may linger behind and become established along the migration or emigration routes.

Less frequent means of distribution of life on land, in streams, and in fresh-water lakes are effected by the agencies of winds, birds, stream piracy, floods, and rafting. Insects, spiders, seeds, pollens, and birds are at times carried hundreds, and sometimes thousands, of miles over barriers that would otherwise restrict their distribution. Seeds of plants, small arthropods, and the eggs of certain animals may be spread by birds over shorter or longer distances. Life in the headwaters of streams may be diverted into other drainage systems by stream piracy. Swift currents in times of flood will transport organisms to other areas. Mammals, land snails, insects, and many other kinds of land life sometimes drift out from the mouths of rivers in times of torrential floods on floating debris or in an acre or more of dislodged mangrove trees drifting out to sea with their entangled roots holding the trees in an upright position. Oceanic islands like those in the West Indies, Madagascar, and Australia have been partly populated when such rafting materials chanced to break up on their beaches or were stranded in embayments near their shores.

Barriers to the distribution of animals and plants are effective in varying degrees, depending on chance and on the ability of the organisms to get across the barriers at certain times. Suitable environments may exist for long periods of time in adjacent or distant areas and will not be occupied by invading forms until populations of the animals or plants change sufficiently, the barrier is physically removed, or a chance dispersal occurs. A barrier to one animal or plant may not be a barrier to another. Some of the most conspicuous land-life barriers are oceans or seas, ice sheets, wide rivers, mountain ranges, deserts, rain forests, and temperature. For marine life, land, subsurface ridges (sea mounts), ocean deeps, strong, steady currents, and temperature tend to restrict its distribution.

Isolation and Speciation. As animals and plants evolve and spread into new areas, eventually their different *populations* tend to favor certain areas more to their liking. Though there are still only minor, seemingly insignificant, ecological differences in the areas occupied by two contiguous populations, they are sufficient to keep the two assemblages apart. This pattern in the distribution of populations extends throughout the geographic range of the initial species. Some areas and populations will be large, others will be small.

There is a multiple of factors, such as the nature of the food, soil, shelter, temperature, and rainfall, that causes these patterns to develop. Of course, the landscape continues to change while the organisms tend to keep pace in their evolution, move on, or become extinct. In time those that live become more and more different from their neighbors, which now could be called *subspecies*, and show some distinction from preceding populations. Unless for some reason the subspecies merge again, they will continue to be different and become progressively more distinct.

Species, Genera, and other Categories. "Species are actually (or potentially) interbreeding populations which are reproductively isolated from other such groups," Mayr, Linsley, and Usinger, 1953. In living plants and animals this can usually be determined, hence a species is thought to be *objective*. It is objective in the sense that all those individuals that will interbreed and give rise to fertile offspring can be included in the species, and those that cannot, or will not, do not belong to the species. The paleontologist must solve the species problem on morphologic characters, since he has no means of knowing about reproductive isolation in any given fossil sample before him. His interpretation is based on the magnitude of the characters and their occurrence in a single quarry or in a stratigraphic unit such as a member or a formation. Thus his species are *subjective.*

The delimitation of the genus and other higher categories are also subjective because the number of groups to be included in each is a matter of opinion. The genus is composed of a group of related species, the family, a group of related genera, the order, a group of related families, and so on through the higher categories. In some cases the category (genus, family, etc.) may be monotypic, i.e., it may be represented by only one species or one genus.

Barriers. The development of rivers, mountains, deserts, and many other kinds of barriers may permanently separate populations. At this stage, or even earlier, they may have become distinct *species* in *geographic isolation* from other related species. Such species living in different geographic areas are known as *allopatric* species, but sometimes distinct species eventually come to occupy the same general area; they are then called *sympatric* species, simply because they live side by side. If individuals of two species do interbreed (mostly they will not), the offspring are *hybrids.*

Other factors also effect isolation between two species. Seasonal breeding, i.e., one species breeding in the late winter and another in late spring, can be just as effective in isolation as thousands of miles of intervening territory between the populations. This is *reproductive isolation.* Also there is functional incompatibility in mating in which fertilization will not come about. Even if it does, as in the case of the horse and the burro, the resultant hybrid mule is almost always sterile. This is *physiological* separation. Psychologically different species usually will not tolerate each other. All of these factors, then, may be instrumental in isolation.

Adaptation. All animals and plants are adapted in a greater or lesser degree to the environments in which they live. This is also well exemplified throughout the fossil record. Hares, rabbits, and rodents, for example, are adapted to various means of locomotion: long-distance running (jack rabbit), short-distance running (cottontail), jumping (jerboa), quick movement in underground tunnel (gophers), climbing (squirrels), gliding (flying squirrels), swimming and diving (beavers). Furthermore, almost every part of the body from the microscopic cells to the limbs and teeth are adapted in some manner to the life that the animal or plant lives. Some of the most intricate adaptations which offer protection and methods of feeding and reproducing are seen in insects.

In some groups such as the marsupials of Australia we see that in their *adaptive radiation* or *divergent evolution* through millions of years they have come to resemble distantly related mammals in other parts of the world in many ways.

Such animals as the dolphin (a mammal), the ichthyosaur (an extinct reptile), and the shark (a fish in the broad sense) have superficially come to look much alike, but when examined closely it is seen that they are distantly related. This is *convergent evolution* in *adaptation*.

The early ancestors of the coyote and fox seemingly diverged from a common ancestor late in the Eocene; they then continued to evolve very much alike throughout the remainder of geologic time until we see the end products of that *parallel evolution* in the wilds today. Usually, as in convergent features, parallel evolution is expressed only in certain parts of the organism, i.e., body form, teeth, limbs, feet, etc.

OTHER EVIDENCE OF EVOLUTION

Comparative Morphology. All organisms are basically alike in possessing a cell-structure organization. All animals are composed of protoplasm, and all higher vertebrate animals have a pectoral limb structure, if it has not been lost secondarily, as in the snakes. These and other structures irrespective of their functions show distant or approximate relationships; though the forelimbs of the bat, the mole, and the whale are adapted to the functions of flying, burrowing, and swimming, each has a scapula, a humerus, a radius, an ulna, carpals (small bones in the wrist), metacarpals, and phalanges. There is a basic similarity in the construction of these parts in which they seem to be so dissimilar. This makes it quite clear that in their early ancestry these limbs were used in terrestrial locomotion, and if one wishes to compare the *homologous structures* further it can be found that even earlier they were used in swimming. Thus the *morphology* of the animals in question displays evidence in evolution.

Homology and Analogy. *Homologous* structures that are so dissimilar in function but of common origin (wing of bird and flipper of whale) contrast markedly with *analogous* structures that are similar in function but different in origin. A classic example is the wing of an insect that is derived from extensions of the body wall and the wing of a bat or of a bird that evolved from the forelimb of a terrestrial animal.

Comparative Physiology. Physiological similarities of the tissues and fluids in different animals show evidence of relationships; otherwise serums used in vaccinations would not work as they do. It is fortunate that we understand these basic relationships in organisms. If we did not, medical science upon which we all depend would be seriously handicapped. Hematin crystals obtained from the hemoglobin in the blood of vertebrate animals show their greatest similarity in the kinds of animals most closely related. *Serology,* then, offers us clues to the relationships and evolution of vertebrates.

Comparative Embryology. The embryonic developmental stages in animals *tend* also to give us an idea of the evolutionary trends in classes and orders. Most multicellular animals originate much in the same way from a fertilized egg or zygote. In the first stages of development the embryos of fish, salamanders, tortoises, chickens, hogs, cattle, rabbits, and man look very much alike. Indeed, structurally they are alike during almost the entire embryonic stage. As they gradually become more distinct, they can be recognized as four groups: (1) the fish; (2) the salamander; (3) the tortoise and the chicken; (4) the four mammals. Later some features make each of the individuals recognizable. *Embryologically* the fish and man start out with essentially the same kinds of circulatory system and heart as well as most of the organ systems in the body. Though the embryos do not show an exact duplication of their adult ancestors, they do show that homologous organs occur in rather distantly related groups at certain stages in their development.

Vestigial Structures. Structures with little or no use are vestigial in many animals. These reduced structures or organs in the process of being lost were not only functional but necessary in the early ancestors.

The splints lying back of the cannon bones (metapodials III) in horses are the vestiges of former prominent metapodials II and IV that bore phalanges with terminal hoofs in their Eocene, Oligocene, and Miocene predecessors. The remnants of the eyes in the molelike Australian marsupial, *Notoryctes,* are no longer visible externally through the hair. In the huge pythons there are two slender delicate bones with three other small bones at their distal ends lying over the ribs in the pelvic region. These are vestiges of a pelvic girdle and hind limbs that once served in quadrupedal locomotion of their distant ancestors. Many similar examples could be cited in all groups of animals.

Vestigial structures occur in both the soft anatomy and in the skeletons. Such features can be explained only as the result of degenerate evolution in these structures while other structures were taking over new and additional functions. Vestigial structures, too, offer evidence of evolution; they cannot be explained satisfactorily in any other way.

Systematics and Nomenclature. Systematics, or taxonomy, is the science of classification. In order to understand plants and animals, they must be arranged in an orderly manner based on their apparent relationships. These affinities are determined from the morphology, physiology, ecology, and genetics of the organisms. No one can master all the information now in print on the biota, but with systematic terminology the information is accessible, meaningful, and communicable. Without an orderly arrangement of this kind it would be impossible to comprehend the places in nature of all of the millions of organisms and their bearings on human problems.

In the fourth century B.C. Aristotle characterized animals according to their way of living, their actions, their habits, and their bodily parts. Knowledge of animals and plants at that time was for the most part restricted to those in local areas, and nothing was known of their predecessors in geologic time. Though these methods in classification were used by scientists for nearly 2000 years, there were numerous errors in the groupings because of apparent similarities. For example, whales were classified with the fish.

Descriptive taxonomy was greatly improved by John Ray, English naturalist (1627–1705), in zoology. He gave separate diagnoses for genera and for species, and though some of his ideas are illogical, in light of our present knowledge, he seemed even at that time to realize that similarities were the clues to relationships.

We owe our present system of nomenclature (a system of names) to the eminent Swedish botanist Carolus Linnaeus (1707–1778), though others also contributed toward this trend in thought. In the tenth edition of his *Systema naturae*, published in 1758, he applied the binomial system of nomenclature, giving a name to each genus and species (the spelling *species* is both singular and plural) of animal and plant known to him at that time. Latin was the scientific literature of the day. It is interesting to note that Linnaeus believed that species did not change but remained as they had been created. In following John Ray he gave precise diagnoses for each species. He also adopted a hierarchy for higher categories, such as order and class. Family names were introduced a little later by other workers, and still more recent was the inclusion of other higher categories, such as suborders and superfamilies, to indicate some of the broader relationships of the groups. It is useful for the student of animals to know that superfamily names end in *-oidea*, family names end in *-idae*, and subfamily names terminate in *-inae*. In botanical names the endings are order *-ales*, suborder *-ineae*, family *-aceae*, and subfamily *-oideae*.

In the earlier editions of Linnaeus' *Systema naturae* a *polynomial* (many-names) system was employed. As an example, the name for a tree frog disclosed much of the description of the animal. *Rana* (arborea), *pedibus fissus, unguibus fibrotundis, corpore laevi, pone angustato* meant frog (tree), feet cleft, toes fibrous, body smooth, back narrowed posteriorly.

Such naming was too cumbersome, and in his tenth edition Linnaeus fully introduced the *binomial* (two names) system that still prevails. However, when subspecies or varieties are recognized the *trinomial* (three names) system is used.

Linnaeus derived either the generic names from Greek and the specific names from Latin: *Elephas*, ελέφος, elephant, *maximus*, Lat., great; or both the generic and the species names from Latin, *Equus*, Lat., horse, *caballus*, Lat., horse. The names now stand as *Elephas maximus* Linnaeus, 1758; and *Equus caballus* Linnaeus, 1758. Frequently generic or specific names are derived from geographic place names or from the names of people. The name of the author and data are optional; when the author thinks that they help to clarify the status of the name, as they frequently do, they are added at the end of the scientific name.

It was soon discovered that the binomial system was applicable to all languages and could be used universally. This eliminated the confusion of common names which can become so difficult. It was found that the same common name was sometimes used for different animals or different names were applied to the same animal. This confusion still prevails and is not at all suitable for scientists. For example, in Florida a gopher is a tortoise, on the high plains it is a ground squirrel, and in most areas it is a pocket gopher. On the other hand, one of the large cats in the Americas is called puma, mountain lion, cougar, or lion, all of which are referred to one species. In paleontology the generic name alone is frequently used in the discussion of a plant or animal, i.e., *Acer* for the maple or *Smilodon* for certain saber–toothed cats.

All nomenclatorial problems (systems of names) were not solved by the binomial system. With many animals being described all over the world it was found that the same scientific name, either genus or species, or both, was sometimes given to different animals. Though many codes of nomenclature were formulated to solve the problem, it was at a meeting of the Fourth International Zoological Congress in Cambridge, England, 1898, that a special committee presented the *Régles internationales de la nomenclature zoologique.* These were adopted three years later (1901) at the next meeting in Berlin. This was, and still is, a sort of international law for scientific names interpreted by the International Commission of Zoological Nomenclature. The botanists have a similar set of rules or regulations.

There are many rules and regulations that cannot be enumerated here, but some of the important considerations were the *designation* of *types*, the *recognition of the law of priority*, and the *relegation* of *synonyms* and *homonyms*.

In the description of a new species a *type specimen* must be designated, though it may not be typical of the species as a whole, because the species may be composed of many populations of individuals that were not known to the original author. Nevertheless, the *type of a species* is a specimen, hence it is an object. On the other hand, the *type of a genus* is a species. It includes all the individuals in the populations of the species designated. Though it has basis in fact, it is a concept. Similarly, the *type of a family* is a genus. The categories above the order are not typified in this manner. A *taxon* is the term used for any taxonomic category, i.e., species, genus, family, order, etc.

The *law of priority* in zoological nomenclature is effective back to January 1, 1758, but subject to certain conditions set forth in the rules. Priority pertains to the relationships between animals or plants previously designated as a family, a genus, or a species for which a name was proposed and the subsequent application of the symbol or name. Thus the objective of the author, as symbolized in a family, a genus, or a specific name, is retained in its original sense and is controlled by the different type specimens (holotype, paratypes, topotypes, etc.), if the taxon, or category, is of the magnitude the author originally thought. This, then, involves a matter of interpretation. If the original interpretation of the materials is correct and the name has not been applied to another family, genus, or species, the name cannot be shifted from its prior taxonomic designation. All subsequent names applied to that taxon are *synonyms*. The law was not formulated to credit any author or the name he proposed for its own sake. The law of priority does not apply to taxons in categories higher than families. Furthermore, plant and animal synonymies are treated separately, i.e., a name for a plant is not homonymous with one given to an animal, though the names may be the same.

Many systematists favor a moderate application of the law of priority. They feel that there should be exceptions to the rule, particularly when a name has been widely used over a long period of time, but they also feel that they should be carefully and fully explained and must have general approval. These exceptions are recommended only in cases in which real confusion is likely to arise in reverting to an earlier name. Thus a name of long-standing usage is sometimes adopted by the Rules Committee, though historically it may have been the second name to be published.

We have seen on a preceding page that Linnaeus named the domestic horse *Equus caballus* in 1758. In 1913 Oliver Perry Hay described a horse skull found along the Kaw River near Lawrence, Kansas, as *Equus laurentius*, but later it was discovered that the skull described by Hay actually belonged to the species named *E. caballus*. Since *E. laurentius* is a different name for the same species described previously by Linnaeus it became a synonym.

Equus caballus Linnaeus, 1758; *Equus laurentius* Hay, 1913 (synonym).

A *homonym* is the same name given to materials not referable to the species, the genus, or the family to which the type specimen belongs.

Elephas maximus Linnaeus, 1758, was the name adopted for the Ceylon or Indian elephant by application of the *Régles internationales de nomenclature zooligique* in 1901. However, Johann Friedrich Blumenbach, a German anatomist, named the African elephant *Elephas africanus* Blumenbach, 1797. Later it was recognized that these two kinds of animals, though they had been given the same generic name, in fact represented distinct genera. Therefore, the generic name *Elephas*, as applied to the African elephant, became a homonym; since Linnaeus' name was typified by the Indian elephant, that species had priority and retained the name *Elephas*. In 1827 Frédéric Cuvier proposed a new generic name, *Loxodonta* (λοξός, slanting; ὀδόντος, tooth), for the African animals; thus the name for the African elephant became *Loxodonta africana* (Blumenbach).

The species name must agree in gender and number with the generic name; therefore the species name of the African elephant should end in *-a* instead of *-us*. Blumenbach's name is in parentheses because the erroneous, all-inclusive substance of the name to include two distinct genera of elephants, as originally proposed by him, had to be altered.

A CLASSIFICATION OF MODERN MAN

Listing the Taxons and Examples

Phylum: Chordata. Acorn worms, lampreys, fish, frogs, birds, reptiles, mammals, including man.

Subphylum: Craniata. Lampreys, fish, frogs, birds, reptiles, mammals, including man.

Class: Mammalia. Opossum, mole, whale, elephant, horse, man.

Infraclass: Eutheria. Mole, whale, elephant, horse, man.

Order: Primates. Lemurs, lorises, tarsiers, monkeys, apes, man.

Suborder: Anthropoidea. New World monkeys, Old World monkeys, apes, man.

Family: Hominidae. Djetis giant man, Peking man, modern man.

Genus: *Homo*. Heidelberg man, Neanderthal man, modern man.

Species: *sapiens*. *Homo sapiens* Linnaeus (modern man).

The system of biologic classification serves as a dictionary for workers concerned with animals and plants. Classifications are greatly affected by knowledge of the fossil history of the organisms and the phylogenetic interpretation of them. A *phylogeny,* then, is *an author's interpretation of the evolution and relationships of a group of organisms through time* (Fig. 250). No one now has all the phylogenetic evidence on any group of organisms, and it is hardly likely that anyone ever will have such complete evidence; therefore, the phylogenies that we use must be interpretative. Even the same author may change his ideas, and usually does, as new and additional evidence is found. Dispersal between provinces and regions frequently offers important clues in formulating a phylogeny of any group of organisms. Phylogenies are usually expressed graphically in *phylogenetic trees.*

5 Stratigraphy, Geochronology, Correlation, and Geologic Time

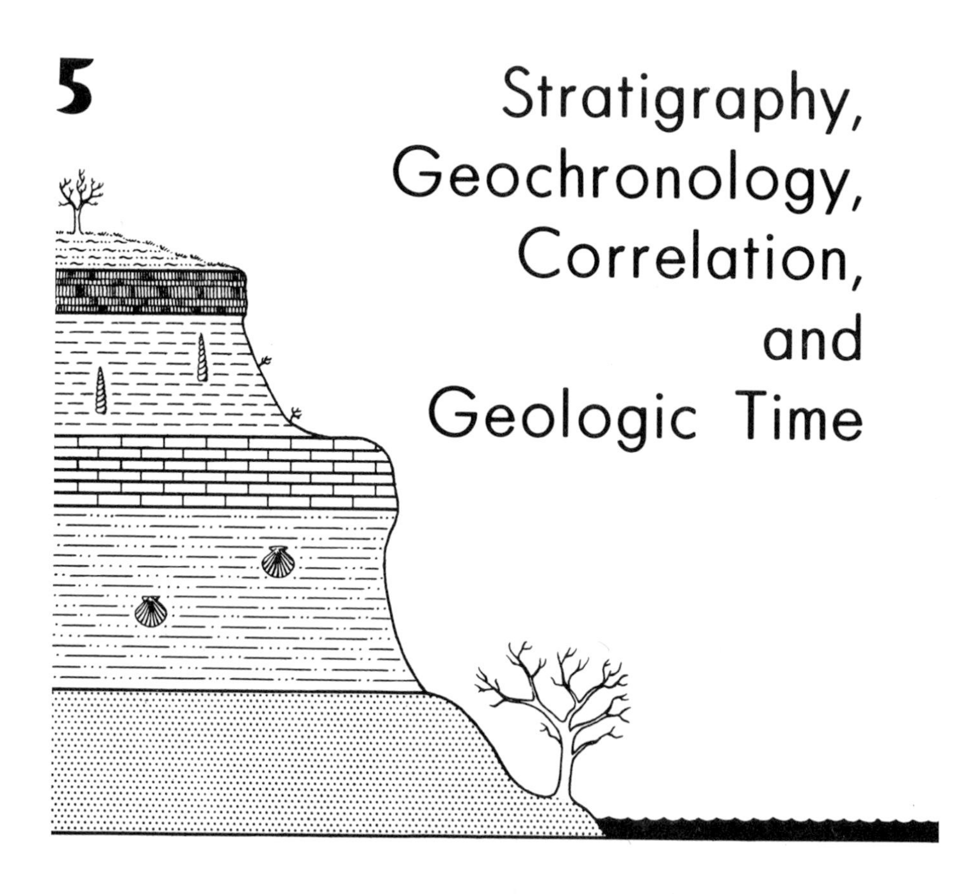

ROCKS OF THE EARTH'S CRUST are igneous, metamorphic, or sedimentary. *Igneous rocks* are solidified from deep-seated molten materials. They are either extrusive or intrusive. Extrusive igneous rocks occur as lava that breaks through the crust and flows over the countryside or erupt as pyroclastics and settle as volcanic ash and related materials. There the lava cools rapidly from a molten state. Lavas seldom contain fossils, but plants, insects, and other remains are frequently covered with volcanic ash. Volcanic ash is also secondarily deposited as a sediment and in this way has become an excellent source for fossils. Intrusive igneous rocks such as granite or diorite well up as molten materials and solidify as deep-seated bodies below the surface rocks. Because of their protection from surface temperatures they cool slowly. Batholiths as deep-seated bodies frequently form the cores of mountains and are devoid of fossils.

Metamorphic rocks are altered from other rocks by heat and pressure. In this way these rocks are delicately adjusted to their environments. Though most rocks are hard and durable when subjected to pressure, they will flow by plastic deformation. This transformation may affect igneous, sedimentary, or other metamorphic rocks. Indeed, some rocks have been metamorphosed several times. This usually occurs in mountainous areas subjected to recurrent crustal disturbances. The kinds of metamorphic rocks formed are determined by temperature, pressure, and the composition of the original rocks. There are many kinds. Some of the more common groups are quartzites derived from sandstones, marbles of limestone or dolomite origin, and schists originating mostly from fine sandy shales. In some poorly metamorphosed rocks, such as slates, distorted fossils can sometimes be recognized. Contact metamorphism occurs in the rocks closely adjacent to igneous masses, whereas regional metamorphism affects rocks more deeply buried and over wider areas.

Sedimentary rocks are composed of the substances deposited by the media of winds, water, or glacial ice. They usually occur as limestones, shales, sandstones, conglomerates, or volcanic ash. The sediments are deposited on land, along stream courses, on flood plains, in deltas, in inland lakes, or in the different environments of the oceans. The great abundance of fossil remains are preserved in sedimentary rocks.

Sedimentary Rocks

Limestones. Limestones containing a predominance of calcium carbonate ($CaCO_3$) are deposited as limy muds and calcareous sands. They are the finer materials that are normally carried far out from the shores or mouths of streams and eventually settle to the bottom. The greatest thickness of these calcareous (Lat., *calx*, lime) accumulations is solidified in association with other marine deposits, though limestones also occur in formations of fresh-water origin. Both inorganic processes and lime-secreting algae precipitate calcium carbonate. This limy mud accumulates in considerable thickness on tropical shoals. Much calcium carbonate, usually in the form of the mineral calcite, is derived from the shells of invertebrate animals. The shells or broken shells of microscopic foraminifers frequently make up much of the limestone.

CHALKS. Chalks also are a porous kind of limestone. They are very extensive in Cretaceous formations. The cliffs of Dover in England and the Niobrara formation in the Great Plains of the United States are well-known chalk formations. In many places chalk formations are made up primarily of the calcareous tests or shells of foraminifers and other invertebrate fossils.

DOLOMITES. Another kind of limy rock is dolomite. It is composed largely of calcium magnesium carbonate ($CaMg(CO_3)_2$). Some Paleozoic marine formations composed primarily of dolomite reach a thickness of 4000 feet. The lip of Niagara Falls is the Lockport dolomite formation.

COQUINA. Coquina is a term commonly applied to poorly cemented shell fragments of the larger invertebrates which when compacted make up another kind of limestone.

Shales. Shales are fine-grained nonplastic materials consolidated from clays and muds. They are usually finely laminated and fracture parallel to the bedding plane. Similar but more massive rocks that do not fracture as shales are called *mudstones*. On the continents shales are solidified from lake sediments or on wide flood plains. Black shales so frequently encountered by the field geologist contain large quantities of organic debris that settled in quiet lagoons. Many beautifully preserved fossils are found on the bedding planes of these shales. Some of the best examples are the leaves of seed ferns and other plants that grew in the old Pennsylvanian swamp forests and the delicate graptolites that floated about so abundantly in Ordovician seas. In the Oligocene Florissant formation in Colorado there is not only a well-preserved flora but an excellent record of insects that can be located by parting the fine layers of shale.

SANDSTONES. Shales grade into sandstones in which there is increase in the size of the sand grains. Sandstones are normally closer to the source of their origin than finer grained materials, as in shales and limestones. Rocks with well-cemented grains of quartz sand are said to be indurated or hard, although almost all gradations of induration are encountered. Argillaceous sandstones contain between 5 and 50 percent of fine materials, such as clay. On the other hand, if there is a strong concentration of calcium carbonate they are referred to as calcareous sandstones. An arkosic sandstone can be recognized by the presence of the mineral feldspar. Ordinarily, feldspars cannot withstand much transportation because of abrasion, so it is usually safe to assume that these sandstones are not far from the mountains or hills from which the sediments came.

A study of sand grains will frequently reveal something of the nature of the local environment at the time the sediments were deposited. If the grains have sharp or rather freshly broken edges, they obviously have not been exposed to much abrasion. Aeolian or wind-blown sands have pitted or frosted surfaces. Sands subjected to considerable water action have well-rounded sand grains and are sometimes well sorted. In other words, they are found in fairly uniformly sized groups. Quicksands, into which animals or other heavy objects will sink, are composed of sand grains of uniform size. Their well-rounded surfaces will not permit them to lodge one against the other to form a cohesive mass, as in sediments of more angular materials, and consequently they yield readily to pressure.

Fossil bones and shells are usually well preserved in sandstones, in which they are less likely to be crushed. Shales, on the other hand, are much more plastic, and the specimens are frequently crushed. Some of the most perfectly preserved skulls have been taken from sandstones.

Poorly sorted, sandy, or gritty sandstones, with a matrix of more than 10 percent clay, deeply buried and firmly indurated or cemented together, are called *graywackes*. The matrix has a salty composition. Different kinds of graywackes may be recognized by the sizes and kinds of sand grains.

CONGLOMERATES. Conglomerates are made up of larger components than sandstones. The particles may range in size from small pebbles no larger than a pea to large boulders several feet in diameter, some of which may be well rounded. These and finer grained particles may be cemented together by calcareous materials, iron oxide, or silica. The coarser the materials, the nearer the sediments are to their source. Only the most turbulent streams will transport massive boulders. Repeated stream or wave action will abrade rocks into a smooth-surfaced stone or cobble. If the component rocks are angular instead of rounded, the sedimentary rock is called a *breccia*. A conglomerate or a breccia is normally a poor media for the preservation of fossils, since turbulent waters are likely to break both bones and shells.

Stratigraphy

Stratigraphy is very closely related to paleontology. In fact it is in stratified rocks that most fossil remains are found. Both sedimentology and paleontology are integral parts of stratigraphy, since the study of *stratigraphy pertains to the origin, the sequence, the description, and the relationships of stratified rocks*. Studies in stratigraphy are primarily concerned with the limestones, shales, sandstones, and conglomerates.

Diastrophic movements (all movements or deformation in the earth's crust or lithosphere) in the form of subsidence and uplift largely control sedimentation. Among the best examples of this phenomenon are the geosynclines and adjacent lands that extended across the continents throughout most of Paleozoic time (Figs. 89–91). As the geosynclines subsided either slowly or rapidly, land areas uplifted at an essentially corresponding rate and supplied sediments to the geosynclines. At times the adjacent lands tended to reach base level, when little or no deposition took place, only to be rejuvenated again and again. Eventually the geosynclines uplifted, folded, and faulted into great chains of mountains.

The now greatly eroded Appalachian Mountains offer in their sedimentary rocks the keys to the history of the Paleozoic Era in eastern North America. When the sediments occur in a great uninterrupted sequence subsidence was continuous. Unconformities (see p. 74) between the stratigraphic units show that the area was subjected to erosion, removing part of the underlying rock and depositing the materials elsewhere. Coarse detritus is indicative of the proximity of land, probably with considerable relief, whereas fine sediments are evidence of deposition in relatively quiet waters. Marine fossils show that the geosyncline was invaded by the sea at that particular time, and fossil plants and evidence of land animals tell the stratigrapher that the basin had filled to the extent that the marine waters had retreated.

In Mongolia the expeditions of the American Museum of Natural History found the nests of dinosaurs in which the eggs had been laid in characteristic *wind-blown sands*. The sands were not only cross-bedded, but the grains were like those found in aeolian deposits, and there were no stream gravels or other evidence of pluvial conditions at the site.

Some of the most important agencies in the formation of sediments are *mechanical* (wind, water, and ice), *chemical* (solution, oxidation, and carbonation), and *organic* (action of plant roots and the accumulation of organic materials). These and many other clues help the stratigrapher to interpret the physical events of the past.

Orogenic disturbances, usually of local or provincial extent which culminate in folded and faulted structures, result in the rapid deposition of coarse clastic sediments. On the other hand, *epeirogenic* movements in the earth's crust in the form of slow emergences over regional areas produce widespread rather fine-grained sediments.

Frequently it is found in tracing the beds laterally that a sandstone will gradually give way to shales; this and similar changes are evidence of change in depositional environments. On the other hand, in widely uniform environments a limestone or a sandstone may extend for more than a hundred miles.

DIASTROPHISM

A trip into the countryside will reveal that rocks in the earth's crust are not resting in a horizontal position, as most of the sediments were laid down originally. They may be folded and faulted into numerous positions. Much to the surprise of the average person, rocks containing shells from prehistoric oceans occur in our lofty mountains. Even so, in some provinces, such as parts of the Great Plains and the Mississippi Valley in the United States, formations have been little disturbed for many millions of years. *These movements or deformations in the earth's crust are referred to as diastrophism.* A movement may occur in the sudden release of pressure built up in the rocks and expressed in a startling earthquake in which considerable displacement takes place at one time; or deformation may be so gradual, something like 20 inches in a century, as measured in the Cajon Pass in Southern California (Gilluly, 1949), that it is not perceptible except by precise measurements made over a number of years. Nevertheless, such seemingly insignificant movements, if continuous, or nearly so, for a million years would elevate a formation from sea level to snow line or higher in our most lofty mountains.

Epeirogeny and Orogeny. Relatively slow, diastrophic disturbances over regional areas, either in the form of uplifts or downwarps, are *epeirogenic.* There were epeirogenic movements of every continent in the late Pennsylvanian and in the early Permian Periods and at the end of the Cenozoic Era; all culminated in glaciation and resulted in profound evolution of plant and animal life.

Orogenies being more intense than epeirogenies are mountain builders. They are local or provincial in extent and are expressed in sharp folds and faults. Displacement along fault zones may be horizontal or vertical or both. These movements are of the utmost concern to paleontologists and stratigraphers in tracing the fossil-bearing formations with which they are concerned.

STRUCTURE. Pressure on rocks in the earth's crust usually produces folded structures. The downwarps of the folds are *synclines* and the upwarps and *anticlines* (Fig. 22).

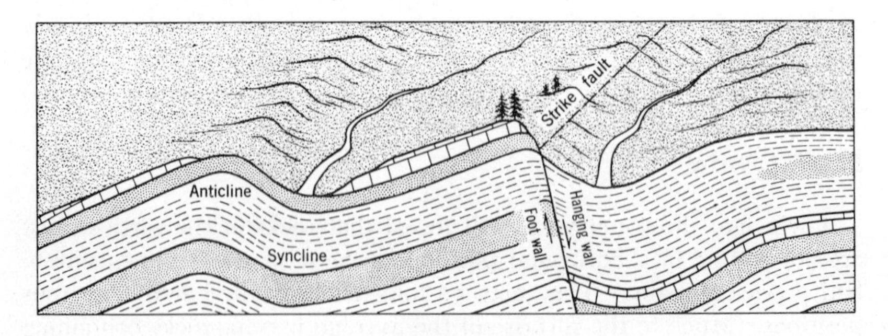

Fig. 22.

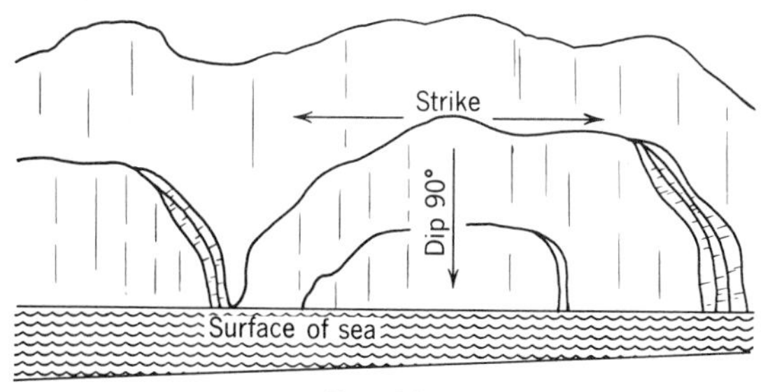

Fig. 23.

DIP AND STRIKE. In all stratigraphic units altered from their original horizontal positions there is a *dip* and a *strike*. *The dip is the angle of inclination of the rock layer from the horizontal plane.* It may be vertical or slightly inclined (only a few degrees). Some rock layers may be folded beyond a vertical position and are said to be overturned. The dip must always be measured along the bedding plane. Frequently a bedding plane in a roadcut or in other crosscut excavations may be difficult to determine because of the angle of the crosscut. The dip is not necessarily constant in all areas of a stratigraphic unit. The angle usually varies when the beds are traced laterally (Fig. 23).

In an area of folded rocks that are eroded, as they all are if exposed at the surface, the more resistant members, such as sandstones, conglomerates, or lava rocks, will stand out in relief. The edges of these formations or members can usually be traced for some distance. Each of these marks the strike of the different beds which at any given place is at right angle to the dip. In other words, *the strike is the direction of the line formed by the intersection of the bedding plane with the horizontal plane.* The so-called "hogbacks" along the eastern foothills of the Rocky Mountains show the strike of the Cretaceous Dakota sandstone formation in that area.

FAULTS. Faults are fractures in rock formations in which displacement takes place along the fracture surface. Many of the larger faults develop a strip called a fault zone in which movement takes place on several surfaces. Typical of this is the San Andreas fault zone that extends from southern California across the Tehachapi mountains, through the Coast Ranges to San Francisco, and finally out into the Pacific Ocean at the mouth of Tomales Bay. It was movement in this fault zone that caused the enormously destructive San Francisco earthquake of 1906.

Displacement may range from 1 inch to more than 100 miles, though it is obvious that all does not take place at one time. Movement may be vertical, at low angles, or horizontal along the fault line. If the rocks in the footwall move upward, it is said to be a *normal fault*, but if the rocks in the hanging wall move up, it is a *reverse fault*. Usually it is difficult, if not impossible, to determine in which direction the movement took place. Low-angle faults, in which older formations are thrust out over younger rocks, are sometimes detected on the evidence from fossils whose age has been determined. Folds or monoclines (strata bent only in one direction) sometimes increase in intensity within a mile or less until they fracture into a fault; these are *hinge faults*. Faults may run in any direction in relation to the strike of the beds. If they are parallel to the strike, they are *strike faults*, whereas *dip faults* or *cross faults* cut at different angles to the strike. These displacements in an area of considerable faulting, followed by erosion, and covered by a soil mantle and vegetation make the tracing of formations through the hills exceedingly difficult. Many features of the faults must be known and evidence of their presence in surface characteristics understood before geologic mapping can be done with any degree of accuracy.

CONFORMITIES AND UNCONFORMITIES. Any continuous accumulation of sediments, such as sands, muds, and volcanic ash, that are laid down on parallel bedding planes are said to be *conformable* stratigraphic units.

After a group of rocks like these has been laid down, diastrophism may occur and the rock formations uplifted and tilted. Then the exposed upturned beds are worn down until a base level is reached at which no more erosion takes place. Later the area subsides below base level, and a fresh set of sediments is deposited on the old erosional surface. The line of contact between the old eroded surface and the base of the later rocks is called an angular *unconformity*. Evidence of this kind is indicative of an orogeny and is demonstrated by uplift and subsequent erosion. The unconformity also represents a hiatus in the time succession represented by the sediments.

On the other hand, in areas like the Mississippi Valley there has been gentle but continuous uplift for a certain length of time without tilting of the underlying formation. An erosional surface is then developed on the uppermost formation in the horizontal sequence already in place. Downwarping of the area again would result in deposition of sediments. The erosional contact between these two groups of rocks would be a *disconformity*. This also would indicate a hiatus but not a disturbance of the intensity, as in the case of an unconformity, though the time interval represented by the disconformity may have been as long or even longer.

BASIC PRINCIPLES

Uniformitarianism. The principle of uniformitarianism was in fact recognized by the premedieval Greeks, though they did not use this term for the phenomena they observed. Indeed, some of them correctly recognized fossil plants and animals as the remains of organisms that once lived, and Herodotus observed the normal sedimentary processes in the Nile Delta. In Italy Leonardo da Vinci came to the same conclusion nearly 20 centuries later.

The principle of uniformitarianism was proposed by the Scottish geologist James Hutton (1795) in his *Theory of the Earth.* From his observations on plutonic rocks, he concluded that the processes then prevailing had occurred in a like manner in the past. Thus Hutton's dictum the "*present is the key to the past*" is uniformitarianism. Hutton believed also that strata were made up of sediments that had been washed into bays by rivers, of sands that had been pounded by waves, or even in part of shells broken into pieces by currents and rolled on sea bottoms.

It remained, however, for Sir Charles Lyell to elucidate uniformitarianism clearly as the natural process of erosion and sedimentation. Rain with resultant floods washed sediments off the hills and mountains onto the lowlands or into the sea. Waves worked continuously at sea cliffs. Great landslides scarred the green mountain slopes. Volcanoes erupted, scattering their ashes near and far, and lavas flowed out over the surface ground and even into the sea. All of these phenomena and more Lyell observed in the British Isles and on the continent of Europe. In these ways, he explained in his *Principles of Geology* (1830–1833), sedimentary rock units were formed in the past as they are today.

Superposition. The Danish bishop of Hamburg, Nicolaus Steno was the first (1669) to propose the *natural law of superposition.* He reasoned that in the accumulation of sediments the oldest or first ones to be laid down would be on the bottom and the last and youngest sediments must be those on top. Of course, he had the mistaken notion that the earth was covered by a universal ocean. Even so, his observation on sequence was correct, and the law of superposition holds for sediments wherever they may accumulate. Even at that early date Steno also realized that strata had been tilted from their original flat-lying position into angular and/or nearly vertical attitudes. He also recognized that such forces, as well as volcanic forces, as he called them, formed mountains, valleys, and plateaus (Fig. 25).

Now we know that the forces in the earth's crust sometimes confuse stratigraphers by the development of overthrusts. In this case the underlying beds are pushed out and over younger beds, but careful studies on the geology and paleontology will disclose these discrepancies in the sequence.

Geochronologic and Stratigraphic Terminology

A terminology is a means of expression. If paleontologists and stratigraphers hope to discuss and write about the things they have observed and interpreted from sedimentary rocks, they must have terms suited for their purposes. These terminologies have been developing slowly for two centuries and will continue to change as we seek greater perfection and clarity in expression. Unanimity of opinion has not been reached and probably will not be attained for some time. Nevertheless, an applicable terminology in expression of the units with which we are concerned may be divided into *time* and *time-rock* units *(geochronologic)* and *rock units (stratigraphic)*. The geochronologic terms include abstract theoretical *time terms* and *time-rock* or *chronostratigraphic terms* (expression of time as based on rocks and fossils). Some authorities maintain that the use of both is superfluous, since an expression of time is the objective of both, but both terminologies are still widely used in literature today.

Time Terms. Time units in geochronologic terminology are the conceptual or abstract divisions of geologic time. It is realized that the time sequence was continuous everywhere and that there was no interruption or overlap of earlier or later units as found in chronostratigraphic sections or stratigraphic units. The fossil record, geophysical dating of rocks, and rhythmic sedimentation have all offered evidence on geochronology. Of these, the paleontologic methods, using fossils, are still the most practical.

The duration of each unit is presently indirectly determined and its extent indefinite because the information how available to us from the rocks and the fossils they contain does not represent all of past time. Yet we have much evidence of it, particularly in the later Eras and Periods. As more facts are obtained, we find that geologic time has been greater, not less, than previously anticipated. It is estimated from radioactive materials that the first Era of geologic time (Archaeozoic) started not less than 2.6 billion years ago. Geologic time is divided into *Eras, Periods, Epochs,* and, in some instance, into *Ages* or even *Chrons*. The term "Chron" has not been adopted by the Commission on Stratigraphic Nomenclature, though it is sometimes used when such small divisions can be recognized. Some geochronologists prefer the term *Zone Moment* or *Secule.* The most accurate method of determining certain points of time within these units throughout most of geologic time is with the radioactive minerals uranium-lead or potassium-argon, though for the last 20,000 years the carbon 14 method is also employed.

Time terms are adjectival words ending in *ian* (Devonian), *ic* (Triassic), *ous* (Cretaceous), *ary* (Tertiary), *ene* (Eocene). The authors of the terms and their derivations are treated in each of the later chapters in which these intervals are discussed.

Time-Rock Terms (also called *chronostratigraphic* or *time-stratigraphic* terms). Time represented as precisely as possible by selected rock sequences, preferably those that are continuously fossiliferous, are time-rock or chronostratigraphic units. A certain amount of time is involved in the deposition of any sequence of sedimentary rocks and in the evolution or change of the organisms during that interval, whether it is of shorter or longer duration. Consequently, the time sequence and duration concept, based primarily on the fossils, in certain lithologic units during the approximate equivalent of a Period are called a *System,* those comparable to an Epoch are a *Series,* those of the magnitude of an Age are a *Stage,* and those equivalent to a *Chron* are called a *Zone.* The term *Terrane* has been suggested for rock sequences of the magnitude of an Era, but this has not been adopted. It has been extremely difficult to find uninterrupted rock sequences equivalent to Eras and Periods. The application of chronostratigraphic equivalents for the larger categories is therefore somewhat impracticable.

A succession of chronostratigraphic units, then, is based on fossils and strata in an almost uninterrupted sequence of rocks, though not necessarily represented in totality in any one section. The units are more or less independent of lithic composition because lithology may change laterally in different environments. Stages and Zones, in particular, are recognized on evidence approximating ideal sections with superpositional and subpositional control (rocks of known age both above and below in a continuous sequence), a continuous and uninterrupted stratigraphic sequence, recognizable paleontologic control both at the top and the bottom of the stratigraphic units, and preferably beds that are continuously fossiliferous. Zones, as the most basic units, are recognized by a *joint occurrence of a number of species.* Each of these species may have a vertical range through rock units that is greater or less than the limits set for the Zone. Thus it is seen that a number of species appearing for the first time in the fossil record but ranging on

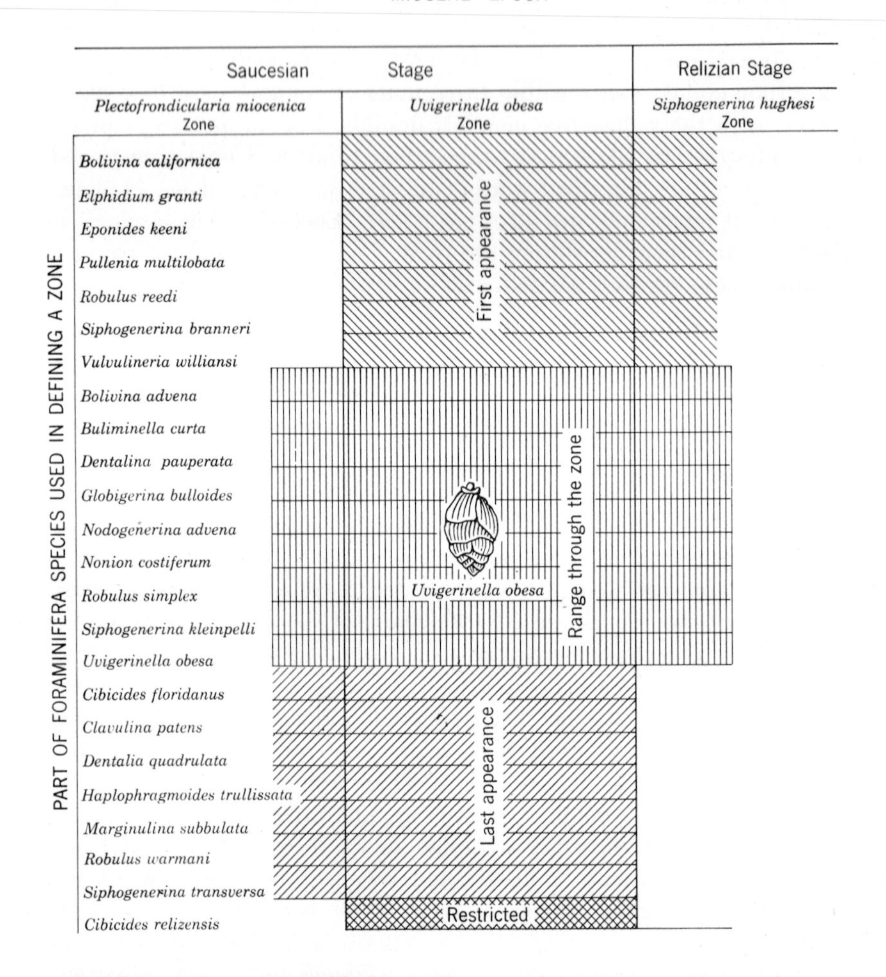

Fig. 24. A Zone as exemplified by rock units and their contained foraminifers.

into the overlying units, a number of species ranging up through the underlying rocks but terminating at that time, and certain species confined to the strata offer evidence for recognition of a Zone. Zones are also supported by the evidence of evolution or change in part of or possibly all of the faunal assemblages in each successive unit. It should be possible to recognize these units in any marine province in which they may occur (Fig. 24).

Zones may be named after any diagnostic or abundantly represented species, and the species need not be restricted to the designated Zone. As an example, in the California Miocene Robert M. Kleinpell used the foraminiferas *Plectofrondicularia miocenica* and *Uvigerinella obesa* for the uppermost Zones in his Saucesian Stage and *Siphogenerina hughesi* for the lowest Zone in his Relizian Stage.

Stage terms usually terminate in the adjectival endings *ian* (Saucesian). As in rock-unit names, the terms are derived from some local geographic name. The Saucesian was taken from the Los Sauces Creek in Ventura County, California. It is preferable to avoid confusion by not using names that have been previously assigned to rock units.

Rock Terms. Rock units as described and mapped on their lithologic (color, size of sand grains, etc., or even the nature of the contained fossils) character and structural position observed in the field are not defined on the basis of geologic time. Furthermore, they do not necessarily coincide with designated geologic time intervals, though in certain cases they may do so. The two major divisions are groups and formations.

The *formation* is the basic unit in stratigraphic classification. Several structurally and lithogenetically related formations may make up a *group*. A formation may be defined as *an observable mappable unit as recognized on its gross lithology and stratigraphic position.* It may be a limestone, a sandstone, a volcanic ash, or any other distinctive lithic unit. Some formations may be composed of members, lentils, or tongues of different sedimentary materials. If several of these units are related

GEOCHRONOLOGIC TERMINOLOGY

Time Terms (abstract theoretical concept)		Chronostratigraphic Terms (time-rock or time-stratigraphic)	
Era	Cenozoic	Terrane	Cenozoic
Period	Tertiary	System	Tertiary
Epoch	Miocene	Series	Miocene
Age	Saucesian	Stage	Saucesian
Chron	*Uvigerinella obesa*	Zone	*Uvigerinella obesa*

STRATIGRAPHIC TERMINOLOGY

Rock Terms	
Group	Vaqueros
Formation	Sandholdt
Member	Lentil, Tongue, and Bed

in their genesis (method of formation or origin), it may be desirable to refer to them as a formational unit. *Members* as distinct units that make up a formation may be of considerable geographic extent, whereas *lentils* are more limited and frequently lens-shaped. *Tongues* which are connected to a larger body of sedimentary rocks wedge out into other sediments in other directions. They are common in deltaic deposits. The term *bed* has a much greater latitude in application. It may be referred to a formation or any of its components.

Group and formation names may be taken from geographic areas, such as a city, town, river, or mountain: Austin chalk; Pinole tuff; Flint River formation; or Frog Mountain sandstone. Members may be designated by letters (A, B, C, etc.), by numbers (1, 2, 3, etc.), by their lithology (bentonitic, tuffaceous, etc.), or by geographic names (Harrisburg gypsiferous members).

Methods in Correlation

Correlation is the recognition of synchrony of biologic and geologic events as demonstrated in the different rock units of the earth's crust. In correlation each facet of earth's history as represented in the lithosphere must be related as precisely as possible in the time of its occurrence to events in other parts of the world. The paleontologist wants to know what kinds of terrestrial animals and plants were present in Australia, South America, and, for that matter, in all parts of the world when the dinosaurs were flourishing in the late Jurassic swamps of North America. It may be important for him to know what the physical conditions were like in other regions when the Himalayan Mountains started their last great orogeny. Petroleum geologists are greatly concerned with precise correlations of marine strata in contiguous areas, for such information may assist in the discovery of a large reservoir of oil. It is only by correlation of events like these that the sequence of life in all parts of the earth can be traced through geologic time.

GEOLOGIC METHODS

Stratigraphic. This is known as the law of superposition, which was recognized as early as 1669 by Nicolaus Steno. It is a very simple law, yet quite fundamental in the study of sedimentary rocks. The oldest sediments were laid down first in a horizontal or nearly horizontal plane, and subsequent materials were deposited on top of them or in a superposition. Though this method in correlation does not give the exact

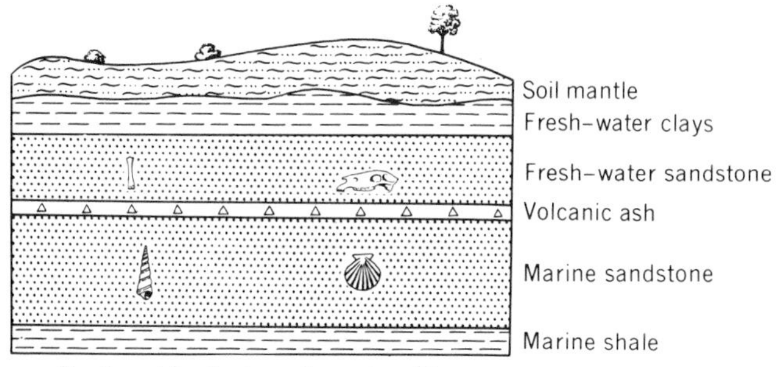

Fig. 25.

age of the stratigraphic units, it does establish their sequence in time in relation to each other. A sequence is shown in Fig. 25 of a clay stone, a marine sandstone with fossil shells, a volcanic ash, a fresh-water sandstone, and another clay stone with an unconformable surface contact with the overlying soil mantle. Normal sequences of this kind are sometimes disrupted by diastrophic disturbances in the earth's crust that may either overturn or even overthrust part of the beds from their former positions (Fig. 25).

Lithologic. Homogeneity in composition, texture, minerals, color, or some other physical characteristic is commonly employed as a basis for lithologic correlation between noncontiguous outcrops in a local basin or even within a province. Thus lithology is useful when superposition or the presence of fossils cannot be employed in establishing the relative age of stratigraphic units in adjacent areas or sometimes in distant areas.

A blue sandstone may be the same age wherever it is found in a local

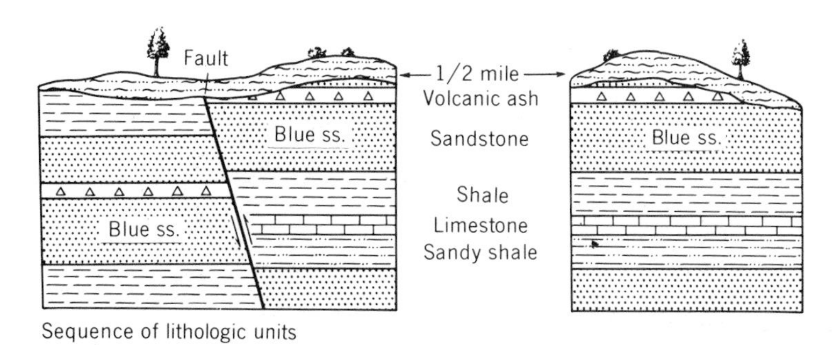

Fig. 26.

basin, though two hundred miles away another blue sandstone formation may be younger or older than the one previously encountered. In this instance correlation by lithology is inaccurate (Fig. 26).

The eruption of a volcano may result in the deposition of a layer of volcanic ash over an area of several thousand square miles. Stratigraphic units like this, if the ash can be accurately identified, are useful in lithologic correlation. Ordinarily, though, because of their heterogeneity most continental and many marine formations are difficult to correlate very far by similarity in the composition of the rocks. A sandstone may thin out and finally give way to shales, whereas at a later time another sandstone quite like it and even derived from the same source rocks may appear. On the other hand, marine formations may be represented by limestones or other rocks that can be traced for several hundred miles. For the most part, though, the lithologic method in correlation is not reliable, or at least it should be used with caution, in comparing the succession of strata in widely separated areas in which the geologic history may be quite different.

Crustal Movements. The results of diastrophic movements in the earth's crust, as shown in local or provincial orogenies or in epeirogenic disturbances effecting the broader configuration of continents, are usually complex and difficult to use in precise correlation. Nevertheless, such disturbances expressed in pronounced unconformities and in old erosional surfaces may be recognized as contemporary events in provincial areas or, infrequently, over most of a continent (Fig. 27).

The Taconian orogeny in the late Ordovician in New England and adjacent Canada is often used as an example of an orogeny used in correlation, but it was not a single episode of relatively short duration and

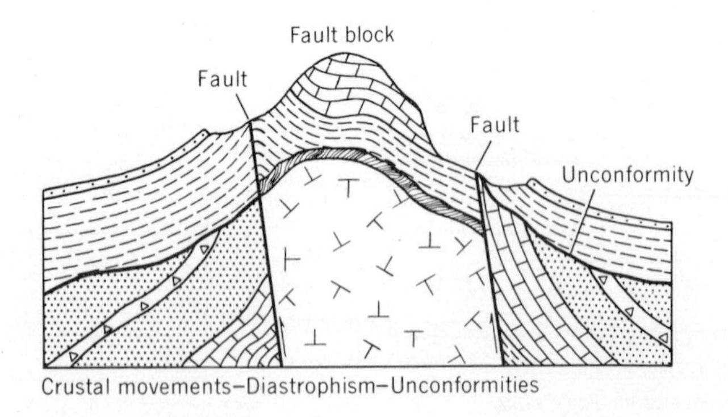

Crustal movements–Diastrophism–Unconformities

Fig. 27.

it had no physical effect as far west as Ohio. Another kind of disturbance was the late Pliocene and Pleistocene epeirogeny which was nearly world-wide in extent. The magnitude of this epeirogeny varied in different areas, and its application in the more refined correlations has become somewhat confusing; the movement did not come about at one time as it has at times been interpreted.

In the nineteenth century and the early part of the twentieth major crustal movements were still thought by many geologists and paleontologists to be world-wide and were used as the clues to intercontinental correlations. Indeed, Thomas Crowder Chamberlain and others called diastrophism the ultimate basis of correlation. We now know that this is not true. Though crustal movements are useful, they have long since lost their front-rank standing as evidence for correlation.

BIOLOGIC CORRELATIONS

Biologic correlations are still the most useful method in our efforts to establish relative contemporaneity of events throughout the geologic past. They are based on the history of life as represented by fossils in the rocks. The evidence is controlled by stratigraphic succession, in which, as is normally true, the overlying beds are younger than those below. Throughout time animals and plants have changed, or evolved, from one species to another. Some have changed rather rapidly and others have existed for long periods of time with little alteration in their structures. In any given line of descent the less specialized kinds occur in the older rocks and the more advanced in later formations. But when a species becomes extinct it never appears again, though some other kinds may come to look surprisingly like it through convergent evolution.

Identity of Species. This method is based on the assumption that identical species occur in beds of the same age. Rapidly evolving and wide-ranging species and genera, frequently called *index fossils,* are particularly useful in this method. As early as 1895 William "Strata" Smith observed that certain species of ammonites were present in the same stratigraphic horizons at different localities in the Mesozoic rocks of England. Later it was found that his correlations could be extended across the English Channel to Europe. This method was soon adopted by all leading stratigraphers.

In applying this method the student must be aware also that there are different geographic species living at the same time in different environments and that they must not be confused with those in geochronologic succession. In the oceans temperature, currents, depth,

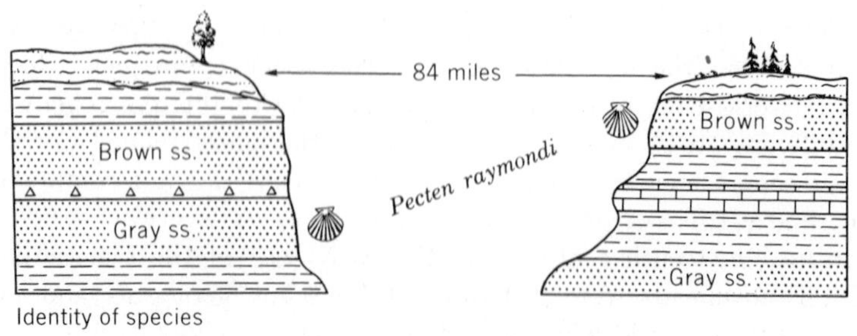

Fig. 28.

nature of the bottom and presence of other organisms are some of the controlling factors in environmental selection, whereas on land we see organisms with preference for everything from lakes and streams to desert conditions. There are, however, certain species or genera with greater environmental tolerances which occupy or spread through different environments. These are the most useful in the identity-of-species method in correlation (Fig. 28).

Stage in Evolution. Another method now widely in use, especially in mammalian paleontology, is the stage in evolution of the fauna as a whole or in any group of animals. This, of course, depends on our knowledge of evolution in the families, genera, and species through time and space. As our information becomes more complete and the phyletic evolution in more groups becomes available, the more precise and useful this method will become. Even now fragments of horse or beaver cheek teeth, showing the height of the crown of the tooth, offer enough evidence to reach Stage or Age refinement in correlation. Other parts of the skeletons of these and other mammals also can be used on a broader scale (Fig. 29).

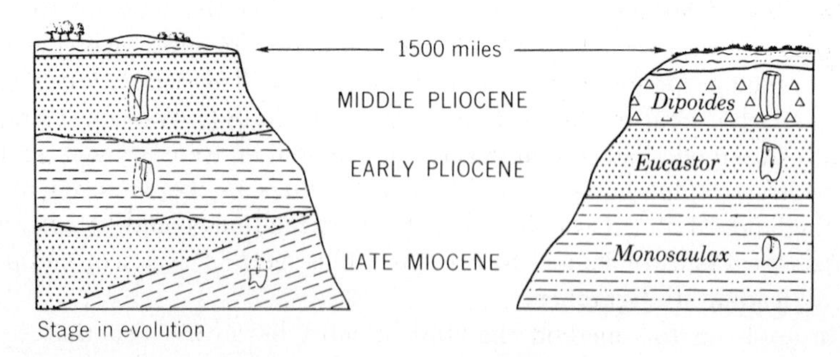

***Fig. 29.** The stage of evolution in the fragmentary beaver teeth to the left can be determined from knowledge of the teeth to the right.*

If we encounter an abundance of scale-tree and seed-fern fossils and though the identity of the genera and species are questionable, we know we are dealing with a formation not older than Devonian and not younger than Jurassic. This discipline is applicable to many other groups of organisms for which we have fairly accurate information on their geologic range. Our knowledge of the sequence of life then helps us to recognize the age of rocks in which fossils occur.

Caution must be used because our knowledge is still incomplete. Every year paleontologists turn up evidence that a group of animals or plants was actually present in a preceding or in a later Epoch or Period than previously recorded. Nevertheless, the practice of precaution in correlation is no indication that a method has little or no merit.

Percentage Method. This method involves a comparison of the species or genera in fossil-faunal assemblages with those living today in similar environments. If a large percentage of the species or genera in the fossil assemblage is still living, the age of the fauna is not far removed from Recent time, but if none or few of the forms are living today then the fossil assemblage is considerably older. One of the prerequisties is to have a fossil collection large enough to be adequately representative of the animals in the region at that time. Its application varies somewhat, depending on the species concept of the author. That is, some scientists allow for more variation in their species than others. The method is of little use beyond the Cenozoic, though it has been extended to the comparison of faunas in the Mesozoic and Paleozoic in which the faunal assemblages have to be compared percentagewise with each other, but it has met with little success.

It is a reasonably reliable method in parts of the world in which little or nothing is known of the fossil record and the living kinds are well described. For the most part, though, the identity-of-species and

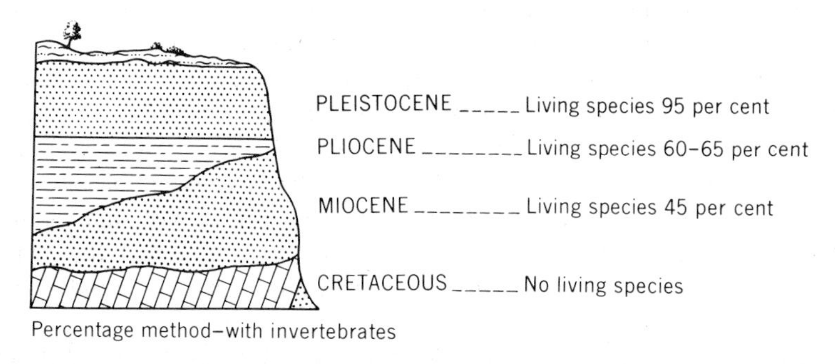

Percentage method–with invertebrates

Fig. 30.

the stage-in-evolution methods have replaced almost totally the percentage method.

The percentage method was employed as early as 1780 by the Abbé Giraud de Saulave in his studies of fossils and stratigraphic succession in southern France. It was also used by Sir Charles Lyell and Gérard Paul Deshayes in drawing up their marine sequence for the Cenozoic of western Europe. This was published in the third volume of Lyell's *Principles of Geology* in 1833. Their evidence was based on invertebrate collections from England, France, and Italy (Fig. 30).

Geologic Time

The student should remember that time was continuous everywhere and that a geologic timetable represents only a human effort to subdivide this continuity into a recognizable sequence of events based on what is known from the rocks of the earth. Careful research through the years has disclosed that earth history did not involve incredible billions of years, as the early orientals would have had us believe, but considerably more than the few thousands permitted by early Christian faith. One of the early estimations was made by Archbishop Ussher of Ireland in 1645. After a careful study of the Scriptures, he concluded that Creation took place on October 26 in the year 4004 B.C., at 9 o'clock in the morning. This was adopted in the King James version of the Bible, and an approximation of not more than 2000 years more was tolerated for the age of the earth. It has always been difficult to overcome such preconceived ideas of long standing. Nevertheless, steady progress was made by which people came to the full realization that the acceptance of factual evidence of earth history really did not conflict with their religion.

Even Herodotus, the early Greek historian, recognized that the sediments in the delta of the Nile River alone required several thousands of years to accumulate. Relative ages were established on the rates of erosion and sedimentation, and such phenomena were utilized by James Hutton as early as 1785. By using these methods Charles D. Walcott in the latter part of the nineteenth century derived an age of something like 45 million years for the age of the earth. But it was soon realized that this figure, as well as those derived from estimates of the time it took for oceanic waters to reach their present degree of salinity, were far too conservative.

Excellent data are now available, and critical analysis toward greater refinement does not militate against their validity. Nothing is gained by ignoring factual information, but few methods are beyond improvement.

Radioactive Minerals. One of the most accurate methods of measuring the millions of years in geologic time involves radioactive elements, especially *uranium* and *thorium*, which occur almost exclusively in pegmatites. After these minerals were formed in certain kinds of intrusive rocks, such as granite, regardless of chemical association, heat, pressure, or any other conditions, they disintegrated slowly into *lead*, invisible radiations of *helium*, and heat. A formula would be $U^{238} = Pb^{206} + 8He + heat$, or ($U^{235} = Pb^{207} + 7He + heat$). Since it has been shown that one gram of uranium yields 1/7,600,000 of a gram of lead in a year, the ratio of lead, then, to the amount of uranium remaining would give the age of the rock. Thorium, the other radioactive element, can also be used. Thus it is possible to derive three independent age determinations from one mineral.

There are, however, certain difficulties to be encountered. Uranium is not found in all granitic rocks. The best samples for age determinations usually occur in pegmatite dikes or other coarsely crystalline plutonic rocks. Once the uranium is located the age of the igneous rock may be determined, but we are usually more interested in the adjacent stratigraphic rocks known to belong to certain Periods or Epochs because of the fossils they contain. The number of years assigned to these rocks must be determined by their structural relationship to the intrusional rocks. Though this interpretation can seldom be exact, it can usually be fixed at least within a few millions of years. The methods used in correlating the two rock units are the ones usually employed in geological correlations.

Caution must be exercised in selecting the uranium-bearing rock sample. Subsurface water may dissolve part of the uranium; hence when

Geologic Age	Number of Years	Locality
Middle Cenozoic	35,000,000	Chihuahua, Mexico
Late Cretaceous or Early Cenozoic	60,000,000	Gilpin Co., Colorado
Early Permian	230,000,000	Oslo, Norway
		Joachimstal, Bohemia
Late Cambrian	400,000,000	Güllhögen, Sweden
Late Pre-Cambrian	620,000,000	Katanga, Belgian Congo
Middle Pre-Cambrian	1,420,000,000	Keystone, South Dakota
Early Pre-Cambrian	1,800,000,000	Careha, Russia
Early Pre-Cambrian	2,300,000,000	Winnipeg River, Manitoba
Early Pre-Cambrian	2,600,000,000	Rhodesian Shield

ABSTRACT HISTORY OF GEOLOGIC TIME

Eras	Periods or Epochs		Time at Beginning and at End	Biologic and Climatic Events Mostly in Middle Latitudes	Geologic and Physical Events	Approximate Duration
CENOZOIC	PLEISTOCENE EPOCH		0 to 600,000 years	Dominance of man and large mammals; marine invertebrates like living ones; modern plants; fluctuating cold and mild climates throughout the world.	High relief of continents; glaciation.	600,000 years
	TERTIARY PERIOD	PLIOCENE EPOCH	600,000 to 12,000,000 years	Abundance of mammals reaching peak in evolution; apelike animals with manlike characters in their teeth and skeletons; large camels and giraffes; antilocaprids and bovines; hyenoid dogs and hyenas; invertebrates more like modern kinds; plants of dryer and cooler environments.	Beginning of continental elevation; retreat of inland marine embayments.	11,400,000 years
		MIOCENE EPOCH	12,000,000 to 28,000,000 years	Rise and rapid evolution of grazing mammals; spread and diversification of mastodonts throughout Northern Hemisphere; temperate kinds of plants; moderate seasonal climates.	Development of plains and steppe grasslands; invasion of marine waters into epicontinental embayments.	16,000,000 years
		OLIGOCENE EPOCH	28,000,000 to 40,000,000 years	Modern families of mammals; primitive apes and monkeys; first saber-toothed cats; oreodonts abundant; true whales; plants trending toward temperate kinds; mild temperate climates.	Holarctic land connection in early part of Epoch; some mountain building; restriction of inland embayments.	12,000,000 years
		EOCENE EPOCH	40,000,000 to 60,000,000 years	All modern orders of mammals present; small tarsiers and lemuroids still common; first horses; first whales; nummulite foraminifers; subtropical forests with heavy rainfall; little or no frost.	Wide land connection between North America and Europe severed by water barrier by middle of Epoch.	20,000,000 years
		PALEOCENE EPOCH	60,000,000 to 75,000,000 years	Archaic mammals as dominant land animals; first tarsiers and lemurs; modern groups of birds; many new groups of marine invertebrates; others continuing from the Cretaceous; plants trending toward subtropical kinds; temperate to subtropical climates.	Culmination of late Cretaceous mountain building; trend toward transgression of marine waters into epicontinental embayments.	15,000,000 years

MESOZOIC	CRETACEOUS PERIOD	75,000,000 to 145,000,000 years	Rapid expansion of flowering plants; extinction of giant land and marine reptiles; last of the toothed birds; rise of pouched and placental mammals; modern groups of insects; ammonites and rudistid clams die out; most foraminiferal faunas cosmopolitan; mild climates at first, cooler later.	Last great spread of oceanic waters over the continents; mountain building toward close of Period.	70,000,000 years
	JURASSIC PERIOD	145,000,000 to 170,000,000 years	Gigantic dinosaurs; large marine reptiles; first mammals; first toothed birds; ginkgos, conifers, and cycadeoides dominant plants; insects of modern aspect; ammonites and belemnites abundant; mild climates.	Much of continents as lowlands; inland seas somewhat restricted.	25,000,000 years
	TRIASSIC PERIOD	170,000,000 to 200,000,000 years	Origin of dinosaurs and many marine reptiles; phytosaurs; diversification of mammallike reptiles; last of scale trees; giant conifers; first hexacorals; ammonites nearly died out at end of Period; last conodonts; arid and semiarid climates.	Elevated continents; widespread deserts.	30,000,000 years
PALEOZOIC	PERMIAN PERIOD	200,000,000 to 225,000,000 years	Great diversification in reptiles; first mammallike reptiles; *Glossopteris* plants in Southern Hemisphere; extinction of many groups of marine invertebrates; development of metamorphosis in insects; last trilobites and tetracorals. Cold dry and moist climates.	Uplift of continents; elimination of geosynclines.	25,000,000 years
	CARBONIFEROUS: PENNSYLVANIAN PERIOD	225,000,000 to 255,000,000 years	Great coal swamp forests; wide specializations in amphibians; origin of reptiles; gigantic insects; scorpions, and cockroaches abundant; fusulinid foraminifers abundant; warm humid climates.	Filling of great geosynclines; formation of coal in swamps; shallow inland seas; glaciation in Southern Hemisphere.	30,000,000 years
	CARBONIFEROUS: MISSISSIPPIAN PERIOD	225,000,000 to 280,000,000 years	Spread of amphibians; sharks and bony fish common; extensive forests in lowlands; insects probably evolved wings; ammonites expanding; crinoids dominant; first fusulinid and first imperforate calcareous foraminifers; warm climates.	Widespread inland seas; mountain building; beginning of great Coal Measures.	25,000,000 years
	DEVONIAN PERIOD	280,000,000 to 335,000,000 years	Reign of fishes; origin of amphibians; first wingless insects; scale-tree forests; brachiopods, corals and bryozoans abundant; blastoids at their peak; earliest ammonites; heavy rainfall and aridity.	Mountain building; filling of some geosynclines; intermontane fresh-water basins; extensive inland seas.	55,000,000 years

ABSTRACT HISTORY OF GEOLOGIC TIME (Continued)

Eras	Periods or Epochs	Time at Beginning and at End	Biologic and Climatic Events Mostly in Middle Latitudes	Geologic and Physical Events	Approximate Duration
PALEOZOIC	SILURIAN PERIOD	335,000,000 to 375,000,000 years	Eurypterids at their peak; tabulate corals of maximum abundance; tetracorals also abundant; fish with lower jaws; lycopsid land plants; scorpions and millepeds, first air-breathing animals; mild climates.	Continents relatively flat; inland seas still widespread.	60,000,000 years
	ORDOVICIAN PERIOD	375,000,000 to 435,000,000 years	Graptolites dominant; tetracorals present; small agglutinated foraminifers; trilobites still abundant; nautiloids of maximum abundance; cystoids at their peak; brachiopods increasing; first ostracods; first conodonts; first vertebrates without lower jaws; seaweeds and other algae; warm mild climates.	Continents remained low with shallow seaways; one of greatest marine inundations of all time.	60,000,000 years
	CAMBRIAN PERIOD	435,000,000 to 520,000,000 years	Abundance of marine invertebrates and algae; first land plants; trilobites dominant; first foraminifers; archaeocyathids abundant; all phyla of plants and animals probably existed; climates mild.	Formation of major Paleozoic geosynclines and adjacent archipelagos.	85,000,000 years
PROTEROZOIC		520,000,000 to 990,000,000 years	Evidence of bacteria, marine algae, fungi radiolarians, worm burrows, sponge spicules and ?brachiopods; all phyla of plants and animals probably recognizable; climates varied from warm moist to dry and cold.	Geosynclines; granite intrusions; volcanic activity, mountain building; glaciation; lowlands and deserts.	470,000,000 years
ARCHAEOZOIC		990,000,000 to 2,600,000,000 years	Earliest forms of life; blue-green algae, fungi and a ? flagellate from Canada, 2 billion years old; carbon in Rhodesian shield associated with minerals dated at 2.6 billion years old by radioactive methods (Holmes, 1954). Graphite and carbonaceous shales in Australia and Canada.	During at least part of this interval granite intrusions, lava flows, erosional and depositional conditions were much as they were in later geologic time; geologic history obscure.	1,610,000,000 years

a sample like this is examined the ratio of lead to uranium will be incorrect. Sometimes extra lead may be added from other sources, which will also give an erroneous ratio. The sample should be taken in deeply buried rock where there is little or no penetration of underground water.

One of the oldest records is 2.6 billion for some rocks in the Rhodesian shield, and there are rocks apparently older than that. The age of the crust of the earth is still not known. One of the most reliable estimates is 4.5 billion. One of the youngest radioactive datings is 35 million, but this is not accurately correlated with stratigraphic rocks with fossil evidence.

Potassium-Argon. The potassium-argon method probably offers the greatest promise of precise dating of sedimentary rocks because potassium-bearing minerals have formed in the depositional environment of these rocks. It has been discovered that when the radioactive isotope of potassium decays it forms two new elements, calcium and argon. In this process an electron from the K-shell of potassium is captured by the nucleus. Experiments have shown that in the decay of potassium 12 per cent changes to argon 40 and 88 per cent to calcium. Thus, with the known quantity of argon emitted from each unit of potassium per second, measurements of these quantities give a rather clear idea of the age of the rocks. In fact potassium-argon measurements of time have coincided well with paleontologic methods in correlation.

Radiocarbon. A sample of bone or plant material from an animal or plant that lived about 25,000 years ago will give an age with an accuracy of plus or minus 1 per cent (years) when subjected to the radiocarbon age-determination techniques. This is one of the most outstanding contributions to archaeology and to paleontologists concerned with late Pleistocene age determinations.

This method evolved when W. F. Libby and his associates at the University of Chicago determined that carbon 14 was formed by neutrons in the upper atmosphere and that it was present in an amount roughly equal to that expected if the isotope were mixed through all living matter and if the cosmic rays had been constant in intensity over the last few thousand years.

Furthermore, it was found that living things contain a constant amount of carbon 14 irrespective of their geographical distribution. It was discovered that when an animal or plant dies the carbon 14 in the organism disintegrates at a rate of one half in 5560 years, one half of the remainder in the next 5560 years, and so on until the quantity is so small accuracy is no longer possible. The limits of possible dating with carbon 14 is in the neighborhood of 40,000 years.

Fluorine. As early as 1892 a French mineralogist detected that the amount of fluorine is greater in fossil bones of the greatest antiquity. In more recent years it has been found that the fluorine content in rocks may vary considerably from one area to another. Nevertheless, fluorine has been very suggestive of age, especially when the comparative age of bones is sought at any given site. Kenneth P. Oakley of the British Museum of Natural History not only uncovered the interesting paper by the Frenchman, Carnot, but in applying the fluorine method disclosed that the remains of the famous Piltdown man did not belong to an early Ice Age, as seemed to be indicated by the alleged contemporary mammalian remains, and that the cranial fragments and the mandible did not belong to the same individual. Indeed, the fragments of the cranium were human and older than the lower jaw, which belonged to a Recent orangutang.

6 History of Paleontology

Comanche warriors and the fossil collector

The evolution of thought in the study of paleontology as well as in other biological and earth sciences in the civilized world has been greatly encumbered throughout much of its history by the beliefs and philosophies of the different societies. These societies in their varied aspects prior to the nineteenth century would not tolerate ideas that did not coincide with their accepted beliefs on the origin of life. With the great advancement in scientific thinking today, it is at times difficult to understand how most of our ancestors could have so ruthlessly ignored facts that are perfectly clear and reasonable to us. Thus history has clearly revealed that society has constantly been the most obstinate obstacle of progress in its own interest. It has taken many centuries to overcome prejudice and skepticism, and this is by no means yet fully attained.

Fossils and Primitive Men. It is not known when man first recognized fossils as the remains of animals and plants. Certainly fossils were picked up and evidently treasured by prehistoric men. Shells of Jurassic brachiopods were among the amulets of Neanderthal men. What they may have thought about these objects is lost in time, but it seems obvious that these primitive men did not realize what they had found.

American Indians utilized the segments of fossil crinoid stems by stringing them into necklaces. The Seneca Indians of western New York had different kinds of invertebrate fossils from the nearby formations in their possession. Evidently they used these fossils as both mystical and ornamental objects. Fossil cup corals were even used for pipes, and fossil brachiopods were placed in the graves of their braves.

Sioux Indians on the northern Great Plains referred to the fossil remains of large extinct mammals, especially those of the brontotheres, as the bones of "thunderhorses." According to the late Captain James H. Cook of Agate, Nebraska, a Sioux legend tells of the thunderhorses that lived "away back" and that these creatures would sometimes come down to earth during thunderstorms to chase and kill bison. The older people of the tribe related how the "Great Spirit" at one time sent the thunderhorses down to drive a herd of bison into the camp of starving Lacota Indians where many of the bison were killed with spears and lances.

George Gaylord Simpson has published on the discovery of parts of jaws of primitive Eocene mammals collected by Indians in New Mexico over 1000 years ago. The attention of the Indians may have been attracted to the dark bony structure and the jet-black, shiny teeth or they may have attributed some mystical quality to the fossils. In any event the fossils were excavated from the ruins of a pit house by E. T. Hall, Jr., in 1941.

Mysticism of Aborigines. All people have had some ideas on the creation of and mysticism pertaining to plants and animals. One of the most interesting tales is told by John W. Gregory in his fascinating book *The Dead Heart of Australia* and is supplemented by other ethnologists and explorers. The versions, as given today, are numerous, but the one related here is essentially like the others.

The aborigines of the Dieri tribe who occupied the area along Cooper's Creek east of Lake Eyre in South Australia had a legend about the giant extinct marsupials known to us as *Diprotodon*. They called them Kadimakara.

The aborigines visualized this desert area as having once been heavily forested. Heavy rains and fertile soil produced a luxurious undergrowth that covered the whole desert area, and in the heavens above were dense clouds that enveloped the tops of three towering gum trees. All above was supported by the strong thick trunks of the trees. Thus the earth below was protected from the blistering rays of the sun.

The Kadimakara dwelt in this cloudy foliage. But the smell of the succulent plants tempted them to venture down the gum trees to eat their fill. One time when these strange monsters were on the ground filling their stomachs with the tasty plants the great gum trees were destroyed. Since there was no way for the animals to climb back to their homes in the high branches they wandered about over the earth, only to become entrapped in bogholes along the streams and in the lakes, where their bones, jaws, and teeth could still be seen.

The holes left in the dense clouds where the tops of the gum trees had been gradually widened until the sky appeared as one hole above. Thus the sky was called "Puri Wilpanina," meaning Great Hole.

It has been said that during periods of drought or during extensive and prolonged floods the aborigines held corroborees at the sites where the bones occurred and offered blood sacrifices to appease the restless spirits of the Kadimakaras that still existed in the sky. Thus it was thought that further disaster could be averted.

Greek Philosophers. The Greeks were less inclined to combine mysticism and religion with their science. This was an extremely important step forward. Though they had some ideas that fossils were documents of earth's history, they treated all objects of natural science with philosophical speculation.

One of the most gifted of these Greek natural philosophers was *Anaximander* who was born about 611 B.C. He visualized an infinite all-pervading primordial substance with inherent powers of movement. Heat and cold derived from this action in the primeval matter he thought gave rise to the earth, the air, and to a surrounding circle of fire. He postulated that stars originated from fire and air and that animals and plants arose from matter on the earth under the influence of the sun. This also included humans, who he reasoned were at first fishlike in form.

Xenophanes of Colophon, a contemporary of Anaximander, is said to have seen molluscan fossils in the mountains of Greece and the impressions of laurel leaves in the sedimentary rocks near Paros. He believed that the sea had previously invaded the area, evidently flooding the mountains. *Herodotus* the great Greek historian and traveler made similar observations in the mountains of Egypt not far from the Ammon oasis some 200 years later. The correct explanation as to why shells of marine organisms occurred in rocks so high above sea level remained a mystery for more than 20 centuries. Nevertheless, Greek philosophers were not inhibited from free thinking and expression, but during and after Aristotle's time they tended to spend too much time dreaming up ideas and not enough on original observations. Consequently, most of the correct observations of the early Greeks were overlooked or disregarded.

The most influential Greek philosopher and scientist was *Aristotle* (384–322 B.C.). He divided the universe into two parts: the earth and planets and the heavens. He believed that there were periods of rejuvenation of the earth in which it had times of growth, maturity, and decay, as observed in all kinds of organisms. In each of these rejuvenations organisms arose from the mud, first as primitive forms, then matured into the more advanced kinds of animals and plants. Each form of life was thought to have had an immutable and intended place in the world and therefore could not evolve and had no history. He believed that fossils were organic but thought that they originated in the rocks. Since Aristotle's ideas were accepted by the Church and since he was held in such high esteem as a scholar, his ideas influenced the thinking of men long after they should have been disregarded. Seven hundred years later a wise man from Persia known as *Avicenna* (980–1037) not only supported but extended Aristotle's ideas. This doctor of medicine, philosopher, and naturalist finally came to the conclusion that the plastic force in the earth's crust worked like a sculptor modeling all kinds of plants and animals, some of which were formed only to fool people.

With the revival of learning in the fifteenth century, some people again turned their attention to the earth sciences and natural history.

This was greatly stimulated by the invention of movable type. Before this, manuscripts were prepared by hand or knowledge was transmitted by word of mouth by the learned men of the time. Universities and learned societies then began to spring up all over the continent, though some institutions were founded earlier, i.e., the University of Bologna in 1088, Cambridge in 1110, the University of Paris in 1257, the University of Uppsala in 1477, and Oxford in the 12th century.

Scientific efforts suffered after the fall of the Roman Empire and through the Dark Ages. This was between 800 and 1300 A.D. Some restricted studies were carried on in monasteries, but the Christian Church would not tolerate any other idea than that of "special creation in six days," irrespective of how this may have conflicted with facts. This was extremely unfortunate, since in the final analysis no one can justifiably deny the truth and that was what sincere scientists were trying to disclose. Efforts directed toward science and religion, then, could have worked together much more advantageously and convincingly for the betterment of mankind.

The Fossil Controversy — 1500–1750. The ideas and explanations of Aristotle, Avicenna, and others on the origins and nature of fossils were widely adopted in Europe until about 1500. Then the observations of an unusually alert and talented man from Italy took up with fervor where the early Greek scientist-philosophers had stopped. This Italian genius was *Leonardo da Vinci* (1452–1519). He was highly respected for his contributions to art, anatomy, engineering, and for other studies he had made.

While traveling and when occupied with canal building in the hills and mountains of Italy, Leonardo observed fossils and molds of fossil shells perfectly preserved in the rocks. He reasoned, as had Herodotus and Zenophanes of Colophon nearly 20 centuries before his time, that the seas had once occupied those areas. These fossils, he argued, could be nothing but the remains of animals that had lived in those waters. He explained at length how the remains of sea animals could not have been distributed so far inland by the Universal Deluge. He also postulated that in the course of time the sea became lower, the salt water drained away, and the mud surrounding the fossils changed to stone. Subsequently, the areas previously occupied by the sea elevated into hills and mountains. Nor did he neglect an explanation of the remains of land animals and plants that were covered over by muds from rivers which at times had flooded the valleys adjacent to their stream beds.

Leonardo ridiculed as absurd and unscientific the idea that such perfect models of living organisms could have taken origin in the rocks under the hypothetical influence of the stars or of plastic forces in the earth. Though Leonardo was held in high esteem for his accomplishment, this interpretation of the origin and nature of fossils was criticized. Obviously people were afraid of condemnation by society and even more of their fate in the hereafter. Thus the fossil controversy was opened in earnest about 1500 and continued with fury for the next 250 years.

Feeling ran so high that scientists were subjected to persecution. The Italian philosopher *Giordano Bruno* (1548–1600) was driven out of the country because of the dynamic and fiery spirit with which he presented his unacceptable views. Bruno had absolutely no tolerance for Aristotle's philosophy, which had been accepted by the Church. When in 1593 he returned by invitation to Venice he was arrested, taken to Rome, and thrown into prison. He was excommunicated seven years later and burned at the stake on February 17, 1600.

Some rather extraordinary explanations and fantastic stories about fossils were formulated by nonscientists of the Renaissance. Yet these were believed by many folk. They interpreted fossils as the remains of legendary monsters, giants, dragons, and basilisks. Some rather weird reconstructions were made. Sometimes the bones in a reconstruction were those of quite unrelated animals. The mayor of Magdeburg, *Otto von Guericke* (1600–1686), was certain that he had the skeleton of the unicorn mentioned in the Book of Job and mounted a fantastic skeleton assembled from a large collection of bones he had uncovered in his search for such monsters. Dragon bones and teeth found in the caves of Europe were later proven to be the remains of cave bears. In other cases supposed dragon, basilisk, and griffin bones were those of mammoths and woolly rhinoceroses. Huge bones and the skull of a mastodontlike animal, now called *Deinotherium*, were believed to be all that remained of the skeleton of Teutobochus, the king of the Cimbri.

Nicolaus Steno (1638–1686), a Danish bishop, directed attention to superposition in sedimentary rocks in 1669. In other words, he noted that the oldest rocks were at the bottom of a section, and all rocks formed later were laid down on top. This was an important contribution and still remains the basic law and the first information to be established in correlation. Unfortunately, religion interfered with the scholarly activities of this gifted scientist to the extent that he essentially refuted his masterful contributions to science.

Other scientists were quite accurate in their observations, but they were confronted by a public that would not yield. In 1696 the skeleton of a Pleistocene mammoth was unearthed at Burgtonna near Gotha in central Germany. *Ernst Tentzel*, a teacher in the gymnasium, was asked to investigate. He found that the bones were well below the soil in rocks of the earth's crust. Comparison with the skeleton of an Indian elephant convinced him of the relationships of the fossilized remains. He suggested that the animal was related to the elephant, but much older, evidently prehistoric. The public was up in arms. Poor Tentzel was brought to trial by the faculty, and though he proved his points by making comparisons, bone by bone, with those in a modern elephant skeleton the whole affair was judged as a freak of nature.

About nine years later when it was realized by many that fossils must be of organic origin *Johann Jakob Scheuchzer* (1672–1733), a doctor of medicine at Zurich, was the chief supporter of the idea that these fossils were the remains of unfortunates who had drowned in the Universal Deluge, as described in the Scriptures. At Gallows Hill near Altdorf in Franconia, Germany, he found what he was searching for. There in the rock were two quite black vertebrae. Scheuchzer pounced upon these with glee and pronounced them part of the skeleton of an ancient sinner who had perished in the Flood. Only 2 years before he had thought of fossils as freaks of nature. Little did he know that these were the vertebrae of a Jurassic marine reptile now known as an ichthyosaur. Still later, a stone workman quarrying in the Miocene rocks at Öeningen, Switzerland, uncovered part of the skeleton of another animal, which was sent to the learned doctor. This made him more positive than be-

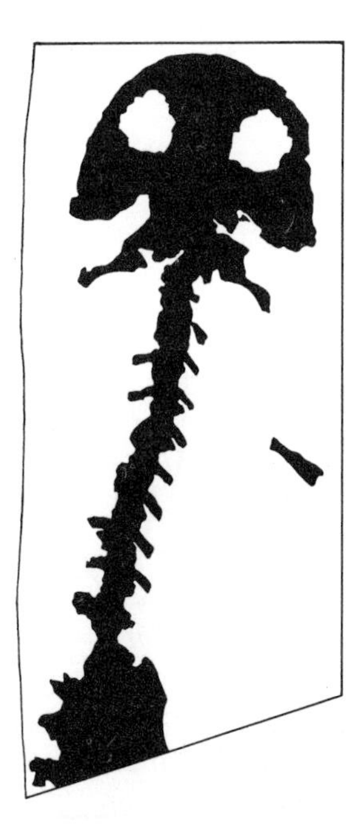

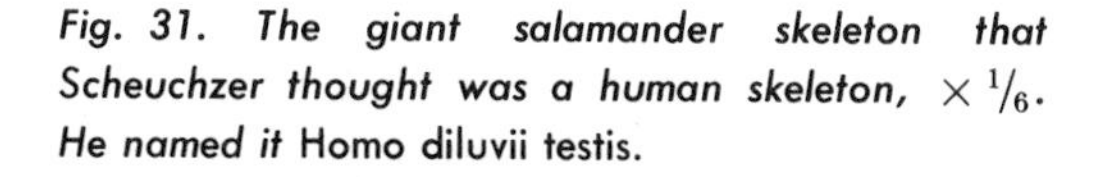

Fig. 31. The giant salamander skeleton that Scheuchzer thought was a human skeleton, $\times \frac{1}{6}$. *He named it* Homo diluvii testis.

fore. This unfortunate he named *Homo diluvii testis.* Fortunately for him, Scheuchzer did not live to learn the true identity of his fossil, for the famous French paleontologist Cuvier nearly 100 years later proved conclusively that the Öeningen skeleton belonged to a giant salamander, which he renamed *Andrias scheuchzeri* (Fig. 31). Nevertheless, the Deluge theory created a great stimulus all over Europe for collecting fossils.

No less misguided in his thinking was *Bartholomaeus Adami Johannes Berringer.* He still believed in Aristotle's and Avicenna's views that a plastic force formed fossils in the rocks. Unlike Scheuchzer, Berringer

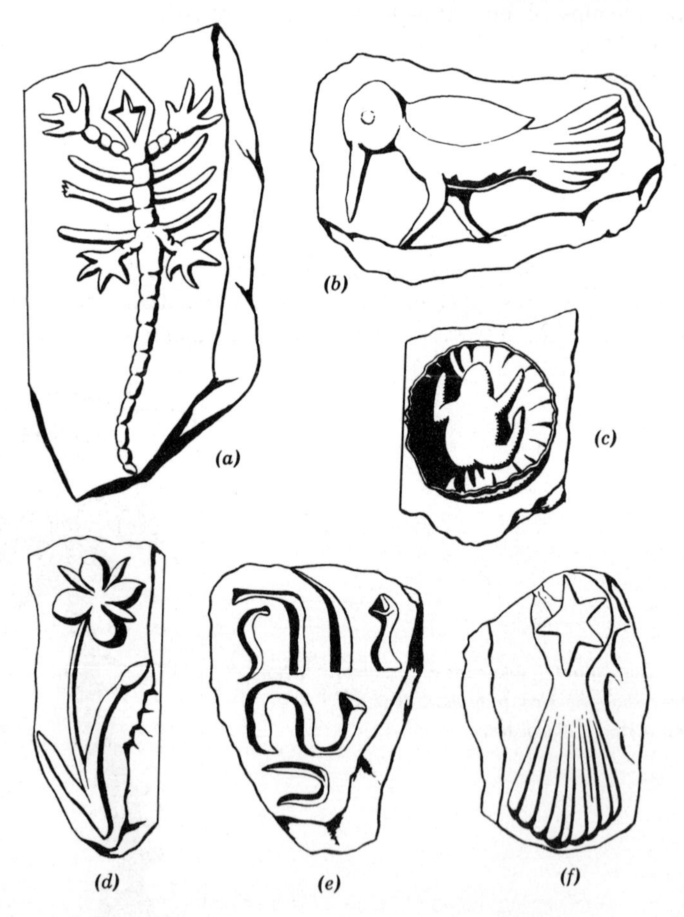

Fig. 32. Berringer's students carved an incredible collection of images resembling many things: (a) *arthropod;* (b) *bird;* (c) *frog in shell;* (d) *flower;* (e) *inscription;* (f) *and shooting star.*

lived to regret his mistake. Being a teacher at Würzberg and an enthusiastic fossil collector, he often took his students to a shale exposure where fossils could be found. As students sometimes do, one of them carved the image of an animal in the shale and left it at the site. When the eager professor found and accepted the carving as a fossil, the "die was cast." Others got into the act until the professor had an incredible collection of all kinds of images, including insects, lizards, frogs, birds, flowers, and astronomical bodies. Berringer figured and described all of these objects and published on them under the title *Lithographica Würceburgensis.* Later he found his name carved in Hebrew letters. The poor professor's reputation was ruined. He spent all the money he had trying to buy up the volumes but finally died in shame and poverty (Fig. 32).

Arguments on the fossil controversy tended to subside near the middle of the eighteenth century, as more scientists and their views were being accepted.

Recovery Period and Rise of Paleontology. Scientists had won the fossil controversy, but this fact could not be confirmed by the opposition and still is not admitted in certain uninformed circles. Nevertheless, it was gradually realized that scholars had more freedom in expressing their views and in reporting their observations.

In 1758 *Carolus Linnaeus* came out with his *Systema naturae,* or system of classification of plants and animals, but it was nearly 50 years before it was used for fossil remains. Linnaeus rigidly adhered to the old Aristotelian concepts of life and believed that each species was created once and gave rise to no other species.

Between 1750 and 1800 stratigraphers made important beginnings to an understanding of stratigraphic sequences. One of the foremost among these was *Giovanni Arduino* (1713–1795). His work was done in northern Italy near Verona. There he recognized a sequence in the major groups of rocks, which he called Primary, Secondary, and Tertiary. The Primary was composed of glassy, micaceous, and strongly folded schistose rocks with veins of quartz. There were no fossils in these rocks. The Secondary was made up chiefly of limestones, marls, and clays, which contained a large number of marine fossils. And finally the Tertiary, consisting of limestones, clays, marls, and sands, was highly fossiliferous.

The *Abbé Giraud de Saulave* in his volumes on the Natural History of Southern France (1780–1784) divided the rocks in the mountains of Vivarais into five parts, much in the manner that Arduino had done in Italy. In addition, he felt confident that if the facts he had observed were confirmed elsewhere a historical chronology of fossils and living organisms could be established. He concluded that the differences in fossils in the rocks were due to their antiquity because they belonged to older Periods and different climates. In the course of time the older species were destroyed and for that reason could not be found in the more recent rocks. Unfortunately, these excellent observations were largely overlooked because the Abbé did not express himself well in writing.

It remained for *William "Strata" Smith* (1769–1839), an English engineer and canal builder, to recognize the full significance of fossils in determining the age of strata. He was the self-educated son of a mechanic. The father died when William was 8 years old, and an uncle encouraged the boy to train himself in geometry and surveying. Consequently at the age of 18 he was employed as a surveyor and canal builder.

Smith was a keen observer and gifted with a remarkable memory. While working on canal construction he made large collections of fossil invertebrates from the excavations and other exposures. He made copious notes and faithfully remembered the stratigraphic position and the nature of the lithology of the beds from which these ammonite fossils came. It soon became known by the geologists in England that Smith had the ability to correlate the stratigraphic units simply by looking at the fossils. Because this unusual man was always willing to communicate orally the Reverend Benjamin Richardson invited him to his home to see his own fossil collection. Smith promptly told him where each specimen belonged in the stratigraphic sequence of Mesozoic rocks.

At first Smith thought that the same succession of strata stretched through England from coast to coast, but his later observations proved this not to be true. Even so, he was the first to make known on incontestable evidence that the stratified rocks of England could be most securely identified and arranged in chronological order according to their organic content. Eventually, he prepared colored geologic maps of Britain in which he used place names as well as the quarry men's names in his stratigraphic terminology. Smith was not a man to make broad generalizations but based his ideas on clear-cut observations. For this reason his views were readily adopted and he was acclaimed later as "the father of stratigraphy." Nearly all of his work was done in England, probably because of his restricted finances.

Meanwhile the geologists of Europe were divided into two groups. One supported the views of the Scot *James Hutton* (1726–1797), who considered all sedimentary rocks to have been derived from igneous rocks. This was the *Plutonic theory.* Furthermore, he held that the geologic processes of the past were the same as those then in operation, more precisely stated as "*the present is the key to the past.*" Other geologists argued for the theory of *Abraham Gottlob Werner* (1750–1817) of the University of Freiberg. The professor for years had lectured convincingly to his students that the rocks in the earth's crust were the result of precipitation through the waters of a universal ocean. This was known as the *Neptunist theory*. Eventually, Hutton's theory prevailed, with some modifications, but only after a vigorous fight.

At the turn of the nineteenth century vertebrate fossils were introduced prominently into the history of paleontology. Explorers and scientists of the French Empire had for years been sending skeletons of the living vertebrate animals to Paris from all parts of the world. Meanwhile *Georges Léopold Chrêtien Frédéric Dagobert*, Baron *Cuvier* (1769–1832), a well-trained comparative anatomist, utilized these skeletons to establish the broader relationships of the fossil vertebrate remains he had been recovering from the Paris Basin and from elsewhere in France. He had the gift of producing vivid descriptions both in writing and in lecturing. He clearly showed that the fossils represented extinct species and that they must have lived in the areas in which they were found as fossils. His observations and comparisons also revealed that some of the plant and animal assemblages were of tropical aspect.

Cuvier's ability to interpret correctly the relationships of fossil vertebrates from a few bones, jaws, or teeth established him as "the father of vertebrate paleontology and comparative anatomy." Once when part of a small skeleton was found he identified it as an opossum and demonstrated his prediction by removing the stony matrix, thus revealing the epipubic, or pouch bones, in the presence of his astonished colleagues. Cuvier soon attained fame and became a favorite of Napoleon I. Fossils were sent to him from all over the world for identification.

Nevertheless, Cuvier was not correct in all scientific matters. He believed that the surface of the earth had been devastated from time to time by violent catastrophic events when all the life in any given country was destroyed. He attributed this mostly to floods. Perhaps the Noachian Deluge was the last one. After each of these catastrophies life was created anew. This was known as the *catastrophic theory*.

No less important were the contributions of *Jean Baptiste Pierre Antoine de Monet, Chevalier de Lamarck* (1744–1829), who has been hailed as "the father of invertebrate paleontology." His accurate descriptions of living invertebrate marine animals and of those from the Tertiary sequence in the Paris Basin established invertebrate paleontology on a systematic basis. Lamarck was even more widely known for a theory of evolution which he was the first to state explicitly. His theory embraced all kinds of organisms. He thought that the living ones had formed by gradual modification from those that previously had existed. He also felt that adaptations played an important role in evolutionary progression. Habits were thought to bring about structural changes by use and disuse in the different parts of animals and plants and these modifications in turn were inherited in succeeding generations.

Though Lamarck's theory of the inheritance of acquired characters through use and disuse was vigorously criticized, especially by Cuvier, Lamarck was much nearer to the truth than his distinguished contemporary in the catastrophic theory. As G. G. Simpson recently stated, Lamarck's theory was a "brave attempt" at the time it was advanced, and though it was not adopted his ideas had a profound effect on the concepts of scientists in the following generation.

In the early nineteenth century more and more young scientists turned their attention to the study of fossils. Fossils were widely recognized as the key to stratigraphy and chronology. Sir *Charles Lyell* and *Paul Deshayes* used fossil invertebrates in their division of the Tertiary sequence, and morphologists like Sir *Richard Owen* (1804–1892) carried the comparative osteological studies initiated by Cuvier to even greater perfection in the classification and restorations of animals of the past. Owen's restorations of the giant marsupial *Diprotodon* from Australia and the huge ground bird from New Zealand, known as the moa, on meager fossil materials were phenomenal. Soon paleontology with its close relationship to life sciences and earth sciences became firmly established as one of the most interesting and rewarding fields of study.

The Origin of Species and Descent of Man. Exactly 101 years after Linnaeus made his major contribution to our system of biologic nomenclature Darwin's equally monumental book *On the Origin of the Species by Means of Natural Selection* appeared. Darwin would not have dared publish it 100 years earlier, but by 1859 society was much more tolerant of scientific endeavor. Even so, no one has ever been more abusively criticized by some and so highly praised by others as was *Charles Robert Darwin* (1809–1882). Being a sensitive person, he was hurt by the comments of his critics. Thirteen years later Darwin's *The Descent of Man* excited more opposition from his critics and rigorous support by his colleagues. All of this stimulated an even greater effort to find fossils as evidence of evolution.

Paleontology in the Americas. The white man's attention was directed to fossil bones in Mexico as early as 1519. Indians showed proboscidean bones to Cortez' army. The Indians believed these were the bones of wicked giants whom their ancestors had slain. It was estimated quite incorrectly that some of the thigh bones were as tall as a man. Some 50 years later in South America fossils of vertebrate animals said to be the bones of giants were taken to the *Reverend Joseph d'Acosta,* a Jesuit missionary. Until rather recently little or no notice was taken of these early discoveries.

Colonists in North America during the eighteenth century had in their possession the bones and teeth of mammoths, mastodonts, and horses. Teeth from the Carolinas were recognized by Negro slaves as being like those of African elephants they had seen. *Joseph Dudley*, Governor of Massachusetts, remarked that mastodont teeth and bones found along the Hudson River were the remains of a giant human who had drowned in the Universal Deluge. The slaves were much better scientists than the Honorable Governor. Another incident of interest was the discovery of the famous Big Bone Lick in the Ohio River Valley. En route from Canada with a troup of French and Indians to relieve the Governor of New Orleans from the attacks of hostile Chickasaw, Major the *Baron de Longueuil* found bones, tusks, and molars of mastodonts in a marsh just above the site of the Louisville Falls. These specimens were later sent to Paris.

Many of the fossils found in the Americas during the eighteenth and the early nineteenth centuries were sent to Europe. Among them were the famous Andean and Humboldt mastodont molars discovered in New Granada (Colombia), South America, about 1800 by the great explorer and geographer Baron *Alexander von Humboldt* (1769–1859). This practice of sending everything to Europe soon changed. The work of Smith, Cuvier, Lamarck, Sedgwick, Murchison, and others stimulated interest in North America. State geological surveys were founded in the United States, and the Canadian Geological Survey was organized. The need for fossils in stratigraphic studies was a boon to paleontologic investigations, particularly in Paleozoic invertebrate paleontology. Paleontology was given an added impulse in the United States by the interest of President *Thomas Jefferson*, who kept a collection of fossil bones of ground sloths, mastodonts, and other specimens at the White House. This greatly influenced society that paleontology was an honorable and learned avocation. The President advised Lewis and Clark to be on the alert for some of these animals on their expedition to the Northwest where he thought some of these huge creatures might still be living.

The most outstanding of these early studies in Paleozoic invertebrate paleontology and stratigraphy was made by *James Hall* (1843) of the New York Geological Survey. He was assigned to the western part of the state where the other state geologists did not wish to work, but young Hall made the most of his opportunity. He was an aggressive person and at times not at all tactful with his colleagues. This resulted in numerous enmities, but his contributions to stratigraphy and paleontology were highly respected. Using stratigraphic, lithologic, and paleontologic evidence, he worked out a sequence of the Silurian and Devonian marine rocks that became a classic. His keen observations in the field and the careful comparisons he made with the paleontologic and stratigraphic sequences outlined by Sedgwick and Murchison in England firmly established the section in New York state. Geologists on other state surveys followed Hall's lead; thus stratigraphic paleontology was launched in North America.

As early as 1846 a few fragments of vertebrate fossils collected by fur traders and explorers found their way to institutions in eastern United States. These were the first of an enormous record of extinct vertebrate animals that were to come from the areas west of the Mississippi River. Between 1853 and 1878 more than sixteen explorations were made in the Western Territories. Most of these were financed by the government of the United States. Six separate Government surveys were functioning in 1874.

Among the foremost early geologists and fossil collectors in the West were *Fielding Bradford Meek* (1817–1876) for invertebrates and *Ferdinand Vandeveer Hayden* (1829–1887) for the vertebrates. These were hazardous times for fossil collectors. The Indians were hostile, and army escorts accompanied all of the early expeditions. Nevertheless, Hayden searched far and wide quite on his own. He never carried firearms, and when at times he was captured by the Indians they soon released him as a demented person whom they dared not harm. Surely, they thought, a man scurrying over the hills with rocks in his pockets must be crazy. Later when Hayden became better known to the Indians he was trusted and established as their friend. Most of Hayden's collections were sent to the Philadelphia Academy of Science where they were described by *Joseph Leidy* (1823–1891) the foremost vertebrate paleontologist in the United States at that time. This was an ideal cooperative team. Vertebrate paleontology based on the rich fossil beds of the West seemed destined to advance under most cordial conditions. *Timothy A. Conrad* a conchologist also of Philadelphia, described fossil invertebrates sent to him from all parts of the United States and its territories.

The atmosphere of cooperation between individuals and between institutions so admirably exemplified by Hayden, Leidy, and others was soon shattered when *Othniel Charles Marsh* (1831–1899) of Yale and *Edward Drinker Cope* (1840–1897) of Philadelphia entered the field. Though these wealthy, competent scientists assembled tremendous collections from the West and threw much light on the relationships and evolution of the vertebrates of the past, they became so involved in a lamentable personal conflict to excel each other that their methods and behavior detracted greatly from their stature as capable paleontologists.

It is always extremely unfortunate for all concerned when men of wealth use their money to amass and hoard fossils, or any scientific materials, exclusively for their own research and pleasure. It is quite clear that much more can be accomplished through cordial cooperative efforts.

By an act of Congress on June 30, 1878, the United States Geological Survey was formed to consolidate and take over the functions of the various surveys of the Western Territories. *Clarence King* (1842–1901) was the first director. By this time the general acceptance of evolution as outlined by Darwin and expounded by Huxley and others had given a new approach to the study of paleontology. Members of the Geological Survey have continued to contribute tremendously to our knowledge of all phases of geology and paleontology.

The first paleobotanist in North America was *Léo Lesquereux* (1806–1889). This well-trained botanist who was a specialist in living plants came to the United States from Switzerland in 1848. He first directed his attention to the large collections of fossil plants from the Pennsylvanian coal measures and later studied fossil plants from all parts of the West. Though Lesquereux had been deaf since early manhood, this eminent scientist laid the foundation for later paleobotanical research in America.

Toward the close of the century more museums and universities took up the study of paleontology. *Henry Fairfield Osborn* (1857–1935) developed the American Museum of Natural History in New York City into one of the foremost centers of paleontologic learning in the world. The discovery of complete dinosaur skeletons in the West and innumerable other vertebrate animals aroused considerable public interest. The evidence assembled by Marsh on the evolution of the horse was carried to greater perfection by *William Diller Matthew* (1871–1930). *Samuel Wendell Williston* (1852–1918), first at the University of Kansas and later at the University of Chicago, described the great marine reptiles of the Cretaceous Niobrara sea and the Permian reptiles and amphibians of the Red beds of Texas. *David Starr Jordan* (1851–1931), president of Stanford University, exerted a tremendous influence in the study of fossil fish, and *John Campbell Merriam* (1869–1945) under the earlier influence of *Joseph LeConte* (1823–1901) established the department of paleontology at the University of California.

Meanwhile an obscure provincial school teacher, *Florentino Ameghino* (1854–1911) and his brother Carlos, were making outstanding discoveries in the pampas of Argentina. Florentino operated a stationery store in La Plata to make their living while Carlos roamed over the interior on horseback collecting fossil bones, skulls, and jaws. Florentino began his descriptions of these unusual South American mammals in 1874. Here were animals that looked like horses, camels, rhinoceroses, and other mammals from other parts of the world, and Ameghino thought they were related. To the contrary, they were later proved to belong to orders that had evolved in South America where, by convergent evolution, they came to look much like the mammals of North America and the Old World. Nevertheless, the descriptions of Florentino and the recognition of the faunal sequence by Carlos were so well done that paleontologists from many foreign countries were attracted to the Argentine. Paleontology in Europe, India, and elsewhere progressed rapidly after the turn of the nineteenth century, and innumerable outstanding contributions have been made.

Progress is also being made toward greater refinement in our studies. Studies in evolution and speciation are now based on population samples. With our increase in knowledge on the spread of faunas and floras throughout the world, our interpretations are becoming much more accurate. Information from all fields of paleontology are employed in the search for petroleum. Of still greater importance, most paleontologists, except in a few unusual cases, are no longer shackled by intolerant philosophies and beliefs in their efforts to give society the greatest benefits to be derived from their studies.

7 Classification

THE MATERIALS with which we deal in science represent *inorganic* (nonliving) or *organic* (living) things. The differences in them, for the most part, are readily recognizable by the average person. A tree is distinguished from a rock, and a whale and other kinds of marine life differ from the oceanic waters in which they live.

Inorganic materials are composed of liquids or crystals. Crystals may be defined as solids bounded by smooth plane surfaces. Inorganic materials grow by external additions of minerals like those that make up rocks. The growth of salt or ice crystals is an excellent example. Many fossil bones when broken will reveal calcite crystals beautifully displayed in the hollow inner cavities.

Viruses. A virus is a substance that is the nearest thing we know to a transitional state between living and nonliving matter, and it is able to reproduce. Inert fragments of tobacco mosaic viruses have been put together to form self-duplicating systems. It is still not known whether viruses are minute living organisms or nonliving complex proteins.

A virus, then, is an agent or substance still invisible when enlarged more than 3000 times. It affects living cells of plants and animals, frequently with disastrous results. How many there may be in the tissues of living organisms having no deleterious effect upon their hosts is not known. Some of the diseases caused by viruses are the dreaded poliomyelitis, influenza, yellow fever, measles, hoof-and-mouth, and the mosaic diseases in plants. When magnified something like 13,000 times with an electron microscope it has been found that viruses are of different outlines but are usually spherical or rod-shaped. They are composed of crystalline particles containing hydrogen, carbon, nitrogen, sulfur, and phosphorus in about equal amounts. They also contain nucleic acids.

Organic materials. Living organisms contain protoplasm and grow by internal addition. *Protoplasm* is the cell substance in plants and animals. It contains water, mineral salts, and organic compounds. These compounds are *carbohydrates* (sugars and glycogen), *fats* (liquids and solids), and *proteins*. Proteins contain many elements, such as carbon, hydrogen, oxygen, and amino acids. Other substances in protoplasm that regulate growth, especially in animals, are enzymes, hormones, and vitamins.

A most essential process known as *photosynthesis* takes place in all but a few plants (fungi, etc.). The energy of sunlight is absorbed and stored in the plant as high-energy compounds in the form of carbohydrates by the aid of chlorophyll and the low-energy compounds carbon dioxide and water. In this process oxygen is released. *Chlorophyll* is the green coloring matter in the leaves and stems. Thus all organisms depend directly or indirectly on plants for their food.

In contrast to inorganic materials, then, living organisms waste away by oxidation of the protoplasm, a kind of internal combustion, but they continually replace this waste by additions between the existing molecules. In this way growth and sustenance from the earliest germination to old age takes place. Even in the lowest single-celled organisms the body functions have progressed a long way from their simple beginning in the Pre-Cambrian.

Living organisms are either plants or animals.

Plants produce food by photosynthesis or absorption. Carbon dioxide, water, and chlorophyll in the plant produces carbohydrates, fats, and proteins.

Animals feed on plants or other animals.

Digestion is the breaking down of raw foods into smaller units, aided by mastication and by chemical action of the digestive system.

Metabolism is the total chemical change in an animal's body.

A Simplified Classification of Plants and Animals

Classification is an orderly arrangement of organisms based on their supposed relationships.

Unicellular organisms included here in the phyla Thallophyta and Protozoa have posed many difficult problems for systematic biologists. Indeed, most of the microscopic forms have characters and characteristics found in both plants and animals. This, of course, should be expected because of the proximity of relationships in these primitive organisms. In the two-fold classification of living things into plant and animal kingdoms many of the unicellular groups have been claimed by both botanists and zoologists. As early as 1866 Haeckel proposed a third kingdom, Protista, to include all of these primitive phyla. Recently, numerous systematists have adopted this as a more realistic classification. Schizomycetes, bacteria; Myxophyceae, blue-green algae; Chlorophyceae, green algae; Chrysophyta, golden-brown algae and diatoms; Pyrrhophyta, dinoflagellates; Euglenoidea, euglenas; Chloromonadina; Cryptomonadina; Euflagellata; Rhodephyceae, red algae; Phaeophyceae, brown algae; Myxomycetes, slime mold; Eumycophyta, fungi; Protozoa, amoeba, foraminifers, radiolarians, etc.

KINGDOM: PLANTAE

Phylum. Thallophyta (Gr. *thalles,* young shoot; *phyton,* plant). The phylum Thallophyta includes the simplest kinds of plants. Related to some group in this phylum are perhaps the first organisms to appear on earth. Thallophytes are mainly water plants, but they also occur in the tissues of other plants and in animals. *Bacteria* (microbes) are very primitive, microscopically minute plants of a low order in organization. There are several kinds of algae. *Green algae* (pond scum) live mostly in fresh water, though some occur in the sea and are lime-secreting. *Blue-green algae* are colonial and are held together by jelly-like materials; some have a blue-green pigment and others have red pigment, as seen in the Red Sea; they also secrete lime. *Brown algae* (kelp) may have trunklike stalks 100 feet long but do not secrete lime. *Diatoms* (plankton) possibly are related to green and brown algae. They inhabit both salt and fresh water. The plant is enclosed in a cell wall with two valves of silica; consequently, they are important in the fossil record. Some formations are so full of diatoms that the matrix is called diatomaceous earth or diatomite. *Fungi, lime molds,* and some *flagellates* are also included in the phylum (Fig. 33).

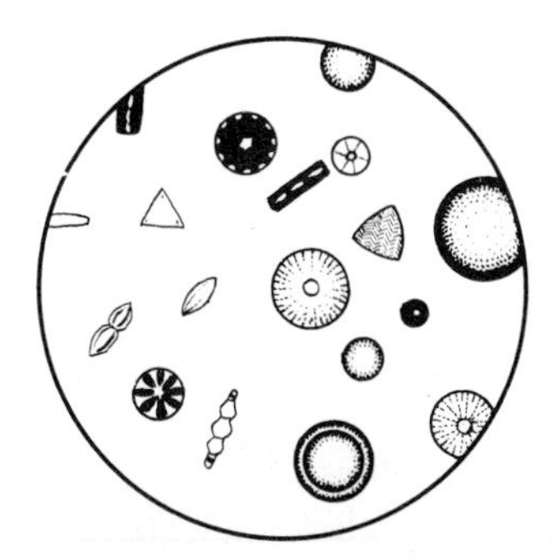

Fig. 33. Diatoms.

In a three-fold classification of organisms the Thallophyta are included in the kingdom Protista.

KNOWN GEOLOGIC RANGE. Proterozoic–Recent.

CHARACTERS. No true roots, stems, or leaves; sex organs unicellular or multicellular; reproduce by spores or by fission.

Phylum. Bryophyta (Gr. *bryos,* moss; *phyton,* plant).

The most primitive of the bryophytes are the liverworts, which may be found in sheltered, moist places, often on damp ground. Also included here are the mosses which form beautiful green carpets wherever moisture permits their growth. Some kinds form deep peat bogs. As the old plants die below, the new ones continue to grow above. Beautifully preserved fossils, such as skulls of the Irish deer and mammoth have been found in old peat bogs (Fig. 34).

KNOWN GEOLOGIC RANGE. Pennsylvanian–Recent.

CHARACTERS. Lacking differentiation into root, stem, and leaf; no vascular system; multicellular sex organs.

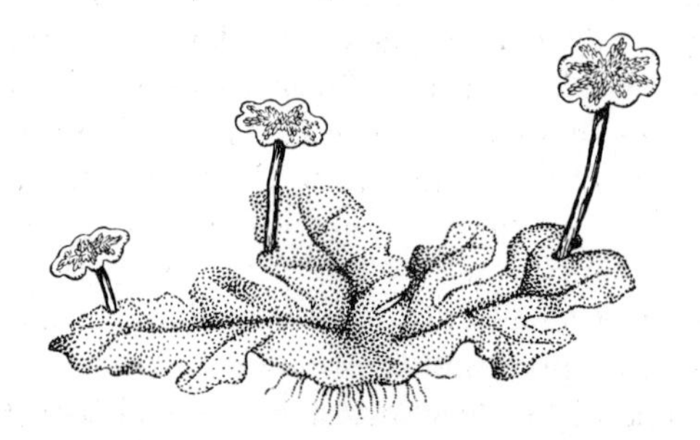

Fig. 34. Liverwort.

Phylum. Tracheophyta (Gr. *tracheos,* pipe; *phyton,* plant).

This phylum includes all of the remaining higher plants: psilophytes, club mosses, scale trees, scouring rushes, ferns, seed ferns, cycadophytes, ginkgos, conifers, and flowering plants.

KNOWN GEOLOGIC RANGE. Cambrian–Recent.

CHARACTERS. Possessing vascular system in body of plant: *xylem,* or woody tissue, which conducts water with dissolved substances up through roots, stems, and leaves; and *phloem* which chiefly conducts food down from leaves into stems after it has been manufactured with the aid of chlorophyll; with true stem and/or roots and leaves; multicellular sex cells.

Subphylum. Psilopsida (Gr. *psilos,* slender; *opsidos,* appearance).

These are the simplest land plants known. The only living represen-

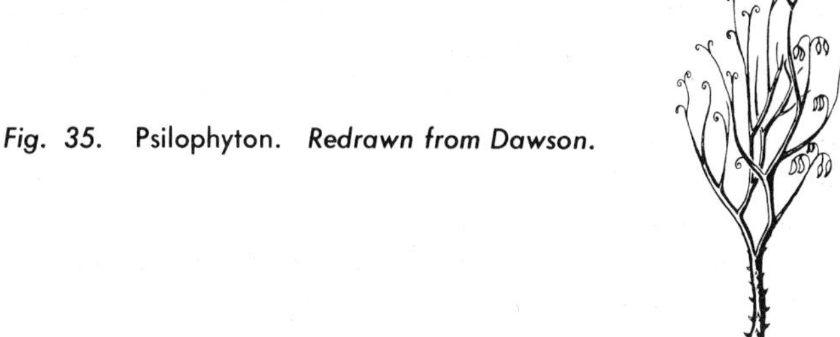
Fig. 35. Psilophyton. *Redrawn from Dawson.*

tatives are *Psilotum* and *Tmesipteris,* which are found in tropical regions. There are several Devonian genera, including *Rhynia, Asteroxylon, Psilophyton,* and *Horneophyton.* If they are representative of the plants intermediate between the lower phyla and the more advanced groups, they must have been in existence long before the Silurian, possibly in the late Proterozoic (Fig. 35).

KNOWN GEOLOGIC RANGE. Silurian–Recent.

CHARACTERS. Vascular stem; rhizomes and rhizoids instead of true roots, and without leaves; reproduce by spores; sporangia terminal on dichotomous or single branches.

Subphylum. Lycopsida (Gr. *lycos,* wolf; *opsidos,* appearance).

These include the living club mosses *(Lycopodium),* which are widely distributed but are most numerous in subtropical and tropical forests.

Fig. 36. Lepidodendron. *Redrawn from Hirmer.*

They are small prostrate plants showing little superficial resemblance to their giant distant relatives of the late Paleozoic. *Aldanophyton*, discovered in the middle Cambrian of Siberia, is the oldest known lycopsid. But much more spectacular are the giant scale trees, *Lepidodendron* and *Sigillaria*, which were so abundant in the Pennsylvanian swamps. Some of these large pithy-trunked trees were 100 feet tall and 6 feet in diameter at the base (Fig. 36).

KNOWN GEOLOGIC RANGE. Cambrian–Recent.

CHARACTERS. True roots, stems abundantly clothed with simple leaves; reproduce by spores; single spore cases on upper surface near base of spirally arranged leaves, in leaf axil, on stem above leaf, grouped into clusters (strobili).

Subphylum. Sphenopsida (Gr. *sphenos*, wedge; *opsidos*, appearance).

In this subphylum are the scouring rushes or horsetails, *Equisetum*, which are found in moist habitats. Because of the silica in the stems they were used in the early days to scour pots and pans; hence their common name. In the Pennsylvanian swamps *Calamites* grew in suitable habitats as dense as "canebrakes." The leaf whorls originally mistaken for petals of flowers were called *Annularia* (Fig. 37).

KNOWN GEOLOGIC RANGE. Devonian–Recent.

CHARACTERS. Stems jointed, ridged vertically; leaves as minor lateral branch systems arranged in whorls out from nodes; reproduce by spores; strobili composed of shieldlike disks on short stalks at right angles from stems.

Subphylum. Pteropsida (Gr. *pteris*, fern; *opsidos*, appearance).

The Pteropsida include the true ferns that reproduce by spores, the gymnosperms, among which are the seed ferns, cycadophytes, ginkgos, cordaites, conifers, and the flowering plants. The leaves were variable

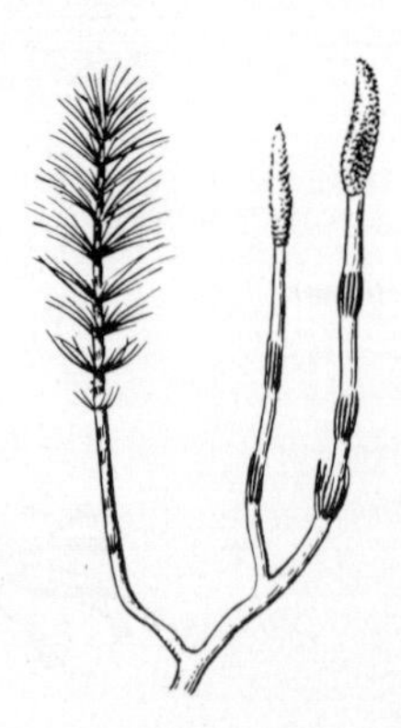

Fig. 37. Equisetum.

Fig. 38. Fern.

in size and complex. Branchlets of the leaves were flattened and expanded or connected by webs of tissue. Spores or seeds were attached to their surfaces or to the basal part of the leaves. The vascular systems as well as the trunks and stems were more advanced than in those of the other subphyla.

Class. Filicineae (Lat. *filix,* fern).

This class includes the ground ferns, tree ferns, and water ferns. Ferns prefer moist habitats, though some, like the bracken ferns, may occur on rather dry slopes. In the late Paleozoic it is difficult to differentiate between the leaves of ferns and seed ferns, unless the spores or seeds are attached (Fig. 38).

KNOWN GEOLOGIC RANGE. Devonian–Recent.

CHARACTERS. Upright leaves of complex form and structure, much larger than other parts of plant, develops axically for long periods of time; rhizomes with adventitious roots; vascular structure much like that in seed-bearing plants; reproduce by spores; spores marginal or on underside of leaf.

Class. Gymnospermae (Gr. *gymnos,* naked; *spermos,* seed).

In this class are the seed ferns, cycads, cycadeoids, ginkgos, cordaites, and conifers. They all reproduce by seeds, in which the micropyle (opening to embryo sac where cross-pollenation takes place) is exposed to the air.

Subclass. Pteridospermae (Gr. *pterido,* fern; *spermos,* seed).

This subclass includes the extinct seed ferns as exemplified by such genera as *Neuropteris, Eospermatopteris,* and *Pecopteris.* Plants like these, though abundant in the late Paleozoic, died out in the Jurassic. *Gangamopteris* and *Glossopteris* the "tongue-leafed" plants from the

Permian and Triassic of the Southern Hemisphere previously have been included in this subclass, but the recent discovery of fruiting bodies attached to the leaves offers some suggestion that they may be sufficiently distinctive to warrant the proposal of another class or subclass near the angiosperms (Fig. 39).

KNOWN GEOLOGIC RANGE. Devonian–Jurassic.

CHARACTERS. Fernlike leaves (fronds) with seeds attached; seeds single, not in clusters, nor in conelike organs; some were trees and others were of sprawling habit.

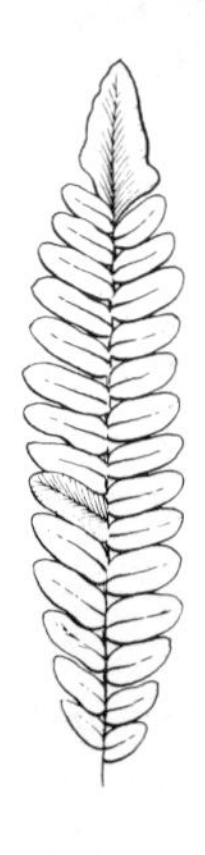

Fig. 39. Neuropteris.

Subclass. Cycadophytae (Gr. *cycad*, coco palm; *phytos*, plant).

The cycadophytes include the living coco "palms," or cycads, and the cycadeoid plants of the Mesozoic typically represented by the genera *Cycadeoidea*, *Bennettites*, and *Williamsonia*. Cycadophytes probably arose from some group of seed ferns in the late Paleozoic. This is suggested by certain leaf forms from Pennsylvanian and Permian rocks.

KNOWN GEOLOGIC RANGE. Triassic–Recent.

CHARACTERS. Pithy trunks surrounded by a thin zone of wood and with a thick irregular surface mantle of leaf-base scars; palmlike leaves and cones.

Order. Cycadeoidales (Gr. *cycad*, coco palm; *eoidos*, resemblance; *ales*, bear).

The cycadeoids formed a conspicuous part of the Jurassic and early Cretaceous floras. Specimens were first identified as cycads, but most of these errors have been corrected. Though there is a marked resemblance to the living cycads, there is no evidence that the cycadeoids gave rise to them (Fig. 40).

KNOWN GEOLOGIC RANGE. Triassic–Cretaceous.

Fig. 40. Cycadeoidea.

CHARACTERS. Trunks short, 3 feet long, subspherical or oval, or 10 to 12 feet long, columnar or forked, as in *Williamsonia;* epidermis of leaves with guard cells of stomata (pore or opening) flanked by a pair of subsidiary cells that originated from the mother guard cell; male and female organs in same floral axis; seed small; embryo nearly fills central seed cavity.

Order. Cycadales (Gr. *cycad*, coco palm; *ales*, bear).

Living cycads are known as coco "palms". They are scattered and in limited numbers in tropical provinces throughout the world. True cycads also occur in Mesozoic rocks, but their trunks are not so well known as those of the cycadeoids (Fig. 41).

KNOWN GEOLOGIC RANGE. Triassic or earlier–Recent.

CHARACTERS. Trunks cylindrical, usually erect; epidermis of leaves usually straight; subsidiary cells are not from the mother cell but de-

Fig. 41. Living cycad.

Fig. 42. Ginkgo. *Redrawn from figure in* Andrews' Ancient Plants, *by permission of* Cornell University Press.

rived independently from guard cells of stomata (pore or opening); male and female organs on separate plants; seeds larger than in cycadeoids; embryo embedded in mass of reserve food.

Subclass. Ginkgophytae (Chinese name).

The ginkgos, or maidenhair trees, which have continued on into Recent time from the Mesozoic, are said to be living in a wild state in western China. Trees now living in botanical gardens and parks were introduced from Japan and China where they were found in temple grounds. Ginkgos were one of the most dominant groups of plants in the early and middle Mesozoic. Their characteristic leaves are well preserved as fossils (Fig. 42).

KNOWN GEOLOGIC RANGE. Pennsylvanian–Recent.

CHARACTERS. Tree habit as in many angiosperms; dichotomous fan-shaped or laminate leaves with nearly parallel fine venation in each half; male and female organs in separate trees; catkinlike strobili in separate trees; microspores carried by wind to mature ovules in female tree; embryo develops into seed.

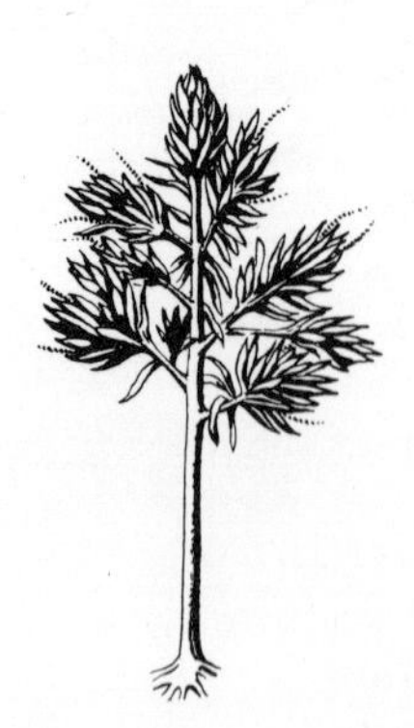

Fig. 43. Cordaites. *Redrawn from Grand'Eury.*

Subclass. Cordaitae (Corda, German botanist).

Cordaites is a late Paleozoic genus closely related to the conifers. The evolution and intergradation between the subclasses recently described by Rudolf Florin is one of the most completely known records in paleobotany. Cordaites were one of the abundant plants in the Mississippian and Pennsylvanian forests that formed thick coal beds in different parts of the world. There is some evidence that they grew in rather dense clusters, as do some living conifers. Some cordaitelike fossils have been found in the Devonian, but their identity is uncertain.

KNOWN GEOLOGIC RANGE. Mississippian–Permian (Fig. 43).

CHARACTERS. Tree habit somewhat like conifers; trunks with central pith surrounded by separate strands of primary xylem; long strap-shaped leaves; seeds borne in racemes on small budlike branches; pollen in sacs attached to long stalks surrounded by clusters of leaves; no resin canals in the wood.

Fig. 44. Pine.

Subclass. Coniferae (Gr. *conifer,* cone).

The conifers include the pines, spruce, firs, redwoods, and many others. They have a long geologic history in the Northern Hemisphere. One of the most interesting recent discoveries was the living deciduous *Metasequoia* in China. Fossil leaves were so much like those in the evergreen redwood, *Sequoia,* that, though common in the Tertiary of the northern part of the world, their identity was not detected until the living trees were found by T. Wang, a forester, and announced by Professor W. C. Cheng. The Araucariaceae, without resin canals in the wood, are somewhat intermediate in their characters between the Cordaitae and the Coniferae (Fig. 44).

KNOWN GEOLOGIC RANGE. Pennsylvanian–Recent.

CHARACTERS. Tree habit usually tall with straight trunks, leaves needlelike or scalelike, mostly evergreen; seeds protected under the scales of compact cones; pithy center of trunks almost totally reduced; all except araucarians have resin canals in wood.

Class. Angiospermae (Gr. *angios,* vessel; *spermos,* seed).

These are the flowering plants which include the multitude of plants we see about us today. There are such different kinds as small grasses and majestic oaks and all the garden vegetables and flowers about the house. Most modern animals depend directly or indirectly on angiosperms for food. Angiosperms have had an intriguing history since they first appeared in abundance in the early Cretaceous in widely separated localities of the world; yet their earlier history and time of origin is still not known. Palmlike leaves have been found in the Triassic of New Mexico, but angiosperms may have become diversified from other plants as early as the Permian.

KNOWN GEOLOGIC RANGE. Triassic–Recent.

CHARACTERS. Reproduce by seeds in which the micropyle is not exposed to the air.

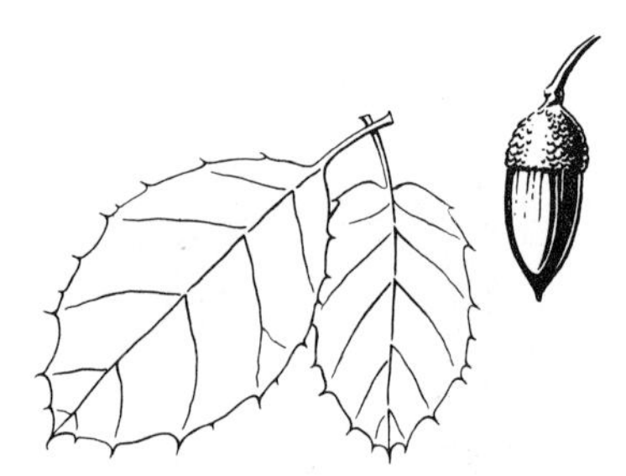

Fig. 45. Oak.

Subclass. Dicotyledonae (Gr. *di,* two; *cotyledon,* cup-shaped cavity).

Most of our trees, such as the elm, oak, willow, and maple, and most of the garden vegetables (beans, squash, melons, and radishes) are in this subclass. Most garden flowers are also dicots. Fossil representatives of these plants have been very useful as climatic indicators in the Epochs of the Cenozoic (Fig. 45).

KNOWN GEOLOGIC RANGE. Jurassic–Recent.

CHARACTERS. Leaf venation netted; vascular bundles of stem arranged in form of cylinder; embryo possesses two cotyledons (two first leaves); parts of flowers usually in fours or fives or multiples thereof.

Subclass. Monocotyledonae (Gr. *mono,* one; *cotyledon,* cup-shaped cavity).

These are the grasses, lilies, corn, and bamboos. The fossil record of monocots is not well known because fruiting bodies have been hard to find and there is little differentiation in the leaves. The best fossil record is that of the spear grass, *Stipidium,* so clearly elucidated by

Fig. 46. Corn.

M. K. Elias on the siliceous husks of seeds found in Miocene and Pliocene continental formations in the Great Plains region of the United States. A palmlike plant, *Sanmiguela,* is the oldest known flowering plant (Fig. 46).

KNOWN GEOLOGIC RANGE. Triassic–Recent.

CHARACTERS. Leaf venation parallel; vascular bundles of stem scattered throughout cylindrical mass of tissue; vascular cambium absent; embryo possesses one cotyledon (one first leaf); parts of flower in threes or multiples thereof.

KINGDOM: ANIMALIA

Phylum. Protozoa (Gr. *protos,* first; *zoon,* animal).

There are many kinds of protozoans, most of which are soft-bodied, microscopic organisms not preserved as fossils. Many of these are parasitic and cause serious diseases in other animals and in some plants. Protozoans include the flagellates, trypanosomes, amoebas, foraminifers, and radiolarians (Fig. 47).

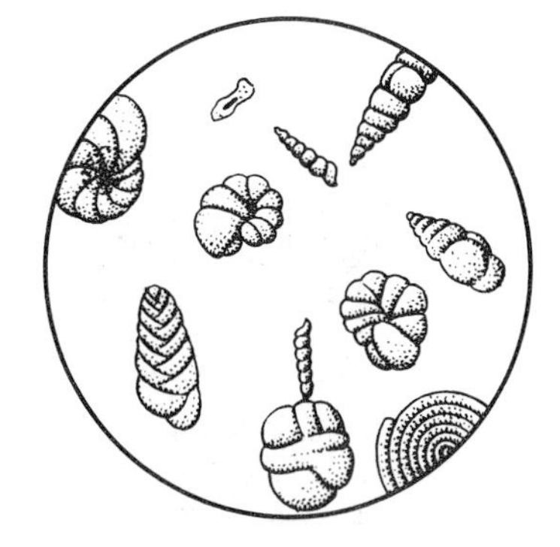

Fig. 47. Foraminifers.

In a three-fold classification of organisms the Protozoa are included in the kingdom Protista.

KNOWN GEOLOGIC RANGE. Cambrian–Recent.

CHARACTERS. Mononucleated (one nucleus) or with several nuclei at some stage in life history; usually unicellular (one-celled); some live in colonies but with no division of labor between the different units; rarely possess chlorophyll; ingestion by engulfing, excretion by egestion; reproduction by binary or multiple fission; tests (hard parts) of foraminifers, chitinous, agglutinated, calcareous, or siliceous; tests of radiolarians, siliceous or of strontium (an alkaline-earth metal) sulfate.

Phylum. Porifera (Lat. *porus*, pore; *ferre*, to bear).

This phylum includes the sponges, the simplest of the multicellular animals. They are usually colonial in habit and occur primarily attached to rocky bottoms in warm, shallow marine waters. Some live in fresh water. The larvae are free-swimming. The different kinds range from 1 millimeter to more than 6 feet in diameter. The flexible skeletons of bath sponges with the living protoplasm removed are of considerable commercial value (Fig. 48).

KNOWN GEOLOGIC RANGE. Proterozoic–Recent.

CHARACTERS. Multicellular (many-celled); radially symmetrical saclike body with two germ layers; water taken in through many small openings, bringing in food and expelling waste materials through a single large opening; reproduction by asexual budding or sexually by eggs and sperm; skeletons supported by calcareous or siliceous spicules or by an irregular spongin network as in bath sponges.

CLOSELY RELATED GROUPS. There are other groups of spongelike, fossils of uncertain systematic status. These include the Paleospongia (Cambrian) and the Receptaculitida (Ordovician-Devonian). The paleosponges seem to be confined to Cambrian rocks. One of the most widely known genera is *Archaeocyathus*. Fossils of these animals have been

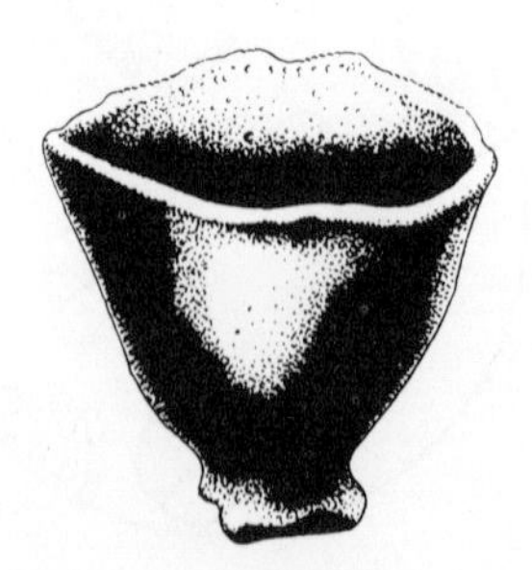

Fig. 48. Sponge.

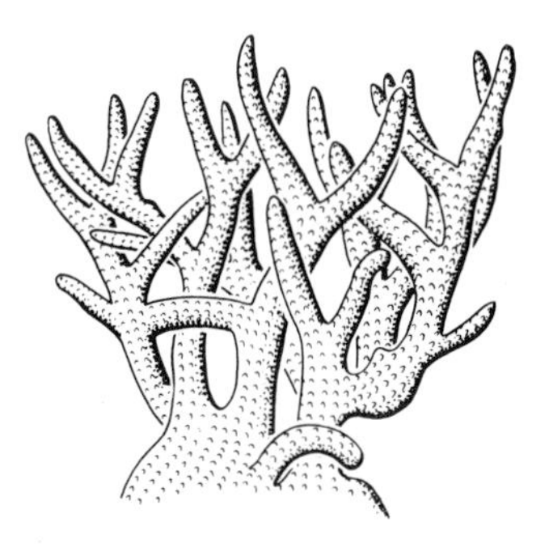

Fig. 49. Horn coral.

found in Europe, Asia, and North America. The equally puzzling receptaculitids are also widely distributed geographically.

Phylum. Coelenterata (Gr. *koilos,* hollow; *enteron,* inside cavity). The many soft-bodied coelenterates include the hydroids, Portuguese man-of-war, jellyfish, and sea anemones which are rarely preserved as fossils, but both corals and graptolites are well represented.

Class. Anthozoa (Gr. *anthos,* flower; *zoon,* animal).
Corals are solitary and colonial in habit. Classification of the corals is difficult, and agreement among the authorities is not fully attained. The tetracorals (or rugose corals), honeycomb corals, and others of the Paleozoic are rather clearly differentiated from hexacorals (or scleractinian corals) of the Mesozoic and Cenozoic. The most spectacular coral reef in the world is the Great Barrier Reef off the northeast coast of Australia (Fig. 49).

KNOWN GEOLOGIC RANGE. Cambrian–Recent.

CHARACTERS. Radially symmetrical body with saclike, central digestion cavity; external epidermis and internal endoderm layers of cells; plexus of nervelike cells in epidermis; food taken through mouth, which is surrounded by tentacles; waste also passes out through mouth; reproduction sexually and asexually; no blood, respiratory, or excretory organs; skeleton calcareous or horny; radiating septa in skeleton.

Class. Graptolithina (Gr. *graptos,* write; *lithos,* stone).
Most graptolites were planktonic, colonial, marine animals that evidently were widely distributed by ocean currents and waves. Some were benthonic (bottom-dwellers). They resembled fronds of plants and colonial animals such as certain corals or bryozoans. Their relationship to chordates has been suggested by R. Kozlowski, who dissolved the cherty matrix about the specimens with hydrofluoric acid. The specimens were then imbedded in paraffin and sectioned with a microtome to reveal structures in the fossils not seen before (Fig. 50).

KNOWN GEOLOGIC RANGE. Cambrian–Mississippian.

CHARACTERS. Bell-like colonial organisms suspended by threadlike rods (nema); individual animals (zooids) in cuplike overlapping structures (thecae) on the branches (stipes); wall structure, method of budding and system of internal tubules resembles structures in pterobranchiate chordates, but otherwise graptolites are markedly different from pterobranchs; reproduction asexually by budding in groups of threes; chitinized skeleton.

CLOSELY RELATED PHYLUM. Small, free-swimming, jellyfishlike marine animals called Ctenophora are frequently classified with the coelenterates. Though there are many features in which they show closer relationships to the coelenterates than to other phyla, they also display more progression in their body organization than in members of that phylum. Ctenophores are not known in the fossil record.

Fig. 50. Graptolite colony.

Phylum. Platyhelminthes (Gr. *platy*, flat; *helminthes*, worm).

This phylum includes the planarians, flukes, tapeworms, and ribbon worms. The ribbon worms are sometimes placed in a separate phylum. Tapeworms and flukes are dreaded parasites in man and in other animals.

KNOWN GEOLOGIC RECORD. There is no positive fossil record of flatworms. It has been assumed that some of the trails and burrows on Pre-Cambrian sea floors were made by these animals, but this has not been substantiated.

CHARACTERS. Bilaterally symmetrical; not segmented in the strict sense of the word; lowest phylum of animals with organ system of construction; body flat dorsoventrally, with head; three layers of cells; small ganglion with a few longitudinal nerves, no sense of sight or sound; food passes through body cells by diffusion, waste passes out through mouth, excretory system connected to excretory ducts; asexual reproduction or with both sex glands in an individual, highly specialized reproductive system; no blood; no hard parts (Fig. 51).

Fig. 51. Tapeworm.

Phylum. Nemathelminthes (Gr. *nematos*, thread; *helminthos*, worm). These are also called roundworms. Most are parasites. In animals they may be found in the intestines, blood, muscles, or other parts of the body. *Ascaris*, the largest of the intestinal worms, hookworms, pinworms, and trichina, are some of the commonest kinds. The plant parasites of this phylum do millions of dollars in damage annually to agricultural plants, usually by attacking the roots (Fig. 52).

KNOWN GEOLOGIC RECORD. Not certainly known as fossils.

CHARACTERS. Body round, unsegmented, mouth and anus at opposite ends, with pharynx and intestine between; no head; ring of nerve tissue around pharynx, ventral nerve cord; excretory system with two lateral longitudinal canals; separate sexes, fertilization internal, eggs microscopic, well protected by chitinous shell; no circulatory or respiratory systems; no skeletal parts.

CLOSELY RELATED PHYLA. The long, slender "horsehair worms," Gordiacea, frequently seen in fresh-water pools, streams, or horse troughs, and the parasitic spinyheaded worms, Acanthocephala, are usually included in one phylum with the roundworms. The ribbonworms, Nemertea, and the marine worms, Kinorhyncha, are less closely related to the nematodes but are not distantly related.

Phylum. Trochelminthes (Gr. *trochos*, wheel; *helminthos*, worm). Rotifers are microscopic organisms less than 1 millimeter long. They may be found in almost any retainer with fresh water and are common

Fig. 52. Whipworms.

in slow-flowing streams, ponds, and pools. Some live in brackish and salt waters. Others are parasitic. Most of them are free-moving. The movement of the cilia around the mouth is suggestive of a rotating wheel, and the name is derived from this movement. Rotifers are approximately on the flatworm level of development (Fig. 53).

KNOWN GEOLOGIC RECORD. No fossils are known.

CHARACTERS. Bilaterally symmetrical, not segmented; three germ layers; ciliated mouth anteriorly and footlike appendage posteriorly; no true head; alimentary tract with mouth, pharynx, masticatory structure, stomach, intestine, and anus; dorsal nerve ganglion and sensory nerve fibers; two nephridia (primitive kidneylike structures); excretory canal; two sexes, reproduction parthogenetic and sexual, usually oviparous; no skeletal parts.

Fig. 53. Rotifer.

Phylum. Bryozoa (Gr. *bryon*, moss; *zoon*, animal).

Microscopic animals living in branched colonies attached to different objects. Their body form led most early biologists to classify them as plants. Bryozoans form carpetlike or thin films as incrustations over rocks, shells, or other kinds of life in marine waters. Only a few kinds live in fresh waters. The structure of the lophophore (mouthlike opening) is suggestive of relationships with the brachiopods. The body structure is more advanced than in coelenterates; currently, there is considerable doubt expressed on the validity of the characters used in the classification of the fossils (Fig. 54).

KNOWN GEOLOGIC RANGE. Ordovician–Recent.

CHARACTERS. Bilaterally symmetrical; not segmented; individuals or zooids as budlike structures at ends of branching colony; U-shaped alimentary tract, with esophagus, stomach, and intestine; opening of anus usually below lophophore; without head; nerve ganglion present; no circulatory system, blood cells in body cavity; no respiratory organs; hermaphroditic, eggs fertilized in body cavity; larva as free-swimming

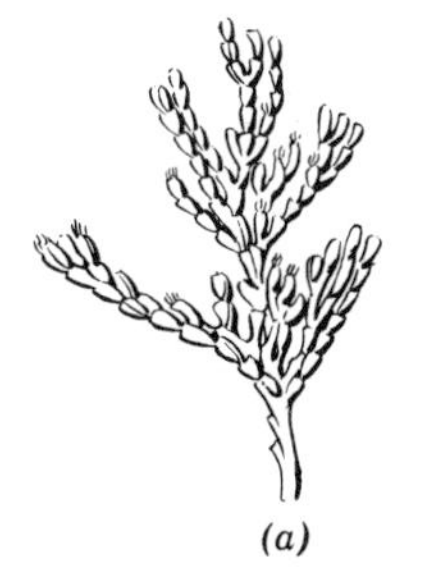

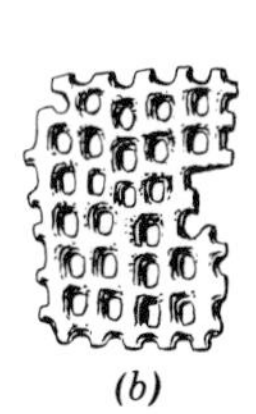

Fig. 54. (a) *Bryozoan colonies;* (b) *part of skeleton,* × *20.*

trochophores (larval stage); colonies developed by asexual budding; skeleton usually slender branching tubes with latticelike perforations.

Phylum. Brachiopoda (Gr. *brachios,* arm; Gr. *pods,* foot).

The brachiopods are one of the most important phyla in the Paleozoic fossil record. Their abundance and diversification in Paleozoic and Mesozoic rocks makes them very useful in correlation. Superficially, they show a marked resemblance to bivalve mollusks (Pelecypoda), but structurally they are much closer to some of the less-advanced phyla. For the most part they are bentho-neritic marine animals, though some live in brackish and estuarine waters. The adults are sessile and may be attached by a pedicle to any object on the bottom. Their dispersal takes place in the free-swimming larval (trochophore) stage. Representatives of the genus *Lingula,* have changed very little since Ordovician time (Fig. 55).

KNOWN GEOLOGIC RANGE. Proterozoic?–Recent.

CHARACTERS. Bilaterally symmetrical, not segmented, large lophophore mouth, centrally located; stomach with digestive glands, intestine, with or without anus; without head; nerve ring around gullet; large coelom (body cavity); small heartlike structure and blood vessels; excretion by two pairs of nephridia (kidneylike structures); no respiratory organs; separate sexes, eggs and sperm discharged into water; not colonial; shells of chitin, with some calcium sulfate, calcium phosphate, and magnesium carbonate; exterior smooth, with concentric growth lines, covered with spines or with radially arranged grooves and ridges.

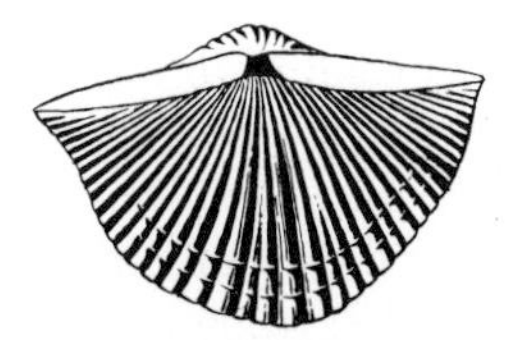

Fig. 55. Brachiopod, Spirifer.

Class. Inarticulata. Unhinged valves held together by muscles; no teeth and sockets; shells usually chitino-phosphatic.

Class. Articulata. Hinged valves; well-developed teeth and sockets along hinge line; shells calcareous.

Phylum. Mollusca (Lat. *mollis*, soft).

The mollusks, sometimes called lamellibranchs, are another very important group of invertebrates to the paleontologist because so many are preserved in the fossil record. They have become widely adapted to life in marine waters, in streams and lakes, and even on land. Aside from the hard exoskeleton in most of them, they have soft, unsegmented, bilaterally symmetrical bodies with a head, a ventral foot, and dorsal viscera.

KNOWN GEOLOGIC RANGE. Cambrian–Recent.

CHARACTERS. Bilaterally symmetrical; nervous system with ganglia for different parts of body; digestive tract well developed; liver aids in digestion; heart, aorta, and other blood vessels, dorsal; respiration by gills, "lung," or by epidermis; excretion by nephridia; sexes usually separate, fertilization external or internal, mostly oviparous. Mollusks are divided into five classes.

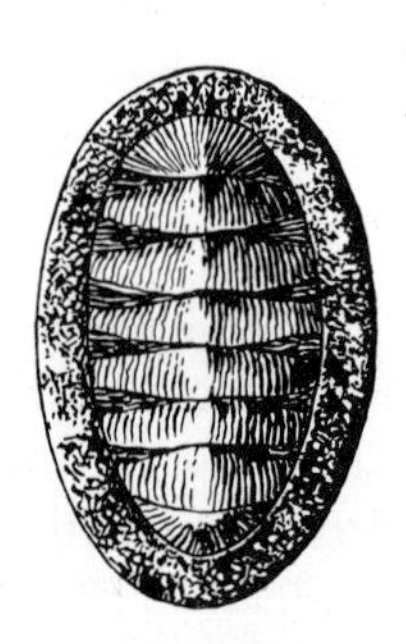

Fig. 56. Chiton.

Class. Amphineura (Gr. *amphi*, both sides; *neura*, sinew).

These slow moving sea-bottom chitons, are frequently found holding fast to rocks; they feed on algae; when disturbed they can roll up like a sow bug. Though they have a long geologic range, the fossil record is not so significant as in the other classes (Fig. 56).

KNOWN GEOLOGIC RANGE. Ordovician–Recent.

CHARACTERS. Head reduced; mouth and anus at opposite ends of body; foot broad; gills usually posterior in position; with or without eight imbricated scales; scales slightly or nonoverlapping in Paleozoic forms.

Fig. 57. Snail.

Class. Gastropoda (Gr. *gaster*, stomach; *pods*, foot).

The Gastropoda compose the largest class of mollusks, in which are snails, slugs, turritellas, whelks, limpets, abalones, and many others; most of the shells are coiled clockwise (dextral); some are coiled in the other direction (sinistral); others are coiled in a horizontal plane; and still others, like the limpet and abalone, show little evidence of coiling. Most gastropods live in water and derive their oxygen by means of gills, but snails and slugs, called *pulmonate gastropods*, have modified the mantle and mantle cavity for breathing air through lunglike structures.

KNOWN GEOLOGIC RANGE. Cambrian–Recent (Fig. 57).

CHARACTERS. Anterior part of body as movable head with mouth, eyes, and sensory tentacles; mouth with rasping, minute, horny teeth (radula) and lateral horny jaws; tentacles not segmented; stomach serves to store, not digest, food; anus opens near mouth; blood colorless, blue, or red; muscular fleshy foot; shells calcareous, with one large central chamber.

Class. Cephalopoda (Gr. *kephale*, head; *pods*, foot).

This is the most spectacular class of mollusks. It includes the squid, the octopus, the pearly nautilus, and the belemnites and ammonites. Some of these exclusively marine creatures are among the largest invertebrate animals known. The living ones are active and alert free swimmers that prey upon other life about them, and the extinct kinds were probably like their living relatives in habits. Though extremely abundant and diversified in the late Paleozoic and in the Mesozoic, they were largely replaced by other classes and phyla in the Cenozoic.

KNOWN GEOLOGIC RANGE. Cambrian–Recent (Fig. 58).

Fig. 58. Ammonite.

CHARACTERS. Prominent head possessing large eyes with well-developed lenses; mouth at anterior end circled with muscular tentacles or arms, with radula and horny jaws; water pulled into and expelled from body through siphon, a tubelike opening below the head; breathes by gills; possesses an ink sac from which a dark pigment is emitted to make a "smoke screen" in water; brachial and systemic hearts pump blood through body; shell of calcium carbonate, divided into series of internal camerae (chambers) by septa; camerae traversed by siphon.

Class. Pelecypoda (Gr. *pelekys,* hatchet; *pod* foot).

Pelecypods consist of clams, oysters, scallops, and many other bivalve shells. Most of them live in marine waters, but many prefer fresh-water habitats. They move about in a vertical position by means of a narrow "hatchet foot." The toothed hinge of the two shells, or valves, is dorsal, or above; this is in contrast to the lamp shells or brachiopods which are fastened to the bottom by a pedicle in a horizontal position. Some pelecypods burrow in mud or even into rock (Fig. 59).

KNOWN GEOLOGIC RANGE. Cambrian–Recent.

Fig. 59. Pearl oyster.

CHARACTERS. Anterior and posterior ends, but no heads; no lophophore; dorsal circulatory system; gills located in posterior mantle cavity; shells symmetrical, composed of three layers, inner one with mother-of-pearl lining; shell grows by addition of materials at margin and in thickness by deposits to inner layer; hinge teeth serve as pivot for shell to open.

Class. Scaphopoda (Gr. *skaphe;* boat; *podoa,* foot).

These are called tooth, or tusk, shells because of their supposed resemblance to slightly curved toothpicklike, miniature tusks similar to those in living elephants. They are all shallow-water marine animals, though some occur at depths of 15,000 feet. The anterior end, which resembles a foot at the wide end of the shell, is buried in the mud to winnow out food while the narrow end, which is open, remains above to extract oxygen from the sea water (Fig. 60).

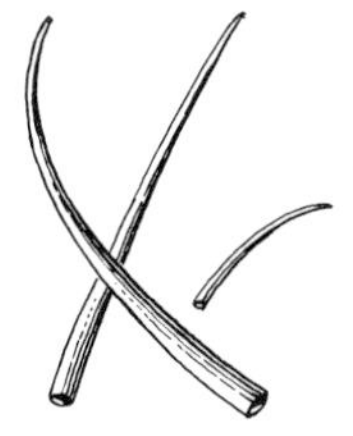

Fig. 60. Scaphopods.

KNOWN GEOLOGIC RANGE. Cambrian–Recent.

CHARACTERS. Mouth lined with several delicate contractile "tentacles;" no head; respiration by mantle; separate sexes; two-layered shell, openings at both ends, tuck shaped.

Phylum. Annelida (Lat. *annelus,* little ring).

The phylum Annelida includes the earthworm, the leeches, and the marine worms. Some are parasites. Earthworms, or angleworms, that frequent damp, rich soils in abundance are a source of food for many animals. Leeches are the scourge of a warm, moist, tropical environment, where they occur in water or in damp, heavily vegetative areas. Some of these repulsive blood suckers render some tropical lowlands almost uninhabitable. The marine worms, or polychaetes, with feet on each segment of the body, are found along sandy seashores or in marine muds. Annelids were well diversified by middle Cambrian time. Several families have been recorded from the Burgess shales in British Colombia. Trails and burrows are also indicative of their presence as early as the Proterozoic. Chitinous, horny, or siliceous jaws are the hard parts usually preserved as fossils, and these structures are present in only some kinds of annelids. Because of certain similarities in their anatomy annelids are thought to be ancestral to the arthropods (Fig. 61).

KNOWN GEOLOGIC RANGE. Proterozoic?–Recent.

CHARACTERS. Bilaterally symmetrical; body elongate, usually divided

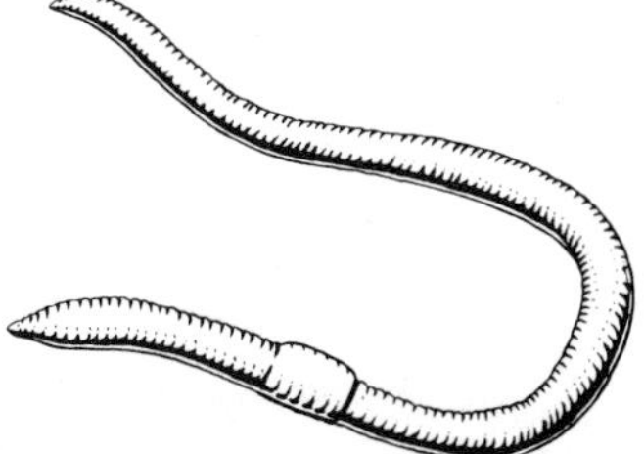

Fig. 61. Earthworm.

into segments (somites), both externally and internally (only the marine gephyre worms are not segmented); minute, rodlike, chitinous setae in each somite; no distinct head; alimentary tract straight, with pharynx, esophagus, crop, gizzard, and intestine; closed circulatory system with five "hearts," colorless corpuscles, and red plasma; nervous system with cerebral ganglia, mid-ventral nerve cord, and lateral nerves for each somite; sensitive to taste, touch, and light; respiration through skin or by gills; nephridia in each somite; separate sexes with trochophore larvae, united sexes, or asexual reproduction by budding.

Phylum. Arthropoda (Gr. *arthro,* joint; *pods,* foot).

This enormous phylum includes the onychophorids, crustaceans, trilobites, spiders, centipedes, millepedes, and insects. They range into nearly all environments at all suited for life, in the sea, on land, and in the air. Insects are the only invertebrate animals capable of flight. Though arthropods as a whole are not so well known in the fossil record as several of the other phyla, all of the classes occur as fossils at one time or another, and some, such as the minute ostracods, the trilobites, and the eurypterids, are abundant.

KNOWN GEOLOGIC RANGE. Cambrian–Recent.

CHARACTERS. Bilaterally symmetrical; body usually segmented and jointed into head, thorax, and abdomen; appendages or legs with hinged joints; digestive tract complete, as in annelids; circulatory system open, dorsal heart pumps blood out through arteries, but blood returns to heart through body cavity; blood nearly colorless; nervous system with pair of dorsal ganglia above mouth and two ventral nerve cords; nerve ganglia in each somite; sensory hairs and antennae, hearing organs, and simple or compound eyes; respiration by gills, trachea (tubelike air duct), book lungs, or through the body wall; excretion by organs called green glands in front of mouth or malpighian tubules in abdomen; sexes usually separate, fertilization nearly always internal, oviparous egg-laying or ovoviviparous, (eggs hatched in body), one to several larval stages.

Class. Onychophora (Gr. *onychus,* claw; *phorus,* bearing).

This interesting class not only has distinctive characters of its own but several features that it shares with the annelids and other arthropods. Some 70 caterpillarlike species of the genus *Peripatus* are still living in the tropical regions of the world. They have changed very little since middle Cambrian where the oldest known specimen, *Aysheaia peduncu-*

Fig. 62. Peripatus.

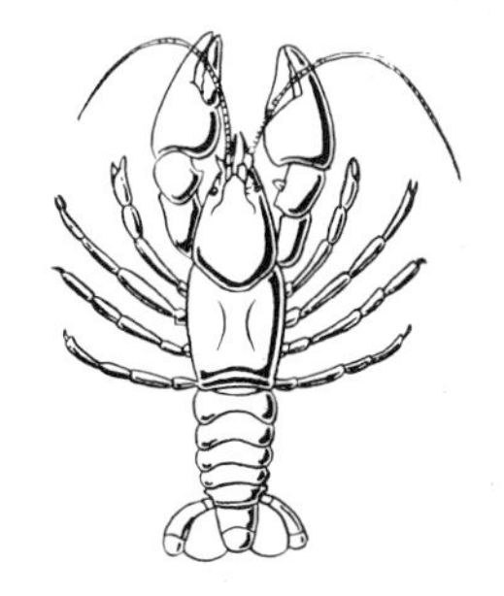

Fig. 63. Lobster.

lata Walcott, was found. The evidence from this class indicates that the arthropods are probably derived from an annelid ancestry, perhaps in the Proterozoic. Some authorities believe that the onychophorids are annelids that have paralleled arthropods in certain features (Fig. 62).

KNOWN GEOLOGIC RANGE. Cambrian–Recent.

CHARACTERS. *Annelidlike:* simple straight digestive tract; photoreceptor cells sensitive to light; coiled tubes as excretory organs, segmentally arranged; reproductive organs with ciliated ducts. *Arthropodlike:* dorsal "heart"; trachea as delicate tubular ingrowth of body wall; general construction of reproductive organs, as in arthropods; jaws derived from appendages. *Distinctive:* nerve cords with no true ganglia along length of body; thick, short legs; skin thin, with little chitin, transverse rings with delicate papillae, each with a spine; tracheal opening distinctive; one pair of horn jaws; ovoviviparous, with each oviduct a specialized uterus.

Sea "spiders," Pycnogonida: Devonian, Recent. Linguatulids, Pentastomida: Recent. Water "bears," Tardigrada: Recent. Garden "centipedes," Symphyla: Recent. Pauropods, Pauropoda: Recent. All of these groups are of uncertain relationship among the Arthropoda.

Class. Crustacea (Lat. *crusta,* hard shell).

Lobsters, crayfish, shrimps, crabs, water fleas, ostracods, barnacles, and others are grouped together in this class. They inhabit both marine and fresh waters. The sow bugs and wood lice live on land. Crustaceans have a long geologic record. The microscopic ostracods, because of their abundance, are useful in correlation and have been employed extensively in petroleum explorations (Fig. 63).

KNOWN GEOLOGIC RANGE. Cambrian–Recent.

CHARACTERS. Head composed of five fused parts, with two pairs of antennae, two pairs of maxillae, and one pair of lateral mandibles; respiration by gills; terminal segment of limbs with two parts, sometimes in form of pincers; separate sexes (except in barnacles and sow

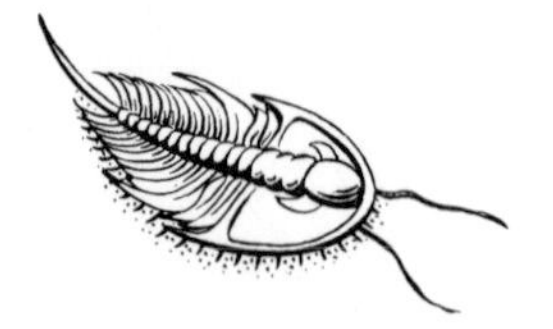

Fig. 64. Trilobite.

Fig. 65. Centipede.

bugs), sex openings separate, parthenogensis (development from unfertilized egg) in water fleas.

Class. Trilobita (Gr. *trilos,* three; *bitos,* lobe).

Trilobites are known to everyone who has read much about fossils. They appeared suddenly in great abundance in the Cambrian, and evolved into numerous phyletic lines in the early Paleozoic, but they gradually dwindled away to extinction at the end of that Era. They are among the most distinctive index fossils of their time. Evidence from the evolutionary progression in the early Cambrian trilobites alone indicates that the phylum Arthropoda was differentiated into some of its known classes as early as the Proterozoic. Evidently trilobites were exclusively marine. They ranged in size from 6 millimeters in *Agnostus* of the Cambrian to 27 inches in *Terataspis* of the Devonian (Fig. 64).

KNOWN GEOLOGIC RANGE. Cambrian–Permian.

CHARACTERS. Body divided laterally into a central (axial) and two lateral (pleural) regions, anteroposteriorly into body (thorax), and tail (pygidium); each somite has a pair of jointed legs, those on the head and body with two parts in terminal segment (biramous); head with eyes above and mouth below.

Class. Chilopoda (Gr. *kilos,* lip; *podos,* foot).

These are the land-living centipedes, or hundred-legged worms, found in slightly moist places in all moderate to warm climates. They are rare as fossils but must have existed much earlier than our present record indicates. Some excellent fossils have been found in Oligocene amber. They are terrestrial and nocturnal in habit (Fig. 65).

KNOWN GEOLOGIC RANGE. ? Pennsylvanian–Recent.

CHARACTERS. Body elongate and flattened, pair of legs on each somite, first pair four-jointed and terminal claw with poison gland; head with pair of jointed antennae, pair of lower jaws, two pairs of maxillae; simple eye; sex openings ventrally located on next to last somite.

Class. Diplopoda (Gr. *diplos,* two; *podos,* foot).

The millepedes or thousand-legged worms live on the ground under decaying vegetation. Millipedes and scorpions from the late Silurian are the oldest known records of air-breathing animals. They were numerous in the Pennsylvanian when some were about 8 inches long. They are now world-wide in distribution (Fig. 66).

KNOWN GEOLOGIC RANGE. Silurian–Recent.

CHARACTERS. Body long, nearly cylindrical, two pairs of legs to each somite; head with pair of short antennae, jaws, and maxillae; two groups of simple eyes; sex openings ventral on third somite.

Class. Arachnoidea (Gr. *arachne,* spider, *oid,* like).

This is another large class of arthropods, which includes spiders, daddy longlegs, mites, ticks, scorpions, king crabs, eurypterids, and many others. Some of the eurypterids reached enormous size in the Silurian. Their earliest ancestors are in Ordovician rocks. It is thought that these interesting predators lived not only in marine and brackish waters, but some kinds may have preferred fresh-water streams. The last of them died out in the Permian. Another group of arachnids which made its first appearance in the Cambrian is the king or horseshoe crab. Even the earliest scorpions in the Silurian and those still living look very much alike. Spiders are common in the fossil record. They first appear in the Devonian and are well represented in Cenozoic amber where they seem to have been attracted by insects struggling in the sticky resin (Fig. 67).

KNOWN GEOLOGIC RANGE. Cambrian–Recent.

CHARACTERS. Cephalothorax (head + thorax) and abdomen; four pairs of legs and two anterior appendages (palpi), with pincers called chelae; no antennae; no true jaws; respiration by book lungs, gills, or trachea; single sex opening on second abdominal somite.

Class. Insecta (Lat. *incised,* into parts).

This is by far the largest class of land-living animals. Included here are the cockroaches, dragonflies, flies, beetles, bees, butterflies, and

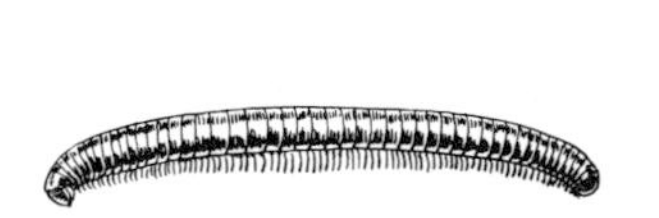

Fig. 66. Millepede.

Fig. 67. Spider.

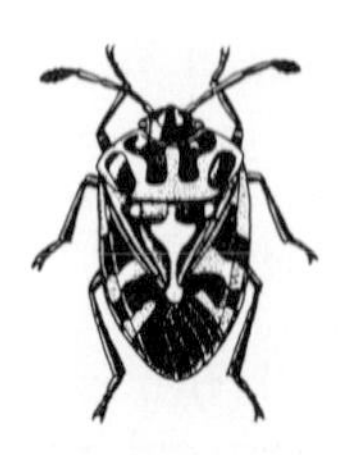

Fig. 68. Harlequin bug.

many others. Most of them are able to fly and have been widely dispersed. Many air-breathing adults have aquatic larvae. They occur abundantly in nearly all habitats, but for some strange reason they have not occupied marine habitats except in a few limited examples. As fossils they are best preserved in amber and in fine-textured shales. The earliest known insects from the Devonian are wingless and thought to be creatures similar to springtails and snow fleas (Colembola), but their direct ancestry is not known. Giant dragonflies and other winged insects appear for the first time in the Pennsylvanian. After that, insects evolved rapidly and increased tremendously in numbers and kinds (Fig. 68).

KNOWN GEOLOGIC RANGE. Devonian–Recent.

CHARACTERS. Head, thorax, and abdomen distinct; three pairs of legs; antennae present; mouth parts for chewing, sucking, or lapping; breathes by trachea; simple and compound eyes; multiple sex tubules with medean posterior opening.

Phylum. Echinodermata (Gr. *echinos,* hedgehog; *derma,* skin).

The echinoderms are so named because their numerous armlike appendages resemble spines. The phylum is divided into two subphyla: the Pelmatozoa, which have stems and are attached to the bottom (sessile), and the Eleutherozoa, which are free-moving. The pelmatozoans include the cystoids, blastoids (sea buds), crinoids (sea lilies and feather stars), and four other classes not so well known. Echinoderms were abundant in the Paleozoic; only the crinoids are still living. The eleutherozoans are the sea cucumbers, starfish, sea urchins, heart urchins, sand dollars, and two classes of uncertain relationships restricted to the Paleozoic.

The echinoderms are known only in marine waters and from marine formations. They seem to have invaded all oceanic environments. They are of particular interest, since they appear to be more closely related to the chordates (including all vertebrate animals) than any of the invertebrate phyla. There are several features in the embryonic development that suggest this relationship; for instance, the skeleton was developed from the mesoderm as in the vertebrates and not from the ectoderm as in the other invertebrates. The larvae of some echinoderms resemble those in the most primitive chordates. The two subphyla were possibly derived from a common ancestry in the Proterozoic.

CHARACTERS. Primitive bilateral symmetry masked by well-developed radial symmetry; no head; podia or tube feet, upper ones for respiration, lower ones for locomotion; radial canals, called ambulacra, leading to mouth; long, coiled alimentary tract; nervous system around mouth, radial nerves to body and appendages; no heart; body cavity with extension to appendages for water circulatory system; water taken in through porous plates (madreporite) or by action of cilia into stone canal; sexes separate, ova fertilized in sea; larvae microscopic, ciliated, transparent, free-swimming, with conspicuous metamorphosis; skeleton of crystalline calcite, with minute passageways through the hard parts.

Subphylum. Pelmatozoa (Gr. *pelmato,* foot; *zoon,* animal). Sessile forms.

Class. Cystoidea (Gr. *cystis,* bladder; *eidos,* form).

The cystoids are primarily an early Paleozoic class. Though the oldest known records are taken from the Ordovician, they must have been present much earlier, since the two main groups were quite distinct at that time. They are among the most primitive of the stemmed echinoderms.

The classes Eocrinoidea, Cambrian, Ordovician; Paracrinoidea, Ordovician; Carpoidea, Cambrian, Devonian; and Edrioasteroidea, Cambrian, Mississippian (Fig. 69).

KNOWN GEOLOGIC RANGE. Ordovician–Devonian.

CHARACTERS. Body or calyx, outline spherical or cystlike; calyx, plates usually numerous, somewhat irregularly arranged, and with porous structure; mostly with radial symmetry; biserial food-gathering appendages.

Class. Blastoidea (Gr. *blastos,* bud; *eidos,* form).

The remarkable radially symmetrical blastoids are suggestive of the bud of a flower soon to burst into bloom. As with the cystoids, they

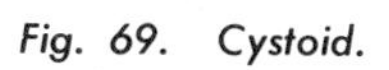
Fig. 69. Cystoid.

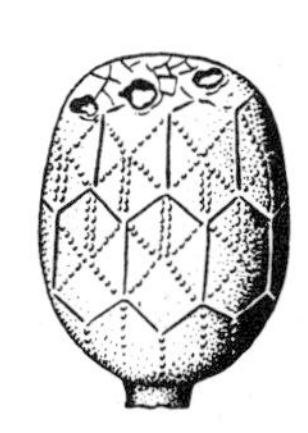

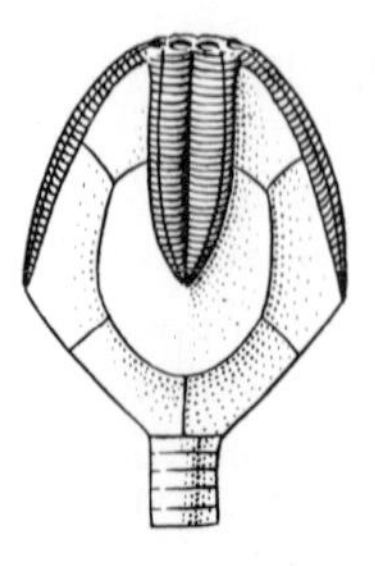

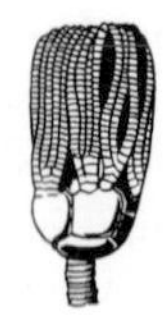

Fig. 70. Blastoid.

Fig. 71. Crinoid.

are restricted to the Paleozoic. Permian rocks on the island of Timor are among the best collecting grounds for these fascinating fossils (Fig. 70).

KNOWN GEOLOGIC RANGE. Ordovician–Permian.

CHARACTERS. Bodies bud-shaped with five symmetrically arranged food grooves (ambulacra); calyx plates uniformly arranged; structure called hydrospire in central body cavity; short armlike appendages along ambulacial grooves.

Class. Crinoidea (Gr. *crinon,* lily; *eidos,* form).

These are the sea lilies and the stemless feather stars. The group as a whole is extremely varied and complex in the different structures. Most of the Paleozoic sea floors must have been blanketed with these animals, since remains of their stems and bodies are found abundantly in most of the marine formations of that time. Most crinoids were small and of medium size, though some were about 60 feet long. The living descendants are limited in numbers and are the only living representatives of the pelmatozoans or sessile echinoderms. The feather stars are free-swimming in the adult stage (Fig. 71).

KNOWN GEOLOGIC RANGE. Ordovician–Recent.

CHARACTERS. Body with symmetrical calyx plates, flexible appendages for food capturing, arms usually branched, with ambulacra on inner surface; calyx encloses soft parts and viscera of body; stem of numerous disk-shaped elements.

Subphylum. Eleutherozoa (Gr. *eleutheros,* free; *zoon,* animal). Free-moving forms.

Class. Holothuroidea (Gr. *holothourion,* waterpolyp; *eidos,* form).

Soft-bodied echinoderms, called sea cucumbers, are sluggish bottom-living animals. They may burrow in the mud or attach themselves to solid surfaces, but they are seldom found as fossils. Some minute hard parts have been found, and other unusual discoveries have been made in carbonaceous or fine-textured shales (Fig. 72).

KNOWN GEOLOGIC RANGE. Cambrian–Recent.

CHARACTERS. Soft, elongate bodies, leathery skinned, with microscopic calcareous plates; mouth surrounded by tentacles; two dorsal rows and three ventral areas of tube feet; large, undivided, fluid-filled coelom; respiratory tree in coelom for respiration and excretion.

Class. Asteroidea (Gr. *aster,* star; *eidos,* form).

Starfish, with their sturdy arms used for crawling about, are familiar objects along seacoasts in nearly all parts of the world. They are usually found more or less clustered together as fossils, but these sites are infrequently found in the rocks. One group known as primitive starfish range from the Ordovician to the Devonian.

Brittle stars are also found in abundance on or near the floor of oceans. They use their delicate, slender arms to propel themselves through the water. There seems to be good evidence for placing them in a distinct class known as Ophiuroidea, Ordovician, Recent (Fig. 73).

KNOWN GEOLOGIC RANGE. Cambrian–Recent.

CHARACTERS. Body with central disc and radiating arms in the shape of a star; skeleton composed of many loosely joined ossicles, with blunt, calcareous spines; ambulacral grooves and mouth on the oral, or lower, surface.

Class. Echinoidea (Gr. *echinos,* hedgehog; *eidos,* form).

This class includes the sea urchins, heart urchins, and sand dollars. They are exclusively marine animals that occupy rocky, sandy, or muddy bottoms, or they may live within the area of high and low tides. Sand dollars are usually found partly buried in loose sand. Echinoids move about by their short spines and tube feet. Some groups have been very useful in Cenozoic correlations. They are divided into two subclasses: Regularia, Ordovician–Recent, and Irregularia, Jurassic–Recent. The

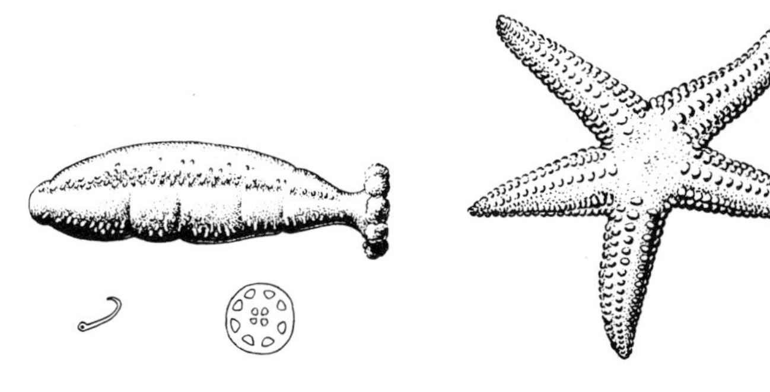

Fig. 72. Sea cucumber.

Fig. 73. Starfish.

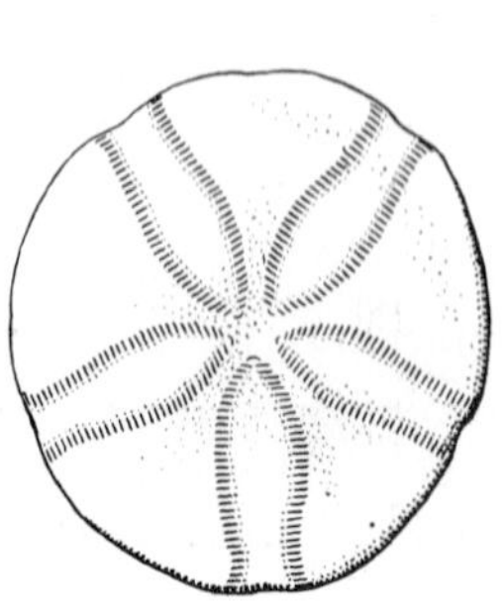

Fig. 74. Sand dollar.

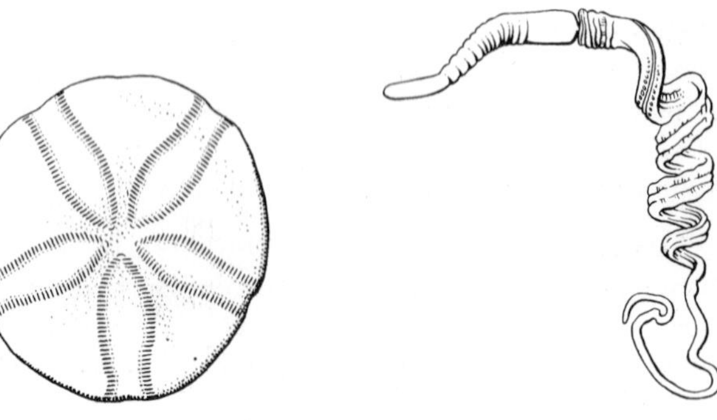

Fig. 75. Acorn worm.

classes Bothriocidaroidea, Ordovician, and Ophiocystia, Ordovician, Devonian, are sometimes included in the class Echinoidea (Fig. 74).

KNOWN GEOLOGIC RANGE. Ordovician–Recent.

CHARACTERS. Body skeleton globular, discoidal, or sometimes subcylindrical, many subcutaneous plates covered with movable spines; five ambulacral and five interambulacral areas. Regularia with anus located opposite mouth in oculogenital ring; Irregularia with anus outside and posterior to oculogenital ring.

Phylum. Chordata (Gr. *chorda,* string).

This phylum includes not only the higher vertebrate animals (lampreys, fish, amphibians, reptiles, birds, and mammals), but also acorn worms, tunicates, and lancelets, frequently referred to as the lower chordates. Chordates frequent almost every environment in water, on land, and in the air.

CHARACTERS. One dorsal, tubular nerve cord; notochord (slender rod of cells for supporting structure) present at some time during life of individual; gill slits or vestiges of them in the pharynx; principal blood vessels ventral to nerve cord when present.

Subphylum. Acrania (Gr. *a,* without; *crania,* cranium).

This includes all of the lower chordates which are currently arranged into subphyla: pterobranchs, Pterobranchia; acorn worms, Hemichordata; tunicates, Tunicata; and lancelets, Cephalochordata (Fig. 75).

Subphylum. Craniata (Gr. *crania,* cranium).

This group is composed of the vertebrate animals. Here are the primitive cyclostomes, sharks, fishes, amphibians, reptiles, birds, and mammals.

CHARACTERS. Body segmented in early stages of development; enlarged brain anteriorly; head with paired sense organs; semicircular canals for equilibrium; closed circulatory system, heart with two to four chambers; skull and vertebral column; ossified or cartilaginous skeleton; notochord from head to tail.

Class. Cyclostomata (Gr. *cyklos*, circular; *stoma*, mouth).

Of the living animals, this class includes the lampreys and hagfish; many but not all of them are parasitic on fish to which they attach themselves to suck the blood of the temporary host. Others continue to feed upon the unfortunate fish until it dies. The lampreys live in both marine and fresh waters; the hagfish are marine animals.

Of greater interest to paleontologists are the extinct ostracoderms. Erik A. Stensiö has shown by a careful method of cross-sectioning that these Paleozoic jawless vertebrates are related to the modern cyclostomes but referable to different orders. Ostracoderms died out in the Devonian (Fig. 76).

KNOWN GEOLOGIC RANGE. Ordovician–Recent.

CHARACTERS. Mouth anteroventral, adapted for sucking or rasping; gills in saclike pouches of pharynx; two-chambered heart; fertilization external; direct development to adult or long larval stage; vertebrae as small imperfect arches; no lower jaws or paired fins; no scales; partial bony armor in ostracoderms.

Class. Pisces (Lat. *piscis*, fishes).

The so-called fishes have many classifications, but for the convenience of the elementary student the acanthodians, arthrodires, antiarchs, true sharks, bony fishes (trout, bass, catfish, etc.), and the crossopterygians (lungfish, etc.) are included in one class. It will be advantageous for students with special interests in vertebrates to familiarize themselves with more precise classifications (Fig. 77).

KNOWN GEOLOGIC RANGE. Silurian–Recent.

CHARACTERS. Mouth variously modified; breathes by gills; heart with two chambers; fertilization usually external (some ovoviviparous or vivi-

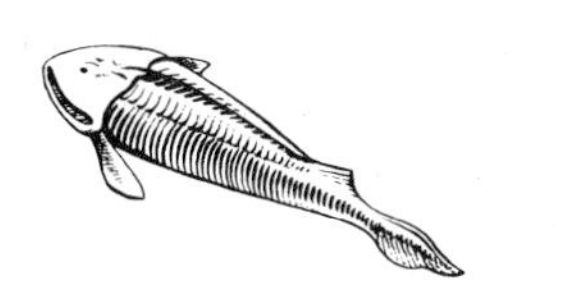

Fig. 76. Ostracoderm. Redrawn from Colbert.

Fig. 77. Tilefish.

parous), eggs without protective covering; vertebral column cartilaginous or ossified; bone dense; one pair of visceral arches modified as jaws; mandible with numerous bony or cartilaginous parts; paired fins; skin smooth or with scales, some with body armor; growth continuous almost throughout life.

Class. Amphibia (Gr. *amphi*, duel; *bios*, life).

Amphibians include the frogs, toads, salamanders, legless caecilians, and extinct labyrinthodonts. They are, of course, adapted to live on land and in the water. None is marine. Labyrinthodonts are the oldest known land-living, or terrestrial, vertebrate animals. The interesting transitional stages between crossopterygian fish and amphibians are now well known. The large bony heads of the late Paleozoic and early Triassic labyrinthodonts were well suited for preservation as fossils. The fossil record of the living groups are not so well known (Fig. 78).

KNOWN GEOLOGIC RANGE. Devonian–Recent.

Fig. 78. Frog.

CHARACTERS. Breathes by gills in larval stage, by lungs and through skin as adults; three-chambered heart; cold-blooded; fertilization external or internal, eggs without protective covering, laid in water or otherwise protected from evaporation; skeleton partly cartilaginous and partly ossified; mandible consists of many bones; four limbs for walking, hopping, or swimming; no paired fins; skin smooth; growth may continue almost throughout life.

Class. Reptilia (Lat. *reptilis*, creeping).

These are the lizards, snakes, turtles, crocodiles, the tuatara *(Sphenodon)*, dinosaurs, plesiosaurs, mosasaurs, ichthyosaurs, pterosaurs, and many other extinct groups. Reptiles are a most interesting group of quadruped (four-footed) animals. They evolved from the amphibians in the late Paleozoic and in turn gave rise to the birds and mammals of the Mesozoic. As four-footed creatures adapted to life on land, they invaded both marine and fresh waters, and some were spectacularly modified for flying or soaring (Fig. 79).

KNOWN GEOLOGIC RANGE. Pennsylvanian–Recent.

CHARACTERS. Breathes by lungs; four-chambered heart; cold-blooded;

Fig. 79. Black snake.

fertilization internal, eggs with tough protective covering, laid on land, hatch from heat of sun; some reptiles give birth to young; skeleton ossified, bone usually dense except in flying forms; coracoid and precoracoid in pectoral girdle; mandible consists of more than one bone; forelimbs for walking, flying, or swimming; skin with scales, dermal ossicles, or bony shields over body; growth may continue almost throughout life.

Class. Aves (Lat. *avis*, bird).

This class contains all of the birds, starting with the reptilelike *Archaeopteryx* of the Jurassic, the toothed bird of the Cretaceous, *Hesperornis*, and all later birds, including gulls, ducks, penguins, owls, hawks, crows, bobolinks, robins, humming birds, and many more too numerous to mention. The ground birds, most of which were large, are the ostrich, rhea, emu, and the extinct moa, *Aepyornis*, *Diatryma*, and the like. Birds have invaded almost all environments, from surface waters to land and the air (Fig. 80).

KNOWN GEOLOGIC RANGE. Jurassic–Recent.

CHARACTERS. Breathes by lungs; four-chambered heart; warm blooded; fertilization internal, eggs with hard-shell covering laid in nests or on ground, incubated by heat of adult; bones hollow, pneumatic, and lightweight, except in large birds; sacrum and pelvis fused; mandible consists of more than one bone; skin covered by feathers; growth limited.

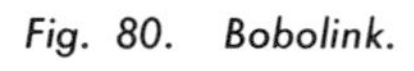

Fig. 80. Bobolink.

Fig. 81. Red fox. Redrawn from Seaton.

Class. Mammalia (Lat. *mamma,* breast).

This is the highest class of vertebrate animals. It contains the opossums, shrews, bats, armadillos, rodents, dogs, elephants, whales, camels, horses, monkeys, man, and many others, including numerous extinct orders and families. This class was so dominant during the later part of geologic time that the Cenozoic has been called the "age of mammals."

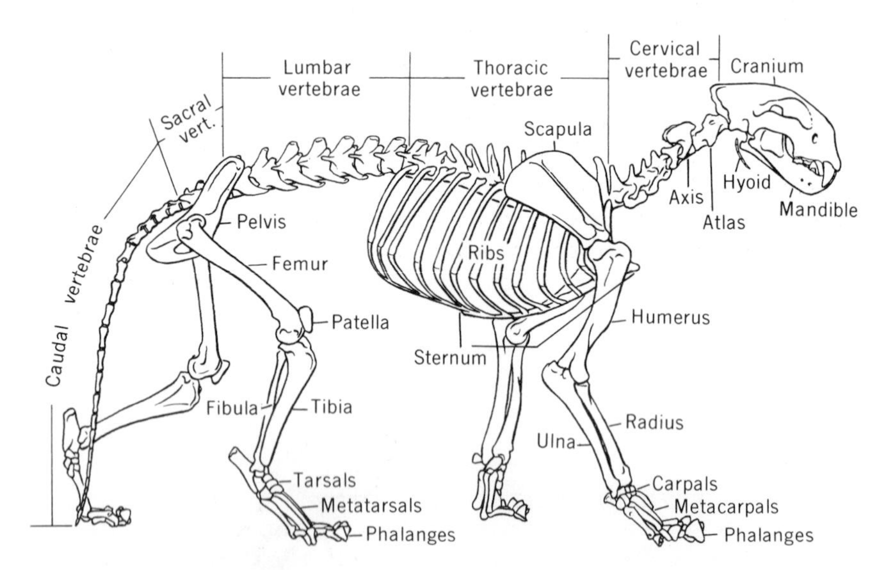

Fig. 82. The skeleton of Felis atrox, the Rancho La Brea jaguar, illustrates the bones of a mammal. (For bones of mammal skull, see Figs. 138, 225.)

Phylogenies of many of the families are well enough known to make even isolated teeth useful in the correlation of the formations in which they occur. Mammals are primarily terrestrial but are variously adapted to live in oceans, lakes, and streams, high mountains, polar regions, deserts, grasslands, forests, underground, and in the air (Figs. 81, 82).

KNOWN GEOLOGIC RANGE. ? Triassic–Recent.

CHARACTERS. Breathes by lungs; four-chambered heart; warm blooded; fertilization internal, eggs protected in mother, embryo nourished in the mother or in external pouch, gives birth to young; nourishes young with milk; bone heavy (terrestrial forms) and even dense in some (sea cows), hollow and light (rabbits, bats, etc.), seven neck vertebrae, except in tree sloths and manatees; mandible consists of one bone in each side; skin covered by hair, except in some adults, some with dermal ossicles or with bony covering over body; growth limited.

8 Pre-Cambrian and the Beginning of Life

The Grand Canyon

Archaeozoic: *earliest forms of life: blue-green algae, fungi, and a ? flagellate from Canada, two billion years old; carbon in Rhodesian Shield associated with minerals dated 2.6 billion years old by radioactive minerals; graphite and carbonaceous shales in Australia and Canada.* **Proterozoic:** *first evidence of bacteria, marine algae, fungi, radiolarians, sponge spicules, brachiopods, and worm burrows; all phyla of plants and animals probably recognizable; climates vary from warm moist to dry and cold.*

A PALEONTOLOGIST is greatly concerned with the origin, evolution, and dispersal of life on earth. His evidence comes from the rocks in the earth's crust, and the nearer the time of deposition of the sediments is toward Recent time the better his record is likely to be. Much has been learned about plants and animals of the geologic past, and some lineages have been traced through millions of years. Unfortunately, however, the rocks older than Cambrian, which represent something like four fifths of all geologic time, have yielded an extremely meager record. Yet, at some time during this long interval life originated, evolved, and dispersed into the environments in which these earliest kinds of organisms could live and reproduce.

It has been estimated that the earth's crust is approximately 4.5 billion years old, or older. If this is true, there were some 1.9 billion years of which we have no certain geologic information because the oldest dated minerals from the Rhodesian Shield in Africa and elsewhere are about 2.6 billion years. Therefore, it is useful and convenient in our terminology to start known geologic time at that point. Pre-Cambrian time has been divided into two Eras, the Archaeozoic (beginning of life) and the Proterozoic (very early life). Other names for these Eras have been proposed as more appropriate, but each of these also has some undesirable connotation. Some authors have strongly advocated the use of Archaean, instead of Archaeozoic, since they have maintained there is no direct evidence of fossils in the oldest known rocks. Nevertheless, Archaeozoic is widely used. It is evident that life did exist at that time, as the name implies, but these academic problems need not confuse beginning students in paleontology.

Occurrences and Nature of Pre-Cambrian Rocks. Pre-Cambrian rocks, for the most part, are deeply buried, but in many areas they have been exposed by uplift and erosion. Consequently, most of them are metamorphic rocks; in other words, they have been altered from their original state by heat and pressure, though some formations in the late Proterozoic are little altered. Pre-Cambrian rocks occur on every continent, and in some countries, such as Australia and Canada, they make up much of the surface exposures. Both epeirogenic and orogenic disturbances have elevated them into positions in which they have been uncovered by erosion and by the surface-stripping of glacial ice. In other areas, like the Grand Canyon, rivers have cut deeply into the surface of the earth to expose underlying Pre-Cambrian rocks.

Thus Pre-Cambrian rocks may be found in the cores of most mountain ranges or as great stable shields over rather wide areas. Some of the more classic shields are the Canadian in North America, the Baltic in Europe, the Angarian in Siberia, the Ethiopian in Africa, the Guianian and Amazonian in South America, and the Western Australian Shield. Numerous blocks of Pre-Cambrian rock groups from widely scattered localities throughout the Commonwealth of Australia have also been described.

Canadian Shield. Study of Pre-Cambrian rocks was initiated about 1842 in the area adjacent to the Great Lakes by Sir William Edmund Logan (1798–1875). Sir William, a Scottish Canadian who organized the Canadian Geological Survey, has been justly called the "Father of Pre-Cambrian Geology" because of his pioneering work in the field. After several years of intensive field work he drew up a broad classification of the formations in Canada as he understood them. Though his classification was improved in the years that followed, it is a compliment to his insight into the problem and to his careful field work that much of it still stands as the foundation of our present terminology on Pre-Cambrian geology.

It was some 30 years later that Andrew Cowper Lawson of the University of California described some of the oldest metasediments (metamorphosed sedimentary rocks) and lavas and discovered that some of the great igneous rocks of pink granite were younger than some of the metasedimentary formations, but his correlations of the metasedimentaries were based on similar lithologies.

Most of the Pre-Cambrian sedimentary rocks are thought to have been of marine origin, though some of the conglomerates may have been deposited on land as fans near the edges of inland seaways. There probably were many more groups of rocks than are currently recognized, and it is quite possible that there were three or more cycles of sedimentation, followed by volcanism and orogenies in different geosynclines or troughs.

Most of the recent and more detailed work has shown that accurate correlation of the rock units from one basin of deposition to another on lithology alone is not accurate, except in a limited number of widely spread distinctive rocks. Correlation from eastern to western Canada is still not possible by this method. The most accurate method of correlating Pre-Cambrian rock units in the Canadian shield is by age determinations from radioactive minerals. It has also been demonstrated that there were many more granitic intrusions, including dikes and sills, than were previously recorded and that many errors were made in attempts to correlate these igneous rocks.

As might be expected over such long periods of time, the geological history is much more complicated than has yet been realized. Perhaps a more satisfactory classification can be made when sequences are com-

CLASSIFICATION OF PRE-CAMBRIAN ROCKS AND APPROXIMATE TECTONIC HISTORY IN SOUTHERN PART OF CANADIAN SHIELD

Canada Geological Survey (Cooke, 1947) slightly altered	Tectonic History (King, 1951) slightly altered
Late Cambrian	Transgression of Late Cambrian seas
PROTEROZOIC Keweenawan; Upper Huronian (Animikie); Middle Huronian (Cobalt); Lower Huronian (Bruce)	Warping and faulting, partly contemporaneous with deposition of *Keweenawan*. Deposition of *Keweenawan* clastics and volcanics; postorogenic. Red beds, flood plains, mudcracks. Orogeny and intrusion of granite, partly contemporaneous with deposition of Huronian. Deposition of *Huronian* geosynclinal deposits in trough northwest of Grenville mountain belt. Glaciation, black slaty carbonaceous shales.
POSTARCHAEOZOIC INTERVAL	Orogeny and intrusion of *Laurentian* granite and other plutonic rocks.
ARCHAEOZOIC Timiskaming; Keewatin (in NW.); Grenville (in SE.)	Deposition of sedimentary rocks in *Grenville* province, including carbonates, in a geosyncline southeast of Timiskaming-Keewatin province. Formation of rocks of *Timiskaming-Keewatin* province. Development of two or more geosynclines at successive times which received clastic deposits and volcanics, interrupted and followed by orogenies and intrusion of granite (including Laurentian and Algoman). Graphite.

piled for each of the local areas. At present any attempt to correlate Pre-Cambrian rock units of different areas and regions cannot be done with the degree of accuracy attained with Paleozoic formations.

Grand Canyon Section. One of the most widely known exposures of Pre-Cambrian rocks is in the lower gorge of the Grand Canyon of the Colorado River which marks the boundary line between Arizona and

Nevada. Numerous scientists, mingled among thousands of tourists from nearly every country in the world, visit this spectacular spot each year. Standing on Permian rocks at Hopi Point or at other vantage points on the rim of the canyon, they can see in one section of superimposed rocks an impressive record of the early history of the earth. Much of the Paleozoic is missing, and later Mesozoic and Cenozoic formations are exposed on escarpments in badlands far back across the desert, but in the deep inner gorge of the canyon are some of the oldest rocks of the earth's crust.

The oldest of these great groups of rocks are the *Vishnu schists,* lying in distorted confusion and intruded by igneous dikes and sills at the bottom of the section. They are so altered by metamorphism that little of their original nature can be detected. This metamorphism must have taken place miles below the surface of the earth; then late in geologic time they were elevated into their present position. The Colorado River has already cut down more than half a mile through these schists. How much deeper they go is anyone's guess. The chemical composition and structure of these recrystalized rocks indicate that they were laid down originally as sedimentary rocks. Indeed, there is even some suggestion of stratification and cross-bedding in some localities. The Vishnu schists are referred to the Archaeozoic Era, but there is no direct evidence for correlating them with any of the rocks in the Canadian shield or elsewhere.

Grand Canyon Group. There is a pronounced unconformity, representing a long erosional interval between the Vishnu schists and the overlying Grand Canyon group of the Proterozoic Era. These formations also are exposed in the inner gorge, deep down in the canyon. The section is over 2 miles thick. These rocks are not so profoundly metamorphosed as those making up the Vishnu schists below, and they lie in an orderly sequence, but before the end of the Pre-Cambrian they evidently were block-faulted into mountains equivalent in height to any we know today. The formations are composed of metasediments of conglomerates, sandstones, sandy shales, calcareous shales, impure linestones, and lava flows. Current and wave action is indicated locally by wave and ripple parks and by cross-bedding. The sediments evidently were laid down in deltas near the sea. In places mud cracks are exposed rather clearly. Several structures have been found in the Grand Canyon group that are suggestive of fossils. Algal deposits and isolated siliceous sponge spicules have been recognized, and traces of other organisms are suggested.

As in many other areas in North America and in Eurasia, there was a pronounced uplift and long erosional interval after the deposition of the Grand Canyon sediments.

Belt Group. In part of Montana, British Columbia, and adjacent areas is another great sequence of Proterozoic rocks. These are named the Belt group after the Belt Mountains in Montana. Actually, they might best be considered as representing several groups of formations. Their maximum thickness is said to be about 6 miles. Like the Grand Canyon group, the Belt has not been greatly deformed but clearly shows a conformable arrangement, mostly of slightly metamorphosed strata. The source of these great beds was evidently primarily westward because the stratigraphic rocks increase in coarseness and the units are thicker in that direction. There is evidence of deposition in an inland sea and over lowlands that were above sea level.

The best evidence of life at that time is the reeflike accumulations of calcareous algae. Other fossils are wormlike burrows and trails.

Pre-Cambrian in Australia. Turning to another part of the world we learn that Australia has been a relatively stable land mass for many millions of years and as such has not been subjected to the profound orogenic disturbances that have so continuously changed the topography of other continents from time to time. Consequently, Australia is one of the best places in the world to study Pre-Cambrian geology. Nevertheless, stratigraphers in their attempts to correlate the sequences of geologic events are confronted with many of the same difficulties that their colleagues have encounted in other parts of the world (Fig. 83).

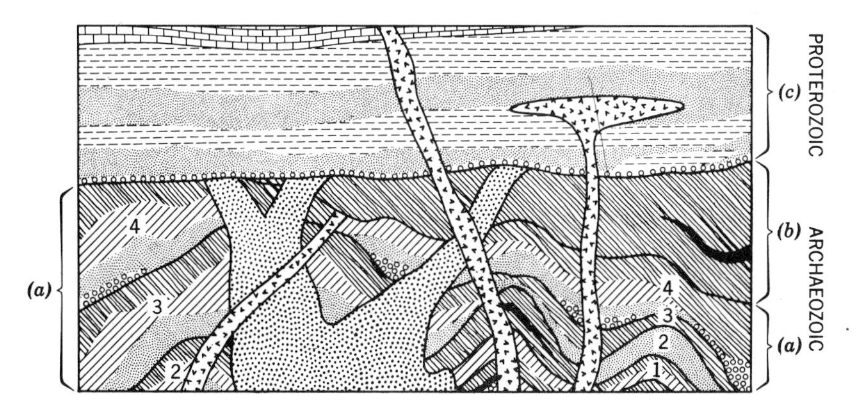

Fig. 83. Generalized Pre-Cambrian sequence for Western Australia. Archaeozoic. (a) Kalgoorlie. 1. Older Greenstones; 2. Black Flag Group; 3. Yildarlgooda Group; 4. Kurrawana Group. (b) Mosquito Group. Proterozoic. (c) Nullagine Group. Slightly modified from T. W. Edgeworth David and W. R. Browne.

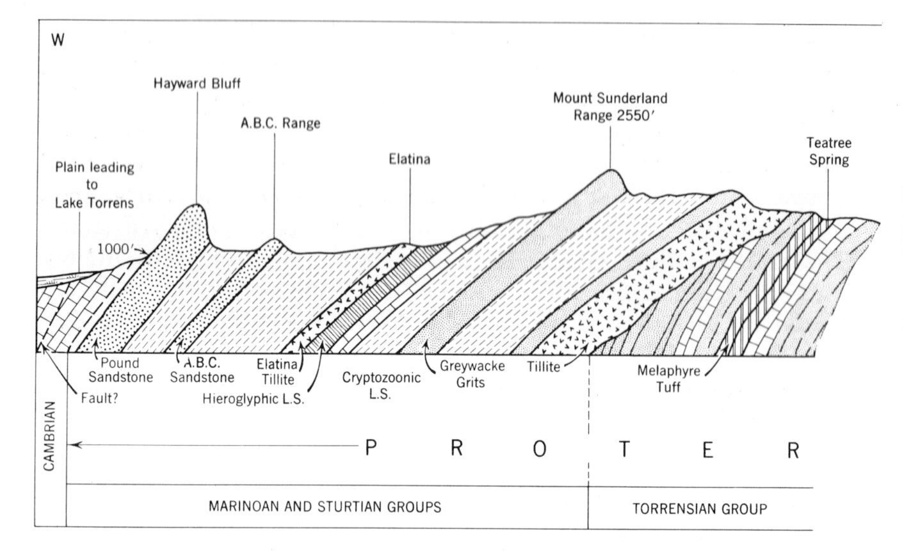

Fig. 84. Cross section of Blinman anticline across the Flinders Range at Wilpena Pound, South Australia, showing conformable contact of Proterozoic and Cambrian formations. Slightly modified after D. Mawson, from T. W. Edgeworth David and

The fossil record, so useful in the later Eras and Periods of geologic time, is either absent or extremely meager. For the most part correlations are based on lithology, degree of metamorphism, and the magnitude of unconformities, though it is realized that in certain localities these may be misleading. Even so, much has been learned about the Pre-Cambrian sequence on the continent "down under."

One of the most significant structural features which played an important role in the geologic history of the continent was the Tasman geosyncline, formed in the Proterozoic. Another remarkable feature of the Pre-Cambrian sequence in Australia is the conformable or nearly conformable position of Proterozoic and Cambrian formations. This is well exemplified in the colorful blue Flinders Range of South Australia. Lying conformably between the Brighton Limestone (Proterozoic) and the calcareous beds, which contain characteristic fossils of early Cambrian, are the "Purple Slates" (Marinoan Group), formerly considered by some to be Cambrian and by others to be latest Proterozoic. Marine fossils have been described from the uppermost formation of the Marinoan Group, the Pound Sandstone, which is now considered by most authorities to be latest Proterozoic (Fig. 84). Moreover, there is no evidence of a temporal break in deposition between the Pre-Cambrian and the Paleo-

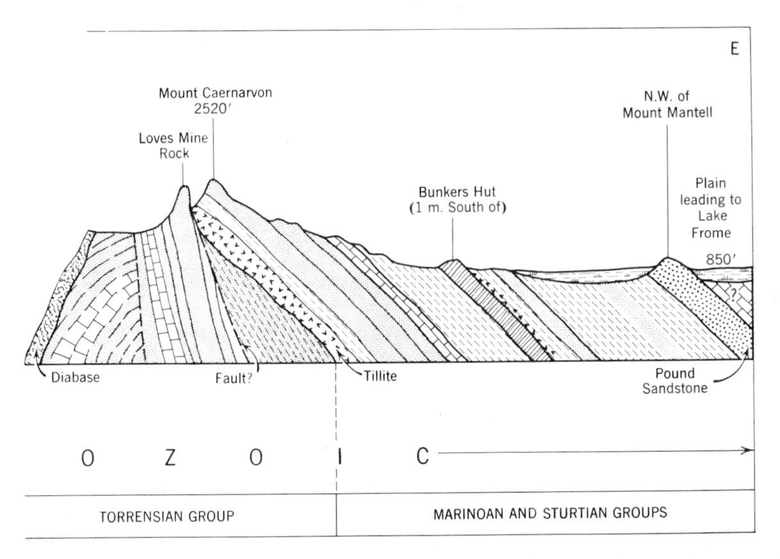

W. R. Browne. Modified by Brian Daily. Note the unconformity between the Sturtian Group Tillite and the underlying Torrensian Group and the conformity between the Cambrian fossiliferous limestones and the Proterozoic Pound Sandstone.

zoic so clearly displayed in many other parts of the world, though in California, also, something like 8000 feet of unfossiliferous, more-or-less conformable formations have been located below early Cambrian fossil-bearing beds.

Primitive Life and Its Possible Origin. After the earth formed as a planet in one of the galaxies of the universe water on it probably existed only in the form of steam because of the high temperatures that prevailed. At that time, it is believed, there was no free oxygen, but simple carbon and nitrogenous compounds may have been present. The earth then cooled, and liquid water formed even before an atmosphere developed. Because of these conditions it is assumed certain chemical reactions could have taken place in that environment that are no longer possible.

The manner in which life originated is not known. It is probable, though, that sometime in the early Archaeozoic the chance association of certain chemicals or compounds effected by ideal conditions of temperature, moisture, and light resulted in the origin of the first living matter that could reproduce itself. This substance, with its amino acids and nucleoproteins, we might have likened to protoplasm. It seems most likely that this reaction occurred in water, or possibly in most places near the surface of the ground. After the origin of living entities and the manifestations of primordial life had evolved into slightly more complex living matter the more progressive living things may have tended to ingest most or all of the organic compounds that had been formed.

Filterable viruses such as the tobacco mosaic virus may be the closest representation we have to such intermediate entities between living and nonliving substances. Indeed, the studies on viruses indicate that the difference between living and inanimate things is as transitional as it is between living organisms. Possibly viruses always existed much in the forms that we recognize today and made their first appearance in the Archaeozoic much in the manner suggested in the preceding paragraph. It has been suggested also that they arose by retrogressive evolution as parasites from bacteria.

Pre-Cambrian Life. The fossil record of life in the Pre-Cambrian is scanty, and it is debatable whether some of the specimens are really fossils. Graphites as well as black shales, schists, and slates are reported from several localities in different parts of the world. In later Periods it has been demonstrated that graphites and carbonaceous shales are derived from plant and animal remains. There must have been an abundance of soft-tissued organisms at certain times in suitable Pre-Cambrian environments to form these carbons.

Among the oldest known fossils are blue-green algae, fungi, and probably a flagellate described in 1954 by S. A. Tyler and E. S. Barghoorn. These interesting specimens were recovered from black flint near the base of the Gunflint formation in Ontario, Canada. There were at least two kinds of fungi and two different algae represented by spores and longer filaments. Another unicellular structure resembles a flagellate. It has been suggested that the flints are about 2 billion years old. Irrespective of their great age, these specimens from the Huronian sequence may be more correctly referred to the Proterozoic.

Fig. 85. Proterozoic algae, Collenia frequens *Walcott. Redrawn from Fenton.*

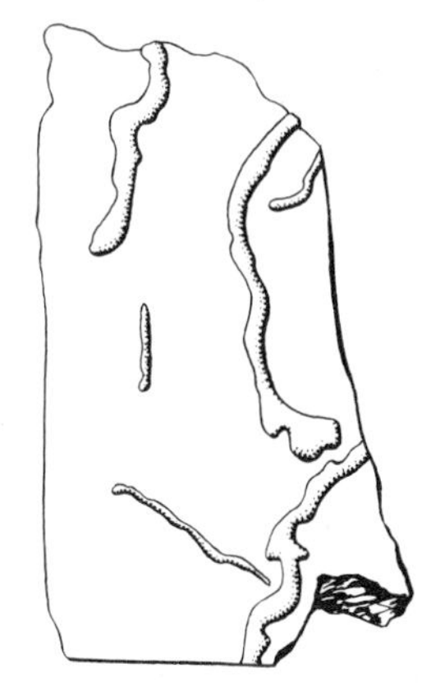

Fig. 86. Fossil trails of Proterozoic segmented worms. Redrawn from Fenton.

Proterozoic rocks have yielded innumerable fossils, especially calcareous algae. Some of the best specimens may be seen in Glacier National Park in western Montana, where they were first reported in 1906 by Charles Doolittle Walcott. They occur there in several of the Belt marine limestone formations, some of which are metamorphosed into fine-grained marble. These colonial algae form fossil reefs as much as 33 feet wide and 25 feet high. Dark gray or black, flattened, spheroidal, isolated colonies are clearly visible in the massive outcrops where they were buried over a billion years ago. Nothing of the original algal structures remain, but the concentric, laminate, calcareous concentrates are quite like those deposited by Pleistocene and living algae. Some of the better known genera of Proterozoic algae are *Collenia* (Fig. 85) and *Newlandia*.

Much of the iron ore of the world is taken from Proterozoic rocks. Since the presence of iron pyrite and glauconite are frequently associated with reducing conditions in the action of bacteria, it is reasonable to assume that these microscopic thalloid plants were abundant in the Proterozoic.

Other evidence of plants are small carbon tubules from Finland that have been named *Corycium enigmatum* Sederholm. The high carbon ratio of C^{12}/C^{13} supports the contention of an organic origin of these tubules. Possibly there were plants higher in organization than the known Thallophyta. Structures in the Belt of Montana, at first thought to be the remains of eurypterids and called *Beltina danae*, are now believed to be film imprints from pieces of seaweed.

Fossils of Proterozoic animals are even more inadequately known than those of plants. Several structures have been found that are strongly suggestive of sponge spicules. These are the minute hard parts in a sponge that tend to strengthen and help hold the shape of the body. Spicules

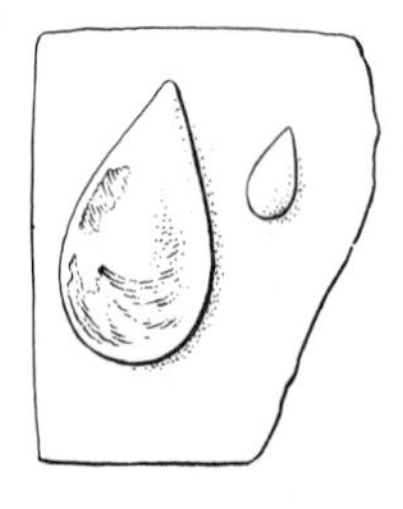

Fig. 87. Questionable molds of Proterozoic brachiopods, Lingulella montana *Fenton and Fenton. Redrawn from their illustrations,* $\times \frac{1}{2}$. *Specimens in the Princeton University collections.*

quite like the structures from the Proterozoic of Montana, the Grand Canyon, and elsewhere are well known in sponges. Radiolarians have been reported from France.

Many traces and fossil trail marks much like those made by segmented worms have been discovered in the sediments hardened into rocks from Proterozoic mud flats. These flexuous annelidlike trails are frequently exposed on ripple-marked surfaces (Fig. 86). Other trails ending in subovate depressions are similar to those made by pelecypods, yet no hard parts of mollusks have been encountered.

One of the most interesting and controversial discoveries in Proterozoic rocks was made by Mildred A. Fenton in 1932. While prospecting the limestone outcrops in the Newland formation of the Belt near Little Birch Creek, Meagher County, Montana, she suddenly recognized a small oboloid form on the rock surface. Ten more of these specimens, suggestive of primitive brachiopods, were found in the vicinity. The fossils were moldlike structures. Unfortunately, no traces of shell could be found, but other molds were found at the same locality and more were discovered higher in the section. Carroll Lane and Mildred Adams Fenton named these specimens *Lingulella montana* and referred them to the phylum Brachiopoda. Some opponents have maintained that this species of *Lingulella* is not a fossil but something resembling one; others have been willing to accept them as fossils but believe they are Cambrian and not Proterozoic in age. Still other competent stratigraphers have confirmed the Proterozoic age of the rocks. Therefore, there is at least a fifty-fifty probability that the Fentons discovered the first evidence of Pre-Cambrian brachiopods (Fig. 87).

The Pound Sandstone jellyfish and associated fauna found at Ediacara and elsewhere is the Flinders Range in South Australia are now believed to be late Proterozoic in age. At Ediacara this fauna is 600 feet stratigraphically below the earliest conventional Cambrian fossils, the Archaeocyatha, of the region. The *Olenellus* fauna is unknown in Australia, but the archaeocyathid limestones which overlie the Pound Sandstone in South Australia are tentatively correlated with similar limestones in

Morocco, where trilobites related to *Olenellus* succeed archaeocyathids at the base of the Cambrian. Further collections made recently at Ediacara have yielded a wealth of new species belonging to diverse zoological groups. Dr. Martin F. Glaessner, University of Adelaide, has recently described a polychaete annelid worm, *Spriggina floundersi*, and an enigmatic anchor shaped organism, *Parvancorina minchami*, of uncertain relationships.

Why Fossils are Rare in Pre-Cambrian Rocks. Pre-Cambrian rocks have been searched diligently for fossil remains, particularily in the middle latitudes, yet the results are extremely meager. This has been difficult to explain, since the marine fossil record is so well represented in the Cambrian. Four reasons are more or less currently accepted for this paucity of fossils prior to Cambrian time.

1. The animals and plants were for the most part soft-bodied, free-swimming or planktonic organisms that left no suitable hard parts for preservation.
2. Life may not have been so abundant as in later time.
3. Pre-Cambrian rocks are usually altered or metamorphosed by heat and pressure, thus destroying evidence of fossils.
4. For the most part Pre-Cambrian rocks, being covered by all the later rocks, can be prospected only in limited areas where their exposure occurs.

These undoubtedly are logical and reasonable explanations, but climatic conditions and also the adaptive potential of Pre-Cambrian organisms may have played an important role in the distribution and kind of life now known from the Proterozoic. Possibly most animals and plants with hard parts to preserve as fossils had not yet evolved sufficiently physiologically to tolerate the colder seasonal extremes in middle latitudes. Thus the greatest populations of life during that Era may have existed in rather restricted, warm, oceanic waters near the equator. Equatorial invertebrate faunas, then, could have been the source of the well-represented, early Cambrian marine faunas that dispersed into middle latitudes with wide-spread warmer climates at the beginning of the Paleozoic.

Pre-Cambrian Climates. Climatic conditions in the Pre-Cambrian are difficult to determine without the evidence of a fossil record. Nevertheless, we know from the composition of the metasediments that there was aridity, rainfall, and erosion much like they occur today. Thick formations of limestone were probably deposited in rather warm waters as climatic conditions varied through the hundreds of millions of years. Rather coarse clastic sediments that accumulated in large fresh-water lakes are indicative of heavy rainfall in certain areas, but at other times there must have been seasonal droughts, as indicated by mud cracks.

The most positive evidence of climatic conditions in the Pre-Cambrian is in the form of tillites or glacial deposits. These formations frequently contain small to enormous striated boulders that received their striations by being transported by ice and scoured across other rock surfaces. Proterozoic tillites have been found in many parts of the world. It is difficult to know how much of the earth's surface was covered by ice at any one time, but it has been estimated that probably half of Australia was covered with ice sometime during the Proterozoic. There is good reason to believe that glaciation was quite extensive on all continents. In India there is evidence of glaciation not far from the equator. The distribution of glaciers so close to equatorial latitudes must have lowered temperatures considerably all over the world at different times during the Proterozoic.

Our knowledge of the distribution of Pre-Cambrian rocks is too incomplete to make an outline of the continents and oceans at that time. There is evidence of mountain building, but the distribution of these as ranges or chains is obscure.

On the whole we have much to learn about the Pre-Cambrian; yet the solutions to many of our unsolved problems are buried somewhere in those rocks.

9 Paleozoic Era

Dominance of invertebrates; origin of fish, amphibians, reptiles, and land plants.

THE NAME PALEOZOIC (ancient life) was originally proposed in 1838 by Adam Sedgwick (1785–1873) to include the rocks and fossils of the Periods Cambrian, Ordovician, and Silurian of our present timetable, though the Ordovician was not named by Charles Lapworth (1842–1920) as a distinct unit until 1837. Then John Phillips (1800–1874) in 1839 and 1841 added the Devonian and Carboniferous as well as the older term "Magnesian limestone," which had been used for certain rocks in the British Isles and western Europe. The Permian, which, in part, includes the "Magnesian limestone," was named in 1841 by Sir Roderick Impey Murchison (1792–1871) and considered by him as part of the Paleozoic Era. Within a period of seven years, then, the Paleozoic came to include essentially the time interval we now recognize for the Era. In America the Carboniferous is divided into the Mississippian and Pennsylvanian Periods, and some authorities in Europe still refer to the Ordovician as early Silurian, whereas others recognize the Ordovician but call the Silurian equivalent Gothlandian.

Geosynclines and Inland Seaways. The seven Periods of the Paleozoic Era mark the dominance of invertebrate life in the inland seaways

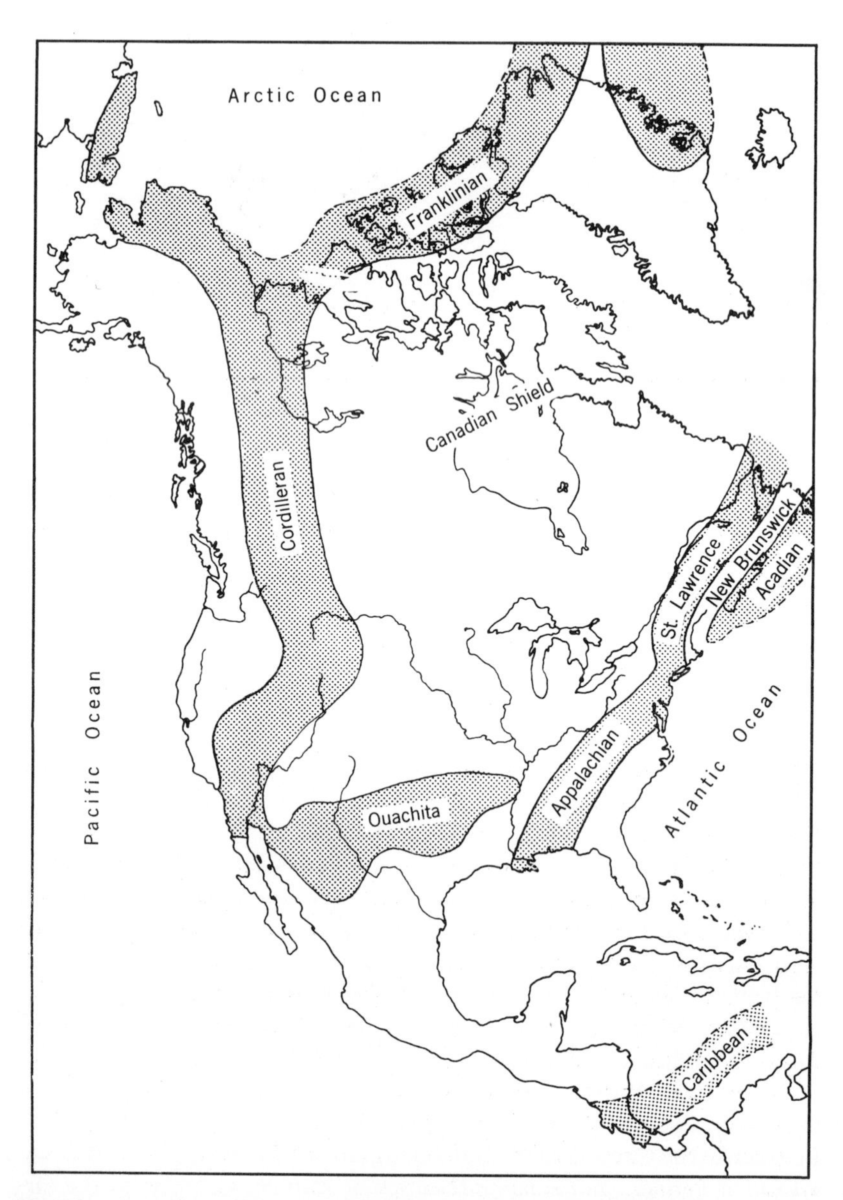

Fig. 88. An outline of the geosynclines and troughs in North America. The patterns of these Paleozoic seaways changed considerably and almost continuously. After Schuchert.

that spread into the geosynclines and marine troughs across the continents of the world. In North America there were four major geosynclines and three troughs: the *Franklinian* in the far north, the *Ouachita* in the south, the *Cordilleran* in the west, and the *Appalachian* in the east, from which the *Saint Lawrence* and the *Acadian* troughs were extensions to the Northeast. A hypothetical trough, the *Caribbean*, probably separated North America from South America (Fig. 88).

A *geosyncline* is an area of long-continued subsidence which receives a considerable thickness of eroded superficial rocks. The depressions in which the deposition takes place are comparatively narrow, linear, structural troughs or inland seaways when compared to the continent as a whole. Most of the time during their development the geosynclinal basins were occupied by shallow marine waters in which subsidence kept pace with deposition. Sedimentation in geosynclines may also take place in troughs temporarily elevated above sea level. They may be restricted to areas like the San Joaquin Valley, California, or may be of regional extent like the Cordilleran geosyncline in western North America. Eventually geosynclines are elevated, folded, and faulted into mountain ranges such as the Appalachians and the Rocky Mountains in North America.

During the Paleozoic in North America, Marshall Kay believes, there was a rather stable *craton*, or central land area, with marginal depressed surfaces called *miogeosynclines* in which carbonate sediments and fine

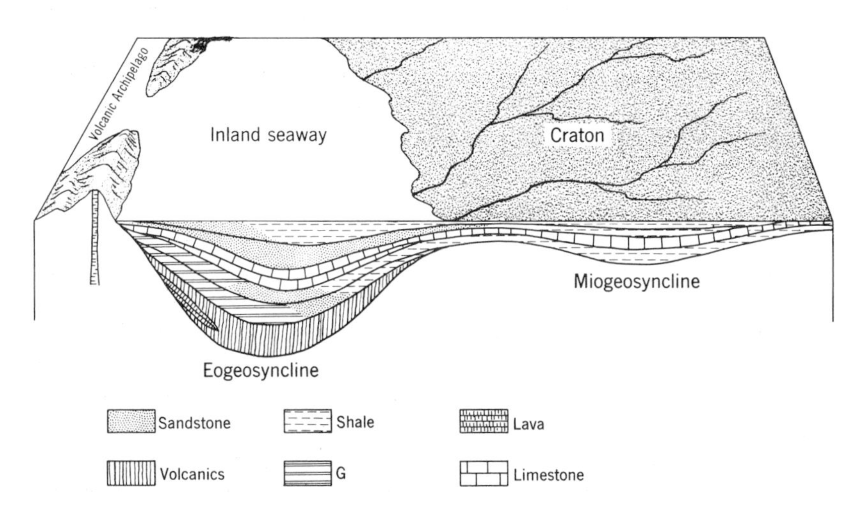

Fig. 89. An ideal section showing the thickness and the nature of the contained rocks in eugeosynclinal and miogeosynclinal (terms proposed by Marshall Kay) structures of a geosyncline. Slightly modified after Eardley.

quartz sands accumulated. These sediments were derived primarily from the craton. Nearer the continental borders where the geosynclines had subsided much deeper they have been referred to as *eugeosynclines.* The continent is thought to have been fringed by chains of volcanic archipelagos which contributed large amounts of volcanic materials, flow rocks (like lavas), and graywackes as sediments to the adjacent eugeosynclinal troughs. The activity in these orogenic belts is indicated further by the numerous and varied kinds of metamorphosed rocks secondarily deposited in the eugeosynclines (Fig. 89).

Similar geosynclines and adjacent archipelagos appeared on other continents at the beginning of the Paleozoic. Except for Europe, they have not been so well mapped nor are the fossils from them so well known as in North America. Many generations of future paleontologists and stratigraphers will be occupied with these problems before the configuration of continents and inland seas of the Paleozoic are adequately understood.

In South America a marine seaway crossed Venezuela and Colombia, then stretched on to the south across Ecuador, Peru, and Bolivia. Before the Era closed the fluctuating patterns of that seaway covered parts of Argentina and most of Brazil.

Paleozoic marine formations and their fascinating fossils have been found in Siberia, China, Burma, and India. There is still not enough evidence to map the seaways accurately in those areas, but at times they contained faunas and floras similar to those in North America, and other assemblages show affinities with the fossils of western Europe. An interesting, yet unsolved, problem of this kind prevails in the early Paleozoic marine faunas of the same age in Burma and India. The Burma fossils seemingly are related to those in Europe, whereas the specimens from India bear a greater resemblance to those in North America. Yet the bulk of the materials from both of these areas is distinctly Asiatic.

There were two major geosynclines in Australia. For the most part offshore animals occurred in the *Tasman* geosyncline which overlapped the eastern shore line from Queensland to Tasmania, and a shallower *Amadeus* trough inundated the center from South Australia out through the Northern Territory. These geosynclines were no more fixed in their positions in Australia than on other continents. Paleozoic fossils of different Periods have been discovered in Western Australia where the *Bonepart Gulf, Desert, Carnarvon,* and *Perth* basins have been mapped by Curt Teichert and others. Indeed, there is good evidence that the waters of the Tasman geosyncline spread westward in the Ordovician from southern Queensland across central Australia through the Amadeus trough northwest into the Indian Ocean (Fig. 90).

The Paleozoic of Africa is not so well known except in the northern

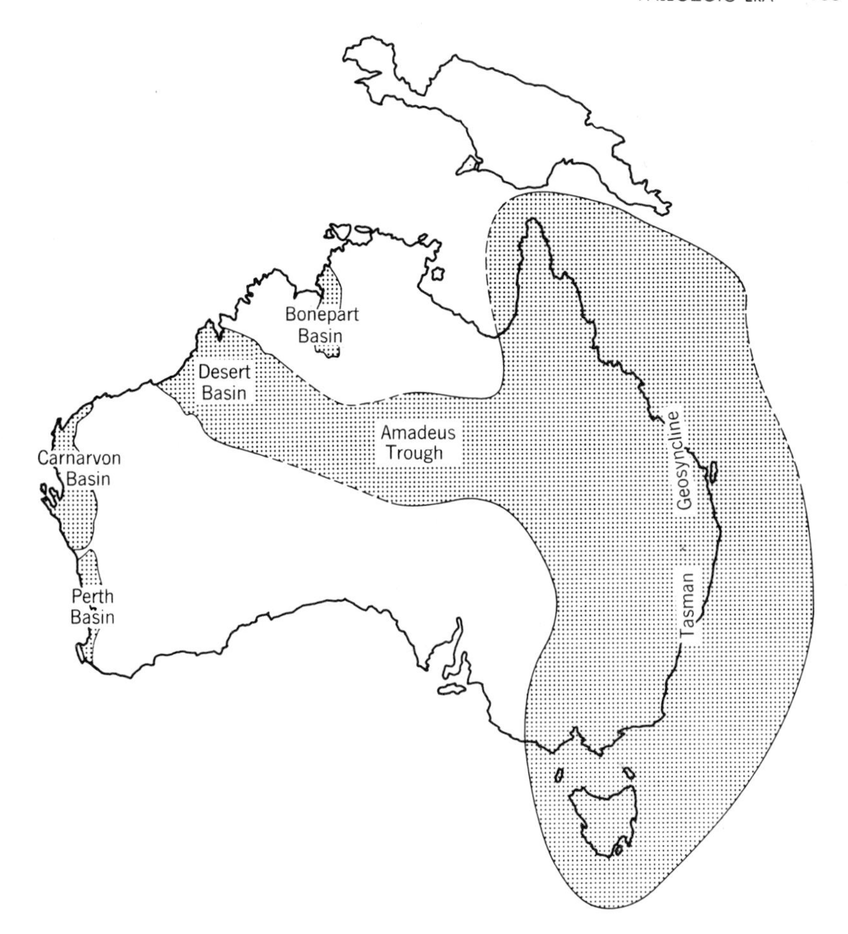

Fig. 90. An outline of the Tasman geosyncline, the Amadeus trough, and other Paleozoic basins of deposition in Australia. As on the other continents, these waterways fluctuated considerably from time to time. After Teichert and from David and Browne.

part, and this ties in fairly closely with western and central Europe and Russia. The marine Paleozoic of Europe has been carefully mapped and studied over a longer period of time than that of the other continents. Orogenies at different times greatly altered the patterns of embayments and troughs. Most of the type sections and, consequently, the typical faunas are in Europe, so for the most part final comparisons for correlations must be made with the fossils from that continent.

Life of the Paleozoic. There is excellent reason to believe that all phyla of plants and animals were present in the earliest Period of the Paleozoic. Some of these changed rapidly from small and simple kinds in the beginning into the giants of their day. The greatest over-all change in the Paleozoic marine invertebrates seems to have come about between the early and the middle Ordovician. Throughout the Paleozoic life continued to evolve and increase in numbers while competition for food and space to live intensified. Lower forms of marginal land plants in the earlier Periods gave rise to rather meager primordial forests in the Devonian, and these, in turn, evolved into the great humid swampland forests of the Mississippian and Pennsylvanian. Toward the close of the Era, when there was a gradual retreat of the inland seas and a general uplift in the continental masses, primitive conifers, ginkgos, and cycadophytes began to blanket the barren landscapes farther back from the borders of the seas and the stream courses.

Once the fishes were established they soon became highly diversified. One of the groups with lunglike structures gave rise to the amphibians. These were the first four-footed animals to walk on the face of the earth. Long before the end of the Paleozoic some of these terrestrial creatures gradually changed into the earliest reptiles. The Era closed with world-wide epeirogenic uplifts in the earth's crust which resulted in the rapid evolution and dispersal or extinction of most plants and animals at that time.

10 Cambrian Period

Olenellus

Abundance of marine invertebrates and algae; trilobites dominant; first land plants; first foraminifers; archaeocyathids abundant; all phyla of plants and animals probably existed.

ADAM SEDGWICK named the Cambrian in 1835. He derived the name from "Cambria," the Roman name for Wales, site of the type area, in which shales and sandstones make up a section about two miles thick. Though these rocks are strongly folded and faulted, some are fossiliferous.

Adam Sedgwick was born on March 22, 1785, in Yorkshire, and died in Cambridge, January 27, 1873. He became Professor of Geology at the University of Cambridge in 1818, where he was an inspired teacher, noted for his enthusiasm and humor. Active in field work, he traveled throughout England and Scotland. In 1831 Professor Sedgwick and Sir Roderick Impey Murchison tackled the enormous task of working out the complicated geology below the Carboniferous and Old Red Sandstone formations in the British Isles. Sedgwick had already demonstrated his unusual ability by disentangling the structure and stratigraphy in another complicated area. In the new project in northern Wales, which many thought was hopeless, the professor started at the bottom of the section where fossils were rare or wanting, while Murchison worked down from the top of the section. Murchison was more fortunate; fossils were plentiful in his formations. Sedgwick named his sequence of rocks Cambrian, and Murchison applied the name Silurian to the upper formations. Eventually their sections overlapped, so that the uppermost part of Sedgwick's column was claimed as early Silurian by Murchison, while the Cambridge professor stoutly maintained that they belonged to his latest Cambrian. The controversy spread as geologists took sides. Forty-four years later Professor Charles Lapworth introduced the name Ordovician for the controversial part of the section.

Geosynclines and Archipelagos in North America. At the beginning of Cambrian time marine seaways spread into the sunken lowlands, bringing with them milder climates and temperate-to-warm kinds of life probably from southern latitudes. Most of the sediments filling these basins were laid down in shallow waters, and, as these *geosynclines* stretching across the continents continued to fill and sink, peripheral archipelagos and a central craton kept pace by furnishing new sediments as the land areas elevated and eroded (see p. 161).

In eastern North America the *Appalachian geosyncline* followed the northeast-southwest trend now occupied by the Appalachian mountains. Its southern entrance was the Gulf of Mexico, and in the north it joined the *Saint Lawrence* and *Acadian* troughs which crossed the New England states, Labrador, and the Gulf of Saint Lawrence. A tectonic belt of volcanic islands and uplifted metamorphic rocks to the east are believed to have supplied most of the sediments that covered the Cambrian and later Paleozoic fossils. This is indicated by the

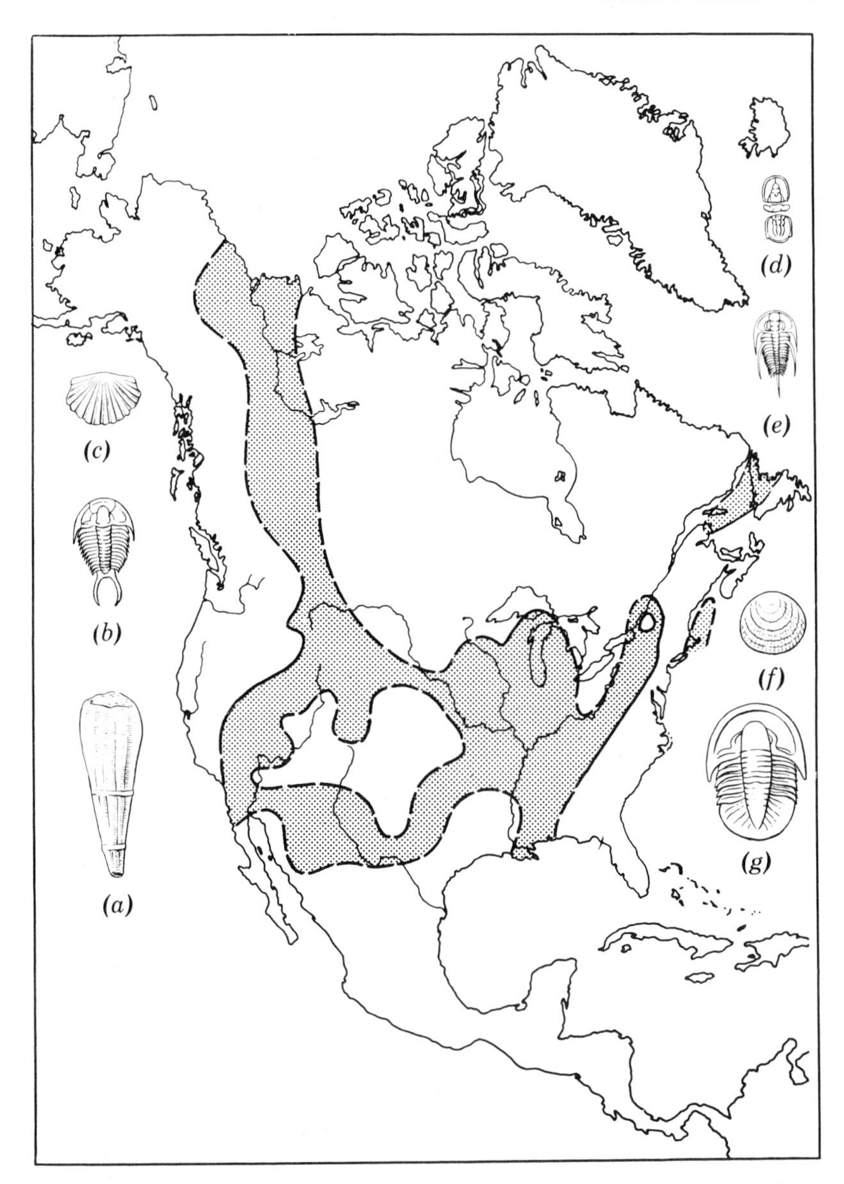

Fig. 91. Greatest inundation of seaways across North America in late Cambrian. After Schuchert. The fossils illustrated come from different parts of the Period: (a) Archaeocyathus rensselaericus; (b) Crepicephalus texanus, *trilobite;* (c) Eoorthis remnicha, *brachiopod;* (d) Agnostus trisectus, *trilobite;* (e) Olenellus gilberti, *trilobite;* (f) Obollela chromatica, *brachiopod;* (g) Cerdaria, *trilobite.* (a), (c) *after Shimer and Shrock;* (b), (d), (e), (f), (g) *after Grabau and Shimer.*

coarser sands and conglomerates in the geosycline and troughs to the east and finer sediments in the same levels farther west.

The Ouachita trough spread across Arkansas, Oklahoma, north and central Texas, New Mexico, southeastern Arizona, and northwestern Mexico. It was at times connected with the Appalachian geosycline to the east and with the southern end of the Cordilleran geosycline to the west.

The *Cordilleran geosyncline* was longer and usually wider than the others. It entered from the Arctic Ocean through the Yukon territory, and extended down across the present site of the Canadian Rockies through Idaho, western Montana, and part of Wyoming into Nevada, where it eventually joined an embayment that spread north from the Pacific Ocean across southern California. Its eugeosynclines evidently derived their coarser sediments from archipelagos to the west in the areas now occupied by California, Oregon, Washington, western Canada, and Alaska. Most of the interior of the continent and the Canadian shield area to the north were lowlands, or a craton, from which sediments washed into the miogeosynclines. The best sections and the greatest thickness of sedimentary rocks in North America occur in the Cordilleran geosyncline, in which all the Epochs of the Cambrian have been recognized. Little is known about the *Franklinian geosyncline* across the northern part of the continent (Fig. 91).

Similar geosynclines occurred on other continents in Cambrian times, but for the most part they are not so well known as those in North America.

Climates. Approximately five sixths of geologic time had passed before the mild climates of the Cambrian, with its widespread calcareous deposits, extended far to the north along the inland seaways. Many Cambrian marine invertebrates ranged from southern to northern latitudes along these marine troughs. Warm currents from the south evidently made it possible for these animals to spread toward the poles. We have no direct evidence of temperatures on land, particularly in inland areas. In contrast to the sparsity of fossils in Pre-Cambrian rocks life was abundant, even in the lowermost Cambrian formations.

Animal Life. Trilobites were the dominant forms of life. The name trilobite was derived from the conspicuous division of the body into three parts; the central, or axial, region and the two lateral, or pleural, regions. Even in the early Cambrian trilobites from seven distinct groups are known, showing that considerable evolution had already taken place. This class of arthropods, sometimes classified with the crustaceans, reached its peak in numbers in this first Period of the

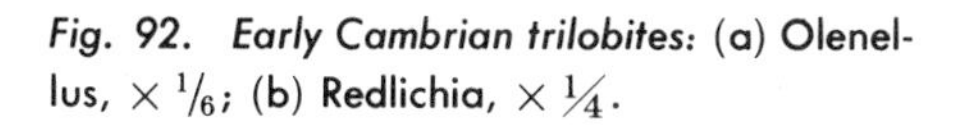

Fig. 92. Early Cambrian trilobites: (a) Olenellus, $\times \frac{1}{6}$; (b) Redlichia, $\times \frac{1}{4}$.

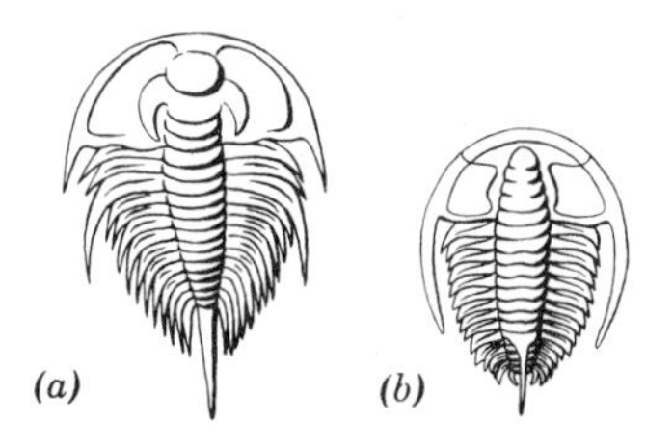

Paleozoic. It then experienced a steady decline in the succeeding Periods until the last of the species died in the Permian.

Most trilobites had a tough casing, usually with sharp spines, that was shed periodically as the animal grew. These old garments, or parts of them, are found more frequently than the complete tests of the original animals. Trilobites were the scavengers and the predators of the Cambrian seas where they groveled on the muddy bottoms, moving along by their paired appendages on the underside of the body. Occasionally they spurted off through the water, propelled by their tail flaps. Sometimes their fossils may be found rolled up into a ball, like a disturbed sow bug. A pair of antennae and the two-faceted eyes on the headpiece were on the alert for unsuspecting prey. Some kinds were blind. Many were ornate with spines and were quite bizarre in appearance. Trilobites inhabited all the marine waters of the world of their time. Their ancestors may have flourished in streams and lakes of fresh water, with one or more groups invading the salty seas near the beginning of the Cambrian. Few Cambrian faunas are found without at least parts of their disarticulated tests.

Several trilobites have been used as index fossils for Zones in the early Cambrian, but two of the commonest genera, though found in different parts of the world, are the little spike-tailed *Olenellus* and *Redlichia*. In the middle Cambrian the large trilobite *Paradoxides* was common in Europe and in the Acadian seaway. *Bathyuriscus*, so characteristic of the Cordilleran inland seaway, spread through the Ouachita trough, and

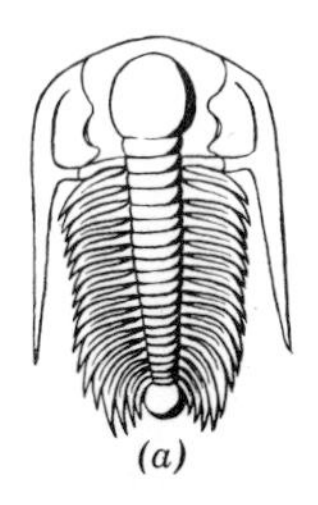

Fig. 93. Middle Cambrian trilobites: (a) Paradoxides, $\times \frac{1}{4}$; (b) Agnostus, $\times 1\frac{1}{2}$.

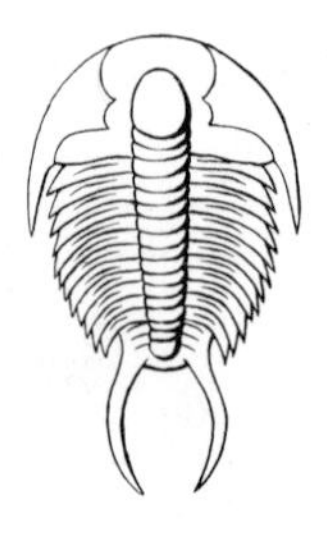

Fig. 94. Late Cambrian trilobite, Crepicephalus; × 1/6.

forms related to it occupied the southern part of the Appalachian geosyncline. During the greatest inundation of the Period, in the late Cambrian, the pincer-tailed genus, *Crepicephalus,* spread far and wide in the evolution of its species. Excellent exposures containing these fossils occur near Sioux Falls, Minnesota, where fine calcareous sediments were deposited from the clearer waters near the middle of the great inland sea (Figs. 92–94).

Other Cambrian arthropods are the numerous trilobitelike creatures and an onychophorid found in the Burgess shales of British Columbia. The onychophorid, related to the living *Peripatus* of tropical forests, is most unusual. It displays some features which suggest a relationship with the annelids. Early ostracodlike fossils have also been found in the Cambrian.

Small primitive brachiopods were second in numbers. Their earliest record came from the early Cambrian, when numerous chitino-phosphatic, inarticulate species and some calcareous, thin-shelled articulate kinds were present. Simple shells were typical of these first relatives of the living lamp shells. Cambrian brachiopods have been found on the bedding planes, lodged in the coral reefs, or as broken parts in a coquina with the hard parts of other organisms (Fig. 95).

Mollusks were not common. The gastropods had simple shells for the most part, but some were coiled. Pelecypods were rare and little is known about them. Cephalopods, some of which were destined to become gigantic in the next Period, were tiny primitive animals with slightly curved or straight conical shells. One genus has been found

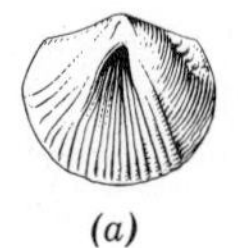

(a) *(b)*

Fig. 95. (a) Eoorthis tatei, *brachiopod,* × 1; (b) *three views of* Micromitra etheridgei, *questionable primitive brachiopod. Cambrian of Australia. After David and Browne.*

in the early Cambrian and another toward the end of the Period (Fig. 96). Some foraminifers and radiolarians have been discovered, and it is reasonable to assume that there must have been other soft-bodied protozoans in abundance.

Sponge spicules representing diversified forms are not uncommon. A group that has caused much controversy in classification is the Archaeocyatha. Recently it has been suggested that the Archaeocyatha is a distinct phylum in a subkingdom of animals which includes the Coelenterata and the Porifera, though it is somewhat closer to the sponges than to the corals. Archaeocyathids have been found commonly in early and middle Cambrian limestones almost around the world and appear to have flourished in shallow seas and gulfs. Excellently preserved specimens are known from the Flinders Range of South Australia.

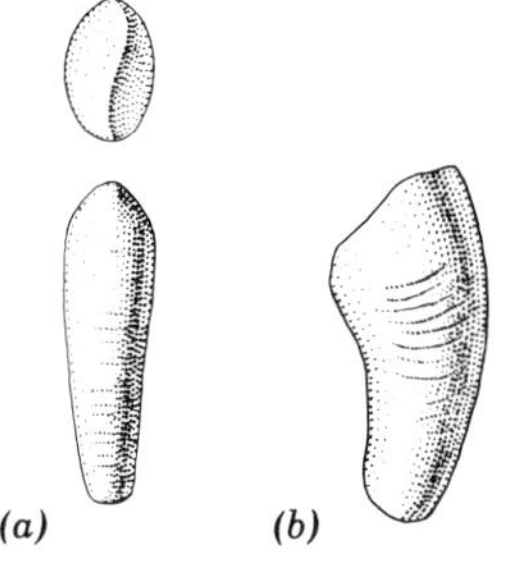

Fig. 96. Questionable primitive cephalopods: (a) Volborthella, × 8; (b) Plectronoceras, × 3. *After Krapinsky.*

Cambrian coelenterates are rarely preserved. A jellyfish is recorded from New York state and a sea anemone was found in the famous Burgess shales of British Columbia. A few graptolites are also known from the Cambrian, and a number of bryozoans have been recognized in early Cambrian faunas from Russia and Canada. Careful prospecting and diligent study will probably reveal more genera in other regions.

There is no Cambrian record of the Platyhelminthes, flat worms, or the Trochelminthes, rotifers, though they, too, may have been present. Nemathelminthes, round worms, and Annelida, segmented worms, are probably represented by the numerous trails so clearly marked in mudstones.

Echinoderms, which were destined to enter so prominently into the later fossil record occur abundantly in the Cambrian in northeastern Australia. Some of these, *Peridionites navicula* and *Cymbionites craticula,* are primitive unattached animals with different calcitic skeletons for which a separate subphylum name, Haplozoa, has been proposed.

Frederick W. Whitehouse has shown that the construction of these interesting echinoderms is similar to that of the larval stages in Recent forms. Except for *Palustrina*, a starfish from Europe, the other Cambrian echinoderms are represented by other classes with stems for attachment to the bottom of the sea. The distinctness of the Cambrian echinoderms indicates a diversification of the phylum as early as the Proterozoic.

It is possible that chordates representing the ancestors of the ostracoderms (vertebrates) will yet be found in Cambrian rocks, since ostracoderms occur in the Ordovician.

Plant Life. Small but widespread calcareous algal reefs have been recorded frequently in Cambrian formations. The detailed structure of these organisms is not well known, but their moundlike appearance and irregular surface is clearly indicative of the structures assumed by Pleistocene and Recent algae. Sections of the colonies are arranged in concentric laminae that were precipitated in layers as the algae grew. Algae with complex structures have been reported from the Cambrian and even from Pre-Cambrian strata. One fossil plant, *Aldanophyton*, from the middle Cambrian of Siberia has been described as a lycopsid. It certainly would not be surprising to see a psilopsid, structurally the most primitive kinds of land plants, turn up in the Cambrian. The appearance of *Aldanophyton* indicates that simple plant life was more abundant than the fossil record might lead one to believe but that suitable remains for preservation as fossils possibly had not yet evolved.

Burgess Shale. One of the most enthusiastic and competent students of paleontology and stratigraphy of the Cambrian was Charles Doolittle Walcott (1850–1927), former chief of the United States Geological Survey. He, of course, concentrated on the faunas and formations in North America, though he did considerable research on the Cambrian of China. He became particularly interested in the Cordilleran geosyncline. Not only did he work on the specimens and information obtained by the earlier surveys, but he spent much of his time in the field, mapping the formations and securing more fossils.

In 1909, as his field party moved slowly along a narrow rocky trail on Mount Wapta in the Selkirk Mountains of British Columbia, it is said that the pack on one animal shifted to one side, as they are prone to do. At that spot Walcott picked up a slab of shale that had been carried down the mountain side by a snow slide. To his surprise it contained the outlines of small creatures, some of which resembled crustaceans. It required considerable careful prospecting before the source of this interesting slab was located farther up the mountain in a middle Cambrian formation. There faithfully imprinted by the carbon-film method of preservation were even the most detailed parts of small,

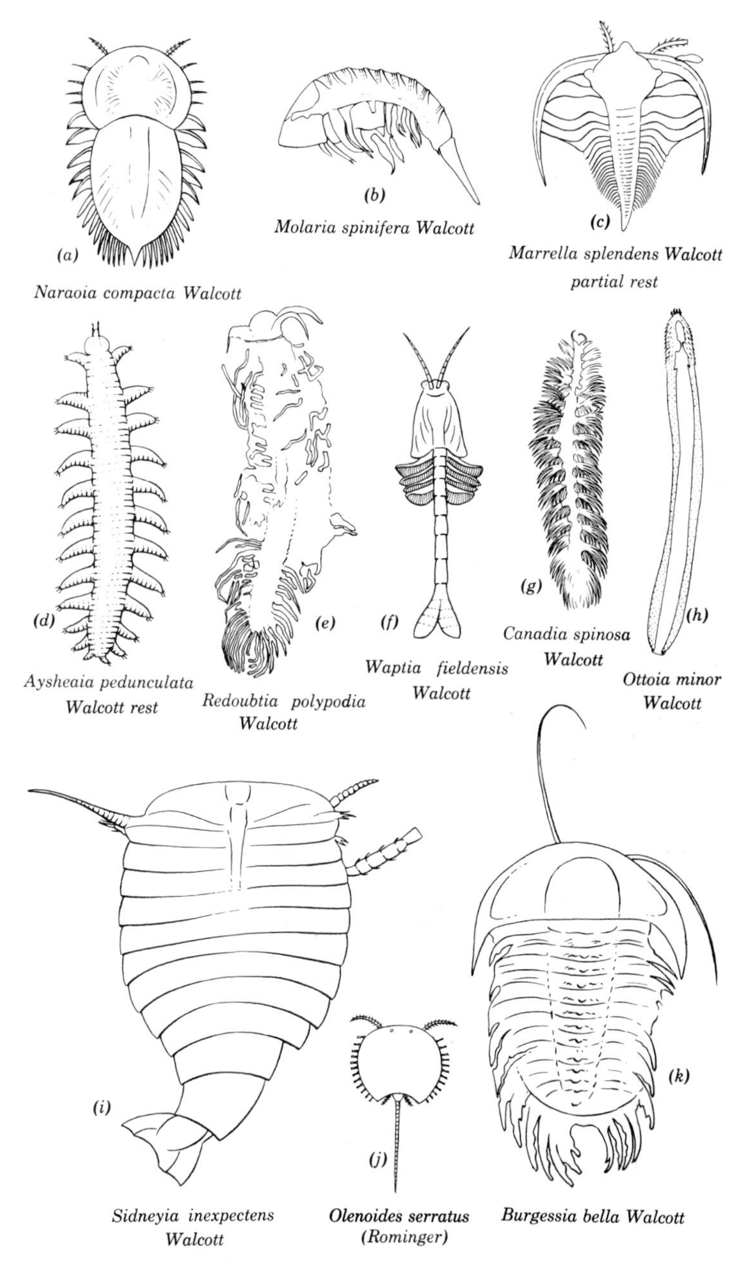

Fig. 97. Fossils from the Burgess shale, partly restored: (a), (c), (f), (i), (j), (k), *trilobitelike;* (b), *eurypteridlike;* (c) *onychophorid;* (e) *holothurian;* (g), (h), *annelid-like worms.* (a) Naraoia compacta × 1; (b) Molaria spinifera × 1½; (c) Marrella splendens × 2; (d) Aysheaia pedunculata × 1; (e) Redoubtia polypoda × ¾; (f) Waptia fieldensis × ¾; (g) Canadia spinosa × 1; (h) Ottoia minor, × 1; (i) Sidneyia inexpectans, × ½; (j) Burgessia bella, × ½; (k) Olenoides serratus, × 2½.

soft-bodied, invertebrate animals. Delicate appendages such as antennae, limbs, and hairlike structures were to be seen on the thin gray shales. Even the outlines of the soft anatomy were observable in some specimens in which traces of the internal organs were shown. This find not only gave a clear insight into the relationships and structures of these little animals of the Cambrian, but demonstrated, as was naturally thought, that there was an abundance of soft-bodied creatures at that time normally not preserved as fossils (Fig. 97).

Unfortunately, most of the Burgess shale specimens are immature and are difficult, if not impossible, to identify with the adults. Nevertheless, about 70 genera and 130 species have been described. These include seaweeds, siliceous sponges, jellyfish, sea anemones, annelidlike worms, trilobitelike forms, a eurypteridlike genus, and the onychophorid, *Aysheaia*. The fossils and sediments indicate a shallow-water environment in a small bay or lagoon not far from land, but with access to the open sea, where poisonous carbonic acid gas escaped from the stagnant muddy accumulations at the bottom. Free-swimming or other kinds of plants and animals that drifted into these relatively quiet waters, both from the sea and from fresh-water sources, perished and sank to the bottom where they became entombed in the soft mud before they could disintegrate. The carbon-film method of preservation fortunately gave us a clear idea of the anatomy of these early Paleozoic invertebrates.

Starting Point of Greater Things to Come. In other seaways and especially in different environments other kinds of soft-bodied or thin-shelled animals must have flourished. Many representatives of these groups may have secreted durable hard parts in later Periods. This is in part exemplified in the Cambrian trilobites and brachiopods in which the hard parts were not so thick nor so massive as those in the middle or late Paleozoic. The protective armor was not only thinner and usually phosphatic, but the spines and other ornate features were less conspicuous in the early animals.

Though the Cambrian gives us the oldest good fossil record, this was not the beginning. Indeed, approximately 3 billion years of geologic time had transpired before Cambrian time, and most of the groups of organisms had already reached an advanced stage in their evolution. Nevertheless, insofar as the record is concerned, the Cambrian was the starting point of much greater things to come. In all probability representatives of some of the phyla were living on land, especially in moist warm places; but without the protection of sheltering vegetation, as we know it today, life on land must have been limited. The fresh-water streams and inland lakes undoubtedly contained animals and plants with less necessity for salinity, but little is known of fossils in fresh-water deposits until later in the Paleozoic.

11 Ordovician Period

Graptolites

Graptolites dominant; brachiopods increasing;
tetracorals present; small aggulinated foraminifers;
trilobites still abundant;
nautiloids of maximum abundance;
cystoids at their peak;
first ostracods; first conodonts;
first vertebrates, without lower jaws;
seaweeds and other algae; warm mild climates.

THE TYPE AREA of the Ordovician, like that of the Cambrian and Silurian, is also in Wales. The name was derived from *Ordovices*, a Celtic people. This Period was proposed, in part, as a compromise by Professor Charles Lapworth in 1879, since the rocks composing the section are between the Silurian and the underlying Cambrian and were claimed by both Murchison and Sedgwick as belonging to the Periods they had named. Nevertheless, the faunas of the Ordovician on the whole have been recognized as distinct throughout the world. In the type section Lapworth found intercalations of volcanic rocks in his sedimentary series which he used in support of his proposal for a separate designation.

Professor Lapworth (1842–1920) was one of the foremost pioneers in stratigraphy. Much of his work was in the highly disturbed and convoluted rocks in the early Paleozoic in the southern uplands of Scotland. There he mapped the aerial geology in minute detail and carefully recorded the stratigraphic sections. Graptolites in the thin-bedded, black, fine-grained shales were his index fossils. The Zones in which these fossils were found ranged from a few inches to several feet in thickness, but the different kinds proved to be good markers and in the same stratigraphic positions in relation to each other whether they were found in Scotland or on the mainland of Europe. Thus he became a specialist in graptolites and again proved the indispensability of fossils in revealing the sequence of stratified rocks where they had been folded and faulted out of their original position.

Inland Seas and Domes in North America. The Ordovician is better represented in North America than it is on any other continent. There was a slight elevation throughout the continent at the close of the Cambrian, but for the most part the continent was low. The shallow seaways not only again flooded the old geosynclines but at times spread over half of the land, eventually inundating most of the Canadian shield. This was called by Charles Schuchert (1858–1942), professor of historical geology at Yale University, the greatest of all submergences. Late Ordovician showed the most far-reaching watermarks of inland flooding in North America, though to the northeast the land was rising continuously even to the extent of eliminating the Acadian trough. Several low, stable domes in the interior tended to outline the inland seas. These were the *Cincinnati dome*, which extended north to south from Ohio and Indiana across Kentucky, the *Ozark dome* in Missouri, and the *Wisconsin dome*. Farther west, most of Kansas, Nebraska, and part of the Colorado area were above water. The domes or lowlands supplied sediments that settled in the seas. The old eugeosynclines continued to sink much deeper than the midcontinent miogeosynclines as they filled with sediments.

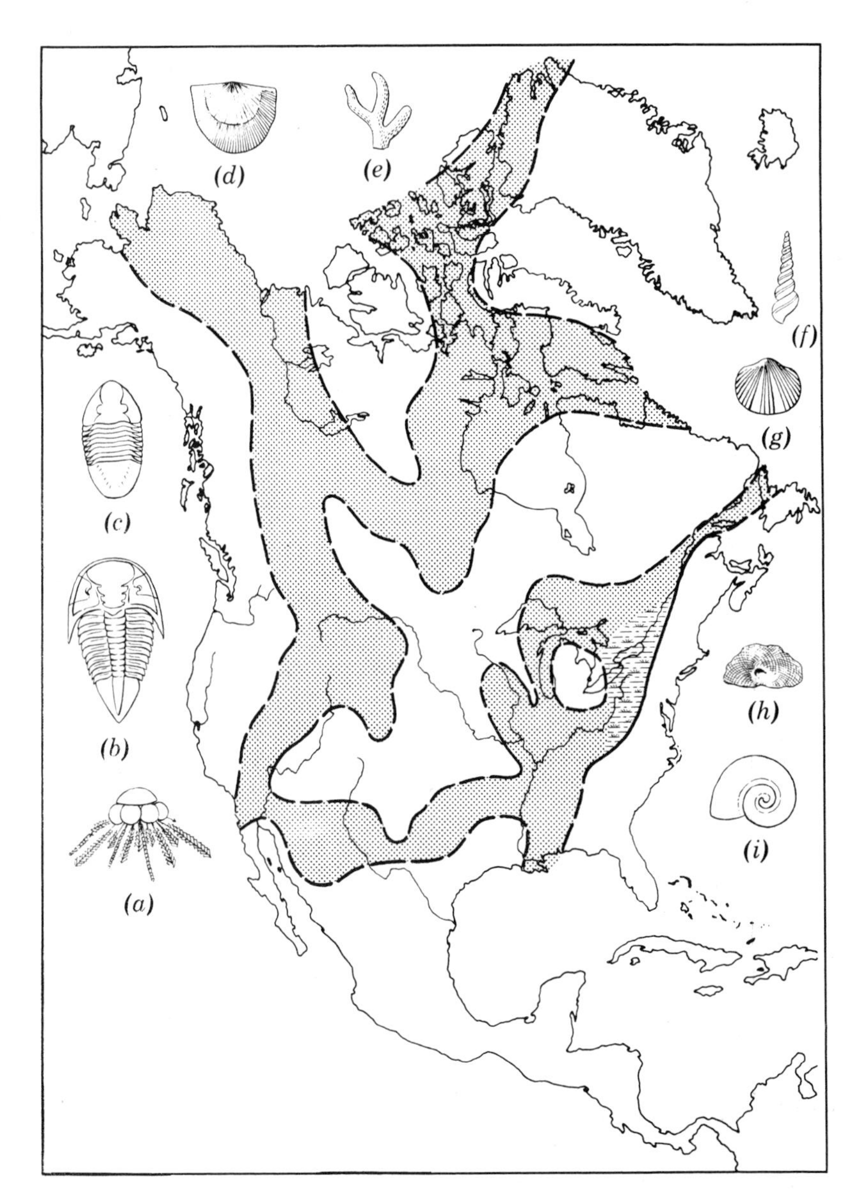

Fig. 98. Greatest inundation of seaways across North America in late Ordovician. After Schuchert. The fossils illustrated come from different parts of the Period: (a) Diplograptus pristis, *graptolite;* (b) Dalmanites achates, *trilobite;* (c) Isotelus gigas, *trilobite;* (d) Öpikina minnesotaensis, *brachiopod;* (e) Hallopora elegantula, *bryozoan;* (f) Horostoma gracilis, *gastropod;* (g) Platystrophia crassa, *brachiopod;* (h) Receptaculites oweni, *sponge;* (i) Maclurites, *gastropod.* (a), (b), (c), (d), (f) *after Grabau and Shimer;* (e) *after Bassler, from* Treatise on Invertebrate Paleontology, *courtesy of Geological Society of America and the University of Kansas Press.*

There were temporary emergences. Two have been used in recognizing a three-fold division of the Ordovician. Toward the close of the Period sediments spread to the west across New York and Pennsylvanian to form the *Queenston delta*. Excellent exposures of the red Queenston shales may be seen in Niagara Gorge below the falls. Sands and muddy silts which settled into the Appalachian geosyncline gave way to fine calcareous silts to the west, where limestones accumulated on the floors of the clear, shallow seas. The last Ordovician formation laid down in that area included continental red beds which showed that the marine waters had retreated at that time. The Period closed with the *Taconian orogeny* which uplifted the land areas and, with pressure from the east, buckled the early Paleozoic formations into high mountains and sharp folds, and in places overthrust Cambrian rocks upon the Ordovician formations.

Animal Life. The climate of the Ordovician was probably milder than that of the Cambrian because of the greater extent of the epicontinental seas. Most of the invertebrates of the early Ordovician were like those in the late Cambrian, but they displayed profound changes before middle Ordovician time (Fig. 98).

Graptolites blossomed out in abundance after their feeble beginning in the late Cambrian. The delicate frondlike colonial structures (rhabdosomes) of these organisms were first observed in 1727. It was over a hundred years before graptolites came to occupy such an exalted position as index fossils in Ordovician formations. At first they were thought to be plants, and this opinion was held by some scientist long after their animal nature was recognized by Walsh in 1771. Though Linnaeus, the great Swedish systematist, divided them into various species as early as 1735, he called them inorganic markings or bodies that simulate fossils. So uncertain were scientists of their relationships that later authors called them bryozoans, corals, cephalopods, or foraminifers. Until 25 years ago, however, they were generally thought to be most closely related to the coelenterates, but most paleontologists saw the chitinous (hard parts) of graptolites preserved as compressed, thin, shining films or markings on the surface of black shales and consequently knew little or nothing of their internal structure. Finally, fully distended larval specimens etched from limestone by Roman Kozlowski have shown that some of the skeletal parts bore a marked resemblance to those in some of the living acraniate Chordata, such as the tunicates and related forms. These similarities were seen in the structure of the cup (theca) walls where the zooids (organic body) formed new zooids, in the method of budding of other zooids, and in the arrangement of the minute internal tubes called stolons. The living acraniate animals (tunicates, lancelets, etc.) have a stiffening rod or notochord as do the higher vertebrate animals in certain stages of their development. Birger Bohlin and Charles E. Decker do not agree with Kozlowski that these characters are indicative of the graptolites being primitive chordates. They point out that these and other features in graptolites are also shared with other phyla of invertebrates. Thus the differences of opinion that began over 200 years ago on the relationships of the fossil remains of graptolites are still with us, but it seems most likely that the graptolites are nearer to the Coelenterata than to any of the other phyla.

Graptolites were colonial animals. Some planktonic kinds drifted in the ocean currents by means of floats; others were attached to seaweeds and were widely distributed; still others were benthonic (sea-bottom dwellers), with restricted distribution of the species. Both the true graptolites (with one cup or theca) and the dendroid graptolites (with three cups or thecae) were common. Usually the leaflike rhabdosomes or parts of them are found in black shales where they settled into muds on the Paleozoic sea floors. Their rapid evolution and wide dispersals have made them most useful in correlation.

Foraminifers were limited in number. Those preserved were agglutinated tubular and globular kinds. They have been placed in five families.

Trilobites were still abundant in the shallower epeiric seas, where they were associated with brachiopods and other invertebrates. They had reached the climax in their evolution. The little shelled ostracods were also there in increasing numbers, and from that time on they played a major role in the drama of aquatic life.

The oldest known true eurypterid was found in early Ordovician graptolite shales in the state of New York. This was the first of a spectacular order of the phylum Arthropoda, some of which were to become the largest invertebrates of the Silurian.

In contrast to the earlier kinds, the brachiopods developed more durable calcareous shells, an adaptive feature that increased in importance as their descendants continued to evolve. These sessile, or attached, animals lived in extraordinary abundance in shallow marine waters in which millions of them have been buried. Their wide distribution in formations throughout the world was brought about by their free-swimming larvae. Beginning students have sometimes confused brachiopods with pelecypods, but a careful inspection will reveal not only a marked difference in the internal anatomy but an inequality in the two shells or valves. The adults were anchored in the mud, attached to rocks, or fastened to other objects by a stem called the pedicle. Thus they were in a position to circulate water containing food and oxygen through the digestive and respiratory systems of the body.

Strange as it may seem, brachiopods are more closely related to the bryozoans than to any other phylum in the animal kingdom. Two classes of brachiopods are recognized: the Articulata have hinged valves on their calcareous shells, and the Inarticulata have unhinged valves and chitinous or phosphatic shells. Many genera have extemely long geologic ranges. For example the living *Lingula* has survived as a genus at least since the Devonian.

Another important group were the cephalopods. Both coiled- and straight-shelled forms were present. These early kinds were more closely related to the living pearly nautilus and the squid than to the ammonites. Some species of straight-shelled cephalopods such as *Endoceras* reached a length of 13 feet and were 8 inches in diameter at the aperture. No animal had reached such gigantic proportions before.

Both gastropods and pelecypods were common in certain environments. Echinoderms appeared in such numbers and diversification as to suggest that more groups should eventually turn up in the preceding Cambrian. All three groups of the class Asteroidea are represented; starfish and brittle stars are fairly common. Though other primitive stemmed echinoderms were present, the cystoids were dominant. Early blastoids and crinoids were also common in the early and middle Ordo-

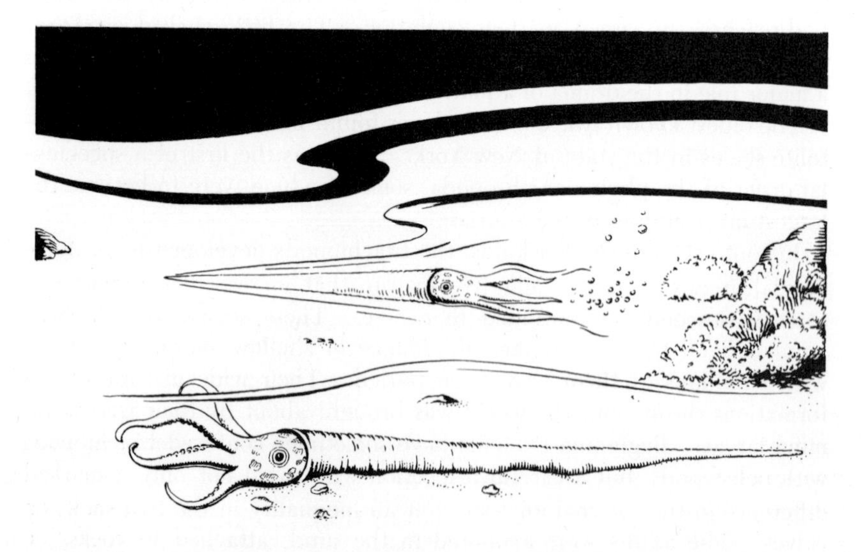

Fig. 99. Giant cephalopods, Endoceras. *Some specimens were 13 feet long.*

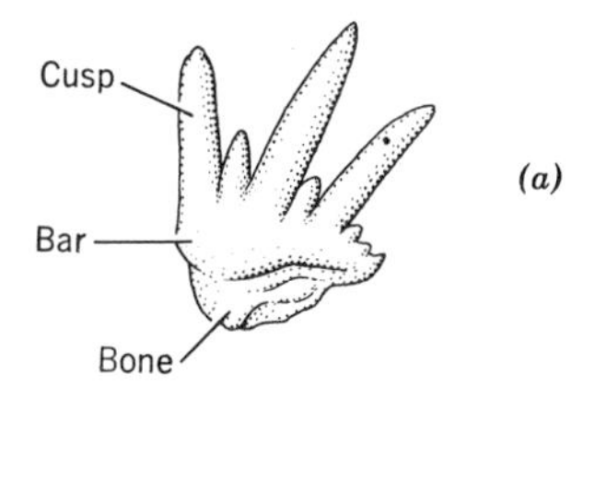

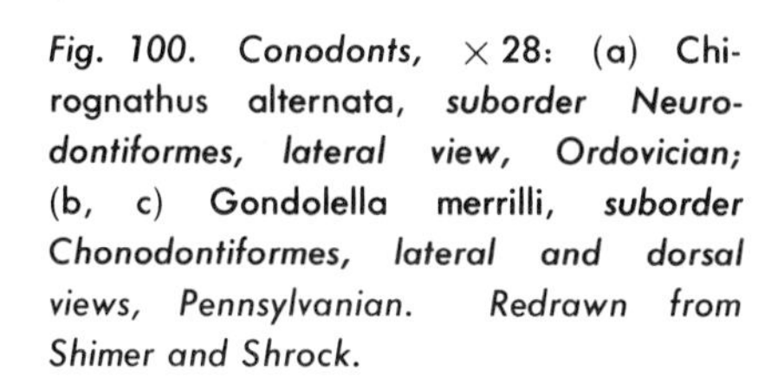

Fig. 100. Conodonts, *×28:* (a) Chirognathus alternata, *suborder* Neurodontiformes, *lateral view, Ordovician;* (b, c) Gondolella merrilli, *suborder* Chonodontiformes, *lateral and dorsal views, Pennsylvanian. Redrawn from Shimer and Shrock.*

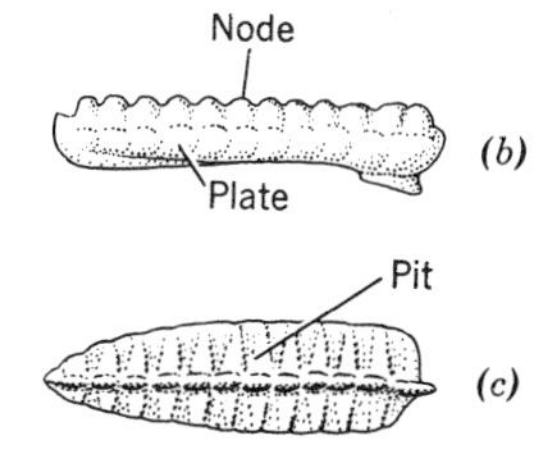

vician. From late Ordovician on the crinoids became one of the most abundant forms of life in the Paleozoic seas. Their discoidal stem segments have become familiar to nearly everyone living in the areas once inundated by marine waters from Ordovician to Permian time.

Primitive honeycomb corals formed low reefs in the early Ordovician. The presence of honeycomb and chain corals among the colonial kinds and the horn-shaped solitary tetracorals attest to the diversity of these animals at this early date. With them in the later part of the Period were countless bryozoans with their lacelike structures enshrouding the corals and the remains of other animals lodged in these *bioherms*. Millions of these and other animals lived and died in these waters, only to become entombed for man to unravel their history in his quest for detailed information of the earth he inherited.

One of the most exasperating yet fascinating and useful groups entered the record at this time. These are the miscroscopic conodonts, represented by teeth or jaw plates. The largest do not exceed 2 millimeters. They are composed of calcium phosphate and occur most abundantly in shales with considerable organic matter. Almost every paleontologic journal or bulletin has a paper dealing with these toothlike structures. Evolution in the different groups has offered convenient evidence for correlation, and these groups together with other microfossils have been useful as marine environment indicators. No record of them has been extended beyond the Triassic (Fig. 100).

The structures assume nearly every shape imaginable but are not difficult to arrange into a scheme of classification. In one suborder, Neurodontiformes, the microscopic internal structure is fibrous, and in the other, Chonodontiformes, it is laminated to show a cone-in-cone arrangement. The known parts of a conodont denticle may be composed of a bar on which the denticles occur or a plate with a platform on which nodes and small pits are arranged. Usually there is a high cusp at one end. Other parts of the anatomy of the animals are not known.

Conodonts are exasperating in that they have not yielded to the numerous attempts to fit them conclusively into any known phylum of animals. They have been classified by different authors as parts of cephalopods, worms, gastropods, crustaceans, and fish. Though conodont teeth are abundant, careful sorting frequently reveals paired elements, which suggests that they belonged to bisymmetrical animals among the lower vertebrates.

Plant Life. Aside from some indications of seaweeds and other kinds of algae, there is nothing known of Ordovician plants. Somewhere in continental shales deposited near the shore, such as might be found at the edge of the Queenston Delta, something may yet be found, since a primitive land plant has been recorded from the middle Cambrian of Siberia.

First Vertebrate Animals. An outstanding milestone was reached in paleontology when T. W. Stanton and Israel Cook Russell of the United States Geological Survey quite independently, while collecting invertebrate fossils, found small bits of bony plates in a Paleozoic formation directly above Pre-Cambrian rocks. In 1891 Charles Doolittle Walcott, Chief of the Survey, identified these bits as the oldest known fossil vertebrate remains. Subsequently, many more broken and abraded parts were uncovered in the middle Ordovician sandstones near Canyon City, Colorado, and named as two genera, *Astrapsis* and *Eriptychius*. Nothing older has been found though similar Ordovician remains have turned up in the Black Hills of South Dakota, in the Big Horn Mountains of Wyoming, and in the Black River formation in Michigan.

Microsections have shown that these fossils were the broken parts of head shields and scalelike coverings of ostracoderms, the most primitive fishlike forms known. Complete skeletons of genera in the Devonian have demonstrated clearly that ostracoderms were jawless, backboned fishlike animals related to the living parasitic lampreys and hagfish.

There has been much speculation as to where these creatures lived and what light they throw on the ancestry of all vertebrate animals. The fossil fragments were found in abundance in Colorado in steeply inclined, colorful, gray and reddish sandstones separated at two levels by shales. Imbedded with the fossil bone fragments were marine gastropods, pelecypods, cephalopods, articulate brachiopods, trilobites, sponges, ostracods, conodonts, and worm borings. The pelecypods and gastropods were abundant, and the other invertebrates were rare. The plates and scales occurred in about 90 feet of the section, but they were most abundant in the reddish sandstones and were widely scattered throughout those beds.

It has been suggested by Alfred Sherwood Romer that the earliest ostracoderms lived in fresh-water streams. This would account for their not having been found in most marine formations of the Ordovician and Cambrian. Romer postulated that the animals died in the streams and their bodies disintegrated. As they washed down to the sea, their armor separated and broke, and, finally, after much abrasion, the fragments were deposited in muds and sands with the shells of invertebrates along the shores of an embayment.

On the other hand, their preferred environment may have been partly sheltered waters along an irregular coastline. Some individual ostracoderms or small schools may at times have ventured into the open seas. Tide action on the beach could have broken and scattered the scales and plates of their bodies before they settled to a permanent resting place.

Robert H. Denison of the Chicago Natural History Museum has written an excellent review of this environment at the edge of the Ordovician sea. After a lengthy discussion on all of the evidence he concluded that the ostracoderms in the Harding sandstones were near-shore marine animals living in shallow water on a bottom of loose, shifting sand or mud.

12 Silurian Period

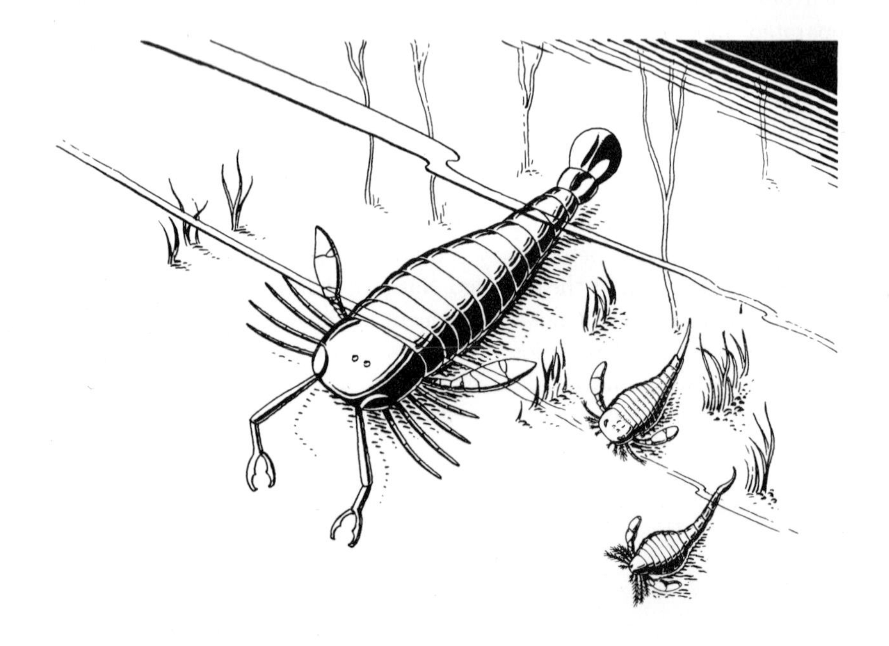

Eurypterids

Eurypterids at their peak;
tabulate corals of maximum abundance;
tetracorals also abundant; fish with lower jaws;
lycopsid land plants;
scorpions and millepeds, first air-breathing animals;
mild climates.

THE SILURIAN PERIOD was named by Sir Roderick Impey Murchison (1792–1871), a Scot who eventually became director of the Geological Survey of Great Britain. The name was derived from the Silures, an ancient Celtic tribe that fought Caesars' legions in the conquest of Britain. Thus this country has more type sections of geologic Periods than any other in the world. In fact, the geologic section across Wales and England represents all of the Periods, from the Pre-Cambrian to the early Eocene.

In his youth Murchison served with the British army in its campaigns against Napoleon's forces. Later, in England, while accompanying Professor Sedgwick of Cambridge and Professor William Buckland (1784–1856) of Oxford on their geologic explorations, he decided to make geology his life's work. After completing short preparatory studies, he worked out a number of geologic problems in Scotland. Then, to broaden his scope, he worked on the geology of Europe with Sir Charles Lyell. He also cooperated extensively with Sedgwick. This led these distinguished geologists to their decision to work out the complicated geologic section below the Old Red Sandstone in Wales. Murchison's classic volume on the Silurian System (1838) gained him world renown. He continued his studies on the Paleozoic rocks in Belgium and Germany and later was invited to Russia by the Czar to work on the formations along the flanks of the Ural Mountains. The work of both Murchison and Sedgwick had a profound influence on geologists everywhere.

Silurian Marine Rocks. Silurian seas spread over parts of every continent. Their distribution is known best in North America and Europe where sediments continued to fill the old geosynclinal troughs as they subsided. Like the ebb and advance of the tide, the marine invasions of the continents spread inland in great inundations, bringing their characteristic faunal units, only to retreat several million years later. This was repeated twice in North America in the Silurian. Except for uplifts in certain areas at the close of the Ordovician, most of North America was relatively flat. Thus slight elevations or depressions on the continent resulted in widespread withdrawals or transgressions now recognized as representing the time intervals of the different Epochs.

Silurian Rocks in North America. The impact of the work of Sedgwick and Murchison in the British Isles stimulated active field work in eastern United States. The New York Geological Survey was formed in 1836. Some states had already formed their surveys and others soon followed. One of the most active workers was James Hall of the New York Survey. Though sometimes rather ruthless with his colleagues, Hall attained such far-reaching results that the Silurian and Devonian of his district in western New York has been looked upon as typical for

North America and far better than the type sections in Wales. When the abundant and beautifully preserved fossils were compared with those of western Europe it was found that many forms were common to the two areas. The evidence indicated intermittent connections of shallow seaways between the two continents.

Formations of early Silurian age in New York have been included in the Albion group. As a result of the Taconian orogeny, sediments from

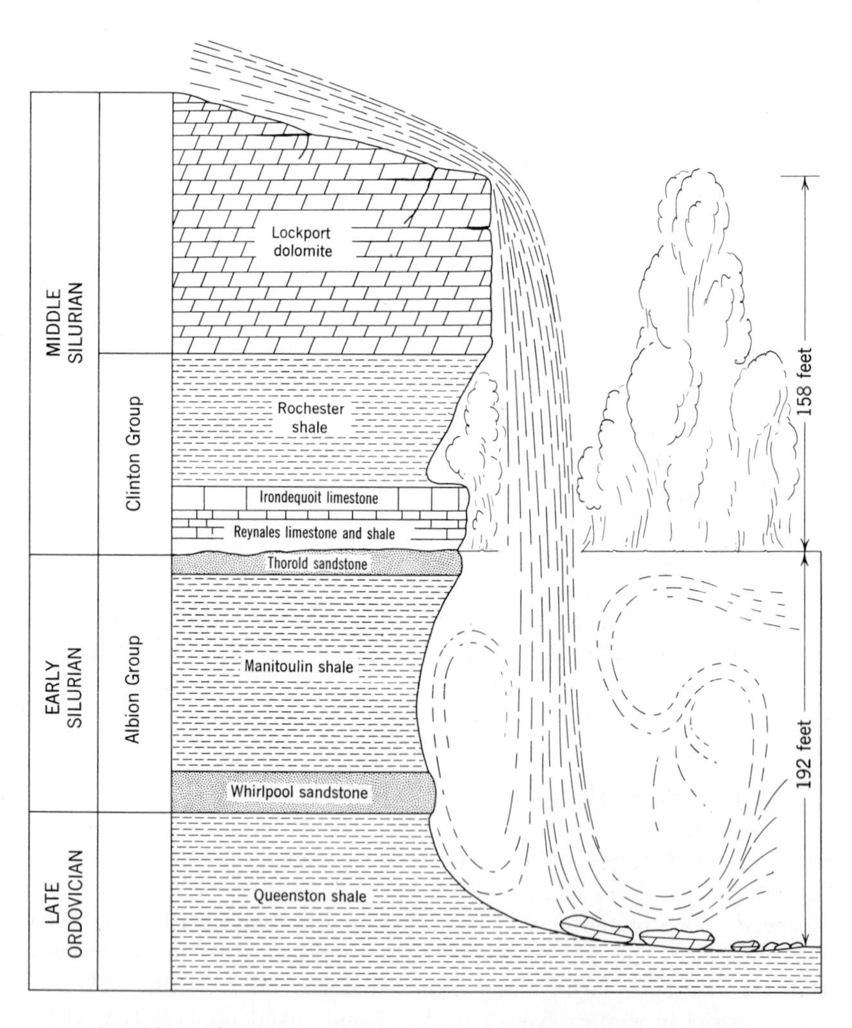

Fig. 101. The section at Niagara Falls. Modified from G. K. Gilbert.

the east were again transported into the Appalachian geosyncline. Some of these materials were derived from the folded and uplifted Cambrian and Ordovician formations. Thus the Albion formations and their equivalents can be traced all the way from Quebec to Alabama, as they grade from coarser sandstones and shales into limestone and dolomites. Conglomerates settled nearer the mountain front, and farther west the fine sandy shales gave way to limestones. West of the Cincinnati dome clear shallow seas spread far across the Mississippi Valley, where dolomites have been mapped in Kansas and Oklahoma. In the limestones west of Cincinnati there are numerous reeflike structures.

One of the best and most accessible sections of Silurian rocks is at Niagara Falls and in the gorge below the falls. There the early Silurian Whirlpool sandstone at the base of the Albion group rests conformably on the late Ordovician Queenston shales. At the site of the falls the Albion rocks lie below the turbulent surface. The water flowing over the Lockport dolomite falls 158 feet past middle Silurian rocks and then plunges 192 feet over early Silurian rocks into the Queenston shales below.

Clearer waters filled the Appalachian geosyncline during the middle Silurian when the highlands in eastern United States were greatly reduced by a long interval of erosion. Much of central United States and Canada was covered by an embayment extending down from the Arctic Ocean, and the Cordilleran geosyncline was again flooded.

At Niagara two groups of middle Silurian rocks are partly concealed by the falls. The lip of the falls is held by the more resistant Lockport dolomite, blocks of which infrequently give away and crash into the whirlpool below with a thunderous roar. The shales and thinner-bedded limestones of the Clinton group below are less durable and tend to undercut the dolomite. Well above the water level of the river below the American Falls Irondequoit limestone forms the floor of the Cave Of The Winds. Lockport dolomite contains a large hingeless brachiopod, a huge clam, and some peculiar gastropods. The Clinton formations are much more productive; great reefs of bryozoans occur in the exposed section below the falls. Elsewhere, corals, together with the remains of other animals, make up extensive reefs or bioherms in the rocks. Corals were some of the most important fossils in the widespread middle Silurian seas.

The late Silurian, as represented by the Salina group, was an interval of aridity in much of northeastern United States. Marine embayments were restricted and even isolated in depressions in which the saline waters evaporated, leaving great thicknesses of salt and gypsum. One of these basins spread over most of Michigan and parts of Ontario, Ohio, Pennsylvania, and New York, an area of about 100,000 square miles. In places wells have been cut through more than 300 feet of solid salt, and mines

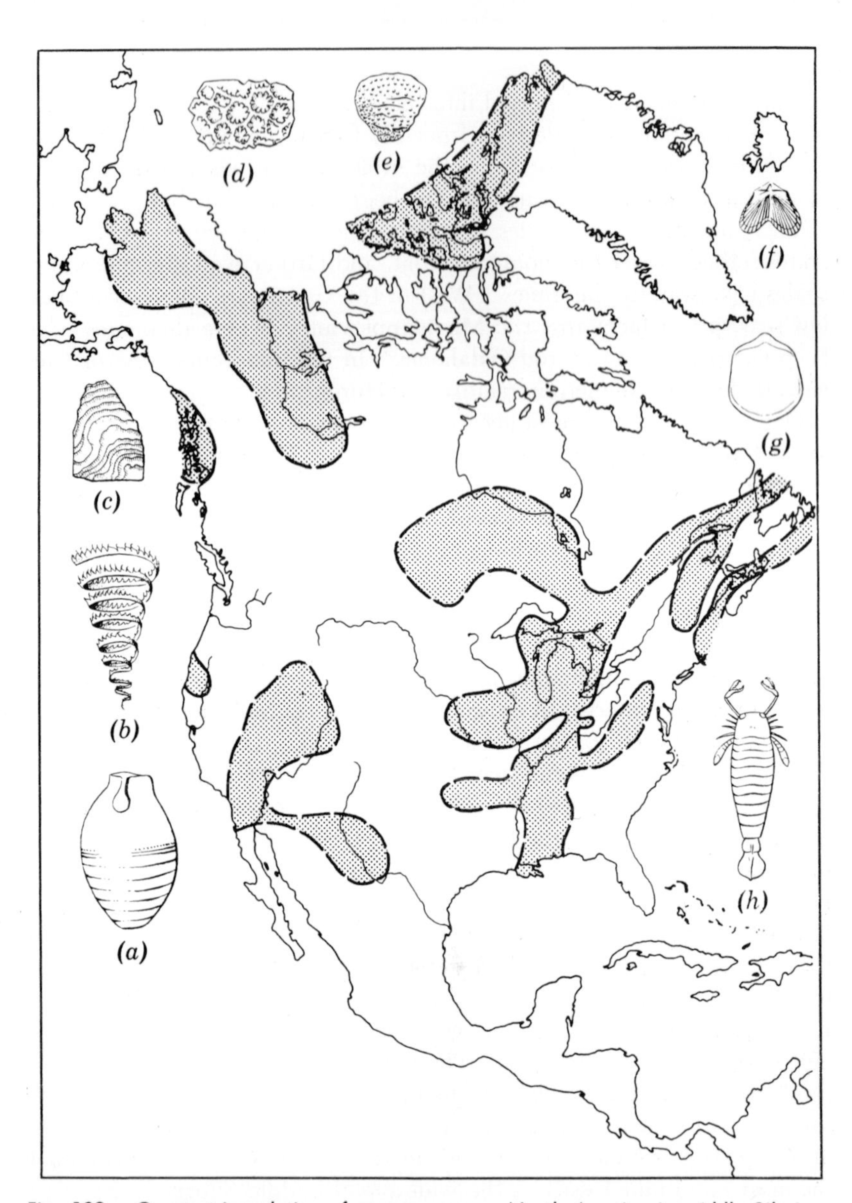

Fig. 102. Greatest inundation of seaways across North America in middle Silurian. After Schuchert. The fossils illustrated come from different parts of the Period: (a) Mandaloceras hawthornense, *cephalopod;* (b) Monograptus turriculatus, *graptolite;* (c) Stromatopora antiqua, *hydrozoan;* (d) Calapoecia canadensis, *honeycomb coral;* (e) Heliolites interstinctus, *tabulate coral;* (f) Bilobites bilobus, *brachiopod;* (g) Pentamerus oblongus, *brachiopod;* (h) Pterygotus buffaloensis, *eurypterid.* (a), (f) *after Shimer and Shrock;* (b) *after Bulman.* (c), (d), (e), (g) *after Grabau and Shimer; and* (h) *after Strømer, courtesy of Geological Society of America and University of Kansas Press.*

have been operated for years from which the salt has been removed in blocks. Along the western edge of the saline basin characteristic dune sands piled up as much as 150 feet. Most of the lower part of the Salina rocks was made up of red and green shales. Some calcareous deposits have yielded the remains of eurypterids and large arthropods, so characteristic of the Silurian, but for the most part fossils are rare from the late Silurian in eastern North America.

The Silurian in western North America is still not well known, though rocks of this age occur in the Cordilleran geosyncline and neighboring troughs. The disclosure of their history is a challenge to the young geologists of this and the next generation.

The Silurian closed without much change in North America, except in the northern part of the Appalachian geosyncline, where slight uplift and some aridity forewarned of the great facies changes of the Devonian. In Europe, on the other hand, the close of the Period marked the initiation of the uplift of the Caledonian Mountains which was to have such a profound effect on life on land and in fresh water.

Animal Life. Eurypterids (chapter heading 17) reached their greatest peak in abundance and diversity in the Silurian and Devonian. It would seem that they were almost antisocial, since their remains are seldom found with the other marine invertebrates. In all probability they were carnivorous, though some of them may have been scavengers. They evidently had an environmental preference to their special liking that was not suitable to most of their contemporaries. Eurypterid remains are found in water lime or calcareous deposits containing much silt.

Some of these creatures can be appropriately called the four-eyed giants of their time, for there are not only two widely spaced, rounded or kidney-shaped compound eyes at the front of the "head" (prosoma), but two tiny ones in the middle of the "head." One of these animals, called *Pterygotus buffaloensis,* from New York, is estimated to have been nine feet long. It was the largest known creature of the Silurian. But not all eurypterids were giants; most species ranged from a few inches to two or three feet in length. Because their remains are usually found in land-locked deposits, apparently near shore, and because these parts are fragmentary it has been postulated that eurypterids lived in fresh-water streams or in brackish waters. They are rarely found with the trilobites, brachiopods, cephalopods, graptolites, or other invertebrates that lived in open-sea environments (Fig. 102).

One of the features that distinguish eurypterids from the crustaceans is a pair of pincers (chelicerae instead of antennae) that stick out in front of the eyes. Five other pairs of appendages are connected to the underside of the prosoma. The body is composed of thirteen segments, with the end one shaped much like a rudder. Most of the eurypterids evidently were rather sluggish bottom-dwellers, but some with paddlelike appendages may have been able swimmers. They made their first appearance in the Ordovician, and the last of them died out with the trilobites in the Permian.

Possibly the *first air-breathing animals* to warm themselves in the sun on land were scorpions. Three genera are known, one of which was found in New York and two in Scotland. Though no gills or spiracles can be seen in these specimens, authorative students think these earliest scorpions breathed air. If this is true, they were among the first animals to have lived at least part of their time on land. They are included in two distinct families which differ from all other families of scorpions. A millepede, also, was found in the late Silurian rocks of Scotland. This thousand-legged worm was also the oldest of its kind to be found. Thus the invertebrate animals seem to have preceded the vertebrates in their invasion of the land, but plants apparently got there first. Otherwise, the animals would have had little or no protection from dehydration. If our record of fresh water or continental deposits were better represented in the first three Periods of the Paleozoic, it is almost certain that we would have a much clearer idea of the ancestry of the arthropods and of the earliest vertebrate animals.

Trilobites had greatly reduced in numbers but still offered important evidence in correlation. On the whole they were showing little more specialization than they had in the Cambrian.

Foraminifers were mostly of the straight- and coiled-shelled agglutinated kinds (see p. 359). At least those with tests (hard shells) had not reached the prominent place they were later to occupy in the fossil record. Little research has been done on Paleozoic radiolarians. Bryozoans were extremely abundant and diversified. There were many reefs in the

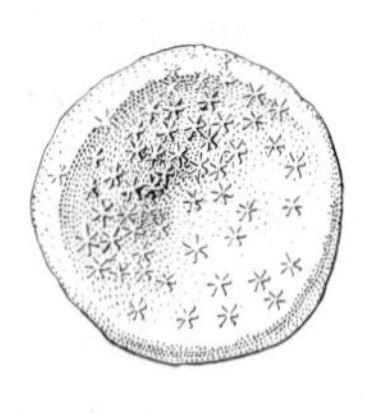

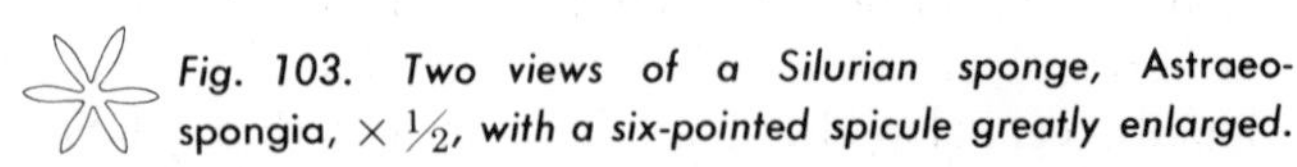

Fig. 103. Two views of a Silurian sponge, Astraeospongia, *× ½, with a six-pointed spicule greatly enlarged.*

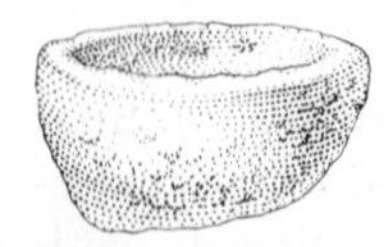

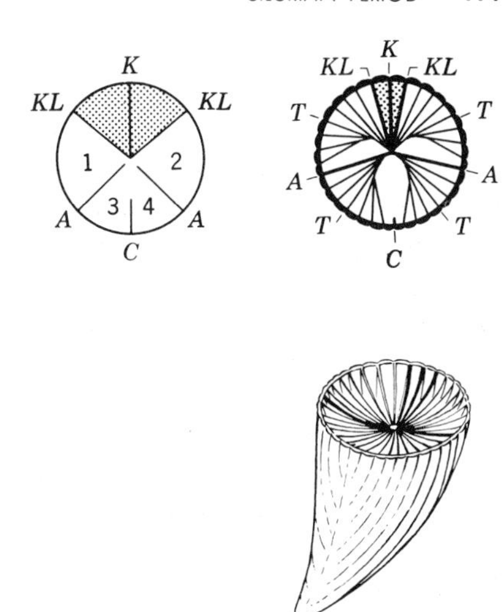

Fig. 104. Ideal sketches with sections at two stages of growth of a rugose, or tetracoral: A, *alar septa;* C, *cardinal septum;* K, *counter septum;* KL, *counter lateral septa;* T, *tertiary septa;* 1, 2, 3, 4, *four divisions of coral. By permission, from* Invertebrate Fossils *by Moore, Lalicker, and Fischer.*

middle Silurian seas in which their lacelike hard parts covered rocks and the skeletal remains of other invertebrates. Sponges were also found in the reefs.

Tetracorals or rugose corals, as they are sometimes called, and the tabulate colonial reef-forming kinds are even more plentiful than the bryozoans. In some cases the reefs are better known as *bioherms* because many other invertebrates lodged among the corals. Reefs of this kind are as much as 75 feet thick and 1 mile long. Many of the coral genera are widely distributed and have been most useful in tracing the connections of epicontinental seaways from one continent to another.

Tetracorals were usually solitary. The skeletons or hard parts assume characteristic shapes in the different genera. Cross sections show that the major septa divide the coral into four parts. The only divisions counted are those in which tertiary septa enter. This is sometimes confusing because the upper or distal part tends to obscure these divisions. Both the honeycomb and the chain corals live in colonies. The hard parts showing the massive polygonal tubes of the honeycomb coral, *Favosites*, have frequently been mistaken for the fossil honeycomb of bees. On the other hand, the tubes of chain corals are joined along their sides in a manner suggestive of the links of a chain. These and still other kinds of corals are characteristic of the Paleozoic seas (Fig. 104).

Brachiopods continued to expand in the Silurian. Some fossil-bearing beds are almost totally composed of their shells. The first of the *spirifers* which played such an important part in later Paleozoic life appeared in

Silurian rocks. Here, too, we find some kinds that were cemented to objects in their feeding positions by long, projecting spines instead of by pedicles as in most brachiopods.

Gastropods and pelecypods are about equal in numbers in Silurian formations. Neither class shows any marked specializations. Two kinds of gastropods are dominant in the Silurian. One is coiled in a vertical spire, and the other is coiled in a single plane. In some formations these mollusks are abundant. The nautiloid cephalopods, both the straight- and coiled-shelled groups, reach the peak of their evolution and start to decline.

Crinoids were the most abundant of the echinoderms. The sea floors in places were obscured by these long-stemmed animals swaying in unison in the disturbed waters. Blastoids were also present in great numbers, but cystoids were on the wane. Other echinoderms such as starfish and echinoids were rare.

Graptolites were still a dominant group, but before the Period closed they started on a decline from which they never recovered.

Vertebrates. As we have seen, marine invertebrate animals dominate the fossil record in the first three Periods of the Paleozoic. This may be more apparent than real. There is no question that the numerous phyla of lower animals were rulers of the oceans and inland seaways at that time, but we have little or no evidence of the kinds of creatures in the fresh-water streams that dissected the mountains and transported sediments to the sea. Possibly the onychophorids and the trilobitelike arthropods in the Cambrian Burgess shales were estuarine or even fresh-water animals that washed into a near-shore deposit where animals of the open seas also died and settled to a muddy grave. Some of the primitive jawless vertebrates could also have had similar living requirements. A situation of this kind is supported further in the late Silurian rocks of Scotland and eastern North America where four orders and ten families of ostracoderms have been recorded. Here, also for the first time, is fragmentary evidence of two orders of placoderm (plate-skin) fish. These acanthodians (spiny fish) (Fig. 219) and arthrodires (joint-necked fish) (Fig. 218) had already acquired lower jaws, a great advance over the ostracoderms (Fig. 217). The lower "jaws" in the placoderms were of different structure and origin from those in most of the later fishes and for this reason are not called true jaws. Frequently associated with these progressive Silurian "fishes" were eurypterids, other smaller arthropods, including the oldest scorpionlike creatures, and millepedes. Some mollusks also could have preferred those environments.

The diversity of these groups clearly indicates a long evolutionary history, probably dating back to the Cambrian or even earlier. Since they were not present in the seas from which we have so much evidence, they must have been in fresh or littoral waters or in both. The most reasonable explanation for not finding them is that deposits of this kind were limited in areal extent and thickness, compared with marine accumulations in the geosynclines and elsewhere. Moreover continental deposits on and along the borders of the land were the first to erode as a result of subsequent uplifts. Then, too, since these rocks are deeply buried and in many places considerably altered, fossils in them are difficult to find. Nevertheless, the answer to these questions is in the rocks somewhere, a reward to the paleontologist who will not be discouraged by days and weeks of fruitless search in his efforts to open new vistas of the past.

Land Plants. It had been assumed with good reason that vascular land plants formed some vegetative cover on land in the Silurian, or earlier, at least in the proximity of water. Indeed, some fossils suggestive of such plants have been found in the late Silurian of North America, and a primitive lycopsidlike plant has been recorded from the middle Cambrian of Siberia.

One of the most stimulating discoveries was made in paleobotany by William Baragwanath, who found parts of fossil plants of this nature in the Silurian rocks of Victoria, Australia, in 1907. Later the remains

Fig. 105. A restoration of the end of a branch of Baragwanathia longifolia *showing needlelike leafy shoots with sporangia at their bases.*

of these plants turned up at four different localities in Victoria. Others have been found in Tasmania and New South Wales. Fortunately, the plants were associated with graptolites of the genus *Monograptus;* frequently both were found on the same slab of rock. Thus the age of the beds has been called middle Silurian on the ever-reliable, wide-ranging graptolites.

Even more startling at that time was the discovery that one of these plants, *Baragwanathia longifolia,* was related more closely to the living club mosses, *Lycopodium,* than to the Psilophytales, the most primitive, structurally, of the vascular land plants. Stems and branches of these lycopsids with thick needlelike leaves, together with the stems of more primitive kinds of plants, evidently drifted out from land before they became water-soaked and sank to the bottom with the dead graptolites. There they were compressed, thin and flat, by the weight of superimposed sediments (Fig. 105).

The preservation is unusually good for such old Paleozoic fossils. Leafy shoots, sometimes with kidney shaped sporangia at the base of the leaves, are shown. Stems with complete leaf bases or leaf scars with or without leaves give much detail of these parts of the plant. Other stems with laterally placed branches show that *Baragwanathia* was dichotomous. Some of the stems clearly display the structure of vascular plants. Here, then, is a member of the subphylum Lycopsida far back in the Silurian which gives us another clue to the possibilities of future discoveries in the early history of land plants.

13 Devonian Period

Eusthenopteron, Protolepidodendron, Asteroxylon

Reign of fishes; origin of amphibians;
first wingless insects; scale tree forests;
brachiopods, corals and bryozoans abundant;
blastoids at their peak; earliest ammonites;
heavy rainfall and aridity.

AFTER MOST OF THEIR FIELD WORK had been done on the Cambrian and Ordovician, Sedgwick and Murchison in 1836 began on the much folded and crushed rocks that underlay the Carboniferous formations in Devonshire and Cornwall. These rocks showed such a marked resemblance to those in the lower part of the section in Wales that they unhesitantly called them Cambrian. It never occurred to them that these undescribed rocks might be equivalent in age to the old Red Sandstone, since excellent sections across Wales and England showed the Old Red Sandstone, of continental origin, to be between the Carboniferous and the Silurian. Difficulties soon arose, however, for when William Lonsdale (1794–1871) studied the fossil corals that had been sent in by local collectors he found them to be intermediate between those in the Silurian and the Carboniferous. He therefore suggested that the graywackes and limestones of South Devonshire occupied the same position in the stratigraphic column as the Old Red Sandstone held elsewhere in Britain.

Indeed Sedgwick and Murchison, who had settled this matter of the age of the Devonshire rocks quite to their own satisfaction, would not at first accept the proposal of Lonsdale. They felt it was unreasonable to assume that these distorted formations with such a marked lithology could be the same age as the Old Red Sandstone across Bristol Bay. Later they took the fossils in hand themselves and discovered to their surprise that the fossils in the lower part of the Devonshire rocks tended to be more like those in the Silurian, whereas those toward the top of the section approached the Carboniferous. The fossil evidence carried the day. The geologists yielded and made these rocks the type of another Period, the Devonian. In many respects this was an unfortunate selection for a type section. Later it was shown that there were less-disturbed beds with much better marine fossils in the Rhine Valley of Germany and even better ones in New York. These have become the standard reference areas for the two continents.

Devonian Record. Rocks of the middle and later parts of this Period give us our first clear insight into life in inland waters and on land. It has been termed the most critical Period in geologic history for the evolution of land-living forms. Though parts of these records come from every continent, the classic one is the Old Red Sandstone of Great Britain, first elucidated in 1841 by Hugh Miller (1802–1856).

One of the finest, if not the most complete, sections of exposed marine formations and faunas of Devonian age in the world outcrops across western New York and Pennsylvania and on to the west. Its description, together with the Silurian section and faunas, was the classic contribution made by James Hall of the New York Geological Survey in 1843.

His field work and that of other early workers was conducted almost simultaneously with that of Sedgwick and Murchison on beds of the same age in Europe. Because of the sound foundation laid by Hall and because of the excellent exposures of Devonian rocks in eastern and central United States the Devonian is perhaps better known than any Paleozoic Period in North America. This research was greatly facilitated by a profound similarity of American and western European faunas and floras, both on land and in the sea.

Though the Devonian of North America was for some time divided into three parts, so much evidence is now available and precise correlations with the European sections so debatable that it has been decided to divide the American Devonian into ten Stages with little effort to recognize three Epochs as has been done so frequently with Periods. Much more work needs to be done on the Cordilleran geosyncline.

Seaways Across North America. North America was very flat at the beginning of the Devonian. There were no lofty mountains to relieve the monotonous landscape. The inland seaways were restricted to the Appalachian and parts of the Cordilleran geosynclines and to the Ouachita trough. The Appalachian geosyncline formed a continuous connection from the northeastern coast of Canada to the Gulf of Mexico, from which an arm extended into the Ouachita as far as Oklahoma. Another branch flooded across the Great Lakes region as far north as the Hudson Bay. Adjacent lands were so low that only fine sediments spread into the geosynclines to produce distinctive formations of limestone and shale. A slight uplift brought in an accumulation of quartz sand, and an almost continuous sheet of limestone blanketed the area except far to the east. Coral reefs were the dominant features in this limestone, some of which have been traced for miles. One of these reefs, of silicified corals crossed the present bed of the Ohio River near Louisville, Kentucky, causing rapids in the river. No other formation has exceeded this Onondaga limestone in the number and variety of corals. Bryozoans and crinoids also played an important part in building the reefs. In the west the Cordilleran geosyncline ran in a rather narrow belt from southern California through Nevada and western Utah north to Canada (Fig. 106).

Toward the middle of the Period an uplift of the land to the east was the source of the first sediments of the Catskill delta that finally closed the Appalachian geosycline from New York north into Canada. But to the south marine waters still opened into the Gulf of Mexico. The Appalachian geosyncline and connecting troughs never again formed a continuous seaway from eastern Canada to the Gulf of Mexico. From this time until the close of the Period one of the greatest shifts of lithofacies from east to west across the Appalachian geosyncline was recorded by the sediments. Conglomerates and red sands gave way to fine gray sand-

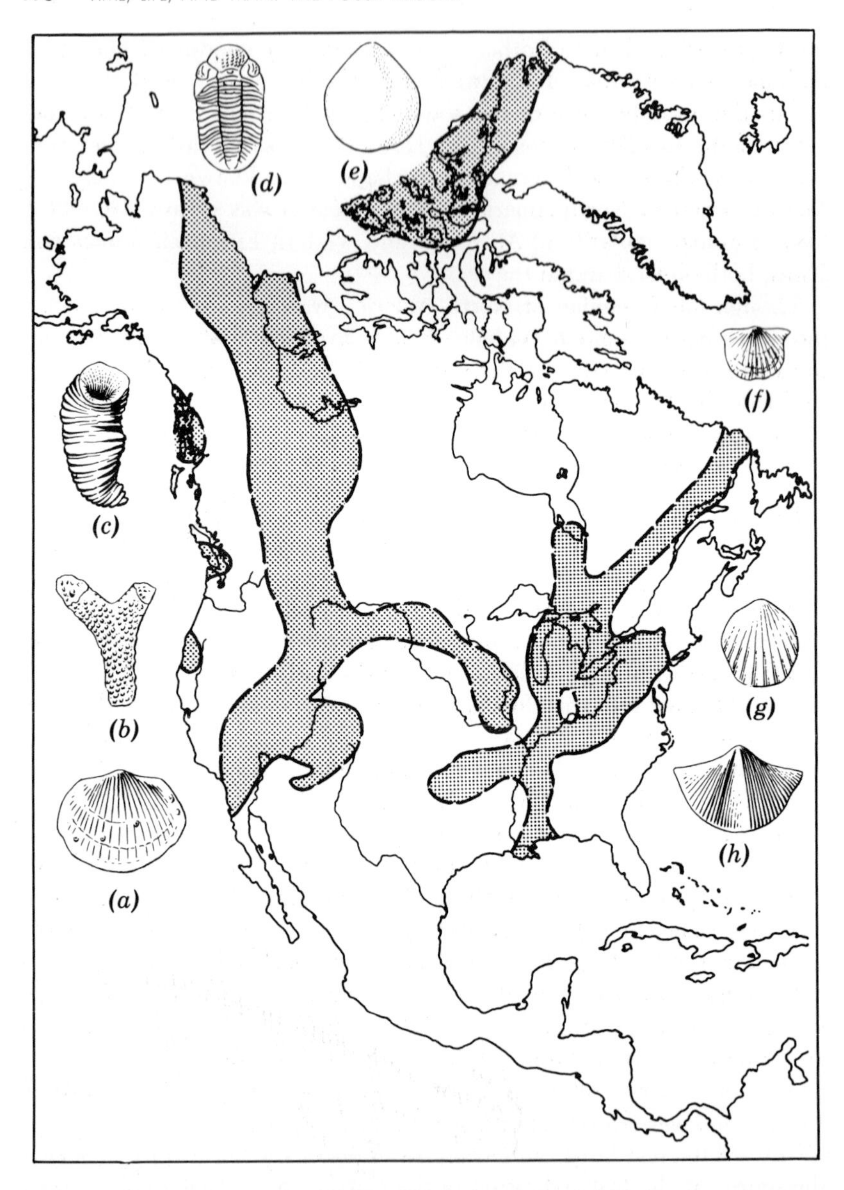

Fig. 106. Greatest inundation of seaways across North America in late Devonian. After Schuchert. The fossils illustrated come from different parts of the Period: (a) Tropidoleptus carinatus, *brachiopod;* (b) Coenites cryptodens, *coral;* (c) Heliophyllum halli, *coral;* (d) Phacops rana, *trilobite;* (e) Cranaena romingeri, *brachiopod;* (f) Stropheodonta erratica, *brachiopod;* (g) Leiorhynchus laura, *brachiopod;* (h) Cyrtospirifer whitneyi, *brachiopod. After Shimer and Shrock.*

stones, then to dark siltstones and black shales, and finally to calcareous shales and limestones with coral reefs and bioherms. Thus the Catskill delta continued to build up and spread out; later the geosycline subsided, and marine waters spread back over parts of the delta only to retreat again before more sediments from the east. This fluctuating condition continued until most of the continent was elevated above sea level. Tree trunks in some localities and mud cracks and fish in others attest to the varied local climatic conditions.

Throughout most of middle Devonian the Cordilleran geosyncline flooded down from the Arctic Ocean in a wide inland sea across Canada. It then branched, one arm spreading over the central states and the other joining the southern embayment in Nevada and California. Other troughs were developed along the Pacific coast. The Sonoran and Mexican troughs also formed to connect the Cordilleran with the Gulf of Mexico. Among the characteristic fossils in the Cordilleran region were the genera of brachiopods *Stringocephalus* and *Spirifer*.

Old Red Sandstone. Two great groups of rocks that arrested the attention of geologists in their early studies of the stratigraphic sequence in Great Britain were the Old Red Sandstone and the New Red Sandstone. Their positions are conspicuous because one is below and the other above the Carboniferous, or coal beds. The Old Red, as it is frequently called, gives us an excellent idea of the animals in inland waters and also the land plants that inhabited contiguous environments. Most of our knowledge of the succession of life on earth in the earlier Paleozoic came from the old marine geosynclines.

At the close of the Silurian the Caledonians, a chain of mountains in western Europe, were folded and elevated high above sea level. The strike of these folds is across Great Britain and Scandinavia, as though a great arc bridged the North Atlantic. This arc, a hypothetical land called Eria, is thought to have connected with folded strata striking in a northeasterly direction from Newfoundland and Nova Scotia. The close relationships of the animals on both continents lend strong support to this hypothesis.

The Old Red sediments were carried down the stream courses from the Caledonian mountains into several intermontane basins. Like the geosynclines, these basins sank as they filled with conglomerates, sands, and muds. Evidently they did not sink to sea level or were barred from the sea by higher land, since there is no evidence of marine waters invading the basins. Deposition in these basins started in the late Silurian and continued until the close of the Devonian.

Red is the conspicuous color, but greenish and gray sands and shales make up much of the section. Careful study of the different formations reveals evidence of varied climates. Heavy boulder conglomerates are indicative of torrential rains that loaded the streams with coarse detritus from the mountain slopes. Coarse sands spread out into the basins where they were laid down in a cross-bedded pattern in the stream channels and at the edges of the deltas, later to be followed by fine-grained sands and silts. At other times long periods of drought are indicated by mud-cracked surfaces when only the deepest parts of the basins contained stagnant water. These climatic conditions are also reflected in the color of the sediments. The red sediments were thoroughly oxidized, but at times of a low ground-water table the gray and green sands and shales show that conditions were reversed.

This, then, was an area of seasonal rainfall and severe droughts that varied considerably from year to year. The droughts never became so severe that the animal and plant life died out, nor did the basins become saline from evaporation. The animal life not only survived, perhaps in restricted areas at times, but it dispersed elsewhere, evolved, and in turn experienced the invasions of progressive creatures from other regions possibly as far away as America and Russia.

The best known of the animals are the fishes. At first the lowly ostracoderms (Fig. 217) continuing from the Silurian were there in abundance. They groveled about on the muddy lake bottoms and were frequently subjected to the attack of the acanthodians (spiny fish) (Fig. 219) and the arthrodires (hinge-necked fish) (Fig. 218). The most interesting of the Old Red fishes are the crossopterygians that possessed lung structures and lobed fins. Rare in the lower Old Red Sandstone, they became abundant and quite diversified before the last Old Red sediments were laid down. One of these groups approached the amphibians so closely that it is considered as ancestral to the terrestrial vertebrates (chapter heading 18). Environments like that of the Old Red, when oxygen in the water was at a premium, evidently forced these fishes to spend part of their time out of water. They then evolved rapidly into air-breathing creatures. This was a momentous event in the evolution of higher vertebrates which marked the first step in the domination of the earth by four-footed animals. The fossils from the Old Red represent an important highlight in the history of life on earth.

Life In The Sea. Life in the Devonian seas reached one of its greatest peaks. This was particularly true of the brachiopods, among which many of the genera and species are Zone or Stage indicators. Some genera spread to two or more continents. Spirifers evolved rapidly and dispersed widely in a short time and are useful tools in correlation. As an example, large brachiopods of the genus *Stringocephalus* have been particularly helpful to the Devonian stratigrapher, since they have told him immediately that he was dealing with late middle Devonian rocks, whether he was working in the arid mountains of Nevada or in the Urals of Russia. North America has been divided into an Atlantic and a Pacific province on the distinctness of the invertebrate faunas. The Pacific forms show their closest relationships to those in Asia, whereas those in the Atlantic disclose an affinity with the western European assemblages.

Corals and bryozoans were almost as abundant as the brachiopods. Never before had coral reefs exceeded those in the middle Silurian Niagaran seas of our central and western states. Wherever the seas were clear, the reefs expanded in colorful profusion as the new colonies grew on the skeletons of their predecessors. Some compound corals measured 7 or 8 feet across. Honeycomb corals lived in the comblike skeletal structures that branched off in different directions like the stems of a cactus. Here under the waters of the inland sea the bryozoans and crinoids helped to form picturesque vistas.

Both crinoids and blastoids were still evolving and spreading into areas where the seas invaded the geosynclines. Starfish and echinoids were still gaining in numbers and diversity, and siliceous sponges occurred in local environments suitable to their existence.

Some trilobites of the genus *Terataspis* were nearly 2 feet 6 inches long. Though they had probably reached their maximum size, they no longer dominated the seas as they had in the past. The genus *Phacops*, with its bulbous beaded head and large crescentic eyes, made a striking appearance (Figs. 107, 108).

Neither pelecypods nor gastropods had reached the point at which they could be considered as abundant, though in certain muddy-bottomed environments pelecypods were common.

The Devonian marked the rapid decline of the nautiloid cephalopods and the spread of the earliest ammonites, a group that was destined to become one of the most important of the invertebrates. These earliest kinds were called goniatites and can be recognized by the simple sutural

Fig. 107. Side view of enrolled specimen of a spectacular Devonian trilobite, Phacops, *in natural size.*

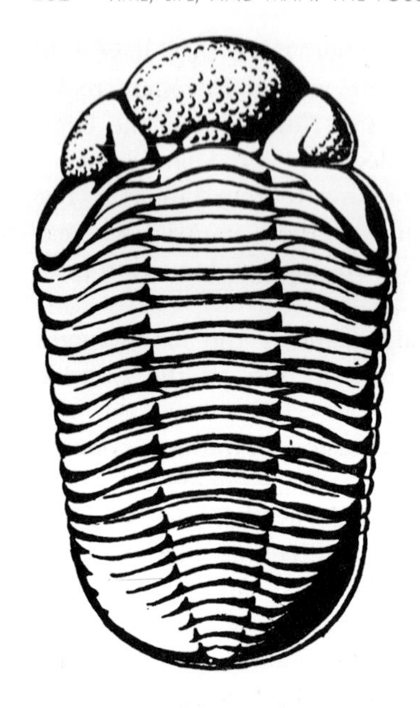

Fig. 108. The Phacops *viewed from above, in natural size.*

lines across the shells. They, of course, were coiled, one compartment (camera) stacked upon another becoming smaller and smaller back to the initial chamber in the center of the whorl. Goniatites evolved from one of the groups of nautiloids.

Among the microfossils ostracods and conodonts were still present in great numbers. Foraminifers and radiolarians, though present, have not been recovered in sufficient numbers to throw much light on their history.

Eurypterids were still the dominant invertebrates in inland-lake and stream deposits. This was particularly true in the Old Red Sandstone of the British Isles. The first insects, apparently referable to the wingless order Collembola (springtails and snow fleas), have been discovered in the Rhynie cherts, middle Devonian, in Scotland.

Reign of Fishes Starts. This is called the age of fishes because they were present in great numbers and because in one group we find the progenitors of terrestrial quadrupeds. Frequenting the fresh-water lakes and streams were the fishlike, jawless ostracoderms (Fig. 217), placoderms (plate-skinned fish) (Fig. 220), sharks, and osteichthyes (bony fishes). The latter include the ancient ancestors of most modern fish, the earliest known lungfish, and the crossopterygians (lobe-finned fishes) one group of which is very close to the amphibians (chapter heading 18). These are the kinds of fishes and fishlike forms from the Old Red and elsewhere which have attracted the attention of scientists for the past 150 years. The ostracoderms were common and had reached the peak in their various specializations. Many had flat heads and a tail designed to keep them on the bottom where their specialized mouth sucked in digestible food from the muddy surface. Other kinds were shaped more like the fast-swimming types of modern streamlined fish and may have relied on living organisms near the surface for food. The Lauge Koch expeditions to Greenland and Spitzbergen discovered such perfectly preserved specimens that Erik A. Stensiö has described the internal anatomy of the head in extraordinary detail. He found that wherever fine silt had seeped into even the most minute cavities and tubes occupied by the blood vessels and nerves in the living animals it was replaced by minerals which made it possible to trace the patterns of the nervous and the circulatory systems. Thus, with a skillful technique of thin sections and painstaking comparisons with other animals, Stensiö worked out a series of plates to reveal these systems and other detailed anatomical features. Stensiö's research established the relationships of the different groups of ostracoderms to living lampreys and hagfish. All ostracoderms died out in the Devonian.

The arthrodires (joint-necked fish) and acanthodians (spine-finned fish) were spectacular animals. They were the predators of the time, and the arthrodires were indeed formidable creatures. Some of the large arthrodires that invaded the seas in the later Devonian reached a length of 30 feet. As a shield they bore two bony armors one over the head and neck

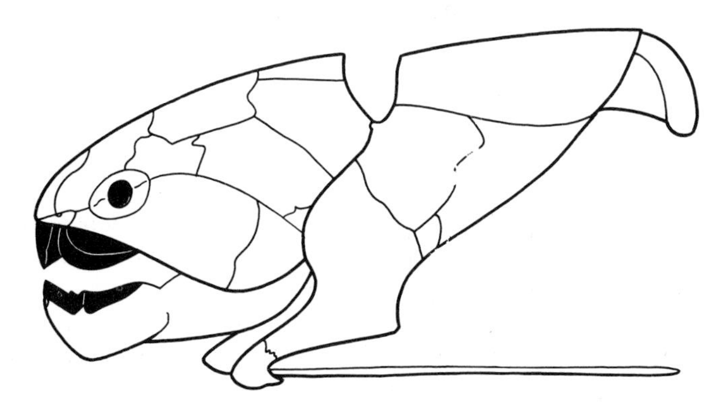

Fig. 109. The head, lower jaw, and neck shield of the arthrodire Dinichthys. *Some of these creatures were 30 feet long.*

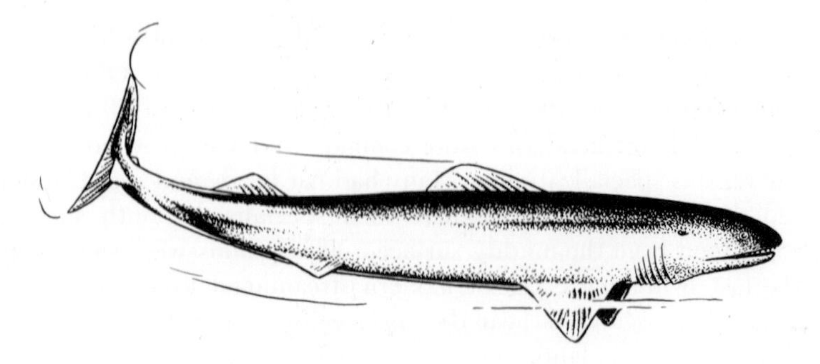

Fig. 110. Cladoselache, a Devonian shark, about 3 feet long. After Bashford Dean.

and the other spreading over the fore part of the body. These protective coverings were connected by a pair of joints, hence the name joint-necked. Armed with sharp bony plates that served as teeth and shears, these creatures must have been the scourge of the seas (Fig. 109).

The acanthodians were about the size of minnows and were less specialized. Since most of the specimens are crushed flat in the shales, their original form has been difficult to determine, but *Climatius* (Fig. 219) from the early Devonian of the Old Red gives a rather clear idea of their appearance. The body was fishlike in outline and protected by small diamond-shaped, bonelike scales. Instead of the normal pectoral and pelvic fins, seen in later fishes, there were seven pairs of ventral fins, an anal fin, and two dorsal fins supported by stiff, sharp spines. In contrast to the heavy armor about the head, as seen in the arthrodires, the head was covered with small plates. The lower jaw was lined with teeth, but there was none in the upper jaw. These little predaceous creatures were abundant in fresh waters in the early Devonian. Some isolated parts indicate that larger forms which evolved from them may have entered marine waters later.

The antiarchs (Fig. 220) were small abundant fresh-water armored fishes that looked much like some of the ostracoderms. Evidently they were bottom-dwelling animals, and many features have shown that they belong with the placoderms. They appeared suddenly in the middle Devonian but died out before the next Period.

True sharks with their modern type jaws and cartilaginous skeletons also came into the fossil record in the middle Devonian where they are found in marine formations. A late Devonian genus *Cladoselache* has been described in detail from numerous specimens found near Cleveland where impressions of the skin and even the muscle fibers have been preserved in the carbonaceous shales (Fig. 110).

Of momentous occasion is the appearance in the middle Devonian of true bony fishes in considerable numbers. This class of fishes, known as Osteichthyes, includes definitely more progressive types quite different from the other Devonian fishes. They had five pairs of gills covered by opercular flaps on either side and just back of the head; and still more important they had lungs or lunglike structures usually in the form of a swim bladder. They had obviously undergone millions of years of evolution somewhere unknown to us, for even at this time they were clearly differentiated into two subclasses, the Actinopterygii (ray-finned fish) and the Choanichthyes (internal nostril fish).

The ray-finned fish as exemplified by *Cheirolepis* were the first rulers in a dynasty of fishes that increased tremendously in numbers and kind and eventually found their way to every body of water on earth. These Devonian fishes were well adapted for swimming in swift currents and in every way were ideally equipped for life in the water. Before the end of the Period they had invaded the sea and were well on their way to becoming the most successful group of all the fishes (Fig. 111).

The Choanichthyes, a name based on the presence of internal nostrils, were destined to give rise to all air-breathing land vertebrates. Even in the earliest species a clear-cut external diagnostic feature was the presence of two dorsal fins instead of one as in the actinopterygians. They were never so abundant as the ray-finned fish but were better adapted to live in the stagnant pools they seemed to prefer. Seasonal rainfall and periodic droughts tended to dry up the water holes therefore the ability to oxygenate their blood by better-developed lungs and, more importantly, to move out of the water on their heavy-lobed fins at least temporarily were adaptations of great importance in their survival.

There are three orders of these fascinating creatures: the lungfish, dipnoans; the fringe-finned ganoids, coelacanths; and the rhipidistians. Lungfish are living today. *Protopterus* occurs in the rivers of Africa;

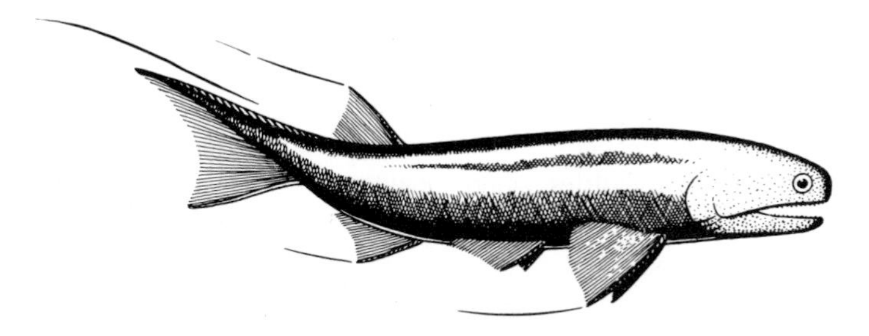

Fig. 111. An early ray-finned fish, Cheirolepis, *about* 11 *inches long. These were near the ancestry of all later bony fish.*

Neoceratodus inhabits the streams and lakes of Australia; and *Lepidosiren* can be found in South America. *Dipterus* is an Old Red Sandstone lungfish very close to the ancestry of all three orders. Until recently the coelacanths were thought to have become extinct over 100 million years ago. Now they have been found in the ocean off the south and east coasts of Africa. These are only cousins of the much more important rhipidistians, but the coelacanths and the rhipidistians are much more closely related to each other than either are to the lungfish. Together they all constitute the order Crossopterygii.

The rhipidistians, typified by genera of wide distribution such as *Osteolepis* and *Eusthenopteron,* show many features much like the oldest amphibians. Three of the more important characters are in the bones in the cranium, the construction of the teeth, and the heavy bones in the lobe-finned paddles. Some bones in the skull have been lost and the pattern of their arrangement is much more like that attained in the amphibians than in the other lunged fish. The pattern of the teeth in cross section shows an intricate labyrinthine infolding of the enamel (Fig. 222) which is also seen on most of the early amphibians. The teeth are spoken of as being labyrinthodont. The third resemblance is seen in the presence of bones in the lobe fin, which correspond with the upper limb bones, and other bones that could readily become reduced and modified into the foot of a terrestrial animal (Fig. 223).

Thus we have seen from evidence in one Period of geological time that much of the ancestry of the fishes is still to be uncovered because many of the groups are well advanced and even specialized. One group has brought us up to the threshold of the terrestrial vertebrates.

First Amphibians. Many years ago Professor Charles E. Beecher found the likeness of a footprint in the late Devonian of Pennsylvania. Much discussion centered around this specimen, which was named *Thinopus antiquus.* Some claimed it was a pseudomorph and not the track of an animal. Be that as it may, there is now no question about the presence of Devonian quadrupeds.

The Lauge Koch expeditions into eastern Greenland found excellent skulls and other skeletal parts of two genera of primitive amphibians, *Ichthyostega* and *Ichthyostegopsis.* There on the slopes of Mount Celsius and in other localities where heavy snows and winter storms now prevail for most of the year late Devonian red and blackish fine-grained sands yielded the remains of amazing animals that are clearly indicative of a mild climate. With these early amphibians were the fossils of groveling antiarchs, joint-necked arthrodires, lungfish, and rhipidistians. Forms closely related to these fish have been found not only in the Old Red,

Fig. 112. Ichthyostega, one of the earliest amphibians, from Greenland.

but in North America, Europe, China, Australia, and of all places, Antarctica. Mild climates must have prevailed almost from the North to the South Pole. But, strange as it may seem, the records of these ichthyostegid amphibians, except for one skull of an even more primitive form, *Epistostege*, from Esuminac Bay in eastern Canada, are restricted to this area far to the north. The transition from fish to amphibians must have taken place in the middle Devonian, since these Greenland forms occur in late Devonian rocks (Fig. 112).

Land Plants. The land plants set the stage for the venture of vertebrates out of the water. As early as middle Silurian at least two groups of vascular plants were already established. Early Devonian records are obscure, but in the middle of the Period psilopsids from the Old Red at Muir of Rhynie in Scotland are well represented. The preservation of these intriguing plants is unusual. They are preserved in a peaty mass in the position in which they grew. The preservation is so good that accurate restorations have been made and even the cell structures have been observed. These are the simplest kinds of land plants (chapter heading 27). Their slender cylindrical branched stems were attached to rhizomes, and they were without true roots and leaves. This kind of plant is suggestive of those in a true intermediate position between some kinds of aquatic algae and land plants.

Most of the ancestors of the plants of the great coal measures were shading the damp ravines and swamps in the middle and late Devonian. Forerunners of the great scale trees (Fig. 208), though not so tall as the giants that followed, towered above the other trees in these primitive forests. Both true ferns and seed ferns (Fig. 208), with their leaves of similar patterns, formed the dense shade nearer the ground, but some tree seed ferns, such as those found in the old Catskill delta near Gilboa, New York, spread their fronds from heights of 40 feet. In certain areas were plants with jointed stems and dichotomously forked leaves. These were the first scouring rushes. On slightly higher and drier ground were slender trees with rather dense wood and no pith in the center. In many respects they displayed a habit intermediate between the seed ferns and the *Cordaites,* sometimes called primitive conifers.

Thus the land plants were well established and widespread through uniform climates over the world. Here was shady protection for vertebrate moisture-loving creatures, if they could adjust themselves to breathing air and to walking on land where they could feed on the numerous arachnids and other invertebrates that had moved into this new realm. Certain rhipidistian fish, such as *Eusthenopteron* (chapter heading 18), met the challenge, and so amphibians soon appeared on land. Essentially, another world had been opened, and through the remainder of geologic time almost every niche was occupied, whether on land, in the air, or back in the water.

14 Mississippian Period

Brachiopods, crinoids

Spread of amphibians; sharks and bony fish common;
extensive forests in lowlands;
insects probably evolved wings; ammonites expanding;
crinoids dominant; spiny brachiopods;
first fusulinid
and first imperforate calcareous foraminifers;
warm climates.

The term "Mississippi group" was introduced by Alexander Winchell in 1869, but it was not until 1891 that Henry Shales Williams of the United States Geological Survey proposed the "series" names Mississippian and Pennsylvanian. These Periods, as they are now called, have not been universally recognized, particularly outside the United States, but their duration and the distinctness of their faunas are becoming more convincing so that they will eventually attain the recognition they so justly deserve.

Carboniferous. Some of the first European rocks to attract attention because of their commercial importance were the coal beds. As early as 1808 the great Belgian geologist Jean Baptiste Julien d'Omalius d'Halloy (1783–1875) called them the Bituminous Terrane. Then in 1822 the Reverend William Daniel Conybeare (1787–1857) and John Phillips used the name Carboniferous to include the coal-bearing beds and the Old Red Sandstone. The type of their Carboniferous was in the Pennine Chain of England. Later Sedgwick and Murchison (1839) recognized the Old Red as the continental equivalent of their marine Devonian. Carboniferous was universally adopted for 69 years before the Mississippian and Pennsylvanian were proposed. The name is still found in some form in all texts and in nearly all of the literature on the subject.

Formations in the Mississippi Valley. In 1891 Williams published a general review of the Paleozoic of North America and checked the correlation of the faunas with those from the European sections. He was greatly impressed with the thickness of limestones, calcareous shales, and a few sandstone formations that intervened between the Devonian rocks and the coal beds, or late Carboniferous. When the faunas of the Midwest were compared it was found that they showed a profound resemblance to the early Carboniferous of eastern United States and Europe. He also found that the time he estimated for the evolution of these faunas was as great as, or greater in some cases, than in other Paleozoic Periods. Consequently, he proposed the name Mississippian and made his type the section of exposures along the Mississippi River in Missouri, Illinois, and Iowa. A fine marine section is exposed at Hannibal, Missouri, where U. S. Highway 50 crosses the Mississippi River.

In most places the lowermost Mississippian rocks rest unconformably on the early Devonian, Silurian, or Ordovician formations. This represents an interval of emergence and erosion between the times of deposition of the sediments. On the other hand, there are localities in which the uppermost Devonian and the lower Mississippian strata form a conformable sequence.

The earliest Mississippian seas were restricted to an embayment that spread north from Mississippi and Alabama across Tennessee, Kentucky, and Ohio into Michigan. An arm of this sea inundated the Ouachita trough. Throughout the remainder of the Period most of central United States was flooded by connections made with the Cordilleran geosyncline, which reached from the Arctic to Central America. The Redwall limestone in the Grand Canyon was laid down in this widespread seaway, a submergence comparable to some of those reached in the Ordovician and Silurian. Characteristic Mississippian limestone is more than 1000 feet thick, and sandstones and shales in the Ouachita trough are as much as 6000 feet thick. These marine formations in North America are divisible into two parts. They correspond with similar divisions of the "Lower Carboniferous" in other parts of the world (Fig. 113).

Appalachian Geosyncline. The filling of the Appalachian geosyncline continued from Devonian time into the Mississippian. Some of these Mississippian sediments were very coarse cobbles, which indicated that land to the east had experienced intermittent rejuvenation. This resulted in deltaic accumulations of silt, sand, and gravel. The sedimentary rocks in the eugeosynclines are known as the Pocono group. To the west they interfingered with marine sediments. Anthracite coal in sufficient thickness to be of commercial value indicates that plants grew in abundance in some of the Mississippian basins. In the later part of the Period marine waters from the west temporarily invaded parts of the eugeosyncline in Pennsylvania and West Virginia but soon retreated or dried up. Mud cracks, raindrop marks, and amphibian trackways have left a clear record of the conditions that prevailed at the time the Mauch Chunk group of formations were laid down on the Pocono. In the late Mississippian of Acadia tree trunks were buried in an erect position.

Close of the Period. There was uplift in several areas in the later Mississippian when the marginal lands were again elevated and folded. An orogeny affecting southern Colorado, the western edges of Oklahoma, Texas, and the northern part of New Mexico slowly pushed up into the old Colorado Mountains. This was simulated in western Europe by the Variscan mountains in southern Wales and England which extended across to the Continent. There were other disturbances in France, Germany, northern Africa, and other parts of the world. The lower Pennsylvanian formations were laid down on the eroded surfaces of these uplifted areas.

Marine Invertebrates. One of the most persuasive arguments for recognition of the Mississippian and Pennsylvanian Periods is the distinctness of the marine invertebrate faunas. This is true not only in North America but elsewhere in the world. The record of marine invertebrates is immeasurably better than that of the vertebrates or plants. The clear waters in the miogeosynclines of the inland seas were well suited for the abundant life that prevailed.

This is appropriately named the *age of crinoids.* Not only are their parts found in nearly every marine limestone formation, but in many places their plates and their segments largely make up the crinoidal limestones. They are more abundant and more diversified in the Mississippian than in any other geologic Period. All three of the crinoid subclasses are present.

Because of their superficial resemblance to plants crinoids (chapter heading 14) are frequently called "sea lilies." None is known to have lived in fresh water. Crinoids are one of the most highly organized and diversified groups of invertebrates. The symmetrical crown, with its body (calyx) surrounded by frilled arms, is attached to a long stalk or stem composed of disklike segments. In some kinds these stems are cemented directly to rocks or other solid objects on the sea floor; others are anchored by radiating rootlike structures. Most of the living crinoids, called feather stars, are free-swimming in the adult stage. The body in crinoids is centrally located and houses the digestive and reproductive systems. The arms bear fringes of small branches (pinnules) on either side which fence in the ambulacral groove on the ventral surface. This groove is floored with tiny cilia which because of their movements direct a current of water with food particles into the mouth. The nervous system, sensitive to touch, particularly in the arms, connects to all parts of the animal. The body and arms are protected by hard parts or ossicles. These are so diagnostic of the genera and species that complete specimens are not necessary for identification.

There is a tremendous range in size in the different kinds of crinoids. Some are only a few millimeters in length, others measure close to 60 feet. They can be visualized as dense, tangled thickets or almost grasslike stubbles on the old Mississippian sea floors.

Another phylum of invertebrates found in abundance in Mississippian seas are the lacy bryozoans. Their fan-shaped or funnel-shaped forms attached to firm bottoms displayed beautiful patterns in the clear, shallow seas. In each of these were colonies of individual animals associated in parallel slender branches which were fastened together at intervals by crossbars. The skeletons of these structures displayed intricate fanciful patterns with two rows of openings along each branch, in contrast to other kinds of bryozoans which have more openings. The individual animals (zooids) had a saclike body and a mouth surrounded by tentacles that circulated water into the mouth.

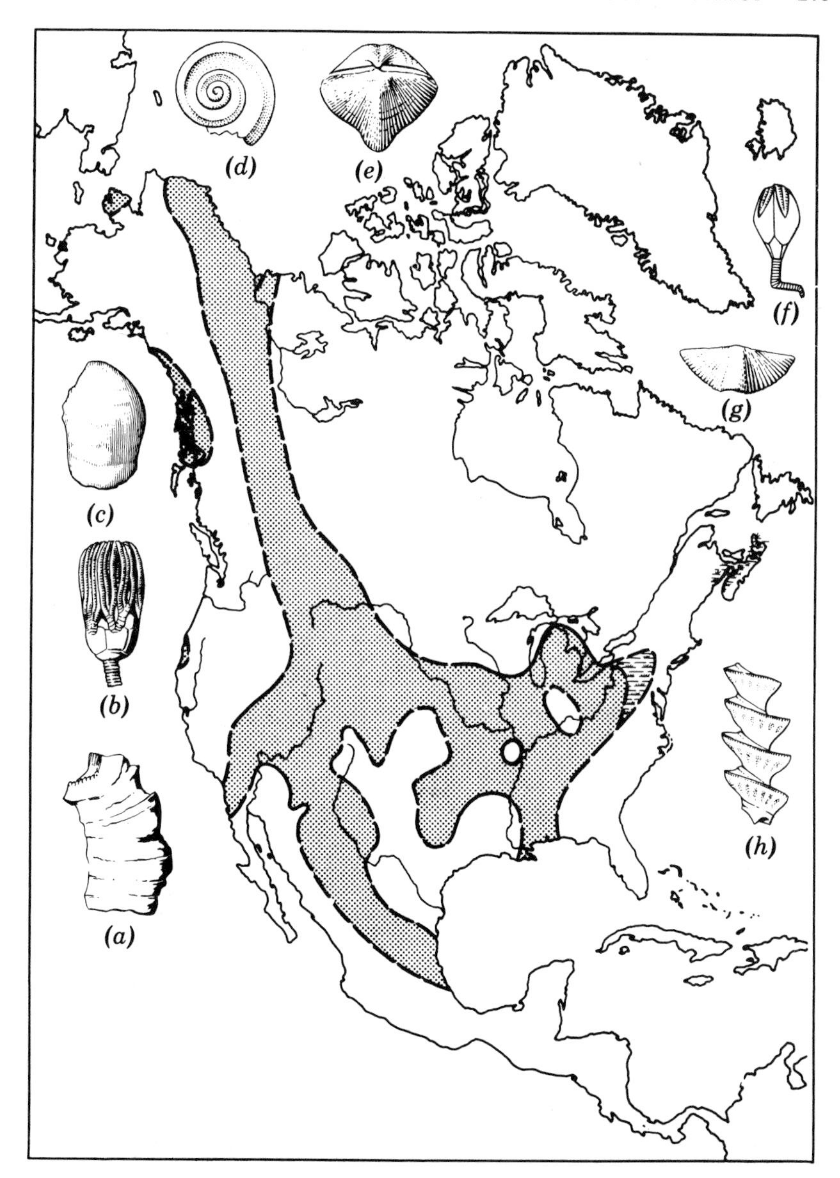

Fig. 113. Greatest inundation of seaways across North America in early Mississippian. After Schuchert. The fossils illustrated come from different parts of the Period: (a) Canina juddi dawsoni, *coral;* (b) Platycrinites symmetricus, *crinoid;* (c) Linoproductus ovatus, *brachiopod;* (d) Euomphalus planidorsatus, *gastropod;* (e) Spirifer grimesi, *brachiopod;* (f) Pentremites pyriformis, *blastoid;* (g) Syringothyris texta, *brachiopod;* (h) Archimedes wortheni, *bryozoan.* (a), (b), (c), (e), (f), (g), (h) *after Shimer and Shrock,* (d) *after Grabau and Shimer.*

There are two methods of reproduction in living bryozoans. After the eggs are fertilized from sperm in the body cavity of the same zooid, they develop and emerge as free-swimming larvae. The larvae may be distributed widely by currents. Finally they become attached to the bottom and then by budding produce another colony.

Though the bryozoans are abundant in Mississippian rocks they are extremely difficult to remove from the limestone because of their delicate structure and the hardness of the stone. Fortunately, the specimens preserved in shales can be weathered out, and sometimes they are found in a satisfactory condition for study. Those in limestone can, in certain samples, be removed by disolving the matrix in acid. Even when the bryozoan skeleton is free from the matrix, its preservation presents another problem because it is so delicate. Identifications are possible from thin sections, if they are properly oriented.

Other conspicuous fossils in the marine faunas of the Mississippian are the productid and spirifer brachiopods. The shells of these invertebrates are distributed throughout the world. Most useful as index fossils are species of the genus *Spirifer*, so named because of the spiral arrangement of the supporting structure (lophophore) of the cilia which direct water currents to the mouth. Though most fossils have their external spines broken off, these spines were conspicuous features in the living animals. Rugae (corrugations) and spines are very characteristic in both the productids and the spirifers. The function of the spines is not clearly understood. They grew out from the outer surface of the shells as hollow tubules. A small depression is made where the spine attaches to the shell, but this does not perforate the shell, as earlier students in paleontology thought when they named the group.

Productids first appeared in the late Devonian but reached their climax in the Mississippian and died out in the Permian. Spirifers ranged from the Ordovician to the Triassic but were dominant in Pennsylvanian and Permian seas.

Among the cephalopods the primitive ammonites assume considerable importance. Their shells show various ornamentations and the sutures display characteristic patterns. For the Mississippian they have become useful in the zonation of formations, and long ranging forms are useful in intercontinental correlations.

Foraminifers, though present, do not play an important role in the marine faunas. The earliest fusulinids (shaped like a grain of wheat) become differentiated from one of the preceding groups earlier in the Mississippian. In certain formations other foraminifers and radiolarians are common, but they have not been studied adequately. A formation near Bedford, Indiana, contains foraminifers, pieces of shells, and grains of oölites in such profusion that the polished surface of the quarried stone has become a choice interior decoration. Other microfossils like conodonts and ostracods are still abundant.

Only four genera of trilobites survived in the Mississippian, when they dwindled almost to the point of extinction. The eurypterids, already greatly reduced in numbers, had spread into the open seas. No good insect faunas have been found, but undoubtedly they were numerous, since they were present in the Devonian and were well diversified in the Pennsylvanian. It seems reasonable to assume that by this time water-living insects by crawling up the stems of plants and gliding back to the water eventually evolved wings.

Fishes. We have seen that some groups of the early fishes did not survive the Devonian, but others carried on and took their places. Sharks dominated the seas. One group developed flat, crushing teeth, and it is assumed that they fed on mollusks and other hard-shelled invertebrates. These sharks first appeared in the late Devonian and died out in the Permian.

The primitive ray-finned fish, called palaeoniscoids, reigned supreme in the fresh waters, both in Europe and North America. Superficially they resembled the minnows of our streams today but differed in their thick, shiny scales and in many other features. Associated with them were the more primitive spine-finned acanthodians. Six genera have been recorded from the Mississippian, and lungfish also are recorded. One rhipidistian genus related to the stock that gave rise to the amphibians was still extant, but this was the sole survivor of a once important group; the more progressive amphibians had taken their place.

Amphibians. Amphibians were well diversified by Mississippian time, since two subclasses can be recognized. Even some of the Devonian ichthyostegids had skulls one foot or more in length and were the largest animals in the stagnant pools at that time.

Fig. 114. Rhadinichthys ornatissimus, a ray-finned fish 10 inches long, from the Mississippian of Scotland. After Traquair.

Unfortunately, amphibians remains from the Mississippian are rare. The best specimens have come from European deposits, where four genera are recorded. These animals evidently spent most of their time in the water, though they were quite able to stagger across land from one pool to another, where they offered the crossopterygian fish keen competition for the available food supply. Fragmentary remains in shales of the Mauch Chunk group have been discovered in West Virginia. These, together with numerous footprints, are indicative of their abundance on the North American continent, and it would seem that sooner or later a deposit suitable to their preservation will be found. A find of this kind could be of the greatest importance, since we are still uncertain whether amphibians as they are now classified came from one or from two groups of crossopterygian fish.

These early amphibians, with their large heads, solidly roofed with thick bone, for which they are sometimes called stegocephalians, show little resemblance to the modern frog, toad, and salamander. For the most part they had heavy bodies and short, stubby legs. When they stopped to rest they dropped down on their fat bellies and wide throats. Though all of them had wide skulls, in some, like the ichthyostegids, the skulls were still rather deep, and some Mississippian and later genera had very flat skulls. They also had a third eye well back on the roof of the skull, but it is doubtful if they could discern more than light and darkness with this organ. Possibly it had another function.

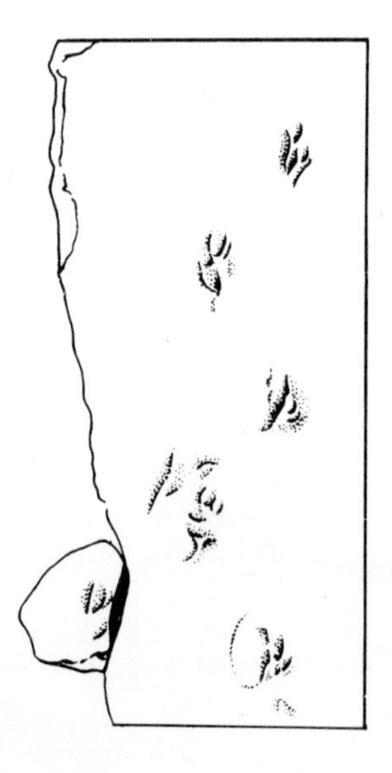

Fig. 115. Amphibian trackway from Mississippian rocks. After Moodie.

In the late Devonian genera the external and internal nares opened close together down on the edge of the skull. In this feature they were not far removed from their fish ancestors. The Mississippian labyrinthodont amphibians had the external nares at the top of the head, and the internal nares were far underneath in the front part of the palatal region. Thus we can see they were better modified for breathing air.

Since the earliest amphibians spent most of their time in the water, their back vertebrae were little modified to function as a strong beam, but in the more advanced kinds the bony parts of the vertebrae became interlocked in different ways to give more rigidity to the body. This offered firm areas of attachment for the pectoral and pelvic limbs and strong support to hold the body off the ground as the creatures ventured across land to the next pool. These different kinds of vertebra simulate the efforts of an architectural engineer in his efforts to devise an efficient structure. Vertebrae have been used by taxonomists in working out a classification of the labyrinthodont amphibians.

Land Plants. The land plants of the Mississippian are so poorly preserved that their details are not well known. Insofar as we know, no new groups of plants came into the picture at this time. Coal beds of economic value, especially in Russia and Europe, offer sufficient evidence for the existance of dense, swampy forests in certain parts of the world. It is interesting to note that the early Mississippian floras showed a marked similarity to those in the Devonian and that plants from late Mississippian rocks were much more like the kinds in the great Pennsylvanian swamps.

15 Pennsylvanian Period

Lepidodendron, Meganeuron, Calamites, Sigillaria

Great coal swamp forests;
wide specializations in amphibians; origin of reptiles;
gigantic insects; scorpions and cockroaches abundant;
fusulinid foraminifera abundant; warm humid climates,
and glaciation at end of period in Southern Hemisphere.

IN THE SAME YEAR that Henry Shaler Williams proposed the name Mississippian for the "Lower Carboniferous," he used Pennsylvanian for the descriptive term "Coal Measures," or Upper Carboniferous of European terminology. The type area was in the Appalachian geosyncline of Pennsylvania where he outlined the section in descending order. Other areas considered were in parts of Pennsylvania and Ohio, West Virginia, and Maryland. In the Mississippi Valley it included the time of deposition of the Coal Measures above the Chester group or late Mississippian. The boundary between the Pennsylvanian and the Permian in many areas in the United States has still not been determined to the satisfaction of all concerned, but there was an interval of uplift and erosion in other parts of North America and in Europe following the Mississippian before sediments of Pennsylvanian age were deposited. Pennsylvanian rocks with their coal beds occur on every continent including Antarctica. There was glaciation at the end of the Period in the Southern Hemisphere. In Africa it extended to the north nearly to the equator.

Appalachian Geosyncline. Most of the Pennsylvanian rocks along the western flanks of the Appalachian Mountains occur at the western border of the old Appalachian eugeosyncline from New York to Alabama. The formations there are neither folded nor in such complex structures as they are farther east where some remnants of the Pennsylvanian can be found. In some folded basins of northeastern Pennsylvania thick beds of anthracite coal are mined. The predominant Pennsylvanian rocks are dark shales and cross-bedded sandstones. Conglomerates are prominent, especially in the lower formations. These materials were carried down by rivers from the east and spread out as deltas to fill swampy areas or were deposited in stream channels. Limestones laid down by infrequent marine invasions are usually thin beds which may extend for long distances. Coal beds, of course, appear frequently in the series, and some units are very thick. The rocks in the Appalachian geosyncline are predominantly continental in origin, as clearly shown by the absence of marine fossils, the abundance of coal and fossil plants, and the nature of the sediments (Fig. 116).

The Pennsylvanian formations in Nova Scotia, New Brunswick, and western Newfoundland were laid down in intermountain basins and are totally continental in origin. These low swampy areas must have been thickly covered with trees throughout the Period. At the Bay of Fundy stumps and trunks were buried in an upright position at twenty different times during the deposition of about 2500 feet of sediments.

Central Interior. The Central Interior was a broad lowland sometimes flooded by marine waters and at times swampy or even dry in places.

This area stretched from the western edge of the Appalachian geosyncline in the east to the middle of the Great Plains in the west and from Michigan and Wisconsin in the north to Oklahoma and Texas in the south. It was so flat that slight warping would elevate shallow seas into land over wide areas; later a depression would again bring in the marine waters from the south and the west. In the Ouachita trough subsidence was much greater. There more than 23,000 feet of sediments were deposited.

Thus alternation of marine and fresh-water deposits has been one of the most interesting features in Pennsylvanian stratigraphy. The cycles can readily be interpreted by reference to columnar sections. Limestones, with their marine fossils, represent deposition of calcium carbonate and fine sediments in quiet marine waters. This may have been followed by slight uplift that brought in more detritus from the land. These sediments settled as silts and eventually hardened into shale. With marine fossils in the base, gray shales giving way to horizons above with no fossils and finally to bits of wood near the top in dark, sandy facies, an elevation of the area was clearly indicated. A layer of coal then showed that land plants had covered the area long enough and in sufficient abundance for great thicknesses of vegetable matter to accumulate. Subsidence was followed by the deposition of more shales until quiet clear waters again prevailed under conditions suitable for the formation of marine limestones. The sequences may be different. For example, marine limestone may appear directly above a coal bed. Thus cyclic fluctuations in local conditions can be readily interpreted by the nature of the sediments and their contained fossils.

Cordilleran Geosyncline. Pennsylvanian formations are found over most of western United States and British Columbia and reach as far north as Alaska. They also extend down through Mexico to Guatemala. Though these are predominantly marine formations, conditions varied considerably in different areas. In central and southwestern Colorado salt and gypsum in the upper part of the section give us evidence of aridity. Coarse detritus, found so frequently in the west, offers evidence of a continuance of a late Mississippian uplift into Pennsylvanian time. One of the most picturesque sequences of Pennsylvanian rocks is exposed by the meandering San Juan River in southeastern Utah. On the whole the climate in western United States tended to be drier and less humid than in the Appalachian province.

Swampland Forest. The soft-tissued trees and seed-fern forests that got their start in the Devonian found climatic conditions well suited to their requirements in the Pennsylvanian. The giant scale trees, *Lepido-*

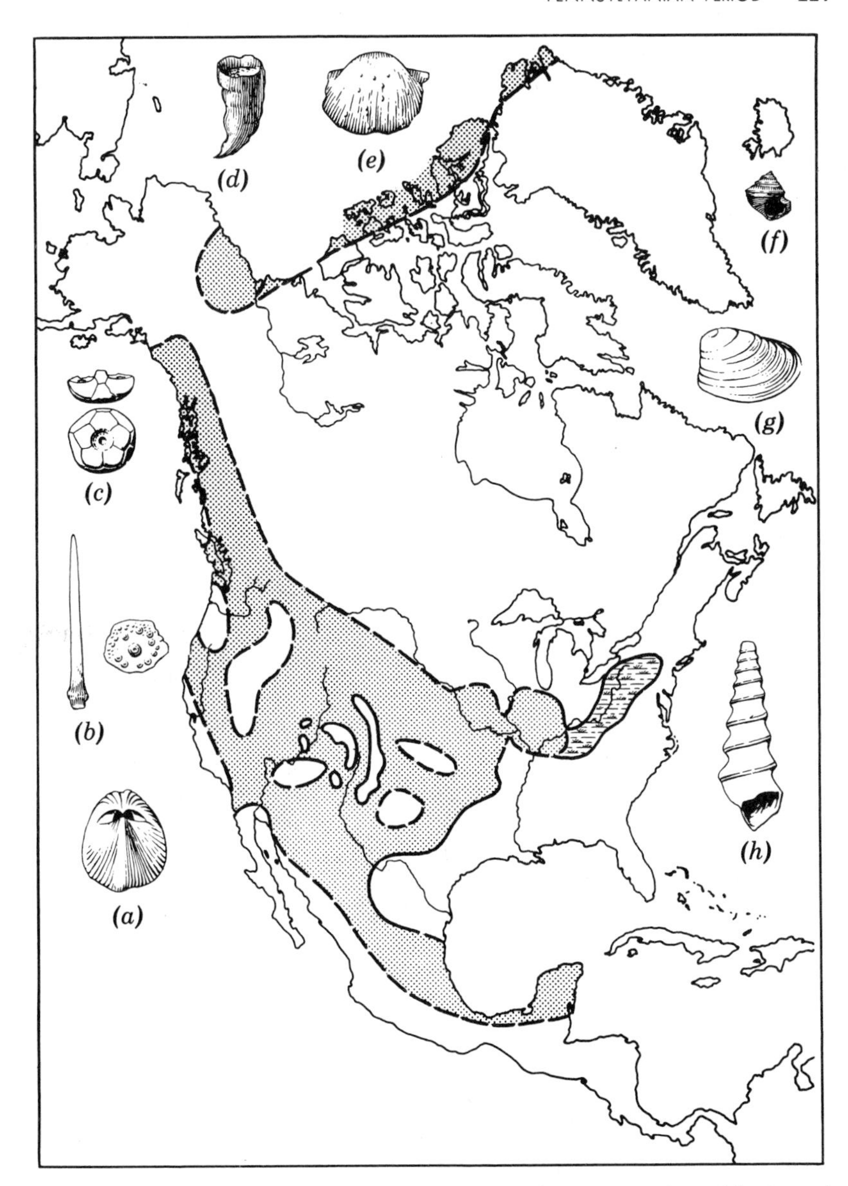

Fig. 116. Greatest inundation of seaways across North America in the middle Pennsylvanian. After Schuchert. The fossils illustrated come from different parts of the Period: (a) Neospirifer texanus, *brachiopod;* (b) Echinocrinus texanus, *spine and plate of crinoid;* (c) Delocrinus nerus, *dorsal and side views of body of crinoid;* (d) Lophophyllidium proliferum, *coral;* (e) Dictyoclostus portlockianus, *brachiopod;* (f) Worthenia speciosa, *gastropod;* (g) Nucula girtyi, *pelecypod;* (h) Goniasma lasallensis, *gastropod. After Grabau and Shimer.*

dendron and *Sigillaria*, towered above everything in the stifling, humid swamp forest. The trunks of some of these trees were 6 feet in diameter, and the trees were as much as 100 feet high. The trunks and limbs had large pithy centers encircled by woody tissue which in turn was covered by bark with leaf scars. As the trees grew and shed their leaves, they left behind permanent leaf scars on the trunks and branches. These scars, from which the name scale tree is derived, look so much like scales that inexperienced folk in uncovering the fossils sometimes insist that they have found the remains of fossil snakes. There were no taproots on these trees for a solid anchor. The roots were relatively short and stubby and spread out nearly horizontally in all directions from the base of the trunk. Small rootlets sprouted out of the main roots into the soft, moist ground. The roots, with their peculiar, pitted rootlet scars, were originally identified as another genus, *Stigmaria*.

Lepidodendron is immediately recognized by its diamond-shaped leaf scars arranged spirally around the trunks and branches. In the higher levels of the forest the branches of *Lepidodendron* spread from numerous forks to give the effect of one of our modern trees that had been topped and had had most of its small branches removed. The leaves resembled giant pine needles nearly 10 inches long and one half inch wide. These spore-bearing plants had conelike spore cases at the ends of the branches (Fig. 117*b*).

In contrast to *Lepidodendron*, *Sigillaria* had a thicker trunk but was not so tall. It had much less tendency to branch, and the leaves were longer and wider. Furthermore, the leaves covered the few sproutlike branches and extended far down on the trunk. Vertically rowed and ovate to round leaf scars were typical of *Sigillaria*. A hundred or more species of these genera have been named, some of which must be synonyms (Fig. 117*a*).

Scouring rushes grew in dense thickets in favorable parts of the swamps. One genus related to the living joint-stemmed *Equisetum* has been called *Calamites*. The tallest ones were about 30 feet high and had a basal trunk 1 foot in diameter. A whorl of slender leaves

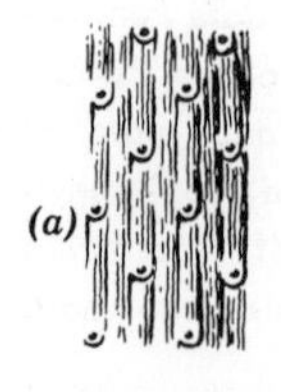

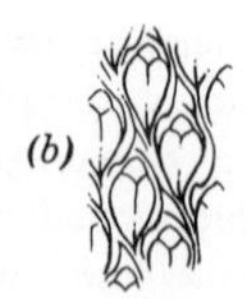

Fig. 117. Leaf-scar patterns on the trunks of scale trees: (a) Sigillaria, × ½; *and* (b) Lepidodendron, × ¼.

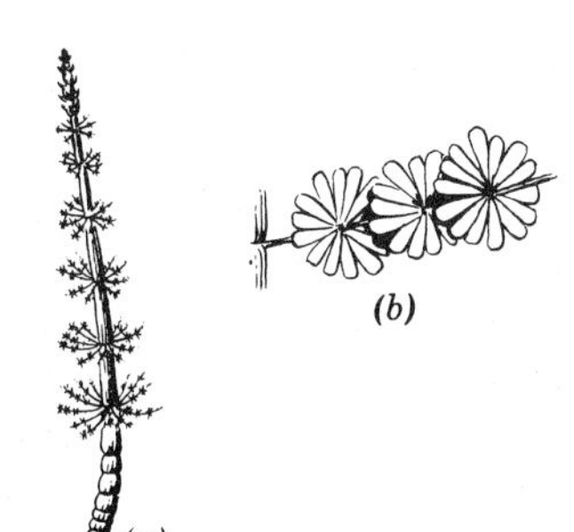

Fig. 118. Scouring rush: (a) Calamites *and its leaf whorls;* (b) Annularia.

radiated out from each joint. Other kinds were much smaller. Unfortunately, in most instances there is little or no association between stems, leaves, and strobili (spore-bearing part) in one plant. The leaf whorls, known as *Annularia*, have been found on a species of *Calamites*. These whorls are very decorative in pattern, since they resemble the petals of flowering plants. Strobili called *Calamostachys* were on the terminal branches of the same *Calamites*. Names like *Annularia* and *Calamostachys* are known as "form genera" in paleobotany because they refer only to certain parts of the plant. The trunks of *Calamites*, like those in the scale trees, contained a pithy center and consequently are found crushed in shales or in coal (Fig. 118).

Seed ferns and true ferns formed most of the dense undercover on the floor of the swamp forest. Their leaves were so much alike that the plants cannot be differentiated unless the fruiting fronds are attached. True ferns, of course, reproduced by spores, but the seed ferns bore nutlike seeds on their fronds. These two kinds of plants may have had a common ancestry somewhere beyond the Devonian.

There is an abundant record of fossil plants from the Pennsylvanian swamps, but little is known of the plants on the higher ground. Plants perhaps tending to bridge this gap were the primitive conifers, known as *Cordaites* (Fig. 209). These were as tall as *Calamites*, the scouring rush, and of equal dimensions. They tend to differ from true conifers in their bladelike leaves which reached lengths of 6 feet, their seeds which were borne in racemes instead of cones, and the pith in the center of the trunk which was much larger. Rudolf Florin has recently described forms showing an evolutionary gradation between the two subclasses. Perhaps other plants more like our modern conifers blanketed the drier ground beyond the edge of the swamps. This is indicated in Kansas, where the smaller leafed conifers *Walchia* and *Lebachia* have been discovered in Pennsylvanian rocks. On the whole Pennsylvanian floras are uniform throughout the world. To fulfill the requirements for this uniformity, there must have been widespread equable environments.

Pennsylvanian Coal. Coal occurs in rocks of all the Periods since the Devonian, but the Pennsylvanian is recognized as the great coal-producing Period of the world. Conditions at that time, with an abundant plant life, were ideal for the accumulation of raw materials. The warm humid swamps overlain by a soggy humus stimulated rapid growth. Most of the plants were large, with soft trunks and a shallow root system. They were easily toppled over by wind storms, but billions of spores and seeds were ready to burst forth with new growths as the old trunks flattened out across each other in the water. Spore cases mixed with yellow and brown spore dust settled in the water and among the fallen plants.

Thus great thicknesses were built up in a relatively short time, only to be covered by another flood of sediments before the organic materials could fully deteriorate. Under these conditions much of the original plant structure, particularly the more resistant parts, was preserved. Deposits like this with partial chemical decay formed *peat*. Compression from the continuous deposition of overlying sediments and heat generated oxygen and hydrogen, which was released as gas as the peat changed to *lignite*. As this process continued, the lignite changed to *bituminous* (soft) *coal* and then to *anthracite* (hard) *coal*. When the heat and pressure became extreme the anthracite metamorphosed into *graphite*, which is pure carbon. Thus the deeper the coal was buried and the greater the compression in the beds due to folding and other processes in mountain building, the higher the carbon ratio became.

Situations similar to that outlined have occurred many times in the same area in Pennsylvanian rocks. This is clearly shown in columnar sections in different parts of the world. Most of the coal in the Illinois and Midwestern fields is not far below the surface. In some areas the coal is dug by surface stripping with heavy equipment. This, of course, is less expensive and not so hazardous as the operation of the deep mines farther east. The famous *anthracite field* in northeastern Pennsylvania, where the coal occurs in a folded structure of the Appalachian mountains, is one of the most interesting and economically important. At one time this field yielded one fourth of the coal mined in North America. The largest is the *Appalachian field* in the western flank of the Appalachian geosyncline under the Alleghany Plateau. As many as sixty coal beds have been recognized in Pennsylvania, but only ten have enough coal to mine.

Petroleum and Natural Gas. The first oil well in North America, called the Dyke Well, was dug in Pennsylvania in 1859. This petroleum was formed in Pennsylvanian rocks from plants that accumulated in the old humid lowland swamp. It might be said that this well started an industry which has done more to revolutionize man's activities and interests than any other. Indeed, it makes it possible for the wheels of every industry to turn. Several billion barrels of petroleum have been taken from Pennsylvanian rocks alone. Natural gas, also derived from organic materials buried in the earth's crust, is used in most cities in the country and can be found on many farms and ranches.

Animal Life in The Swamps. Plant remains are much more abundant in Pennsylvanian formations than the fossils of animals. Nevertheless, the kinds of animals that existed are fairly well known, and the picture would be very inaccurate without the animals that were sheltered by the swamp forest.

Insects appeared for the first time in the record in abundance. The humid swamps with no cold winters and much rain were ideal for their propagation. About 1500 species have been named (Fig. 119).

A most interesting group, Palaeodictyoptera, is close to the ancestry of all orders of insects. These primitive, generalized insects were usually of slender build and had four similar membranous wings with independent movement. It is thought that they had aquatic nymphs. Superficially, the adults resembled earwigs with widespread wings. This order and others that have been proposed but are not yet established with adequate evidence and probably close to the ancestry of the insects in the Permian, in which relationships to the living orders can be recognized. Another order, distantly related to our modern mayflies, made their appearance in these Pennsylvanian forests.

Cockroaches of the order Blattaria were so common that the Pennsylvanian has been called the age of cockroaches. In the insect world they dominated the swamp border around the decaying logs and over the wet ground. Some were three or four inches long, but most were smaller. More than 800 species have been recorded from the Pennsylvanian. Though they were diversified and had experienced much evolution, they were, for the most part, rather conservative and looked not unlike the modern kinds.

Fig. 119. Palaeodictyopteran insect, Stenodictya lobata. *After Handlirsch.*

The rustle of dragonfly wings blended into the familiar sounds of the swamp, as these insects flashed from one resting place to another or held themselves in a stationary position as if suspended by an invisible thread. They were masters of the airways. Nearly 500 kinds have been described, but of the countless numbers that must have existed most will never be found. The giant of them all was *Meganeuron*, with a wingspread of 29 inches. This largest insect of all time turned up in the coal fields of Belgium. Though the average size of Pennsylvanian dragonflies is thought to be about 2 inches, there were gradations from the small primitive order Megasecoptera to the huge order Protodonata. The waters of swamps and streams must have teemed with their predaceous nymphs. No young fish or freshly hatched amphibian could have been safe from these voracious immature dragonflies. Even the eggs of the vertebrates resting in warm waters to hatch were probably greatly depleted by these active foragers.

Delicately constructed creatures like insects are rare in the fossil record. Most of the Pennsylvanian specimens are represented by parts of wings. The most productive locality is in the famous plant-bearing beds at Mazon Creek in Illinois, where they occur in ironstone concretions imbedded in shale. These and other animal remains are usually discovered by fossil plant collectors. It has been estimated that one insect specimen is encountered in every thousand nodules that are broken, and if fossil leaves and sometimes other rare animal remains had not been found in 999 of the other concretions to encourage the collectors few of these insects would have been found.

Other arthropods in the Pennsylvanian include scorpions, centipedes, and spiders. Land-living gastropods have been found where they crawled into hollow stumps which eventually were deeply buried by the muds and sands of the Nova Scotia deposits. Fresh-water pelecypods were common in some formations and ostracods were still plentiful in marine waters.

Fishes. The ray-finned palaeoniscoid fishes, as in the Mississippian, were the most abundant vertebrates in nonmarine waters. They not only fed on other life in the warm waters, but their great numbers of eggs and young offered an abundant food supply for other water-inhabiting creatures. Both the fringe-finned ganoids (coelacanths) and lungfish (dipnoans) were present, but the number of discoveries indicates their relegated position in the environment. Sharks continued to abound in the marine waters.

Amphibians. The most conspicuous animals of the time were the amphibians. Lying in ambush under fern fronds or along the decay-

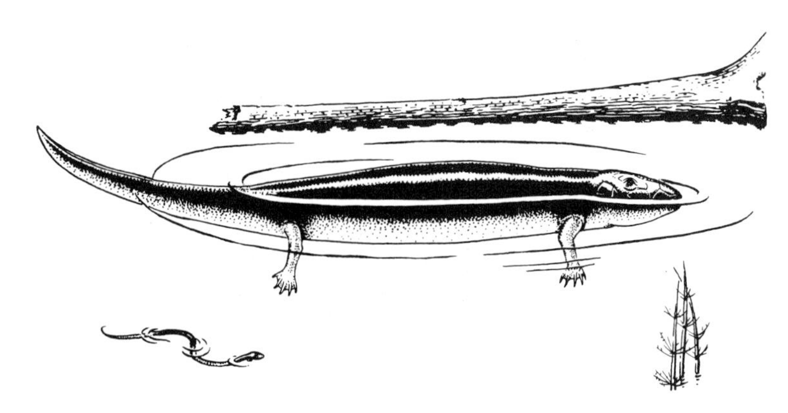

Fig. 120. The gigantic emblomere, Eogyrinus, *about 15 feet long, and a snakelike aistopod,* Ophiderpeton, *about 2.5 feet long.*

ing logs of scale trees were huge *rhachitomes.* For amphibians, they possessed strong legs and were ready and able to move on land as well as in the water in search of food.

The numerous young of these creatures had brachial arches and were formerly thought to represent another order. They evidently remained in the water until they were nearly adults. Other land-frequenting kinds were smaller. The *seymouriamorphs* showed many characters in common with the most primitive reptiles. Indeed the better-known genera, such as *Seymouria* of the Permian, are almost exactly intermediate in their characters between the reptiles and amphibians. Some may have evolved far enough to lay their eggs in warm moist sand instead of in the water, thus affording them better protection.

The giants in the water were the *embolomeres.* Though their limbs were reduced, they had powerful tails to propel them in their pursuit of unlucky fish or other amphibians. Some of these monsters reached a length of 15 feet. Smaller and more agile were the snakelike *aistopods.* These amphibians were extremely specialized in that the limbs were lost. In their length of about 2.5 feet they had as many as 100 vertebrae. They could easily squeeze into places of safety in the submerged vegetable matter. Even smaller were some eellike *nectridians* which had vestigial limbs and flattened tails. The bottom-dwellers in this amphibian world were creatures with flat, flaring wedge-shaped heads, known as *diplocaulids.* The pygmies of them all were angle-wormlike amphibians, unfortunately called *microsaurs.* Their limbs were also greatly reduced, and the skull was solidly roofed over with bone, as in some later reptiles; even the pineal eye was lost. These

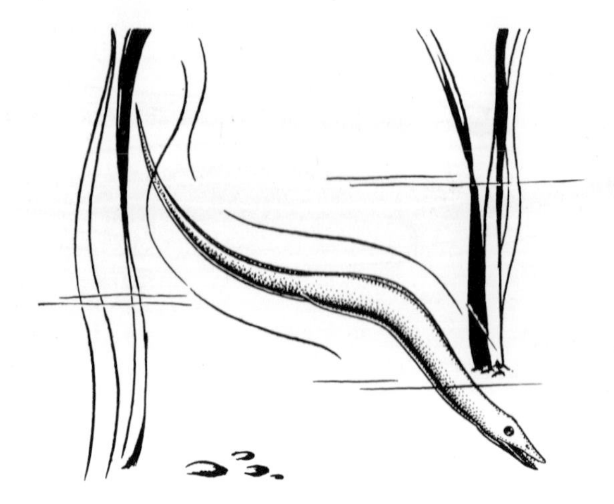

Fig. 121. Sauropleura, an eellike nectridian, about 7 inches long.

features caused the original describer to consider them as reptiles, hence the name Microsauria. In some of the iron concretions at Mazon Creek, Illinois, small amphibians with characters like those found in the heads and vertebrae of frogs and toads were found. Superficially, they looked more like salamanders.

Thus we see a tremendous advancement in the Pennsylvanian amphibians over those in the two preceding Periods. Something like eight orders, twenty-three families, and seventy-two genera from the Pennsylvanian alone were known a few years ago. Many of these were highly specialized (Figs. 120, 121).

Other amphibians have been discovered in the late Pennsylvanian of Texas, where parts of the skeletons of *Diplocaulus* and the large rachitomes related to *Eryops* occur in rocks of the Wichita formations.

First Reptiles. While the amphibians were at their peak the group that was destined to replace them had already appeared. Possibly in the early Pennsylvanian, seymouriamorph amphibians gradually gave rise to certain kinds with more elongate necks, longer and more slender feet, and a skin better adapted to retain moisture in the body. These were the first reptiles. Most of them did not exceed 1 foot in length. They were much more agile than any of the land-frequenting amphibians and must have found an excellent food supply of cockroaches. The land of the earth was theirs to occupy and rule for the next four Periods of geologic time.

Reptile remains are rare in most Pennsylvanian formations, as might be expected. They were just getting started in their tremendous evolutionary history and were few in numbers and variety. Then, too, their habitats probably were on dryer land away from the sluggish borders of the swamps. Infrequently, they did get into these areas or their remains were transported there by streams in flood. Fragmentary remains have been recognized both in Europe and North America. For many years the most complete Pennsylvanian reptile was the posterior part of the body skeleton of a small creature from the famous collecting locality near Linton, Ohio. Unfortunately, the head and petoral girdle were missing. Though poorly preserved, the vertebrae and feet clearly showed its reptilian affinities. This animal is known as *Tuditanus punctulatus* (-synonym *Eosauravus copei*) (Fig. 122).

More recently, parts of more than a dozen individuals representing four genera of reptiles were found in late Pennsylvanian shales near Garnett, Kansas. One of these genera, showing different growth stages, has been called *Petrolacosaurus kansensis*. Since all of the bones in the skeleton except those across the roof of the skull were preserved,

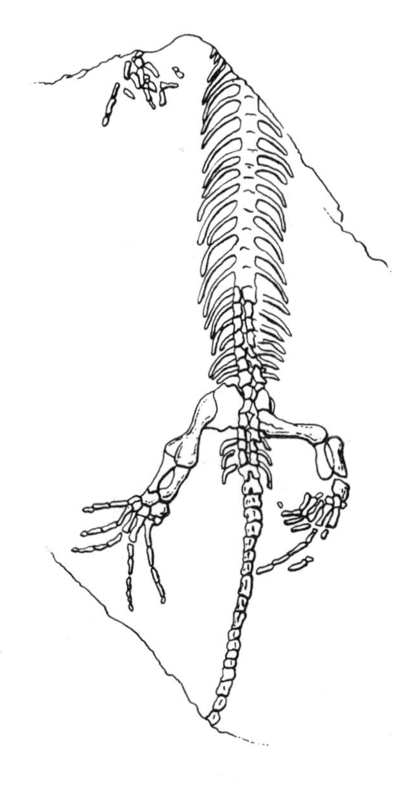

Fig. 122. The hindlegs, ribs, most of the back and tail vertebrae, and part of one front foot preserved of Tuditanus punctulatus, *one of the oldest reptiles known, natural size. From a middle Pennsylvanian concretionary nodule near Linton, Ohio. After Case.*

Frank E. Peabody of the University of California at Los Angeles, then at the University of Kansas, was able to make an excellent restoration of this little creature. He found that it was related to all of the most primitive reptiles, yet retained some characters of the seymouriamorph amphibians.

The nature of the sediments and the associated fauna and flora demonstrate rather clearly that the local area was an embayment or lagoon isolated from the open sea by a sandbar. Intermittently it was flooded by sea waters which brought in cup corals, crinoid stem segments, bryozoans, and brachiopods. At other times inland floods brought in parts of land plants, insects, scorpions, a coelacanth fish, and the remains of other reptiles and buried them with brackish water mollusks. Sometimes the site was a mud flat on which large and small amphibians left their trackways. Peabody pictured *Petrolacosaurus* as an agile terrestrial form that lived among early conifers, *Walchia*, primitive ginkgos, *Dichophyllum*, and seed ferns, *Alethopteris*, *Taeniopteris*, and *Pecopteris*. Accidental deaths in or near a river resulted in their carcasses, together with the remains of cockroaches, small dragonflies, *Parabrodia* (Megasecoptera), scorpions, *Garnettius*, and plants, being carried

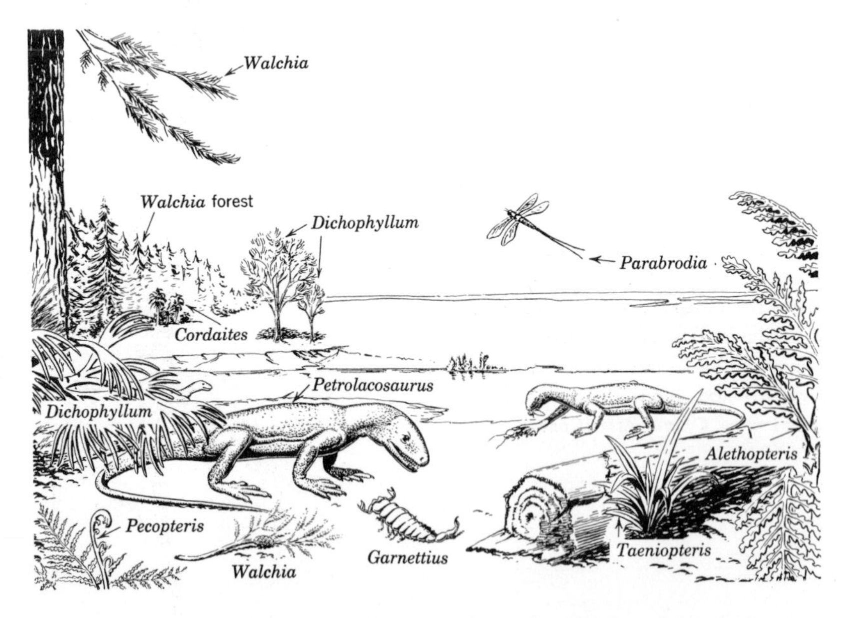

Fig. 123. A Pennsylvanian environment in the mid-continent of North America. Fossils discovered near Garnett, Kansas. Redrawn from Peabody with additions by H. Newcombe.

downstream into the embayment. There the fine sediments, transported at least in part by the stream, fouled the lagoon but simultaneously excluded scavangers and preserved both the mollusks in the lagoon and the organisms washed in by the river (Fig. 123).

This discovery shows an environment quite different from that in the coal swamps. Though *Lepidodendron* and *Cordaites* occur in limited numbers the trees for the most part are much more like those that followed in the Permian.

As with the amphibians, the Wichita formations of Texas have also yielded reptiles that were forerunners of those in the great deltaic faunas of the Permian.

Marine Invertebrates. Marine invertebrates continued their important role in the life of the marine waters. Some groups were expanding while others were declining.

Corals were abundant in certain areas, but most were the solitary kinds. Compound coral reefs were limited. Bryozoans were still abundant in most of the seaways. The productid brachiopods, descended from their Mississippian ancestors, were as important in intercontinental correlation as they had been previously.

Crinoids and blastoids were common but not so well preserved as in the preceding Period. Blastoids died out in North America in the early Pennsylvanian but survived until Permian in the Timor area of the East Indies. Nautiloids were still on the decline, but the goniatites were increasing in numbers and kinds. Ostracods were still common, but trilobites were reduced to two genera.

Pennsylvanian rocks are speckled with benthonic fusuline foraminifers. Some rocks are made up almost exclusively of these bottom-living organisms. The tests or shells are an elongate lense shape, somewhat resembling a grain of wheat. The internal structure is coiled planispirally around a longitudinal central axis. Diagnostic characters are seen in sections of the specimens exposing the internal chambers. Two major groups of these important foraminifers are recognized. The *fusulinids* have four layers in the wall structure, whereas the *schwagerinids* are made up of two layers. These foraminifers which evolved so rapidly are distributed throughout the world and are useful as Zone indicators. The oldest known fusulinids appear in the late Mississippian and after a rapid evolution died out in the late Permian.

16 Permian Period

Dimetrodon

Great diversification in reptiles;
first mammallike reptiles;
glossopteris plants in Southern Hemisphere;
extinction of many groups of marine invertebrates;
development of metamorphosis in insects;
last trilobites and tetracorals;
cold, dry, and moist climates.

After Sedgwick and Murchison had finished their work on the Cambrian, Silurian, and Devonian and were still on friendly terms, they went abroad together to see if the sequence they had worked out in England was represented on the Continent. The rocks they found seemed to conform with those in England and Wales but were more disturbed. At this time Murchison was told about extensive Paleozoic exposures in Russia. Eager to confirm his results in England in sections in other parts of the world, he went to Russia, at the invitation of the Czar, accompanied by a Russian and a French geologist. In contrast to the situation in Britain, Murchison found that the marine and continental rocks above the Carboniferous were thick and richly fossiliferous. In 1841 he proposed the name *Permian* and typified it with the sequence in the province of Perm on the western flank of the Ural Mountains. This he referred to as the youngest part of the Paleozoic.

As early as 1808 d'Omalius d'Halloy had called beds of the same age, and others in Europe somewhat younger, *Terrane Peneen,* with the intention of directing attention to the paucity of fossils in the rocks. The Magnesian limestone and Red conglomerates above the Coal Measures in England were also included in the Permian. Though the Permian was not at first recognized in America, a thick section was soon established in India and later in South Africa and elsewhere in other parts of the world. In Australia the rocks were called Permo-Carboniferous for a time because the marine faunas showed both Permian and Carboniferous affinities.

Epeirogeny and Mountain Building. The Permian was a Period of profound changes in the earth's crust on every continent. These were epeirogenetic movements that elevated the continents into greater relief and orogenic disturbances that folded and faulted the old geosynclines into mountains. Among the mountains formed at this time were the Appalachians in eastern United States, the Ouachitas in Arkansas and Oklahoma, the Variscan chain across England, northern France, and Germany, the Urals in Russia, and other disturbances in India, Australia, and South America.

Ocean currents and atmospheric conditions were altered by these changes in the configuration of the continents. This resulted in most provinces in extreme, dry, cool to cold climates that followed late Pennsylvanian glaciation in the Southern Hemisphere. The temperate to warm climates which had prevailed for such a long time in middle latitudes gave way to colder conditions that had a profound effect on animals and plants. Thick beds of coal in China and Australia show that there must have been heavy rainfall in some areas in the late Permian.

Some animal and plant groups died out; others continued to evolve at a moderate rate but shifted to more suitable climates, whereas the more adaptable ones changed rapidly to fit into the new environments. It was a critical Period in the history of life.

Inland Seas. Epeirogenic movements in the earth's crust forced the inland seas out of most of the old geosynclinal areas they had occupied during most of the Paleozoic. This was particularly true in North America and Europe. In the early Permian, though, a trough extended up through Mexico into Arizona, New Mexico, and Texas, thence to the north across Oklahoma, Kansas, and Nebraska. A temporary inundation spread as far east as Ohio. These early Permian sediments in mid-continent were laid down conformably on late Pennsylvanian rocks. The fossils and in certain places the lithology have been used to locate the boundary between the two Periods. Even in the early Permian local areas were isolated when the water evaporated, leaving a saline residue. These areas were repeatedly flooded and dehydrated. This resulted in thick concentration of rock salt and gypsum, but as the Period continued there was a general retreat of the sea to the southwest. Early Permian formations occur also in northern California and Alaska (Fig. 124).

In the middle Permian the waters in the southwest made connections with the northwest across Utah and Nevada into California and by another trough across Montana and Idaho into Alaska. Some of these formations, the Hermit shales, the Coconino sandstone, and the Kaibab limestone, form the upper rim of the Grand Canyon. In the late Permian the inland waters were restricted to the north in New Mexico, overlapping the eastern border of Arizona; these waters extended from Yucatan up through Mexico. The Carlsbad Caverns in New Mexico were formed probably in late Pliocene and Pleistocene time by limestone being carried away in solution from one of these late Permian formations.

The marine formations in Europe are for the most part disturbed or metamorphosed and are not so fossiliferous as those in other regions. The type area in Russia is much better represented; nearly flat formations with an abundance of fossils occur there. This is an excellent type section, since both marine and continental formations are represented. One of the best Permian marine sections in the world is in the Salt Range of India, and perhaps the greatest evolution in the Permian marine invertebrates is shown in the faunas of India and the East Indies.

One of the finest collecting areas in the world for Permian marine fossils is in the strongly folded and faulted formations on the island of

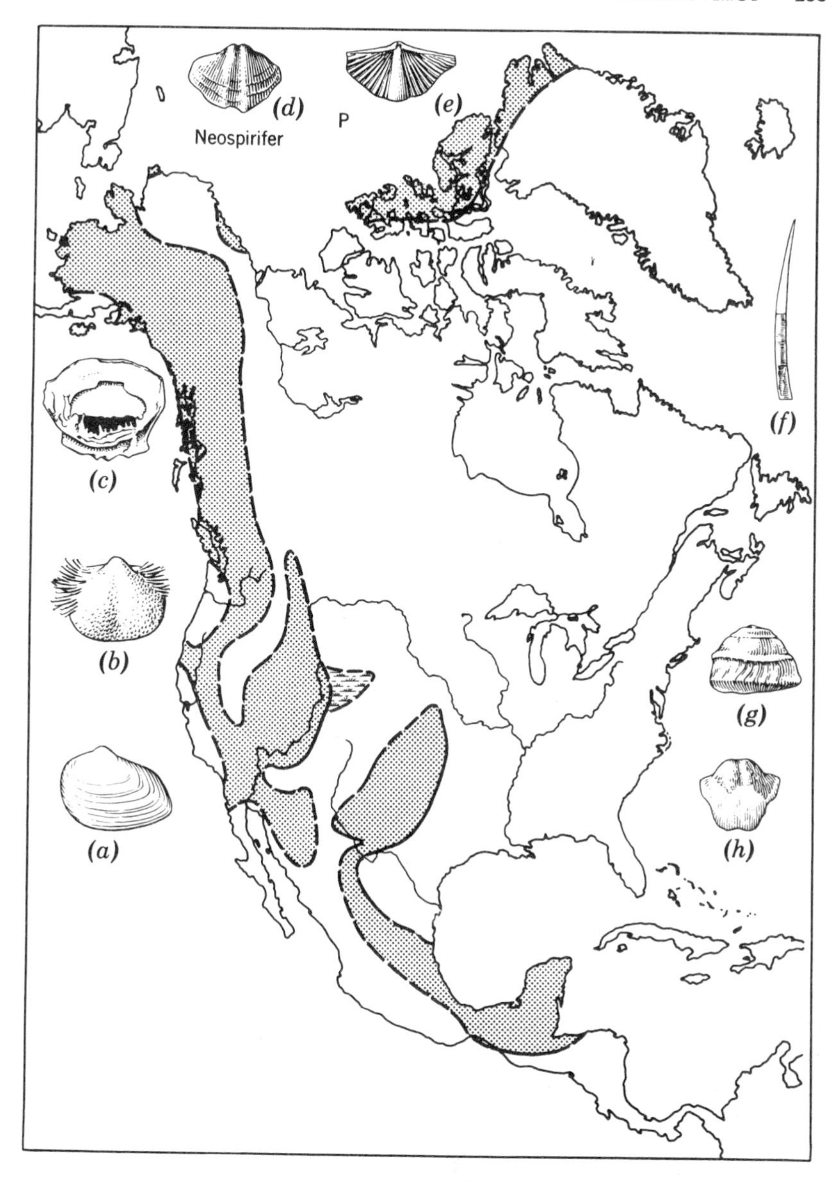

Fig. 124. Greatest inundation of seaways across North America in the middle Permian. After Schuchert. The fossils illustrated come from different parts of the Period: (a) Schizodus wheeleri, *pelecypod;* (b) Waagenoconcha montpelierensis, *brachiopod;* (c) Prorichthofenia uddeni, *brachiopod;* (d) Neospirifer condor, *brachiopod;* (e) Punctospirifer pulcher, *brachiopod;* (f) Dentalium canna, *scaphopod;* (g) Omphalotrochus ferrieri, *gastropod;* (h) Dictyoclostus bassi, *brachiopod. After Grabau and Shimer.*

Timor, which lies east of Java and north of Australia. In some localities the limestones are made up almost exclusively of fusulinid foramifers. The abundance of other invertebrates is indicated by more than 320 kinds of crinoids and at least fifty species of blastoids. Among the other invertebrates ammonites and nautiloids are common. These interesting fossils find their nearest relationships with genera and species in rocks of the old Tethyan geosyncline which stretched across southern Europe to India. Of even greater interest is a marked resemblance in the Timor faunas to those in the early Permian in Texas. The full significance of this distribution pattern is not yet understood. In Western Australia an early Permian marine invasion brought with it a large fauna of more than 100 species of invertebrates. The ammonites there display affinities with those in Timor.

Marine Invertebrates. Only slight differences were seen in the latest Pennsylvanian and the earliest Permian faunas, but before the Permian closed several Paleozoic groups were extinct and others greatly reduced. Invertebrates were abundant in the mid-western and southwestern parts of the United States as long as conditions were favorable.

The larger fusulinid foraminifers which experienced such a rapid evolution and wide dispersal in the Pennsylvanian were still abundant in Permian rocks but none survived the Period.

Tetracorals and honeycomb corals, so important in the earliest Paleozoic Periods, were still extant in the open seaways, but evidently they could not tolerate the changes in climate. Other groups that died out were the blastoids, some groups of crinoids, and some bryozoans. The trilobites that strated with such a flourish in the Cambrian had dwindled to about four genera in the Permian. These were the last of one of the most spectacular groups of animals in the fossil record.

Among the cephalopods, the ammonites with goniatite sutures were abundant. During the Permian these goniatites developed the first complicated sutures, so intricately displayed in their descendents in the Mesozoic Era. The wide-ranging ammonites have been most useful in correlation, but except for a few rare genera the goniatites died out at the end of the Paleozoic.

Appalachian Piedmont Delta. Sediments from the east spread out in a great alluvial flood plain over the old Appalachian geosyncline. This group of rocks is known as the *Dunkard*. In some places the red and gray shales and sandstones reach a thickness of 1200 feet. They differ from similar strata in that they were deposited under drier conditions. The beginning of uplift and pressure from the east gave the entire area better drainage. Locally, plant remains were abundant and accumulated in some basins in sufficient depth to form thick seams of coal. In the later part of the Period the old Appalachian geosyncline was folded into a chain of mountains that permanently eliminated it as an inland seaway. Like the pages in a crumpled manuscript, the eroded, folded, and distorted formations of the Appalachian Mountains contain the story of the Paleozoic in eastern North America.

Texas and New Mexico Red Beds. Sediments derived from the old Colorado mountains and elsewhere eventually filled the mid-continent sea and transformed the area into a desert with saline basins. Farther south the wide, flat, deltaic floodplains offered an excellent environment for terrestrial vertebrates. The climate was moderate, with enough rainfall to fill the ponds and supply the streams. Though there were times of rather severe drought, there was sufficient moisture at all times to grow ferns, seed ferns, and equisetales near the water and conifers on the dryer ground.

Reptiles had evolved rapidly since the Pennsylvanian and in the early Permian dominated the deltas of the southwest. Most abundant were the cotylosaurs (Fig. 224), primitive reptiles, with solid-roofed skulls (anapsids), that had already become diversified into carnivo-

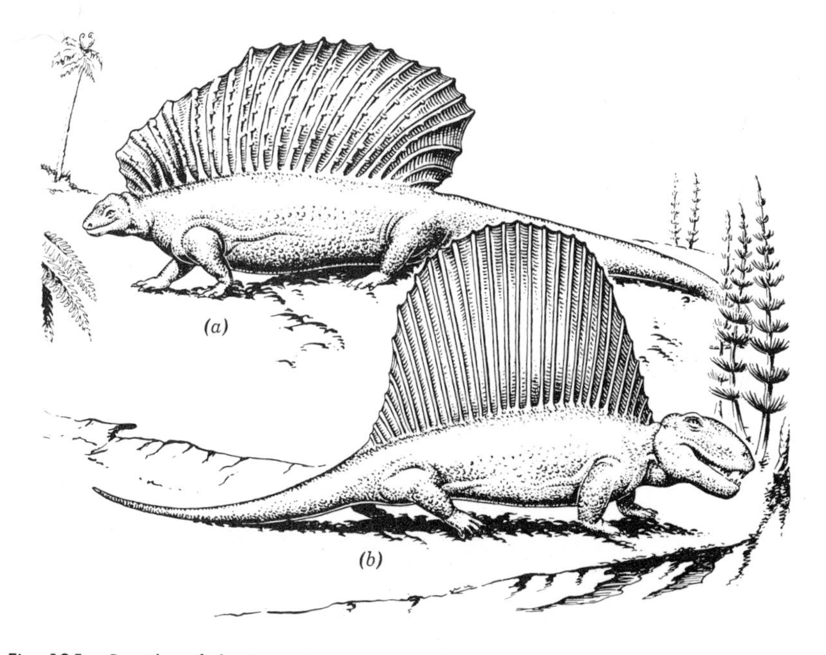

Fig. 125. Reptiles of the Texas Permian: (a) Edaphosaurus, *a plant eater, and* (b) Dimetrodon, *a flesh eater. Length about 9 feet.*

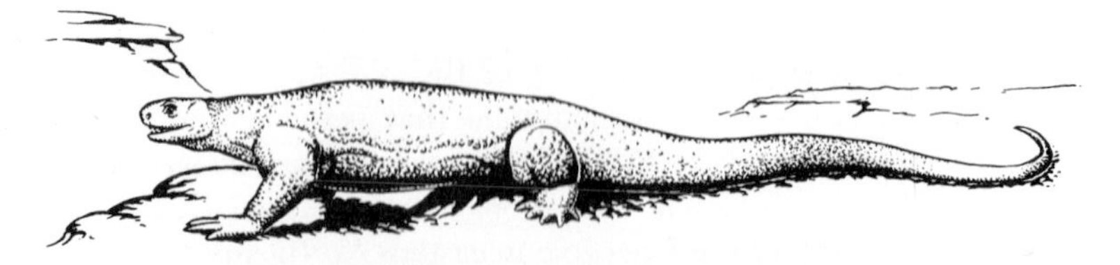

Fig. 126. The elongate-bodied Cásea, a plant-eating reptile, was about 4 feet long.

rous, insectivorous, and herbivorous animals. Equally numerous were the pelycosaurs which had an opening on the lower side of the skull back of the eye (synapsids). These reptiles were even more specialized in their different genera than the cotylosaurs. They included the amazing flesh-eating sail reptiles, *Dimetrodon,* and the plant-eating ship reptile, *Edaphosaurus.* Both had high spines on the vertebrae of their backs which may have supported saillike membranes. In *Dimetrodon,* the dorsal spines were smooth and nearly straight, whereas in *Edaphosaurus* the spines were decorated with peculiar short crossbars. Both the colytosaurs and pelycosaurs with their strong legs, were dry-land and lake-border inhabitants. One of the bulkiest of the early Permian reptiles was the barrel-bodied *Cotylorhynchus,* a pelycosaur estimated to have weighed almost 700 pounds. It was closely related to the much smaller elongate-bodied *Casea* (Figs. 125, 126).

At the lake margins and in the streams were giant labyrinthodont amphibians called *Eryops* and the animals that were intermediate in their characters between amphibians and reptiles known as *Seymouria. Eryops* was a formidable creature quite able to compete with the reptiles around him. Other smaller amphibians, such as *Cacops,* had heavy, bony armor on their backs. These armored creatures may have wandered farther away from the lakes and ponds at night than the other amphibians (Figs. 127, 128).

Fig. 127. The rhachitome amphibian, Eryops. *Huge labyrinthodonts were 6 feet long.*

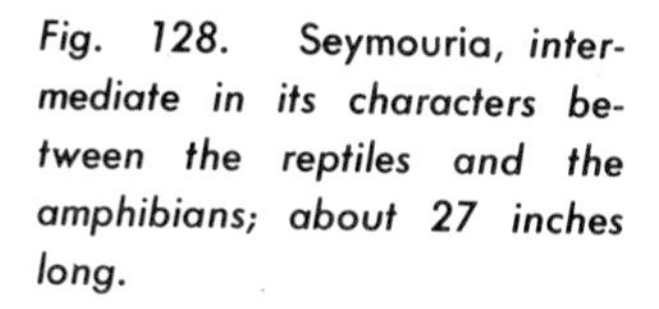

Fig. 128. Seymouria, intermediate in its characters between the reptiles and the amphibians; about 27 inches long.

Predaceous sharks and lungfish also lived in the streams and lakes, and it is thought that small palaeoniscid fish of minnow size must have been present. Spectacular bottom-dwellers were the wedge-headed amphibians known as *Diplocaulus*. These highly specialized forms had flat heads with long, flaring posterior corners. Fresh-water invertebrates and insect larvae must have supplied an important part of the food of this animal world in the early and middle Permian of Texas and New Mexico. These deltaic red beds from the Clear Fork group have yielded the best early Permian record of fresh-water and land-living animals yet found in the world (Fig. 129).

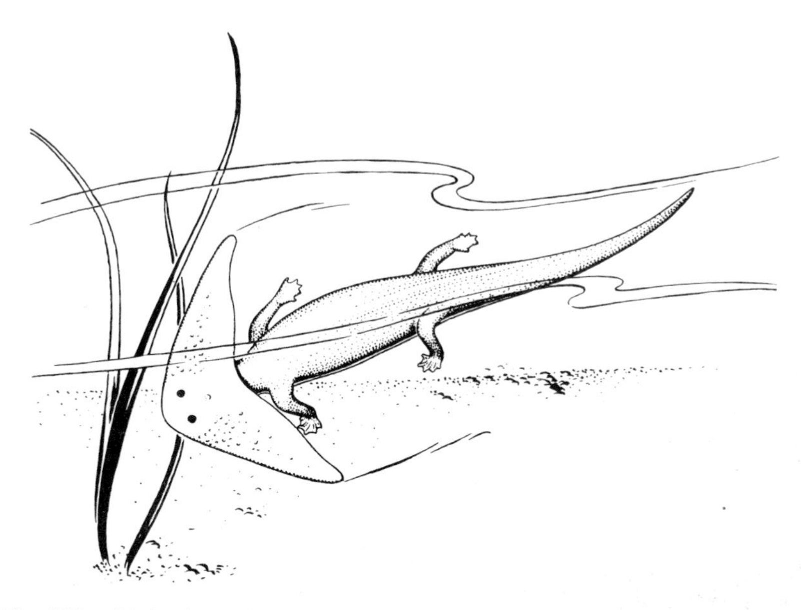

Fig. 129. Wedge-headed, bottom-dwelling amphibian, Diplocaulus, *2 feet long.*

Early Karroo of South Africa. For the land life of the late Permian and more evidence of the early and middle part of the Period we shall turn to South Africa. There in a sequence of formations are the remains of mammallike reptiles called therapsids, which were even more advanced than the pelycosaurs from Texas. They are for the most part clearly the descendents of earlier pelycosaurs. They, too, are specialized in many directions, but those of greatest interest to us are some of the smaller ones that continued into the Triassic and eventually gave rise to the mammals. Similar forms have been found in Russia, South America, and in a few other places, but there is nothing comparable in numbers to those in South Africa.

The total time involved in the deposition of the nearly horizontal formations of Karroo sequence ranges from late Pennsylvanian to early Jurassic. The rocks are divided into four groups, recognized by their lithological character and by their fossils. The first of these are the *Dwyka* tillites (glacial moraine) and shales with *Gangamopteris* plants. The Dwyka shales have yielded fishes much like those in the Pennsylvanian of Europe and small fresh-water aquatic reptiles about 18 inches long. These peculiar reptiles, known as mesosaurs, have long slender snouts and long, delicate, needlelike teeth and are also found in southern Brazil. The Dwyka is followed by the *Ecca* sandstones and shales of the Permian which include the coal beds. In these beds are a few reptilian remains, fish scales, a conifer, and both of the "tongue-leafed plants" *Gangamopteris* and *Glossopteris*.

Superimposed on the Ecca are the *Beaufort* sandstones and shales which contain an abundance of fossil reptiles. These rocks form a continuous sequence from the middle Permian to the close of the Triassic. Robert Broom (1866–1950) divided the Beaufort into six Zones, as typified by different reptiles (Fig. 139). The first three of these Zones are of Permian age. The reptiles, though diversified, differ rather conspicuously from those in the early and middle Permian of North America. For the most part they are mammallike but not so close to the mammals as those in the Triassic and early Jurassic rocks higher in the section. Amphibians, a few fish remains, and fresh-water mollusks also occur in these beds. The last of the *Glossopteris* (Fig. 210) plants occur in the lowermost Beaufort Zone. The overlying *Stormberg* group are considered under the Triassic.

Fig. 130. The skull of Mesosaurus, a late Pennsylvanian genus of reptiles found in South Africa and in Brazil. The skull is about 3 inches long. After von Huene.

Robert Broom was the foremost student of the Karroo reptiles. He was born in Scotland in 1866 and took his degree in medicine in 1889 with commendation. He was actively interested in anatomy and botany. Of the numerous problems to be solved, the origin of mammals appealed to him most. In an attempt to uncover information on this subject and to improve his health he sailed for Australia in 1892, for still living there were the most primitive of all mammals (if indeed they are mammals), the egg-laying monotremes [duck-billed platypus and echidna (anteaters)] and the pouched marsupials (kangaroos, koalas, etc.). Broom found a bone breccia of Pleistocene age near the Wombeyan Cave from which he described interesting small mammals. After four years his interests turned to the Karroo of South Africa and its mammallike reptiles. These rocks, with their abundant reptilian remains were first discovered in 1856 and by the time Broom arrived in South Africa had attracted considerable attention because of the mammalian characters in the reptilian fossils. Broom was appointed Professor of Geology and Zoology at Victoria College in Stellenbosch. Later he gave up his academic position and resorted to the practice of medicine to make his living, while at every opportunity he searched the badlands for fossils. He doctored both man and beast and at one time said, "A horse doctor must be much smarter than other doctors because the horse cannot say what is ailing him." Broom's collections and studies astounded scientists everywhere. Not only did he reveal the origin of mammals from mammallike reptiles, but after a trip to America he pointed out the relationships of the pelycosaurs in the Red Beds of Texas to those he had studied in South Africa. He also made outstanding discoveries in anthropology and with Raymond A. Dart turned the attention of the world to South Africa with discoveries of primitive man in the cave breccias. Broom was a skilled and fascinating speaker who reproduced the extinct creatures on the blackboard as he lectured. He was active in field work and research until the last. His death at the age of 86 closed one of the most illustrious careers in paleontology.

Fossil Insects. The Permian insects showed a marked change and diversity over those in the Pennsylvanian. Many orders, so familiar to us today, made their appearance for the first time. There were a few large ones, but for the most part they were small, some tiny. One of the best representations of early Permian insects was discovered in the Wellington shales near Elmo, Kansas, where more than 10,000 specimens have been recovered. Other localities have turned up in the Dunkard deltaic rocks of Ohio and in the early Permian of Russia. The best late Permian locality is in New South Wales, Australia, where the insect fauna shows a greater resemblance to those that followed in the Mesozoic than to those recorded from the early Permian in North America and Europe. Indeed, most of the earlier groups are rare or absent in the New South Wales collection (Fig. 131).

The giant dragonflies (Protodonata) and the smaller, primitive dragonflies (Megasecoptera), with wings of a more complicated construction, ranged from the Pennsylvanian into the early Permian. These evidently had spread over most of the world, for specimens have been discovered in North America, and Eurasia. They were rare in the late Permian of Australia. Cockroaches also were widespread, several hundred species having been described, but like the dragonflies they were rare in the late Permian of New South Wales.

Of interest to us are the early grasshopper and cricketlike insects (Protorthoptera), which show long antennae and elongate hind legs for jumping. Usually only parts of the wings are found. Another order (Plectoptera) much like our mayflies, is known to have belonged to at least seven families of Permian age. They were much more abundant then in the Pennsylvanian.

The oldest known beetlelike insects (Protelytronoptera), had elytra (hard wing covers) homologous with front wings in other insects and appeared for the first time in the early Permian. They were common both in North America and Europe. Though it was thought at first that these insects were surely ancestral to the living beetles (Coleoptera), it was later discovered that the first true beetles made their appearance in the late Permian of Australia. The Australian beetles are much smaller than those in the early Permian order, and the structure

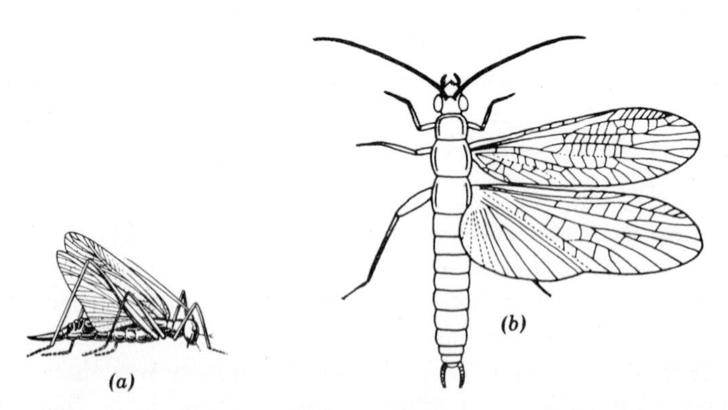

Fig. 131. Permian protorthopteran insects: (a) Metoedischia, *protorthopteran,* × ½, *from Russia, after Martynor;* (b) Protembia, × ¼, *from Elmo, Kansas, after Carpenter.*

in the second wing is entirely different. They are associated with another order (Protocoleoptera) with equally distinct elytra. In fact, the early Permian beetlelike Protelytronoptera are more closely related to the earwigs (Dermoptera) than to any other order of insects.

Other insects appearing for the first time were the book lice (Corrodentia), lacewings (Neuroptera), stone flies (Plecoptera), scorpionflies (Mecoptera), and some insects very much like true flies (Diptera).

This showed a remarkable evolution in the insect life. The development of a complete metamorphosis, perhaps in response to Permian climatic extremes, permitted the larvae to live in a whole new set of habitats (plant and animal parasites, soil, etc.) There were many extinct orders, but with them were the basal stocks of many living orders, though no existing families had yet appeared. The climates of the Permian set the stage for radical changes; that the insects responded is indicated in the large number of new groups coming into the record for the first time.

Land Plants. The flora in the late Pennsylvanian near Garnett, Kansas, gave a clear indication of the predestined fate of the coal swamp plants. As the continents were uplifted and Permian climates became colder, the scale trees, calamites, and seed ferns were able to survive only in local areas in which the temperature was still mild. The *Cordaites* gave way to the smaller leafed, true conifers *Walchia* and *Lebachia*. These true conifers became the dominant plants in the Permian landscapes of the Northern Hemisphere.

A cool flora characterized the late Pennsylvanian and early Permian in the Southern Hemisphere. The dominant plants, with creeping stolons and thin, narrow, elongate leaves, have been termed "tongue-leaved ferns." The genera *Glossopteris* and *Gangamopteris*, so frequently seen in the literature, were not related to the ferns, but they evidently had saclike reproductive structures representing seeds because these have been found in association with and in certain specimens attached to the leaves (Fig. 210). These plants were widely dispersed in South Africa, South America, Antarctica, Australia, and India. Their northernmost records were taken in Russia and Greenland, where they occasionally turn up in the Angara floras. There is no evidence of their ever having been in western Europe or North America. In Australia *Glossopteris* and other genera confined to the Southern Hemisphere were the coal-producing plants of the great Permian coal field at Newcastle.

Hypothetical Gondwanaland. The Gondwana system was proposed by Henry Benedict Meldicott (1829–1905) in 1865 for some 21,000 feet of nonmarine strata in India. These rocks were intercalated with coal beds containing abundant fossils of *Gangamopteris* and *Glossopteris* but no true conifers, as in the beds of the same age in North America and Europe. Since similar deposits have been recorded from South Africa, South America, and other areas widely separated by oceanic waters, it has been suggested that the land masses in the Southern Hemisphere were connected in the Pennsylvanian and the Permian. This was called "Gondwanaland." Gondwanaland was supposed to have existed until early Jurassic, but other geologists have extended its duration well into the Cenozoic. Some paleontologists have directed attention to closely related, near-contemporary Triassic reptilian faunas in South America and Africa that have no counterparts in the Northern Hemisphere. These include mammallike and other reptiles. To account for the distribution of these terrestrial and fresh-water animals, some authorities have maintained that a land connection must have existed across the South Atlantic or via Antarctica in the late Paleozoic and early Mesozoic. Others, employing the theory of "Continental Drift," have suggested that the continents drifted apart from one large land mass. Most scientists, though, who are concerned with the problem now admit that the evidence is overwhelmingly against such a connection in the Cretaceous or in the Cenozoic.

Others believe that the positions of the continents and oceans have been stable; therefore, the distribution of the *Gangamopteris* and *Glossopteris* floras and the faunas showing a similar distribution pattern must be explained in some other way. Geophysicists and oceanographers employing gravity data and the seismic refraction methods now feel that evidence from the crustal structure beneath the ocean basins does not support either continental drift or the foundering of land bridges in the deep ocean basins. Some paleobotanists believe that the double-winged spores of *Glossopteris* could be carried by high winds over long distances, and thus the chance of transportation across oceans within a million years becomes possible. On the whole, we still need to learn much about the dispersal of spores and seeds of plants. Moreover, those who favor permanency of the ocean basins believe that the negative evidence of certain reptiles in the Northern Hemisphere, over which they had to pass if the ocean basins in the past were much as we see them today, can be explained. It is thought that when formations are found to represent the right environments the intermediate faunas will be discovered in North America and elsewhere. After all, our knowledge of life that existed on earth is antagonizingly incomplete in certain groups, and the farther back we go into geologic time the truer this becomes.

17 Mesozoic Era

Dominance of dinosaurs, marine reptiles, ammonites, belemnites, cycadeoides, ginkgos, and conifers; origin of birds, mammals, and angiosperms.

As originally proposed by John Phillips (1840–1841), the Mesozoic (middle + life) included what we now know to be Cretaceous, Jurassic, Triassic, and part of the Permian. This was altered by Murchison, who considered the upper part of the New Red Sandstone to be Triassic and the lower part, with the underlying "magnesian limestone," to be Permian.

John Phillips, a nephew of William Smith, was born in 1800 at Marden in Wiltshire, England. He was greatly influenced by his illustrious uncle. Aside from his work in the field, Phillips arranged and identified different museum collections. He was appointed Professor of Geology at King's College in London in 1834 and later accepted a professorship in Dublin. However, he finally succeeded Professor Buckland at Oxford. Phillips carefully revised the stratigraphic and paleontologic evidence of previous investigators on the Paleozoic and proposed terms still in use in our geologic timetable.

Inland Seas and Invertebrates. In contrast to most of the Paleozoic, when the marine invertebrate faunas made up the major part of the fossil record, the most spectacular groups in the Mesozoic were land animals. The extreme climates of the Permian had a profound effect on life everywhere at the end of the Paleozoic. Many of the older groups of plants and animals became extinct, but others with the potential evolved into new groups. Both plants and animals were more modern in appearance.

Except for the great Tethys Sea across southern Europe, northern Africa, and southern Asia, the inland seas were more restricted in the Mesozoic than in the Paleozoic. An excellent sequence of marine sediments was laid down in Great Britain and in western Europe, though it fluctuated considerably because of intermittent deposition of continental sediments. In North America marine embayments were largely restricted to the mid-continent, though they were not so extensive as before; but in the Pacific province some excellent fossils can be compared with those on other continents. Though good marine fossils have been found in Africa, South America, and Australia, for the most part the Mesozoic seas were not so widespread in those regions as in the Paleozoic.

Modern groups of foraminifers in the Jurassic and Cretaceous apparently assumed the role of the fusulinids of the late Paleozoic. Hexacorals replaced the tetracorals, and pelecypods and gastropods increased in numbers and kinds, but the most outstanding invertebrates were the ammonites and belemnites. The ammonites evolved rapidly and dispersed widely to become the most useful organism in the correlation of Mesozoic rocks, but they died out suddenly at the end of the Cretaceous.

Vertebrate Life. The last of the big cumbersome labyrinthodont amphibians became extinct at the close of the Triassic, and the modern groups gradually came into the record. The first frog was found in the Jurassic, but amphibians never again were dominant.

Reptiles having evolved from amphibians in the late Paleozoic rapidly changed into animals well adapted to a terrestrial existence. But in the Triassic some groups, primarily the early ancestors of the ichthyosaurs and plesiosaurs, moved back into the seas. These groups became gigantic in the Jurassic and in the Cretaceous. The phytosaurs, confined to the Triassic, were crocodilelike in appearance and habits but were in fact more closely related to the dinosaurs. The first crocodilian appeared in the early Jurassic. Other reptiles took to the air. The most conspicuous were the gliding pterosaurs. One of the Cretaceous pteranodons had a wingspread of 27 feet. Another reptile-like creature, though still possessing teeth and other reptilian features, developed feathers and has been classified with the birds. The ichthyosaurs, plesiosaurs, pterosaurs and toothed birds became extinct before Cenozoic time.

Among the persistently terrestrial reptiles certain bipedal, light-boned, fast-running saurians gave rise to the two spectacular orders of dinosaurs. Though specialized in many ways, the dinosaurs were either terrestrial or amphibious in habits. One genus seems to have been partly arboreal. The huge amphibious sauropods were the largest land animals ever known. Other herbivorous dinosaurs were armored, horned, or duck-billed kinds, all of which had to be prepared for the sudden attacks of the vicious carnivorous dinosaurs. This Era was truly the age of reptiles. All of the dinosaurs passed on with the Cretaceous.

Still another branch of reptiles of much more importance insofar as we are concerned, though several groups fall into this category, comprised the mammallike reptiles. Remains of their forerunners are buried in the Permian rocks of Texas and South Africa. In the Triassic some of the earlier groups disappeared, but there still were several kinds much more like the mammals than anything previously seen. Thus it has been suggested with good reason that the most primitive orders of mammals arose from different groups of late Triassic mammallike reptiles.

The first undoubted mammals showed up in the middle Jurassic of England as three distinct orders. In the late Jurassic there were four orders in both North America and Europe. Small and inconspicuous during the Mesozoic, the mammals, together with the birds and insects, rose to dominate the earth after the giant reptiles were gone.

Plants. Perhaps the most momentous event of the Mesozoic was the rise to dominance of the flowering plants, or angiosperms, in the early Cretaceous, though they are first recorded in the Triassic. The great scale trees survived for awhile in the Triassic when conifers grew to tremendous heights. The cycadeoids and ginkgos then came into dominance, only to be outnumbered in the Cretaceous by the angiosperms. Insects, birds, and mammals all responded to the appearance of these new plants with their abundant sources of food in nectar, flowers, and fruits. This set the stage for the prolific life of the Cenozoic and a new chapter in the history of life on earth.

18 Triassic Period

Phytosaur

Origin of dinosaurs and many marine reptiles;
phytosaurs; diversification of mammallike reptiles;
last of scale trees;
giant conifers; first hexacorals;
ammonites nearly died out at end of period;
last conodonts; arid and semiarid climates.

SOME TWENTY YEARS after the turn of the nineteenth century European geologists became concerned with the formations later to be called Triassic. The *Bunter*, continental and marine sandstones, the *Muschelkalk*, marine limestones, and the *Keuper*, bright-colored marls and clays of continental and marine origin, were recognized in central Europe. It was discovered that the New Red Sandstone in the British Isles graded into this threefold division when its equivalents were traced across to the Continent. In 1834 von Alberti suggested the name *Trias* for these well-marked subdivisions in Germany. He united them into a single formation, but it was soon abandoned. On the other hand, Trias, later modified to Triassic, was widely adopted and replaced the term New Red Sandstone.

Friedrich August von Alberti was born in Stuttgart in 1795 where he studied mining and finance. He never became an academician, though he published important geological and paleontological contributions as a result of his investigations in the field. When 20 years of age he began his career at a saltworks. Eventually he became manager of the Friedrichshall saltworks and was a Counselor of Mines. This is an excellent example of the part economic geologists played in the advancement of knowledge at this early date. Von Alberti died in 1878.

Marine Formations. The Triassic, like the Permian, is a Period of limited continental marine embayments. Almost everywhere there is an unconformity between the marine Permian and the marine Triassic. The salt-water invertebrate assemblages for the most part show marked changes either by extinction or by evolution in the different groups. No marine beds occur in eastern North America. Marine formations in the Far West first appear in California, Arizona, Nevada, Utah, Idaho, Wyoming, and Montana. Later, parts of these areas filled with sediments, but similar embayments spread into other areas of the west and northwest, and a narrow trough extended south into Mexico. Rather extensive formations have also been found in Alaska and in the Yukon Territory (Fig. 133).

In central Europe marine beds were intermittently laid down between deposits of continental origin. This is well exemplified in the type section. To the west the marine rocks give way to formations mostly of nonmarine origin. The Tethyan geosyncline, clearly outlined for the first time, extended across the Mediterranean area into India, where some of the sections reached great thicknesses. There had been no marine embayments in that part of India since the Cambrian. Marine Triassic rocks also occur in northern Europe, other parts of Asia, Australia, and Africa. They seem to be most limited in South America.

Marine Invertebrates. The marine invertebrates in western North America, abundant in many places, show relationships with those in Asia, India, and southern Europe. Other European faunas are more distantly related.

Foraminifers are rare in the Triassic, though some collections have been made in Europe. Those that are known show relationships with Jurassic genera. The fusulinids, so abundant in the Pennsylvanian and Permian, are replaced by genera of small imperforate and perforate calcareous foraminifers which had been distinctly in the minority during the great radiation of the fusulinids.

New kinds of corals replaced the old Paleozoic tetracorals and honeycomb corals. These were the hexacorals, or scleractinian corals, related to those in the seas today. These corals are an important part of the invertebrate assemblages in the old Tethyan geosyncline of southern Europe and southern Asia. They are also common in the seaways of western North America but were not of the reef-forming habit until middle Jurassic time (Fig. 132).

In contrast to the tetracorals, sections of the new kinds show the major septa divided into six parts. Scleractinian corals include both solitary and colonial kinds. The origin of these modern corals is still unsolved. Some coral specialists maintain that the scleractinian corals are the descendants of a common ancestral stock with the tetracorals back in the Cambrian and that during the Paleozoic the scleractinians had no hard parts to become preserved as fossils. Others believe they are derived from a group of tetracorals in an accelerated evolution in the late Permian and early Triassic. This view they feel is supported by the presence of additional septa in certain late Paleozoic forms which tend to approach the structure seen in the scleractinians. Much light may be thrown on this interesting problem when more early Triassic genera are sectioned and studied in detail.

Sponges and bryozoans are essentially not known in Triassic rocks, but they should be found somewhere, since both groups occur before and after Triassic time.

Fig. 132. Hexacoral, Margarosmilia zieteni; × ½. After Vaughn and Wells.

Brachiopods, though still common in the Tethys Sea where many spirifers have been found, experienced a decline elsewhere, particularily in America. They never again reached the diversity of groups in the life of the seas that they experienced in the Paleozoic, though two of the groups that are still living increased greatly in numbers in the Jurassic. As the numbers of brachiopods subsided, the gastropods and pelecypods increased, and from the Triassic on they played an increasingly important role in life in marine waters.

The Triassic seas were alive with ammonites. The relatively simple sutured goniatites of the Permian had given way to the complex sutured ammonites. Their buoyant shells were suggestive of different specializations. Some probably were free-swimming (nektonic), and others may have drifted with the ocean currents (pelagic). The shells were smooth in most of the early Triassic genera, but in the middle and later parts of the Period shells with ribs, keels, nodes, and other surface features appeared. Near the close of the Triassic for some unexplicable reason the ammonites nearly died out, but several species of the genus *Phylloceras* survived and it was from this diminished stock that the Jurassic groups arose.

Another group of cephalopods, the torpedo-shaped belemnites, were widespread in Triassic seas. These peculiar squidlike creatures must have had a much earlier origin, since one specimen has been described from the Mississippian of Oklahoma, and another species is common in the Permian of Greenland.

Other marine invertebrates are modern echinoids and starfish. Among the crustaceans were, of course, the ever-present ostracods. Of particular interest are the first lobsters and other crayfishlike or shrimplike forms. Here, too, were the last of the conodonts which had been so abundant during the Paleozoic.

Marine Reptiles. The Triassic throughout the world shows a strong trend toward different groups of reptiles, not only going back into freshwater and estuarine environments, but some which spread out into the seas. Some of these were the forerunners of the fusiform-bodied ichthyosaurs and the long-necked plesiosaurs so well known in the Jurassic and the Cretaceous. Other less specialized groups, like the nothosaurs and rhynchocephalians, were abundant but did not develop into such spectacular forms. Some ichthyosaurs in the inland seas of California and Nevada reached gigantic size. Reptiles related to the plesiosaurs, called placodonts because of their flat teeth, are assumed to have fed on mollusks. Here in the Triassic, too, the first turtles are represented by the peculiar *Triassochelys* that still had teeth in the roof of the mouth.

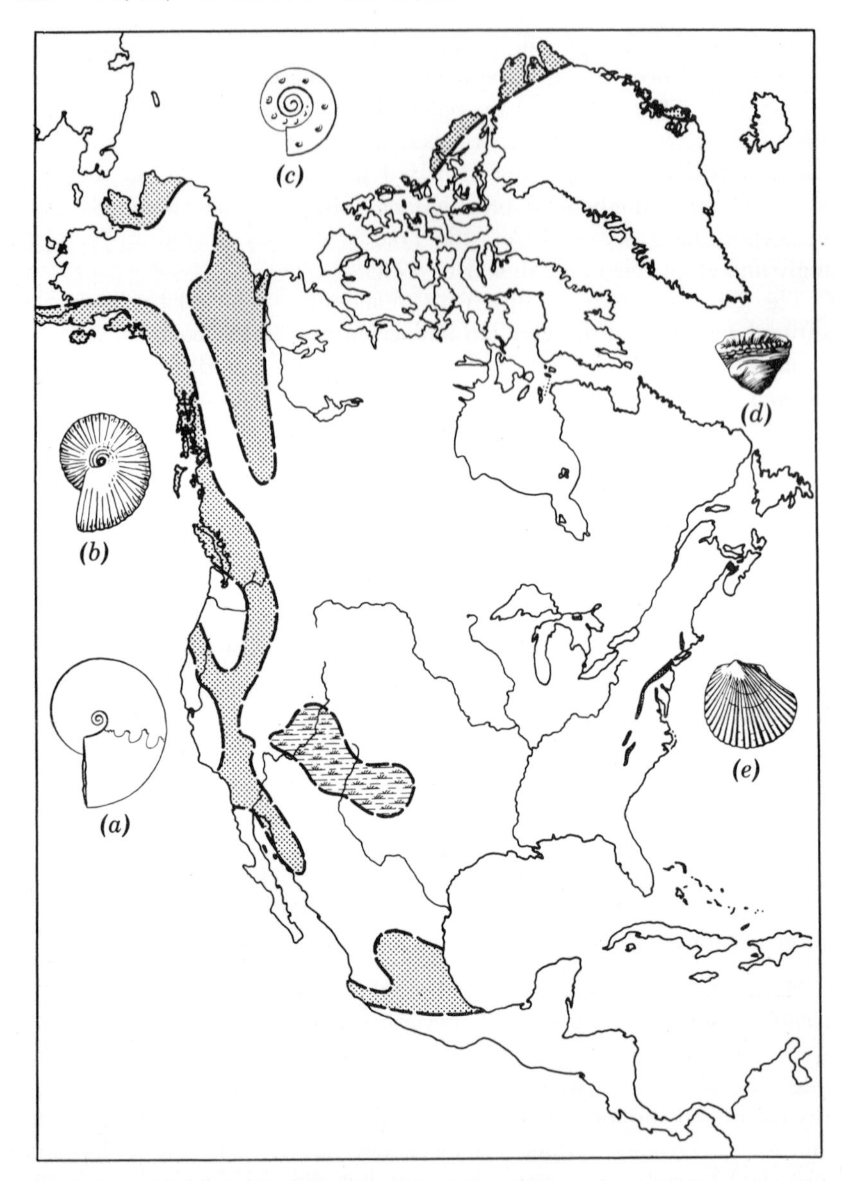

Fig. 133. Greatest inundation of seaways across North America in late Triassic. After Schuchert. The fossils illustrated come from different part of the Period: (a) Meekoceras gracilitatis, *ammonite;* (b) Phyllopachyceras infundibulum, *ammonite;* (c) Tirolites pacificus, *ammonite;* (d) Oppelismilia rudis, *coral;* (e) Pseudomonotis subcircularis, *pelycopod.* (a), (e) *after Pirsson and Schuchert;* (b), (c) *by permission, from* Invertebrate Fossils *by Moore, Lalicker, and Fischer;* (d) *after Wells, courtesy Geological Society of America and University of Kansas Press.*

Continental Deposits. The emergence of continental areas throughout the world resulted in a greater deposition of nonmarine formations than during any previous Period. Even in most areas in which marine rocks were deposited they intergrade or are replaced from time to time by land-laid red beds, which were continuations of sedimentary cycles initiated in the Permian.

Connecticut Valley. After the Appalachian revolution had folded and lifted the Appalachians into a high chain of mountains in late Permian time, these mountains were subjected to erosion during the early and middle Triassic. In some areas faulting formed basins, or grabens, adjacent to block mountains later in the Period. The basins then filled with sediments as much as 2.5 miles deep. A Triassic trough through central Connecticut and Massachusetts was one of these basins. Animal footprints found there in profusion, and mostly leading in one direction on the old sandy mud flats, were some of the first fossils to attract the attention of early Americans. Because of the three digits so clearly displayed in many trackways they were thought to be bird tracks. In 1812 Pliny Moody referred to them as the tracks of "Noah's raven," which gives some idea of the folklore pertaining to these imprints at that time.

The old Connecticut Basin was periodically filled by heavy rains which partly dried up to leave wide stretches of mudflats. The water was then restricted to the deeper parts of the basin and was populated with fresh-water fishes, water insects, and fresh-water mollusks. Ferns, cycadophytes, and conifers fringed the lakes and ponds and outlined the stream courses. Even *Lepidodendron*, which occurred only in certain environments elsewhere along the Atlantic coast, probably showed up in certain swampy areas. As the waters of the lake receded over the mud flats, kept soft by moisture below, showers of raindrops left their marks in the soft surface, and terrestrial animals moved across it seemingly headed toward the same destination. For some reason a few died on the mud flats where their bones were preserved with the raindrop imprints and the trackways of the creatures that wandered away to distant places.

One of these animals was a late Triassic forerunner of the great saurischian (amphibious) dinosaurs which reached such tremendous sizes in the Jurassic. *Yaleosaurus*, as it was called, was not so large. It was only about 8 feet in length from the end of its nose to the tip of the last vertebra in its tail. Despite its small size, the hind limbs were relatively short and the front limbs were not greatly reduced as in the more strictly bipedal forms. Other features in the skeleton also suggested its affinities with the sauropods. Another, *Podokesaurus*, was little more than 3 feet in length. These small creatures must have been fast runners because of their slender proportions and the lightness of their bones. Insofar as

Fig. 134. Triassic dinosaurs: (a) Podokesaurus holyokensis *and* (b) Yaleosaurus colorus; × $^1/_{24}$.

we know dinosaurs made their first appearance and spread to most parts of the world in the late Triassic (Fig. 134).

There were other four-footed reptiles called phytosaurs with a superficial resemblance to crocodiles and probably with habits much like the living saurians. The phytosaurs, which had long snouts and a nose opening far back on the head, preferred the streams and ponds where they lived on fish or other unfortunate creatures that came within their reach.

Continental Triassic of Southwestern United States. Sandstones and shales of continental origin are widespread over much of the Southwest. To the north and east they reach Wyoming, western South Dakota, and western Nebraska, and to the southeast there are excellent exposures in New Mexico and in the panhandle of Texas. The source of the sediments was the old Colorado Mountains which also supplied the detritus for some of the Permian formations. The Triassic formations thicken to the west where they interfinger with the marine sequence in Nevada.

Moenkopi Early Triassic. The early Triassic of the Southwest is exemplified in the Moenkopi formation which reaches a maximum thickness of 2000 feet in western Utah and Nevada. A study of these sedimentary rocks and their fossils indicates fluctuating, moist, temperate to semiarid and at times arid climates with stream beds, lagoons, playas, flood plains, and tidal flats or shallow sea floors at the edge of the sea. Some of the best exposures occur in the valley of the Little Colorado River where variegated sandstones and shales can be traced for miles along the escarpments. The bedding planes in the different units are characterized by the presence of many reptilian and a few amphibian trackways. Occasionally, bones or skulls are found. Near the middle of the section in the Little Colorado Valley are gypsiferous shales in which no trackways or bones have been found. Elsewhere throughout the formation gypsum appears in lenses and layers. These saline evaporites occur in old lagoon or lake deposits a few feet to a mile or more in diameter. This together with shrinkage cracks in mudstones and siltstones is rather clear evidence of periodic drying up.

One of the most characteristic features of the Moenkopi is the abundant reptilian trackways across the ripple-marked tidal and flood-plain mud flats and a more limited number of amphibian tracks in the sand-spits parallel to the streams. The most abundant of the reptiles is *Chirotherium*, the "hand animal," so named because the imprint of its hind foot resembles the digits of a hand with the opposable thumb. There were predinosaurlike creatures as well as lizardlike reptiles. The many species of primitive saurians seemed to cross the mud flats in a direction parallel to the stream courses and did not wander aimlessly over the barren tide flats and flood plains. Reptilian remains are rare, but in quarries like one near the Meteor Crater, Arizona, Samuel Paul Welles and his field party found a number of amphibian skulls. These flat-headed and short-limbed labyrinthodonts evidently stayed in the streams or in close proximity to water (Fig. 135).

The record of plants is scarce, but pieces of wood as well as impressions of stems and reeds are preserved. Though these fossils are ex-

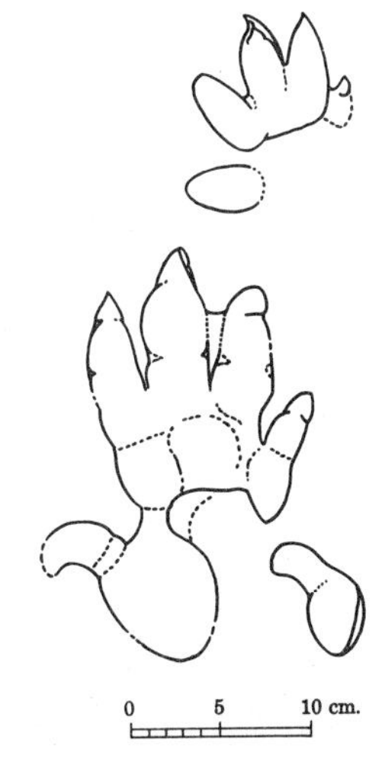

Fig. 135. Footprints of Chirotherium barthi, *the "hand animal," from the Moenkopi formation.*

tremely meager, there is evidence of reeds and horsetails along the streams and primitive conifers, such as *Walchia*, must have partly shaded the higher ground.

Invertebrates left their trails and remains in many places. They ranged from fresh-water and brackish kinds to the marine crinoids and ammonites, which give a picture of the encroachment and retreat of the sea that persisted farther west. The ammonites *Meekoceras* and *Tirolites* have been utilized as Zone markers and have been most useful in age determinations.

Chinle Late Triassic. Well back from the rim of the Grand Canyon are the colorful landscapes of the Painted Desert of Arizona. Variegated colors in reds, yellows, whites, purples, and blues in the low, rounded hillocks or sandstone-capped mesas in the Chinle formation leave a permanent impression of unexcelled beauty. At the base of the Chinle and unconformably on the underlying Moenkopi is the Shinarump conglomerate, considered by some as part of the Chinle formation.

Sediments were carried onto this wide flood plain by meandering streams from mountains probably to the south. Interspersed with the streams were many tracts of swampy terrain. Ferns, club mosses, and scouring rushes grew in considerable numbers. Farther back on the higher and drier slopes were the giant conifers, *Araucarioxylon*, some of which are thought to have been 120 feet high. Their huge logs, together with those of the ginkgos and other plants, drifted in by torrential currents, became stranded, and were covered by the sediments carried by the floods that followed. Some of these logs were partly destroyed by Triassic forest fires because they are still coated with altered charcoal. Others show evidence of wood-boring insects. The forest had different kinds of conifers that find their nearest relatives today in the monkey puzzle pine, *Araucaria*, of the Southern Hemisphere (Fig. 136).

Though thirty-five genera of plants have been recorded from the late Triassic of Arizona, the Petrified Forest got its name from the abundance of immense *Araucarioxylon* logs beautifully preserved with silica. Unfortunately, many of the fine specimens have been destroyed. The broken pieces were carried away as souvenirs or sold to tourists.

These huge fossil logs first attracted serious attention in the middle of the nineteenth century. In 1853 an expedition sent out to explore the resources of the Southwest and to map a route for a railroad to the Pacific coast secured pieces of the logs that had broken up and were scattered over the slopes. Another expedition was dispatched from Fort Wingate in 1879 to collect fossil logs for the United States National Museum, a branch of the Smithsonian Institution.

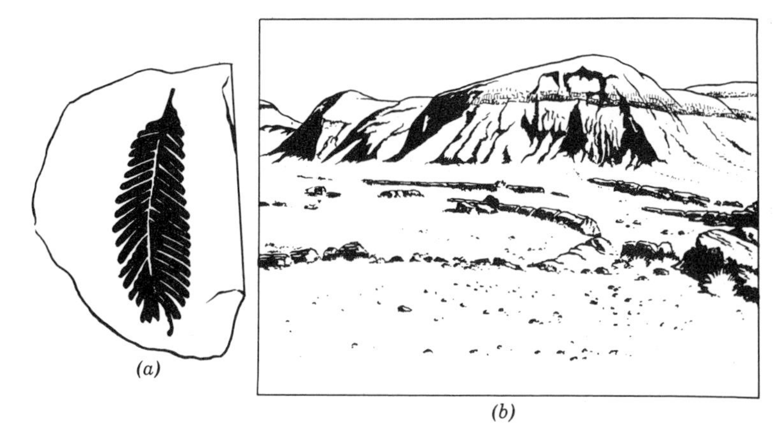

Fig. 136. (a) *The leaf of a cycadeophyte,* Otozamites powelli, $\times \frac{1}{2}$, *from the Chinle formation;* (b) *logs of* Araucarioxylon *in the Blue Forest area of the Petrified Forest National Monument, Chinle formation, Arizona. Sketched from photographs after Daugherty.*

Lungfish and large garlike fish lived in the stagnant pools and streams. These waters and the shores were heavily populated with the giant flat-faced labyrinthodont amphibians, *Eupelor*. These 6- to 8-foot, flat-bodied creatures crawled slowly across the muddy banks or moved more quickly through the water after fish and other forms of life that were swallowed in huge gulps. But they, too, had to contend with the even more terrifying phytosaurs which were so characteristic of the Triassic. The name phytosaur (plant + reptile) has been one of the most misleading misnomers. The first specimen found was the anterior part of the lower jaws where they were fused together. The specimen displayed the rather cylindrical alveolar casts of tooth sockets which were mistaken for the crowns of teeth, and because of the shape of the casts it was assumed that these peculiar reptiles ate plants. Quite to the contray, these predaceous reptiles whose jaws were lined with sharp, serrated teeth, must have been the scourge of Triassic streams and ponds everywhere. Many excellent skulls and the platelike armored body coverings have been found in the Chinle shales and sandstones.

In the forests and around the ponds were large mammallike reptiles called dicynodonts (double + dogtooth). They were big, clumsy, heavy-boned animals with only one tooth in each upper jaw and a beak much like a giant tortoise. Of course, they were not related to the turtles, nor were they much like early mammals. These big *Placerias* were specialized offshoots of an early basal stock of mammal-like reptiles. A quarry near Saint Johns, Arizona, has produced a large

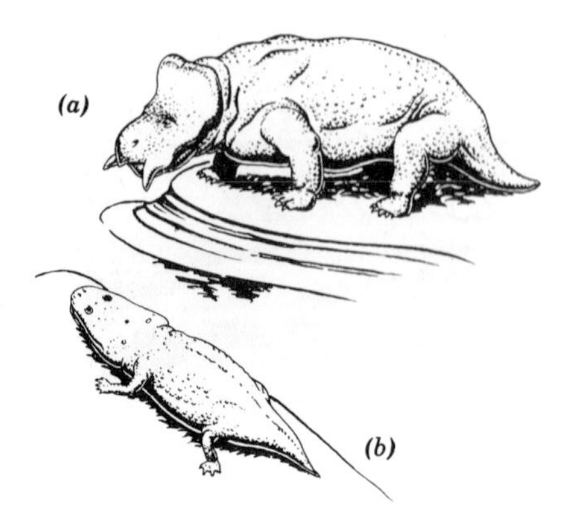

Fig. 137. (a) Mammallike dicynodont reptile, Placerias, *and (b) the large flat-faced labyrinthodont amphibian,* Eupleor, *found in the Chinle formation, Arizona;* × $^1/_{48}$.

collection of broken bones, many of which have been reconstructed by Charles Lewis Camp and Samuel P. Welles into a nearly perfect skull. Related genera have been found in South America, South Africa, China, and Indochina (Fig. 137a).

In the uplands were slender bipedal reptiles 6 to 8 feet in length. These were some of the first dinosaurs. They were agile, fast-running flesh eaters with short front limbs equipped with sharp claws for holding their prey. A quarry, in which a large number of these dinosaurs called *Coelophysis* has been found, lies in the Chinle formation on the famous Ghost Ranch in New Mexico.

Mammallike Reptiles of South Africa. The reptiles from the three upper Zones in the Beaufort of South Africa approach mammals in many features. Some groups are more like mammals in the construction of the cranium, and others show nearer relationships in the reduction of bones in the mandibles or in the feet. This has led to the supposition that warm-blooded, hair-covered, and milk-producing animals, as we know them, may have descended from different groups of mammallike reptiles. Indeed many of these Triassic groups may have had warm blood, hair, and milk glands and may have differed from mammals in retaining more than one bone in the lower jaw or by having more bones in the feet (Fig. 139).

Several reptile groups continued from the Permian lower Beaufort beds into the middle and upper Beaufort. One of the most interesting is the genus *Lystrosaurus* from which one of the Zones gets its name. Like the other dicynodonts, this animal had two large daggerlike teeth in the upper jaw, but the large eyes and the nostrils were located near the top of the skull. These features, together with the reduced limbs, have led to the logical conclusion that the lystrosaurs were aquatic reptiles. They dominate the *Lystrosaurus* Zone almost to the exclusion of terrestrial types. Terrestrial dicynodonts in the upper Beaufort were larger and bulkier than a rhinoceros.

Of much greater interest, though, are some other animals of the South African Triassic: *Thrinaxodon, Diademodon, Ericiolacerta,* and *Cynognathus.* Most of these creatures showed a differentiation of the teeth in incisors, canines, and cheek teeth. In some the digital formula (number of phalanges in each toe) was 2-3-3-3-3, as in mammals, and the others tended in that direction with some of the phalanges reduced to narrow disklike bones. The extra bones in the back part of the lower jaw were greatly diminished in size and no longer important as jaw elements in the more advanced forms. Moreover, the development of a diaphragm as a muscular partition between the visceral and thoracic cavities to faciliate breathing seems to have occurred. A. S. Brink thinks it likely that the high blood temperature, mode of birth, and milk glands were acquired gradually and at different stages of evolution, as were many other features in the mammallike reptiles. Indeed, the dif-

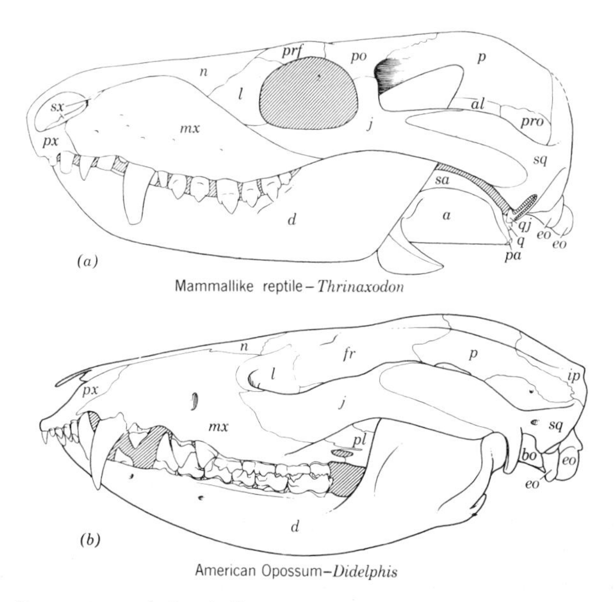

***Fig. 138. Comparison of the skulls in* (a) *a mammallike reptile,* Thrinaxodon, *and* (b) *an American opossum,* Didelphis.** **a, angular; al, alisphenoid; bo, basioccipital; d, dentary; eo, exoccipital; fr, frontal; ip, interparietal; j, jugal; l, lacrimal; mx, maxillary; n, nasal; p, parietal; pa, prearticular; pl, palatine; po, postorbital; prf, prefrontal; pro, prootic; px, premaxillary; q, quadrate; qj, quadratojugal; sa, surangular; sq, squamosal; sx, septomaxillary.**

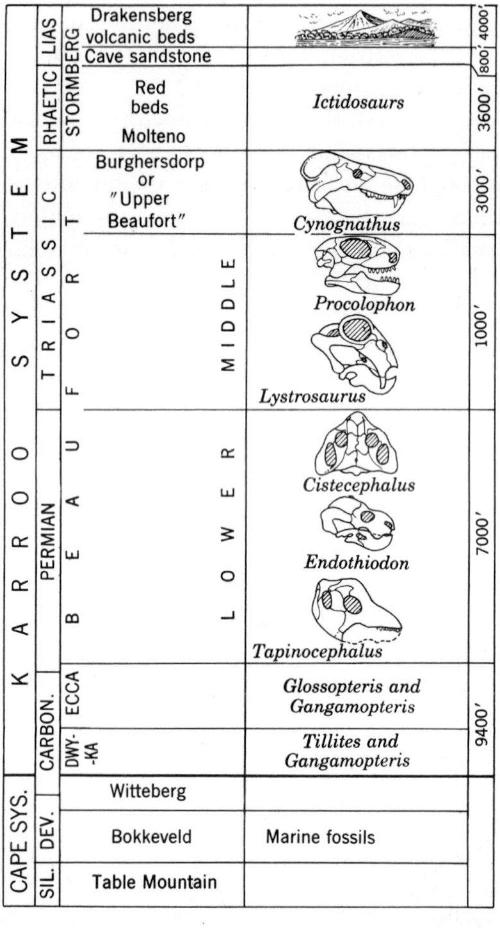

System	Period	Group	Beds	Fossils	Thickness
KARROO SYSTEM	LIAS	STORMBERG	Drakensberg volcanic beds		4000'
	LIAS	STORMBERG	Cave sandstone		800'
	RHAETIC	STORMBERG	Red beds Molteno	*Ictidosaurs*	3600'
	TRIASSIC	BEAUFORT	Burghersdorp or "Upper Beaufort"	*Cynognathus*	3000'
	TRIASSIC	BEAUFORT	MIDDLE	*Procolophon* *Lystrosaurus*	1000'
	PERMIAN	BEAUFORT	LOWER	*Cistecephalus* *Endothiodon* *Tapinocephalus*	7000'
	CARBON.	ECCA		*Glossopteris and Gangamopteris*	9400'
	CARBON.	DWY-KA		*Tillites and Gangamopteris*	
CAPE SYS.		Witteberg			
	DEV.	Bokkeveld		Marine fossils	
	SIL.	Table Mountain			

Fig. 139. The Karroo sequence of South Africa. Modified after William King Gregory.

ferent groups of these interesting creatures display remarkable parallel evolution in different parts of their skeletons, an observation made by David Meredith Seares Watson over 40 years ago (Fig. 138).

In the Stormberg group, part of which is frequently referred to the early Jurassic, fossils are rare, but those found show that some are so close to mammals there is a strong possibility that a number may be classified as mammals.

Rhaetic Mammallike Reptiles in Europe. Recently a composite skeleton was reconstructed by Walter G. Kühne from the parts of perhaps hundreds of individuals which were taken from a fissure deposit in England. These small mammallike reptiles with elongate bodies and very short legs have been called *Oligokyphus*.

Similar fossils occur in beds of Rhaetic age (referred to the Triassic by some authorities and to the Jurassic by others) in Europe where hundreds of isolated mammallike teeth have been found. It is unfortunate that some well-preserved skulls and limb bones of these fascinating little creatures have not been found to settle their fate in our scheme of classification, but as we continue to advance the frontier of our investigations in all parts of the world the relationships and dispersal of the animals and plants are gradually falling into place. That terrestrial animals have spread to every part of the globe where there was opportunity to go is indicated by the presence of Triassic forms in South America much like those in Africa. As new kinds continue to be found and the structures of their skeletons described, it seems that their adaptive radiation must have been nearly as complex as that in the placental mammals, which spread over the world in the Cenozoic.

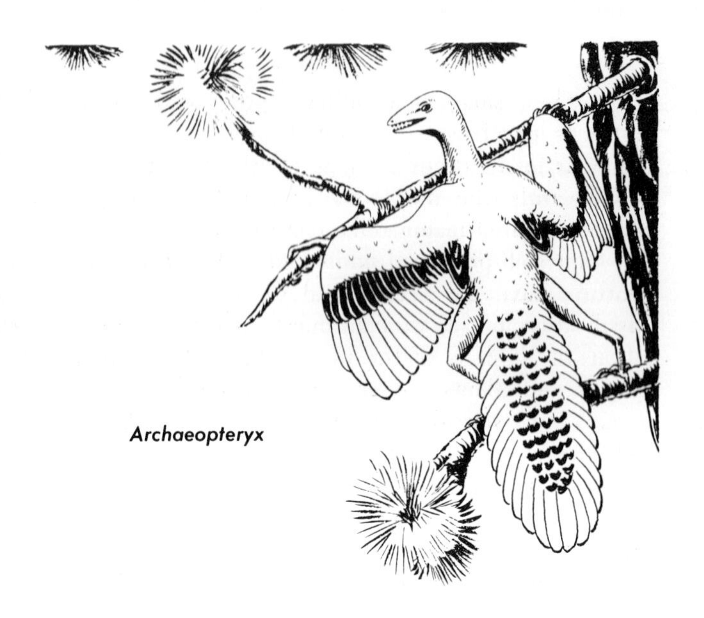
Archaeopteryx

19 Jurassic Period

Gigantic dinosaurs; large marine reptiles; first mammals; first toothed birds; ginkgos, conifers, and cycadeoides dominant plants; insects of modern aspect; ammonites and belemnites abundant; mild climates.

Brontosaurus

THE TYPE SECTION of the Jurassic was eventually established on a sequence of formations between what we now know as the Rhaetic and the lower Cretaceous rocks in the Jura Mountains of Switzerland and France. The name *Jura limestone* was first proposed by Alexander von Humboldt in 1799 for the massive limestones he saw in that area when he traveled from France to northern Italy four years before. Unfortunately, his observations covered nearly all of the rocks that are now known as Triassic and only part of the Jurassic. This was rectified by later workers to include the Liassic and Oölitic (Dogger and Malm) formations. Alexandre Brongniart applied the name *Terrane Jurassique* to these sedimentary rocks.

It is unfortunate, for many reasons, that the type section was not selected in England where the Jurassic section was so securely established by William Smith on uniform lithology and an extraordinary abundance of fossil invertebrates. But the importance of type sections in interregional correlation was not realized at that time. Nevertheless, the uniform stratigraphy and wealth of widely dispersed ammonities has made England a focal point for comparison with Jurassic marine fossils wherever they occur.

The famous explorer, geologist, and geographer *Alexander von Humboldt*, of an old aristocratic family, was born in 1769. He was appointed Superintendent of Mines in Germany but after four years resigned to follow independent lines of research. His wide interests took him to Central and South America to obtain more information on the physical geology and botany of tropical regions. In Colombia he was impressed with the zonal distribution of plants and animals from lower to higher altitudes and thus initiated the study of *life zones*. For this and, more importantly, for his writings on the physical description of the earth he is known as the "Father of Geography." From South America he brought back to Paris mastodont teeth, later described by Cuvier, which are important types from that continent. The expenses of these travels consumed von Humboldt's entire fortune, but in his later life he was generously supported by the King of Prussia. He was ninety years old when he died in Berlin, the city of his birth.

The other establisher of the name Jurassic, *Alexandre Brongniart* (1770–1847), worked closely with Georges Cuvier, renowned vertebrate paleontologist. Brongniart described the marine invertebrates from the Tertiary deposits in the Paris Basin, while Cuvier was concerned with the vertebrates. Their research was contemporary and said to be independent of that of William Smith on the Jurassic in England. Brongniart was one of the most active pioneers in the application of paleontology in stratigraphy. *Terrane Jurassique* appeared in his 1829 attempt to describe all of the rocks in the earth's crust in chronological order. These units as proposed were quite independent of their lithology. The *Terranes* were supposed to have been deposited in one "geologic period." They were divided into formations, which in turn were separated into subformations, each with its distinctive plants and animals. His recognition and descriptions of fossils in the subformations was his most important contribution. Most of Brongniart's other terms were not adopted.

Zones, Paleogeographical Maps, Lithofacies, and Paleofacies. It was from Jurassic rocks that the first Zones were named, the first paleontologic and rock facies were recognized, and the first paleogeographic maps were made. All of these practices are widely used today.

The word Zone (see pp. 77–79) was introduced by Alcide Dessalines d'Orbigny (1802–1857), but it was clarified in application by Carl Albert Oppel (1831–1865) in his detailed work on the early Jurassic (middle Lias) rocks in Swabia. Later Oppel visited localities in France and England in an effort not only to compare the major divisions (Series) of the Jurassic but to see if he could recognize smaller divisions (Stages and Zones) on the basis of the fossil species. Though he realized this was a difficult task, he felt that it was by these methods that an accurate comparison of the different systems could be made. His studies were based on all of the fossil species in the different faunal assemblages, from the sequence of rock units with no reference to the lithology of the beds. At that time he recognized eighty Stages and thirty-three Zones, each of which was typified by certain combinations of ammonite species. This biostratigraphic work in paleontology related the fossils to their containing strata. It was an outstanding contribution, but it is now recognized that such Zones cannot always be traced from continent to continent as Oppel may have thought, though he did not say so.

A contemporary and ardent supporter of Oppel, Jules Marcou (1824–1898), also recognized the limited distribution of Jurassic fossil assemblages. He compared them with the restricted ranges of Recent faunas as representing definite ocean depths. In using these methods Marcou drew *paleogeographic maps* to show not only the position of land and water but also the chief geological provinces in the Jurassic ocean of central and western Europe. In all he distinguished eight provinces and correlated their geographical positions with three climatic zones.

About 16 years earlier, Amanz Gressly (1814–1865), working in the Solothurn Jura of Switzerland, traced the horizontal extension of members in the Jurassic rocks. He thus discovered that the petrographic changes he encountered were accompanied by different faunal assemblages. These he called *facies*. He recognized mud facies, coral facies, sponge facies, pelagic, subpelagic, and littoral facies, among others. Some geologists, even today, have become confused with the presence of differences in these faunal facies at the same stratigraphic level, hence have raised doubts about the ability of paleontologists to correlate the age of the beds with which they are concerned. If the critics had been fully familiar with the nature of the evidence on faunal facies and the nature of their environmental requirements as they exist today, their doubts would not have been voiced.

Marine Jurassic in Great Britain. There are few better sequences of marine Jurassic rocks in the world than that in England. It was in these formations between 1795 and 1799 that William "Strata" Smith first recognized the stratigraphic position of the different units by their fossils. This initiated studies in stratigraphic geology and paleontology which were adopted enthusiastically by geologists and paleontologists in England, on the Continent, and about 30 years later in North America. From a humble beginning of fourteen strata dictated by Smith and written down by his friend the Reverend Benjamin Richardson, the Jurassic marine series (Lias, Dogger, Malm) was divided into sixteen Stages, forty-five Zones, and 102 Subzones. These Zones and Subzones are recognized primarily by the species in the wealth of ammonites which had multiplied tremendously after nearly dying out at the end of the Triassic. There were broad connections with the inland seas across the European continent and north toward the Arctic. Most of the British Isles was covered with Jurassic seas, though there were adjacent lands in which brackish to lagoonal deposits were laid down at or near the coast line. The presence of marine reptiles, oysters, fossil insects, plants, and disseminated iron pyrite indicates that this continental sea at times was relatively shallow.

Stonefield Slates. One of the most fascinating fossil localities occurs in middle Jurassic rocks. It is a bed of sandy slate 1 to 1.5 feet thick under the village of Stonesfield, England. As early as the middle of the seventeenth century the houses in Oxfordshire were roofed with these slates. The accessible slates near the surface were soon used up. Then mine shafts were put down, some to a depth of 70 feet. Before this industry was discontinued and the mines abandoned, startling fossil discoveries were made by the men who carefully parted the slabs.

A decorative little pelecypod, *Trigonia impressa,* though not found elsewhere in the Jurassic, was at Stonefield in abundance, but the amazing discovery of small mammal jaws and teeth so startled the scientific world that they were not accepted as such for nearly 20 years after the first specimen reached the hands of men competent to know. In fact,

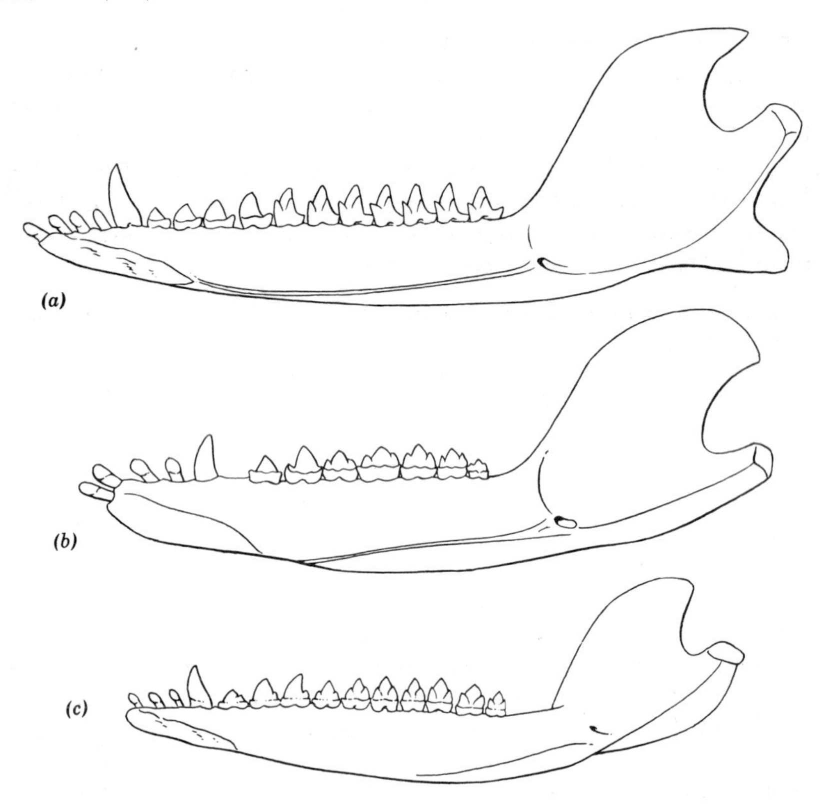

Fig. 140. Lower jaws of three Jurassic mammals. After Simpson. (a) Amphitherium prevostii, *pantothere,* × 2, *from Stonefield slate, middle Jurassic;* (b) Phascolotherium bucklandi, *triconodont,* × 1¼, *from Stonefield slate, middle Jurassic;* (c) Spalacotherium tricuspidens, *symmetrodont,* × 1¼, *from Purbeck beds, late Jurassic.*

the first of these little mammal jaws was found much earlier, in 1764, but its full significance was not suspected. About 1812 a young law student at Oxford, William John Broderip (1789–1859), took two lower jaws to his professor, William Buckland. The specimens were secured by one of the old stonemasons who collected for Broderip. Both Broderip and Professor Buckland thought the jaws were mammalian. But at that time it was generally conceded that mammals were Tertiary in age, and since these came from Mesozoic rocks they had to be something else. Though in 1818 Cuvier confirmed their identification as mammals, he thought they were opossums, probably distantly related to those he had been studying from the early Cenozoic of the Paris Basin. It was 6 more years before one of the little jaws was described by Buckland. Then opposition arose in earnest, with fierce discussions on the question. Finally Sir Richard Owen, later director of the British Museum, settled the issue by directing attention to their numerous mammalian characters (Fig. 140).

Among the fossil remains that drifted into this shallow embayment were parts of the volant reptiles called pterosaurs, dinosaur fragments, long-snouted crocodiles, a plesiosaur, a turtle, sharks and other fish, insects, ferns, conifers, cycadeoids, and a dicotyledonous angiosperm, as well as a few cephalopods and an abundance of pelecypods and gastropods. All of these gave a rather clear picture of life in this shallow sea and on the adjacent land. Most important were the mammals, for among the numerous specimens recovered were three orders: Multituberculata, Triconodonta, and Pantotheria. There is some doubt that the creatures with many tubercles or cusps on their molars were mammals, but, if not, they were reptiles very close to the multituberculate mammals of the late Jurassic. There is no question about the carnivorous triconodonts and the insectivorous pantotheres being mammals.

Purbeck Beds. Other important discoveries of this kind in which a correlation of marine and land faunas can be made is in the late Jurassic Purbeck beds, also in England. Fossil cycadeoids, as well as silicified stumps and logs of conifers, have been found encased in tufa. At another locality in the Purbeck, near Drulston Bay, insects were blown or washed into the water in abundance. This appears to be a rather temperate insect fauna of dragonflies, locusts, grasshoppers, butterflies, beetles, ants, and aphids. The modern aspect of these insects is much more apparent than in earlier Periods. Ostracods are dominant nearly everywhere in the Purbeck. But again the fossils of greatest interest are the teeth and jaw fragments of small mammals found in Beckel's Mammal Pit. There are nineteen species representing all five orders of Jurassic mammals. Multituberculates from these beds are mammals. There is a marked similarity in these mammals to those of the same age in the Morrison formation in Wyoming and Colorado. Other animals found in the Purbeck are unique dwarfed crocodiles and a few fish.

Holzmaden Black Slates. In the early Jurassic the Tethys Sea again spread northward over part of central Europe where it had been during the middle Triassic. The warm waters were relatively shallow. There were archipelagos of small islands made up of the older resistant rocks, and coralline reefs formed other low islands. In the area of Holzmaden, about 20 miles east of the city of Stuttgart, Germany, a nearly isolated embayment had filled with fine carbonaceous sediments that settled as black mud. There was little or no bacteria in the mud, but poisonous hydrogen sulfide gas formed in considerable quantity.

The Holzmaden embayment was occasionally inundated by marine waters and invaded by great numbers of ichthyosaurs. These pelagic reptiles were killed by the poisonous gas and sank to the bottom where they were entombed by the hundreds in the carbonaceous mud. Eventually the mud changed to Lias black slate in which the outline of the soft parts of the ichthyosaurs can be detected by a difference in its porosity. One of these dolphin-shaped ichthyosaurs contained some 200 belemnites in its stomach region, rather clearly attesting to the food habits of these reptiles. Another was found with unborn young in its body cavity and with one evidently in the process of being born. This gives evidence that even at that time some reptiles were viviparous.

Solnhofen Limestone. This fine-textured limestone was quarried near Solnhofen and Eichstätt in the Altmühl Valley, Bavaria, for many generations and sent to all parts of the world for use in the preparation of lithographic prints which played such an important role in the first publication of books. Immediately the presence of fossils and impressions of others were recognized. For here were not only the hard parts of extinct animals, but also the outlines of their soft parts so rarely preserved as fossils. Over 450 kinds of animals have been taken from these quarries.

In the late Jurassic these fine sediments were laid down in a lagoon evidently on an island in the Tethys Sea. The lagoon must have been separated from the sea by a reef formed of corals and sponges much like those in certain parts of the Pacific Ocean today. It was near shore because it appears at times to have dried up and have been flooded again with sea water. This would account for the assemblage of land-living vertebrates and marine life that became entombed in the sticky, limy ooze formed primarily from fine, coralline debris that had settled on the floor of the lagoon. Animals were buried so quickly that their skeletons were not disarticulated and in most instances the soft parts did not deteriorate before their outline was preserved.

Insects were abundant and, in some specimens, well preserved, giving a rather clear picture of their numbers and variety. There were moths and flies not known from older rocks. Soft-bodied animals like jellyfish, so rarely preserved as fossils, also were found. No fresh-water invertebrates were present but, numerous marine crustaceans, belemnites, and ammonites were discovered, together with an abundance of marine fishes. An unexpected find was a small dinosaur, no larger than a rooster, and its young, which somehow found their way into the lagoonal ooze. Pterosaurs also were trapped in the sticky substance, perhaps as they attempted to swoop down on some struggling creature (Fig. 141).

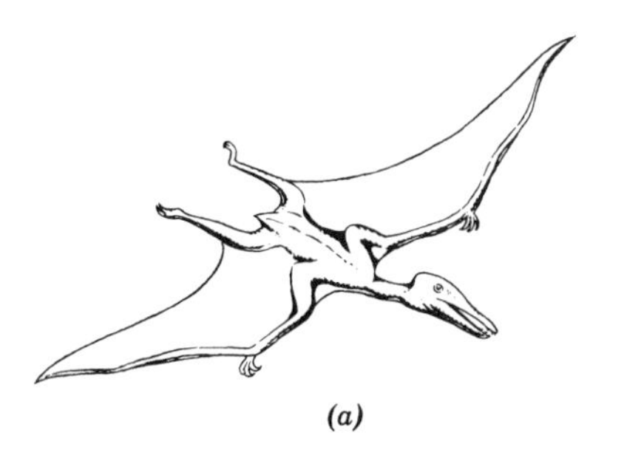

(a)

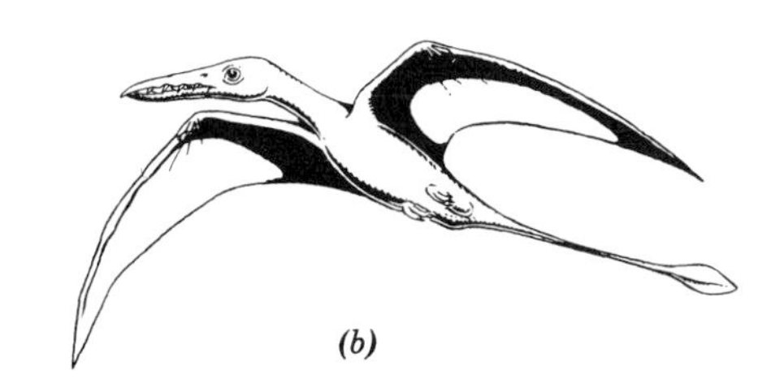

(b)

Fig. 141. Solnhofen pterosaurs: (a) Pterodactylus suavicus, *about the size of a sparrow;* (b) Rhamphorhynchus gemmingi, *about 2 feet long. After Abel.*

Twenty-nine kinds of pterosaurs, ranging from the size of a sparrow to others 4 feet in length, have been found. These interesting fossils showed how the delicate gliding membranes were supported by the greatly elongated fourth digit. One pterosaur had a long tail with a steering rudder at its end, and another was almost without a tail. But the most remarkable of all these creatures at Solnhofen was the oldest feathered animal known. This was the first bird, *Archaeopteryx* (Figs. 241, 242), about the size of a crow. Two most welcome specimens represented by their skeletons and an isolated feather would surely have been called reptiles if clear-cut impressions of their feathers had not been preserved. Though there were birdlike features in the skull and pelvis, the skeletons were more like those in reptiles. The bones were solid, not hollow and pneumatic (with air sacks), as in most modern birds, and the jaws were lined with sharp lizardlike teeth. Even a cast of the brain was more like that in reptiles than in birds.

These unique animals and their unusual preservation in the old Solnhofen lagoon have made valuable contributions to our knowledge of little-known extinct animals not found elsewhere in the world. The discovery of *Archaeopteryx* settled the long-debated question on the origin and relationships of birds, for it is now clear that they arose from a primitive group of diapsid reptiles, a group that would also include the ancestry of the dinosaurs and phytosaurs.

Jurassic in North America. Jurassic rocks are confined to the western and southern parts of the North American continent. Deep wells in Louisiana, Texas, and Arkansas show that a marine embayment extended that far up in the Mississippi Valley in the later part of the Period. As in the Triassic, most of the marine formations are in the Pacific province. One great embayment in middle and late Jurassic time extended down from the Arctic across Canada as far south as Colorado and Utah. Other inland seas came in from the Pacific all along the western coast of Canada into Washington, Oregon, California, and Nevada. Marine life was abundant in these Jurassic seas. As in the Jurassic elsewhere in the world, there was a wealth of ammonites, and one of the most characteristic mollusks was the little pelecypod, *Aucella* (Fig. 142).

We have seen that some reptiles moved back into the seas during the Triassic and continued into the Jurassic and Cretaceous. Among the most conspicuous were the plesiosaurs, which developed such long necks, and the dolphin or whale-proportioned ichthyosaurs. In the late Jurassic marine rocks of central Nevada, where the formations have been elevated 7000 feet above sea level, the largest ichthyosaurs of all time have recently been found. The immense size of these monsters is indicated by their skulls, which have been estimated to be eight feet long, and a vertebra that measures 13.5 inches in diameter. The total length of these denizens of the Jurassic inland seaways must have been about 50 feet. The partly disarticulated skeletons were found where the bodies drifted together and eventually were buried in the soft mud and ooze on the shallow sea floor.

Jurassic continental deposits superimposed on the Chinle Triassic and its equivalents spread over most of the Rocky Mountain and southwestern states, where in some areas great thicknesses of cross-bedded desert sandstones accumulated. The huge stone arch, Rainbow Natural Bridge, in southern Utah, has weathered out of these desert sandstones in the Navajo formation, and the picturesque sandstone cliffs in Zion National Park, also in Utah, are Jurassic.

An ancestral crocodilian, *Protosuchus*, was discovered in Dinosaur Canyon in Arizona by a Navajo Indian who led paleontologists to the site. Another recent find of even greater importance, which shows how much can be expected from these early Jurassic or late Triassic formations, is a quarry containing abundant remains of a mammallike reptile very close to one of the oldest known orders of mammals. These advanced reptiles called tritylodonts, with previous records restricted to Eurasia and South Africa, have multituberculate teeth. Many other features in the skulls and lower jaws are also quite close to those in the

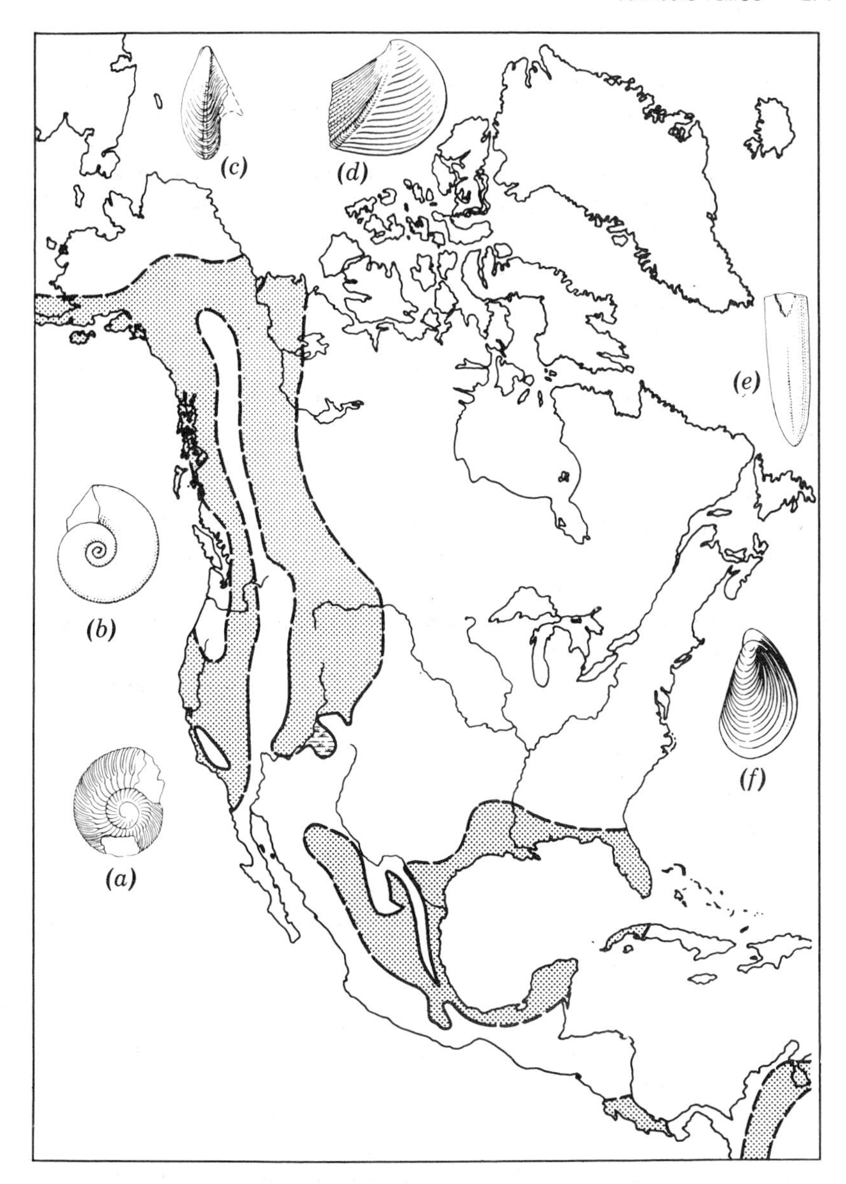

Fig. 142. Greatest inundation of seaways across North America in late Jurassic. After Schuchert. The fossils illustrated come from different parts of the Period: (a) Seymourites loganianus, *ammonite;* (b) Harpoceras propinquum, *ammonite;* (c) Gervillia montanaensis, *pelecypod;* (d) Trigonia americana, *pelecypod;* (e) Pachyteuthis densus, *belemnite;* (f) Aucella pallasi, *pelecypod;* (a-e) *after Shimer and Shrock;* (f) *after Pirsson and Schuchert.*

mammals. Stimulating finds like these tend to fill important gaps in our knowledge of the world-wide distribution of animals thought previously to be restricted to certain parts of the globe.

While erosion was going on in eastern United States evidence of much Jurassic history was being buried in the West. The inland seas with outlets to the Pacific were being continuously filled with sediments. These hardened into limestones, sandstones, and shales. Locally there were accumulations of volcanic detritus. A subsequent deformation metamorphosed the shales into the Mariposa slates with their gold-bearing quartz dikes.

Near the present Pacific coast deposits of silica accumulated locally and later hardened into the reddish radiolarian cherts of the Franciscan group. These rocks may be seen in the San Francisco Bay area and at the top of Mount Diablo. At least some of the Franciscan rock units have been demonstrated to be Cretaceous.

The batholithic intrusive rocks that are exposed in the high crests of the Sierra Nevada, and probably extend from Alaska and British Columbia into Mexico and to the tip of Baja California, are one of the greatest igneous intrusions since the Pre-Cambrian. Previously this was thought to have occurred in the Jurassic. More recently, however, information from radioactive minerals has revealed that the molten materials welled up under the overlying formations in both the late Jurassic and the middle Cretaceous.

Sundance and Morrison Formations. In the late Jurassic a great marine embayment spread down from the Arctic Ocean across Canada into the United States. It covered most of the Rocky Mountain states and extended as far east as the Black Hills in South Dakota. This was the relatively shallow Sundance Sea. It was typified by an abundance of belemnites, *Pachyteuthis densus*, a few ammonites, *Cardioceras cordiforme*, other mollusks, and some ichthyosaur remains. The fauna was boreal and related to those in Greenland, England, and Russia. Throughout most of its area the Sundance deposits did not exceed 300 feet in thickness, but toward the west in Idaho there was a much greater accumulation of sediments. A slight regional uplift and subsequent silting up caused the waters to retreat slowly to the north.

An alluvial plain crossed by streams and spotted with lakes then occupied most of the area previously covered by the Sundance Sea. This marked the deposition of the Morrison continental sediments and its world-famous dinosaur fauna. It was named from the exposures at Morrison, Colorado, but originally spread over an area of about 100,000 square miles. The discovery of dinosaur bones in the Morrison formation was made almost simultaneously by three men in 1877 in Colorado and Wyoming. Until then the huge bones of these creatures had been mistaken for fossil wood. The most important of these localities was at Como Bluff near Medicine Bow, Wyoming, just south of the old Union Pacific Railway station. Eventually several hundred tons of fossil bones were taken from the Morrison formation in this area and shipped to educational institutions in eastern United States. Among them were skeletons of the giant amphibious dinosaurs *Diplodocus* and *Brontosaurus*, remains of other saurians, and from one level came the teeth and jaws of the oldest mammals ever found in North America. About ten miles north of Como Bluff on Little Medicine River the bones of huge reptiles had weathered out in such numbers that a Mexican sheep herder had made the foundation and part of the walls of his cabin from the vertebrae. The excavation opened nearby was known as Bone Cabin Quarry. The first digging uncovered a six-foot femur, the tibia and fibula, and the complete foot of *Diplodocus*. The quarry gave an almost complete picture of the late Jurassic inhabitants of the region (Figs. 143, 145).

At Como Bluff, skeletons, frequently nearly complete, were found 20 to 100 hundred feet apart. This evidently was along the shoreline where

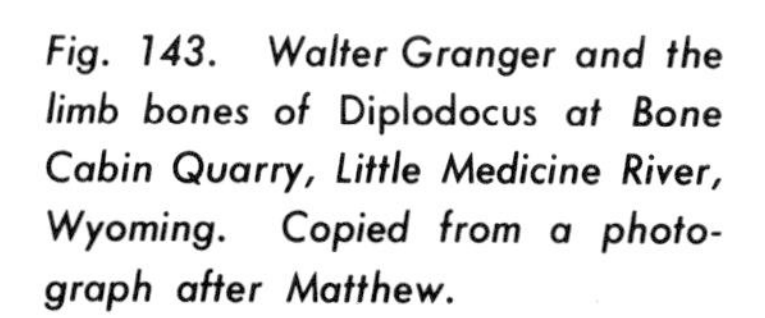
Fig. 143. Walter Granger and the limb bones of Diplodocus at Bone Cabin Quarry, Little Medicine River, Wyoming. Copied from a photograph after Matthew.

the dinosaurs bogged down in a muddy lagoon, whereas at Bone Cabin Quarry the bones of nearly all kinds of dinosaurs were buried together. In addition to the enormous amphibious sauropods, there were the carnivorous, or flesh-eating, *Allosaurus*, the armored *Stegosaurus*, the primitive duck-billed *Camptosaurus*, the small, lightly constructed bipedal *Coelurus*, and crocodiles and turtles. The skeletal remains in the quarry may have been carried downstream by rapidly moving currents in flood and finally buried together in a deeper pool.

There was much excitement when John Bell Hatcher (1861–1904), collecting dinosaurs at Como Bluff, sent a small mammal jaw to Professor Othniel C. Marsh at Yale University. Dinosaurs were nearly forgotten in an eager search for more of these little mammals. There was one thin, dark, shaly member near the east end of the bluff that yielded nearly all of the mammal remains found by Hatcher. It was discovered that this unit did not extend very far, but before the possibility of finding more was exhausted two dozen rare specimens were secured. As in the late Jurassic of England, five orders were represented. It would seem that they lived in the shelter of vegetation in the marginal swampland. Aside from this unique locality, only two specimens have been found elsewhere in the North American Jurassic. These mammals were among the vanguard, small and inconspicuous at first, which several million years later replaced the greatly specialized reptilian giants.

There were dense growths of scouring rushes in the swamps through

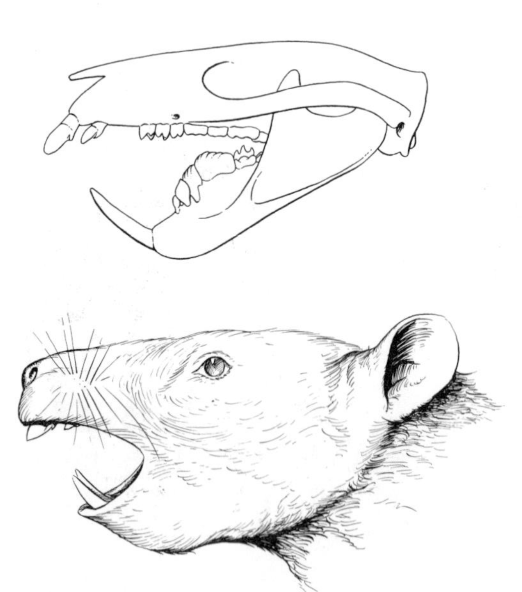

Fig. 144. Restored skull and head of a Jurassic multituberculate mammal, Ctenacodon. *After Simpson.*

which the big amphibious dinosaurs waded and at times partly floated as they fed on soft vegetation. On the bottom and frequently underfoot were fresh-water pelecypods and gastropods. Other water inhabitants were the crocodiles, turtles, and fish. Species of duck-billed dinosaurs, called *Camptosaurus*, frequented the swamp borders and drier terrain. The borders of the swamps were shaded with tree ferns, conifers, and ginkgos and had a ground cover of ferns. This was the rendezvous of ferocious carnosaurs, *Allosaurus*, *Coelurus*, and their allies. It was here that the little mammals found shelter, possibly unobserved because of their size or nocturnal habits, or they may have found protection in the boughs and foliage well beyond the reach of the reptile world below. Here also could have been the reptilelike birds, *Archaeopteryx*, for it is not likely that they were confined to the Old World, as the record now stands. They could well have been the worst enemy of the little mammals. The diet of both was probably predominantly insectivorous. It can be visualized that the plated *Stegosaurus* (Fig. 237) and also some of the flesh-eating dinosaurs ranged over the drier slopes among the clusters of cycadeoids. The cycadeoids, with their short barrel-shaped trunks and long palmlike leaves, apparently dominated the drier terrain.

One of the best places in the world to see dinosaur skeletons in the rocks is the Dinosaur National Monument near Jensen, Utah. The site was discovered by Earl Douglass in 1909. Twelve species of dinosaurs have been recognized, ranging in size from the small duck-billed dinosaur, *Laosaurus*, 6 feet long, to the great *Diplodocus*, 87 feet from the tip of its snout to the end of its tail. Since 1922 twenty-three mountable skeletons have been removed from the quarry, but now the bedding planes of the steeply inclined shale and sandstone are being prepared under the direction of Theodore E. White as a great museum to show the skeletons in their final resting places. A surface 200 feet long and 40 feet high will form the north wall of the museum. There the dinosaur bones can be seen in the rocks in the same positions in which they were buried in late Jurassic time. Work on the museum was started in November 1953 and should be completed in 1960.

Tendaguru in East Africa. Far away in East Africa is an isolated hill on a wooded plateau. This is Tendaguru, from which the famous dinosaur beds in this area take their name. The section is about 500 feet thick and shows three soft shale horizons with dinosaur bones separated by marine sandstones and conglomerates with an abundance of marine invertebrate fossils. Traced laterally, the marine units grade into the continental beds. Evidently there was an alternation of shallow seas that were filled with detritus and eventually changed into mud flats

Fig. 145. Comparative sketches of three large amphibious dinosaurs. These were the largest known land animals. Diplodocus *was at least 84 feet long, and* Brachiosaurus *was thought to have weighed about 50 tons.*

flooded by the rivers. The marine invertebrates are useful in correlating these rocks with the standard European section, and the dinosaurs can be compared with those in the Morrison formation in North America.

Most abundant at Tendaguru were the big amphibious dinosaurs. One of the genera, *Brachiosaurus*, was also found in the Morrison. They were massive brutes built somewhat like a giraffe, with the front limbs longer than those behind and with a small head perched on the end of a long neck. Single neck vertebrae have measured 3 feet in length. Since the nostrils opened on the top of the head, these reptiles were perfectly adapted to wading about in deep swamps where they fed on plants. The surrounding water helped control their body temperature. These were probably the largest reptiles that ever lived (Fig. 145).

Another genus, *Tornieria*, is quite like the long-necked and long-tailed *Diplodocus* from North America. Similarity is shown in many parts of the skeleton. The other amphibious dinosaur at Tendagaru is smaller and of more normal proportions, but it is peculiar in having the dorsal spines on the neck and thoracic vertebrae as two, instead of one, spinous processes.

The only armored dinosaur is *Kentrurosaurus*. This small dinosaur is referable to the same family as *Stegosaurus* from the Morrison, but the platelike armor, greatly reduced in the African form, is partly supplanted by very strong spines. These spines may have offered more protection against the voracious predator *Megalosaurus*. In any event the predaceous dinosaurs probably levied more heavily on the young of any dinosaur, even their own, than on the adults.

Also of much interest is one jaw of a small mammal (order Pantotheria) from Tendaguru. It is closely related to genera that Hatcher found at Como Bluff in Wyoming. These little Jurassic forerunners of all later placental mammals must have spread to all corners of the earth even at that early date.

Uniform Mild Jurassic Climates. Plants and reptiles alike show that mild climates again had spread over the earth almost from the North to the South Pole. Plant communities showed a marked resemblance everywhere, except for the seed ferns lingering on in the Southern Hemisphere. Big amphibious dinosaurs even reached Australia in the early Jurassic.

Pelagic invertebrates such as the free-swimming ammonites were extremely widespread. This is also true of the new groups of foraminifers which replaced the fusulinids of the late Paleozoic. When the marine ichthyosaurs and plesiosaurs are better known they too should show similar distributional patterns. The evidence indicates that mild temperatures also prevailed in the oceans.

Pteranodon

20 Cretaceous Period

Triceratops

Rapid expansion of flowering plants;
extinction of giant land and marine
reptiles; last of toothed birds;
rise of pouched and placental mammals;
modern groups of insects;
ammonites and rudistid clams die out;
most foraminiferan faunas cosmopolitan;
mild climates at first, cooler later.

Elasmosaurus

CRETACEOUS ROCKS were first studied in the London and Paris basins. Fossils were not so abundant as in the marine Jurassic; therefore it was some time before the significance and extent of the Period was fully recognized. In 1822 J. B. J. d'Omalius d'Halloy, a Belgian geologist, traced the formations from France into Belgium and called them Terrane Crétacé (Lat. *creta*, chalk), later altered to Cretaceous. Typical were the white cliffs on both sides of the English Channel. Though most of the white limestones of the world, known as chalks, are late Cretaceous in age, it should be remembered that the predominance of formations in the Period are limestones, shales, and sandstones. This was the last of the great inundations of the continents by marine waters. Of course, there were marine embayments during the Cenozoic, but they were never so widespread as those in the Paleozoic and the Cretaceous.

Continental formations and their contained records of life on land were extremely important. These sedimentary rocks were originally laid down in fresh-water lakes along stream courses and over wide alluvial plains. In many ways the evolution and extinction of life on land in the Cretaceous has been of much more interest to us than that in the sea. With the rise to dominance of the angiosperms and the appearance of modern groups of mammals, birds, and insects, the stage was set for a great transformation of life on land. Thus the rule of the reptiles gave way to the more progressive mammals and birds of the Cenozoic.

Jean Baptiste Julien d'Omalius d'Halloy, one of the most active geologists in Europe during the early part of the nineteenth century, was born at Liége, Belgium, in 1783. He was greatly influenced by the research of and personal contacts with Alexandre Brongniart in France and William Smith in England. One of d'Omalius d'Halloy's major contributions was a geologic map of the adjoining areas in France, Belgium, Germany, and Switzerland. Being an energetic man with much perseverance, he walked over the exposures and thus gained first-hand information in the field. D'Omalius d'Halloy held several important positions, such as Governor of the Province of Namur, Member of the Belgian Senate, and President of the Academy of Science in Brussels. He died at the age of 92.

Atlantic Coast. Sometime during the late Jurassic a proximity to the outline of the present east coast of North America took shape. The Cretaceous formations overlapping the present shore line from New Jersey to Florida derived their sediments from the Appalachian Mountains to the west. The lower part of this sequence was made up of continental sands and clays. Fossil land plants and the remains of crocodiles and dinosaurs, as well as the nature of the sediments of dark lignitic clays have afforded evidence of flood-plain deposits and local

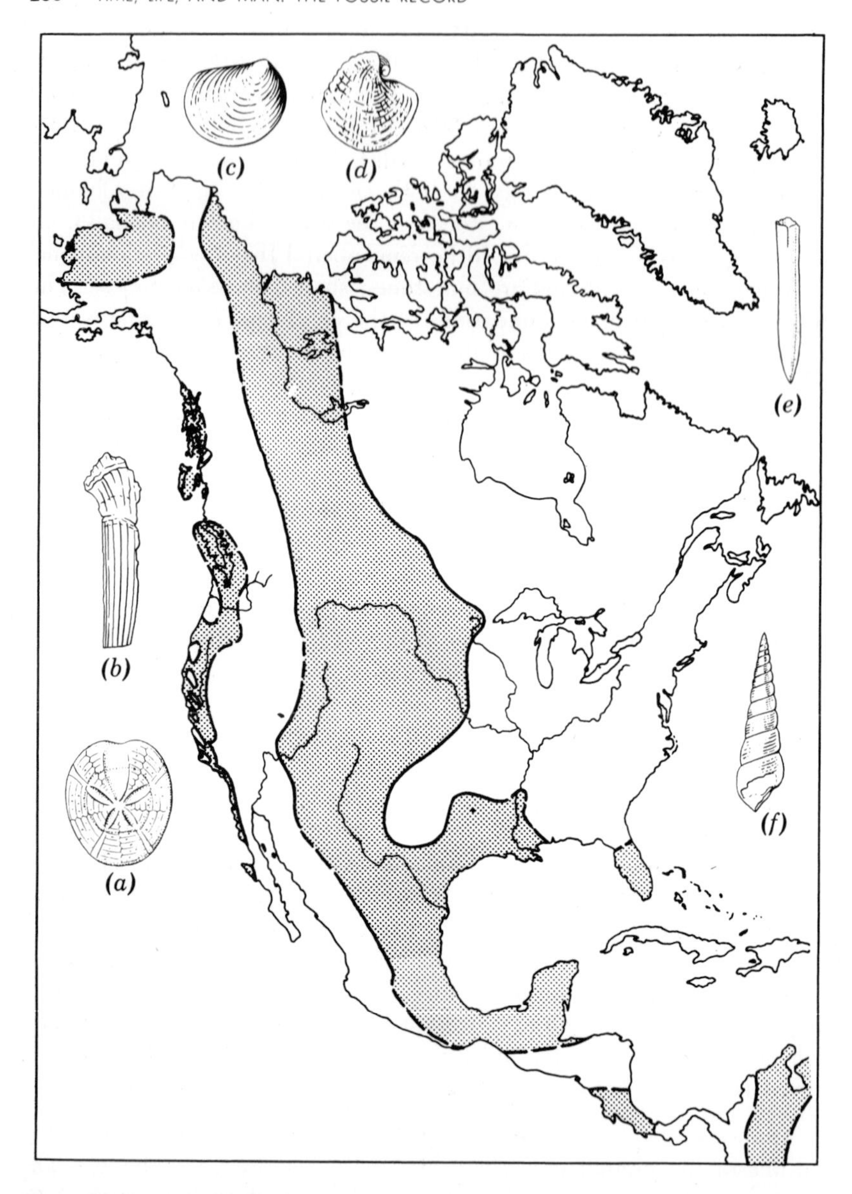

Fig. 146. Greatest inundation of seaways across North America in late Cretaceous. After Schuchert. The fossils illustrated come from different parts of the Period: (a) Hemiaster texanus; *echinoid;* (b) Radiolites davidsoni, *pelecypod;* (c) Inoceramus barabina, *pelecypod;* (d) Exogyra costata, *pelecypod;* (e) Belemnitella americana, *belemnite;* (f) Turritella whitei, *gastropod;* (a), (b), (e) *after Shimer and Shrock;* (c), (d), (f), *after Pirsson and Schuchert.*

swamps along a coastal plain. It was in one of these formations that the first dinosaur was found in North America. This duck-billed dinosaur, *Hadrosaurus*, was discovered while some excavations were being made at Haddonfield, a suburb of Philadelphia. Unfortunately, many of the bones were broken and carried away by curio seekers and lost. These early Cretaceous deposits were followed by the spread of marine waters over the coastal plain in the later part of the Period.

Gulf Coast. In the early Cretaceous the waters from the Gulf of Mexico spread northwest to connect eventually with the Rocky Mountain geosyncline. An arm of this sea extended northeast from New Mexico and Texas across eastern Colorado and Kansas into Iowa. Bioherms made up of the peculiar pelecypods, known as chamids and rudistids, associated with larger foraminifers, corals, and other organisms, were formed. Though the central states were never flooded as in the Paleozoic, marine waters extended up the Mississippi Valley to the southern tip of Illinois. Numerous pluglike salt intrusions in the later Cenozoic rocks were probably derived from extensive salt deposits beneath the coastal plain in eastern Texas. Much oil has been found in structural reservoirs around these salt plugs (see similar fault-trap accumulations in Fig. 201). As in other parts of North America, continental beds were laid down when the inland seas retreated; consequently early Cretaceous rocks in Texas have yielded a few mammal teeth, the remains of enormous amphibious dinosaurs, and an excellent angiosperm flora.

Rocky Mountain and Great Plains Areas. In the early Cretaceous an inundation of marine waters again spread down the Rocky Mountain geosycline from the Arctic. In the middle Cretaceous this had connected with the inland sea from the Gulf coast and Mexico and spread out a thousand miles in width from eastern Idaho and Utah to Minnesota and Iowa. Even the granitic domes of the old Colorado Mountains had eroded and were now covered by this epicontinental sea. Just to the west of the geosyncline was the Mesocordilleran geanticline which formed a chain of mountains extending from Alaska far down to the western part of Mexico. Thus for a time the continent was completely divided by a great seaway. Then the marine waters retreated both north and south. As the mountains to the west continued to rise, sediments were carried off to the east to fill the Rocky Mountain geosyncline. A vast swampy lowland with dense vegetation sprang up over the area previously occupied by the old sea floor. In many areas this luxurious plant life resulted in the formation of thick beds of coal. Intermittently, oceanic waters spread back over much of the area, bringing with it an abundance of marine life.

Particularly notable is the Niobrara formation with its large marine fauna of plesiosaurs, mosasaurs, pterosaurs, giant fish, toothed birds, and an abundance of invertebrates. The widespread, underlying Dakota sandstone is one of the finest sources of artesian water in the world. The Dakota is a land-laid formation, with fossil leaves perfectly preserved in small concretions. Abundant horned dinosaur remains occur in the Lance formation in the late Cretaceous, where they are associated with small mammals representing the multituberculates, marsupials, and insectivores. Most of the fossils of the little mammals have been found as isolated teeth, gathered in by big red ants and dumped on their mounds. There were many kinds of small opossumlike animals. Some

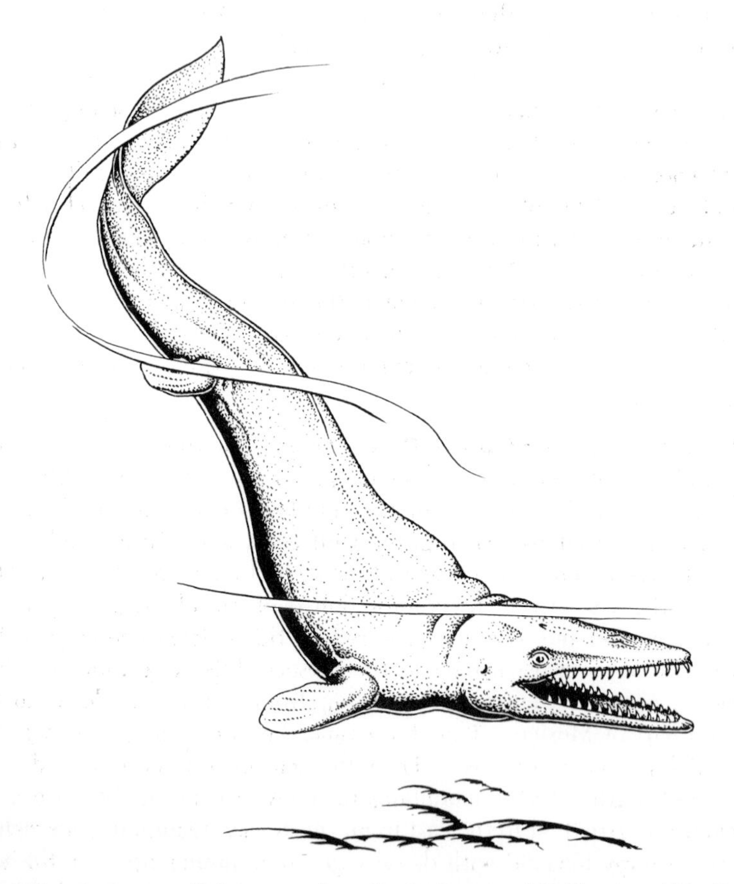

Fig. 147. A mosasaur, Kolposaurus, *from the Cretaceous of California. After Camp. These marine reptiles are related to the lizards. Some of the mosasaurs were over 25 feet long.*

nearly complete jaws have been found. To date these earliest marsupials have not been found in Cretaceous rocks outside North America.

Pacific Coast. Marine Cretaceous rocks in the Pacific coast province outcrop in Baja California, California, Oregon, Washington, and Alaska. There is reason to believe that these areas were not connected by a continuous inland sea. In central California a land-locked embayment extending from the San Francisco Bay area at least as far south as the lower end of the present San Joaquin Valley was bordered by land to the west. Elsewhere, the evidence indicates that the embayments opened directly into the Pacific Ocean, but much more field work must be done before the outline of these seas can be accurately mapped.

In California the Cretaceous record is better known than in other areas along the Pacific coast. During the early Cretaceous the area occupied by the present Coast Ranges was covered by a shallow sea. The eastern margin of this geosynclinal structure is near the center of the present San Joaquin and Sacramento valleys, and the western border lies west of the present coast line. Maximum thickness of these early Cretaceous beds in Tehama County is 22,000 feet (over 4 miles).

The old basin of deposition was gradually widened and deepened in the late Cretaceous. In Merced County 29,000 feet (nearly 6 miles) of sediments were laid down. Sediment accumulation kept pace with subsidence; consequently the water was always relatively shallow. The ancestral Sierra Nevada and the land to the west were the sources of most of the sediments. In the late Cretaceous the basin extended farther east than in the early part of the Period.

Invertebrate fossils suggest that the early Cretaceous seas in California were cooler than in the late Cretaceous. For the most part marine invertebrates are rare, but in some localities ammonites, pelecypods, foraminifers, and radiolarians are abundant. Remains of mosasaurs, plesiosaurs, and duck-billed dinosaurs also have been found along the western edge of the San Joaquin Valley. Other indications of vertebrate animals include numerous records of fish and one bone of a pteranodon from Oregon (Fig. 147).

In the late Cretaceous diastrophic disturbances resulted in an initial uplift in the Coast Range area in central California. This continued intermittently into the early Cenozoic as orogenic disturbances that were more or less contemporary with those in the Rocky Mountain province.

On Other Continents. The spreading of marine waters in the Cretaceous over continental margins and far over the surfaces of the present continents has been considered one of the greatest marine transgressions in geological history. Most of Europe was covered at one time or another during the Period. The Cretaceous seas of Europe apparently connected with the Indian Ocean across southwest Asia and the Persian Gulf. Evidence from invertebrate fossils indicates that the Indian Ocean was present but that much of India was covered with water and a marine connection separated Australia from the Asiatic mainland. East African and Australian ammonites are closely related. Some parts of equatorial Africa and eastern Brazil also were inundated. In Australia a marine embayment from Darwin in the north to the lower end of Lake Eyre in the south nearly divided the eastern from the western part of the continent. A similar geosyncline reached from Colombia to Argentina in South America, and other parts of that continent were also inundated. Continuous land connections between North and South America evidently were severed in the second half of Cretaceous time. Volumes have been written about the Cretaceous and its life, but much more work needs to be done.

Mid-Pacific Seamounts. Cretaceous corals, bryozoans, mollusks, planktonic foraminifers, and a sand dollar have been dredged from the slopes and bases of *guyots,* flat-topped volcanic seamounts 500 to 1100 fathoms (3000–6600 feet) below the surface waters of the northwest Pacific. These fossils are related to those in the old Tethys Sea which extended around the world at that time. These invertebrates lived about 300 to 510 feet below the surface in tropical to subtropical waters. Paleocene foraminifers deposited on the flat surfaces of three of the guyots indicate that volcanos were truncated by wave erosion in the Cretaceous. These are the oldest known rocks in the Pacific Basin (Hamilton, 1956).

Foraminifera. Foraminifers were abundant in chalky and calcareous marine formations throughout the world. They have been collected in tremendous numbers by petroleum geologists and studied in detail for their value as guide fossils. After the late Paleozoic, fusulinids died out, and a radiation in new groups of foraminifers started in the Triassic. Many groups including the calcareous lagenids were well established in the Jurassic, and the strongest relationships in the early Cretaceous samples are with those in the Jurassic. In the late Cretaceous, however, there was a marked change in the microfaunas when nearly all of the modern foraminiferan families made their appearance. Other older groups died out at the end of the Cretaceous. Planktonic (floating) and benthonic (bottom-dwelling) kinds were so prolific that their tests make up much of the rocks in which they occur. The species of *Gümbelina* were the most important natural chalk former in the Cretaceous rocks

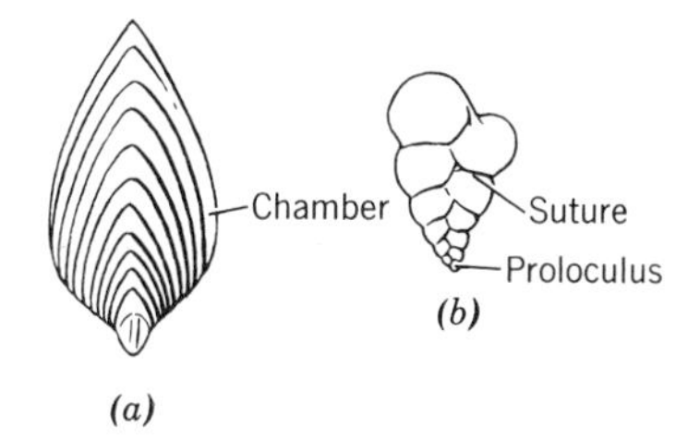

Fig. 148. Cretaceous foraminifers: (a) Frondicularia goldfussi, × 9, *after Cushman, family Lagenidae;* (b) Gümbelina globulosa, × 60, *after Cushman, family Heterohelicidae.*

of the world. They were thought to be planktonic because they accurred in extraordinary numbers and because their outer chambers are globular in contrast to the inner ones which were formed in tiny coils. Many of the smaller foraminifers have proven to be excellent index fossils in delineating even the smallest divisions of Cretaceous time. Phyletic lineages can be traced through two or three Stages (Fig. 148).

Among the commoner families were the arenaceous Verneulinidae, the trochospirally coiled Rotaliidae, and the elongate, biserially chambered Buliminidae. The calcareous benthonic genera *Bolivina* and *Uvigerina* were so widespread and had such long geologic ranges that their species are used extensively in correlation. The economic importance of Cretaceous foraminifers, as well as those in the Cenozoic and late Paleozoic, can hardly be overestimated.

Megainvertebrates. The evolution of mollusks had progressed by Cretaceous time to the extent that most of the later families were established. Many genera which made their appearance for the first time in this Period ranged far up into the Cenozoic. The most conspicuous of the Cretaceous groups to die out were the ammonites and the rudistid clams (Fig. 146).

Marine gastropods are extremely varied in Cretaceous rocks but are not so abundant locally as the pelecypods. Among the most useful groups in determining the age of stratigraphic units are the narrow high-spiralled nerineids. They appear suddenly in the Jurassic from an unknown ancestry and disappear as quickly after evolving into two dozen genera in the Cretaceous. Their characteristic features are found by making vertical sections through the shell to expose the internal structures.

Another genus of gastropods, *Turritella,* bears a marked resemblance to the nerineids, but the growth lines and other features are different. These shells make their first appearance in early Cretaceous fossil assemblages. Many species occur in Cenozoic formations, and others are still living. Their remains often are found as masses of high-spiralled, pointed shells or shell fragments where they have accumulated in depressions or sheltered pockets on the sea floor. Turritellas are gregarious scavengers frequently found together in great numbers.

The bivalved bottom-dwelling pelecypods are common in Cretaceous rocks. Not only are they more varied than in earlier Periods, but in some localities their shells make up most of the rocks. In the Pierre formation heaps of the little shells of *Lucina* are referred to as "Tepee buttes." Another oysterlike bivalve Zone has been traced 2500 miles from New Jersey, where it is 4 feet thick, to Texas where it exceeds 200 feet. These shells, so common in the Cretaceous, are *Exogyra.* A closely related genus is *Gryphaea.* In some localities in Texas and elsewhere, where the calcium carbonate cementation of the sand grains has broken down, the loose shells of *Gryphaea* cover the exposures. No less important is the genus *Inoceramus,* sometimes called the "pearl oyster." Some of these are 3 or 4 feet long. True oysters typified by their thick distorted shells are also abundant Cretaceous fossils. Some are excellent index fossils, and in some the shells are decorated with radiating ribs and tubercles. Among the many other bivalves is the genus *Pecten,* used as a symbol on every Shell Oil gasoline station.

Two other genera so characteristic of Cretaceous rocks are the coral-like chamaceans and rudistids. They are gregareous forms and are particularly noted for forming reefs. Thus they build the foundations for extensive bioherms which include the remains of other organisms. The sessile rudistids originated in the Jurassic but evolved and spread more widely in the Cretaceous. The attached valve continues to grow upward by addition at the base, and the other valve is a lidlike structure sitting on top. They are frequently thought to be fossil corals or even rams' horns by people unfamiliar with their structure. Some specimens are two or three feet in diameter.

The last great radiation of the ammonites started in the early Jurassic from one family that survived when other earlier groups became extinct. Many groups derived from this family existed for only a short time in the Jurassic, but others continued into the Cretaceous where they reached their maximum in size and specialization. Some of the gigantic coiled shells of *Pachydiscus* from Westphalia, Germany, were 7 feet or more in diameter and if uncoiled would have been about 35 feet long. Ammonites of all sizes were represented in the Cretaceous seas; some were no more than an inch in diameter. Some shells were partly and others almost totally uncoiled. Genera such as *Turrilites* and *Cochloceras* were very suggestive of helicoid (vertically spiraled) gastropods, but the sutures marking the margins of the internal chambers (camerae) have clearly disclosed the affinities of these invertebrates to ammonites. For the most part the sutures were extremely com-

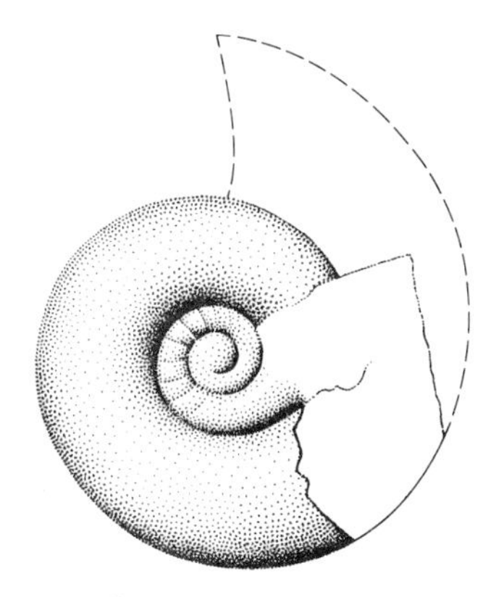

Fig. 149. The largest known ammonite, Pachydiscus suppenradensis, *from Westphalia, Germany, said to be 7 feet or more in diameter.*

plicated in the Cretaceous genera, but many secondarily evolved simple sutures, as found in the earlier Paleozoic genera. Those with retrogressive simplified sutures tended to uncoil the shells. At the close of the Cretaceous ammonites became extinct all over the world, thus culminating one of the most dramatic evolutionary histories among invertebrate animals (Fig. 149).

Dying out a little later in the Eocene was the last of the torpedo-shaped belemnites, which were abundant in the Cretaceous. These, of course, were more closely related to the squid, the cuttlefish, and the octopus than to the ammonites or the nautiloids. Belemnites were seldom perfectly preserved as fossils, since the softer parts at the mantle and head end tended to deteriorate. The elongate, conical, chambered portion of the internal shell was durable and was readily fossilized. The remains of these free-swimming belemnites were buried by the millions in Cretaceous sediments. Their torpedo shape suggested thunderbolt, or belemnite, the name which they bear, to Georgius Agricola (George Bauer of Saxony 1494–1555) in 1546.

Sea urchins and starfish were also well represented in the Cretaceous. Some of the most perfectly preserved starfish to be found anywhere came from the Austin chalk in Texas. Coral reefs were also frequent in the inland seas. Bryozoans expanded considerably, and brachiopods of the modern groups were plentiful.

Among the crustaceans, the ostracods and barnacles were abundant in certain areas. Because of their microscopic size and tremendous numbers ostrocods have been very useful in correlation. Larger crustaceans such as crabs and lobsters also have been found well preserved in numerous localities.

Marine Vertebrates. One of the best records of the vertebrate animals of the Cretaceous seas has been found in the Niobrara formation in the Great Plains area of the United States. Some of the first fossil bones to be uncovered by exploration parties were fish and reptile bones from the Smoky Hill drainage in Kansas Territory. Within a few years the collecting parties of O. C. Marsh and E. D. Cope, in their keen competition to excel each other in startling discoveries, were taking bones of fish, reptiles, and birds from the chalky white exposures along the borders of the ravines and washes. Collectors prospected with a pick in one hand and a rifle in the other. It might be said they had one eye on the ground for fossils and another on the horizon for Comanche warriors.

The Niobrara Sea flourished with life. Fish were there in great numbers. The teleosts or bony fish were large and small. *Hypsodon* (synonym *Portheus*), the giant of them all, attained a length of 15 feet or more. Schools of fish were pursued by the abundant swift-swimming mosasaurs which must have had insatiable appetites. Floating about were the plesiosaurs, with huge fat bodies and wide paddlelike limbs, ever ready to launch their heads and long necks deep into the water for passing fish. Larger fish ate smaller fish, for some have been found as fossils with complete skeletons in the stomach region. Huge marine turtles, *Archelon*, basked in the warm surface waters. Another highly specialized fish catcher was the wingless diving bird *Hesperornis* (Fig. 243). This bird was 3 feet long and showed many primitive features, the most conspicuous of which were teeth in its reptilelike lower jaws and in the back part of its beak.

Overhead glided the volant pterosaurs ready to scoop up fish that were frightened to the surface by fish-catching reptiles. *Pteranodon* had an almost unbelievable wing spread of 23 to 25 feet. Its birdlike bones were hollow and lightly constructed. The first piece of a bone found by Marsh where a bison trail had cut through one of the gully banks was thought to be a bird bone until he examined it in detail. No one has been able to determine for certain whether these long-crested reptiles were able to flap their wings or whether they were only capable of gliding.

Another bird, *Ichthyornis*, was about the size of a small seagull. This bird was quite unlike its contemporary, the flightless *Hesperornis*. The attachment areas for strong pectoral muscles have shown that this archaic bird was capable of long-sustained flight over the Niobrara sea. Birds like these have not been found elsewhere in the world.

A Cretaceous plesiosaur worthy of notice has been discovered in Australia. Details of the body proportions are not known, but the enormous skull was nine feet, 6 inches long.

Land Plants. It is now known that flowering plants were in existence long before Cretaceous time and that the most primitive kinds may have existed in some restricted highland areas as early as the Permian. The rapid development and wide distribution of these plants in the Cretaceous was in marked contrast to later floras. The difference between the early Cretaceous and middle Cretaceous floras was so conspicuous that Léo Lesquereux, who still believed in special creation, described the difference as modified by a new creation on the globe. The dominance of ancestral conifers, cycadeoids, and ginkgos of the earlier Mesozoic gave way to the spread of hardwood forests in the Cretaceous. From Greenland to South Africa and from Texas to Japan to Queensland,

Fig. 150. An early Cretaceous scene in England: Hypsilophodon, *small arboreal dinosaurs, 4 feet long;* Iguanodon, *the first dinosaur named, 23 feet long; and a ginkgo tree.*

Australia, the ancient plant world changed. By the middle of the Period the forests were essentially modern. There were magnolias, figs, sassafras, oaks, poplars, dogwoods, grapes, and other plants so familiar to us. Wherever the evidence was available it was clear that the Cretaceous species differed from those in the early Cenozoic, but most of the genera were alike. Grasses and cereals were also present.

These new plants, in abundance throughout the world, effected a great transformation in all life on land. There were flowers, nectar, and fruit for animals to feed upon. This was especially true in the insect world, where so many kinds depended on flowering plants for food, which cross pollinated a large percentage of the plants. A recent discovery of insects in Cretaceous amber in Alaska should give us much valuable information on a critical stage in the evolution of winged insects.

One of the most famous fossil plant localities in the world was discovered about 1865 in the Dakota sandstone of Ellsworth County, Kansas, by Judge E. P. West. For several years Charles H. Sternberg, collecting bag over his shoulder and pick in hand, took hundreds of small ironstone concretions from the soft yellow sandstones exposed along the hills. Every concretion seemed to contain a fossil leaf. Léo Lesquereux identified and described species of birch, poplar, linden, cherry, sarsaparilla, cinnamon, persimmon, sweet gum, cherry, fig, sassafras, and magnolia among these important collections. Leaves from this old coast-line forest with their abundant angiosperms were in marked contrast to the Jurassic vegetation.

Later in the Cretaceous more temperate conditions were clearly reflected in the plants. These environments accompanied a general uplift in most of the continent. Equally affected were the insects, birds, and mammals that found a new and plentiful food supply. All three classes evolved rapidly into numerous kinds, almost all of which depended directly or indirectly on the more modern kinds of angiosperms. Without angiosperms most of today's food could not be supplied.

Land Reptiles. The conspicuous land and inland lake animals of the Cretaceous were still the dinosaurs. Rather inconspicuous were the much smaller crocodiles, turtles, snakes, lizards, and occasional rhynchocephalians. The monstrous amphibious dinosaurs had nearly died out, though some very large types still inhabited the swamps in lower latitudes (Fig. 151).

Toward the end of the Cretaceous movements in the earth's crust which eventually culminated in our great chain of Rocky Mountains changed the northern Great Plains from an inland sea to a flat lowland marked by meandering streams and adjacent lagoons. Sycamores, wil-

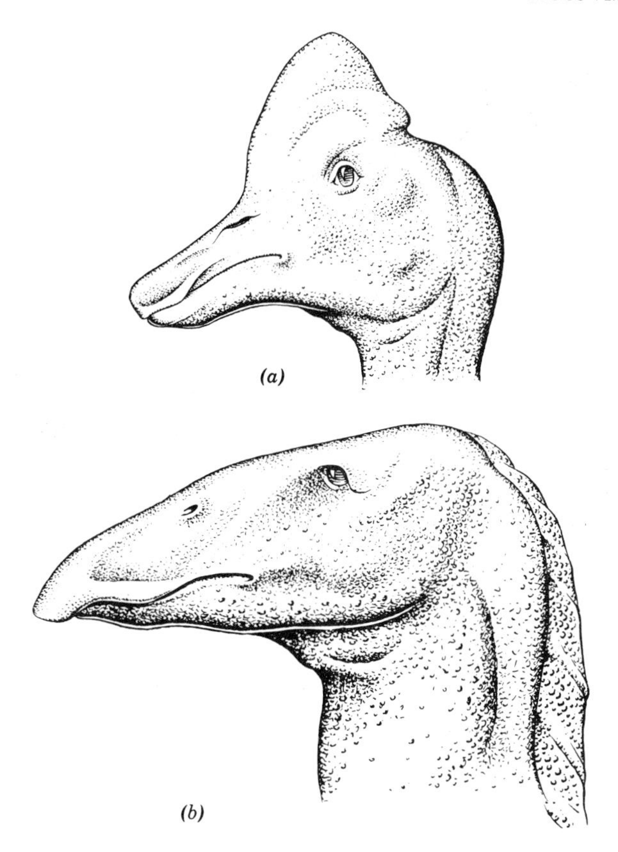

Fig. 151. A crested dinosaur and a broad-billed dinosaur: (a) Corythosaurus excavatus, *and* (b) Anatosaurus annectens. *These are called duck-billed dinosaurs.*

lows, and other plants of modern aspect fringed these bodies of water. There, too, were different kinds of saurians. The swamp-frequenting duck-billed dinosaurs were abundant and varied. Wading about on their strong hind legs or swimming gracefully, propelled by their webbed feet, they fed on the soft-tissued swamp vegetation. Frequently partly or totally submerged, they lived upon the succulent tubers and young plants loosely anchored in the muddy bottoms. The broad-billed *Anatosaurus*, the fan-crested *Corythosaurus*, and the long, bow-horned *Parasaurolophus* splashed about in their daily activities, while along the shores the huge armadillolike *Palaeoscincus*, with armored body and tail, forced his way through tangled undergrowth in search of plants to his liking. In the background was the peculiar ostrichlike *Ornithomi-*

Fig. 152. A restoration of the head of the horned dinosaur, Triceratops horridus. *The skull was found by cowboys in Wyoming and collected by John Bell Hatcher; about $^{1}/_{16}$ natural size.*

mus. It has been suggested that the toothless *Ornithomimus* ate the eggs of other dinosaurs. His long hind legs and slender build enabled him to outdistance any predator.

While J. B. Hatcher was collecting in eastern Colorado he came upon a peculiar horn. This was dispatched immediately to Yale University for Professor Marsh's attention. The only thing comparable to it were giant bison horn cores found at Big Bone Lick in Kentucky. But Hatcher was not satisfied because he was certain that the horn had come from a Cretaceous formation. When he heard of a similar horn in the possession of a Wyoming rancher he set out immediately to investigate. There he learned that some cowboys had located the head of a huge creature on a ledge and to get it down had thrown a lasso over its horn, which broke off. The horn was borrowed and also sent to Yale. Return orders were to get the skull at all costs. Since it weighed nearly 700 pounds, this proved to be a monumental task. The rancher insisted that the horn his cowboys had lassoed be returned. This was done, but as has been so often the case, the horn was subsequently lost at the ranch and never recovered (Fig. 152).

This was the first skull of the large-horned dinosaur *Triceratops*. Its two long horns projecting forward from above the eyes, a shorter one on the nose, and a wide bony shield over the neck made it a formidable opponent of the huge flesh eaters. These gigantic carnivorous creatures, like the *Tyrannosaurus*, were the largest of their kind. Moving about in a bipedal manner, they were 12 to 15 feet tall; their heads were 4 feet long, and their jaws powerful. Their forelimbs, however, were greatly reduced and probably of little or no use. They have been called the most terrifying beasts of all time.

Falming Cliffs of Central Mongolia. The American Museum's Central Asiatic expedition of 1922, led by Roy Chapman Andrews, had finished their field work for that season. Walter Granger (1872–1941) and Wang, his Chinese assistant, had already uncovered dinosaur remains at Gurbun Saikhan, but they were little aware of the tremendous discovery soon to follow. In the late afternoon on the second of September the party had stopped along an old caravan trail to seek information from some Mongols about a shortcut that led north to the old Chinese post road. In the meantime J. B. Shackelford, the photographer, wandered away from the fleet of cars into the purple shadows and deep red exposures where erosion had dissected a labyrinth of ravines, gullies, and sheer cliffs. There he discovered a beautifully preserved white skull about 8 inches long, resting on the tip of a sandstone pinnacle. He hurried back to his comrades with the specimen which was exposed enough to be identified as a peculiar reptile skull. Two hours of exploration revealed bone fragments all over the place. Walter Granger found the first fragments of dinosaur eggs, though they were not recognized as such at that time. This place of brilliant red sandstones capped with black lava was named the Flaming Cliffs.

In the following summer Granger was back again with a team of the best collectors from the American Museum: Albert F. Thompson, Peter C. Kaisen, and George Olsen. Olsen found the first dinosaur nest, and though most of the eggs were badly broken the shell surfaces were perfectly preserved. Their identity was then recognized because Granger received a cable from the American Museum stating that the peculiar reptile skull found by Shackelford had been prepared and that it was a primitive ceratopsian dinosaur. It was named *Protoceratops*. In the sandstone four inches above the nest were the bones of a small toothless dinosaur which was named *Oviraptor philoceratops* (egg seizer + fondness for ceratopsian eggs). Many more and much more perfect nests of *Protoceratops* were found. In all about seventy skulls, fourteen skeletons, and about thirty eggs were recovered. Many more were found that could not be taken for lack of time and space to transport them.

Fig. 153. A restoration of Protoceratops, *the egg-laying dinosaurs found in the Flaming Cliffs of Mongolia. Modified after Charles R. Knight.*

These small ceratopsian dinosaurs differed from the American kinds in not having horns and were considered as ancestral types. They lived in an area of drifting sand in which they laid their clutches of eggs to hatch in the warm sunshine. Sand driven by the wind covered them so deeply they finally became fossilized. This has been visualized as a landscape with sçattered shrubs and occasional pools of water resulting from thunderstorms. Though *Protoceratops* was particularly abundant, there were many other animals about, including smaller and larger dinosaurs (Fig. 153).

Some one in that 1923 party picked up a small concretion with evidence of bone inside. This was not prepared until after the 1925 party was in the field. A hurried message to Walter Granger from William Diller Matthew, back in New York, informed the men in the field that the specimen was one of the rarest of fossils, a small Cretaceous mammal skull. The urgent request to get more of them resulted in Granger's having another within thirty minutes. History repeated itself—dinosaurs were forgotten while everyone carefully examined concretions for mammals. Several poorly preserved skulls were discovered. Here were multituberculates and insectivores, one of which is possibly close to the ancestry of all later placental mammals. The curiosity of man to seek information from remote corners of the earth once again had paid off.

Other mammalian remains had been found in the Cretaceous of North America, but nothing as complete as these little treasures from the heart of Asia. In the Lance formation in Wyoming Hatcher first found teeth of these little creatures in ant hills to which big red ants had carried them as they did grains of sand. Field parties under the direction of Donald E. Savage and William A. Clemens of the University of California recently recovered hundreds of teeth and dozens of jaws of these fascinating little mammals by washing and screening the sedimentary rocks in which they occur. Most of a lower jaw and pieces of the cranium of an opossum, *Eodelphis,* were discovered in the dinosaur beds in Canada. In America multituberculates, marsupials, and insectivores have been found. Somewhere on this continent, sooner or later, someone will find complete skulls and perhaps skeletons which will reveal pertinent information on the ancestry of all later mammals.

Laramide Orogeny. In the later part of the Cretaceous and continuing on into the Cenozoic there were profound movements in the earth's crust which extended all the way from Alaska to the southern tip of South America. Great batholithic masses welled up under the

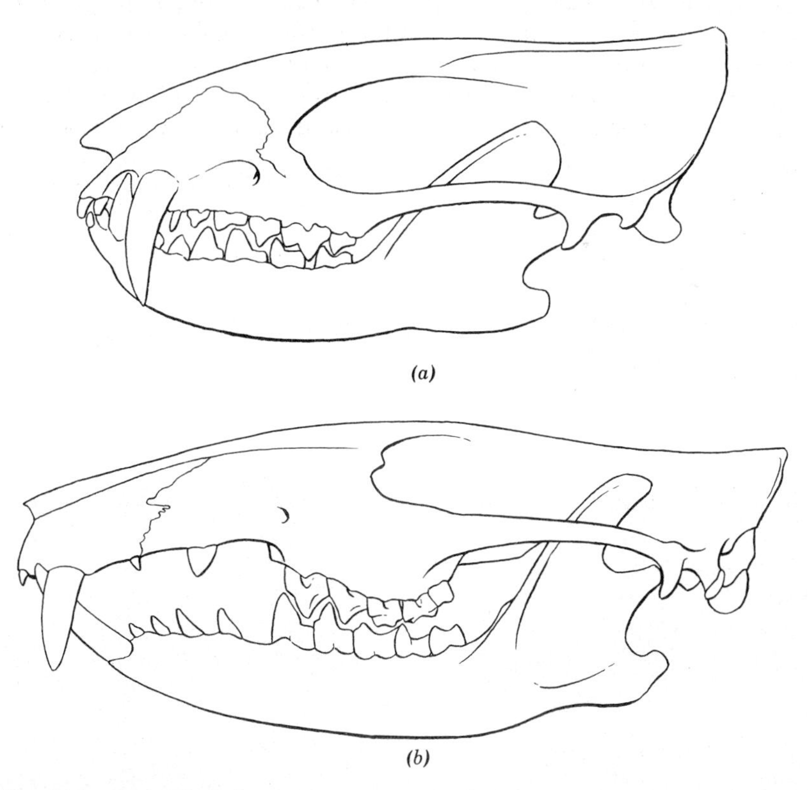

Fig. 154. Restored skulls of small Mongolian Cretaceous insectivores: (a) Deltatheridium pretrituberculare *and* (b) Zalambdalestes lechei. *After Simpson.*

present Rocky and Andes mountain chains. The Sierra and other areas were similarly affected. Not since the close of the Proterozoic had the Western Hemisphere experienced such a disturbance. The sedimentary rocks in the old geosynclines were folded and faulted, and in some places underlying formations were overthrust upon later rocks. These movements in the lithosphere not only altered surface features and climatic conditions but indirectly accelerated evolution in some groups of organisms or extinction in others. The angiosperms, insects, birds, and mammals experienced pronounced diversification and change. On the other hand, the dinosaurs, ichthyosaurs, plesiosaurs, mosasaurs, pterosaurs, toothed birds, ammonites and rudistid clams, some other mollusks, several genera of echinoids, and some foraminifers failed to survive beyond the Cretaceous.

All of these animals did not die out suddenly everywhere. Some must have lingered long after others were gone. Nevertheless, within a few million years many groups that had made up important parts of Mesozoic environments faded out of the picture. Among the remaining dinosaurs it seems to have been rather sudden, for in the lowermost part of some formations their bones and tracks may be found, but in the higher strata they were totally replaced by the earliest Cenozoic mammals.

21 Cenozoic Era

Dominance of mammals, birds, insects, angiosperms, gastropods, pelecypods, echinoids, and agglutinated and calcareous foraminifers.

CHANGES IN TOPOGRAPHY, climate, and environmental conditions in general in the late Cretaceous resulted in profound changes in animal and plant life throughout the world. On land the large reptiles, so conspicuous in the Mesozoic, and other groups like the toothed birds were not able to adjust themselves to these new conditions. They gave way as an abundance of new mammals, modern birds, and the groups of insects with which we are familiar today took over the new environments. For the most part the remaining reptiles were lizards, snakes, crocodilians, and turtles. These new radiations in life were made possible largely by the abundance of angiosperms which preceded the later animals and provided them with a new and plentiful supply of food.

After the remaining geosynclines into which inland seas had spread were folded and faulted into mountains near the end of the Cretaceous, the continents began to assume their present outlines. Throughout the Cenozoic there were invasions of the sea over certain rather restricted continental areas. These were of greater extent in some Epochs than in others, but they never reached the magnitude of those in the Paleozoic or the Cretaceous. As stated previously, some invertebrates so typically Mesozoic died out before the Paleocene. Others ranged on into the Cenozoic, not even diminishing in numbers, whereas some groups which had appeared in small numbers in the Cretaceous increased tremendously in early Cenozoic faunas.

The names "Kainozoic" or "Cainozoic," here called *Cenozoic* (recent life), were introduced by John Phillips (1840–1841) for the Era following the Mesozoic. This included the Tertiary and Quaternary Periods, named previously by Arduino and by Desnoyers. On the whole the Cenozoic corresponded essentially to the "Angeschwemmtgebirge" of Johann Gottłob Lehmann and of G. C. Füchsel of the eighteenth century. Some 17 years earlier Sir Charles Lyell had started the field work on his classic three volume *Principles of Geology.* It was in Lyell's third volume, published in 1833, that he divided the Tertiary into three parts: the *Eocene* (dawn + recent), the *Miocene* (minor + recent), and *Pliocene* (major + recent). The Pliocene he separated into *Newer Pliocene* and *Older Pliocene.* Later he restricted Pliocene to his Older Pliocene and coined a new name *Pleistocene* for the Newer Pliocene.

Charles Lyell was a Scot by birth. He was born in 1797, the son of a rich proprietor, about the time William Smith and Georges Cuvier were attaining the results which later made them famous and when Napoleon was approaching the peak of his career. Lyell later lived in London and was educated at Oxford. Though always happy to exchange ideas with others, he was an independent worker and clear thinker. He disliked being tied down with teaching. Having the means to travel, he did so extensively, assembling information from the field. He soon discovered that Cuvier's catastrophic theory was untenable. By adopting the dictum of uniformitarianism, "the present is a key to the past," he carried this idea much further than did James Hutton and supported it day after day with original observations. Thus he noted that the erosional processes, the deposition of sediments, the ways in which animals and plants become buried today, etc., were the same processes that had taken place in past geologic Periods.

Lyell recognized the sequence of Cenozoic rocks in the London and Paris basins which had been classified by Georges Cuvier and Alexandre Brongniart in 1810. To this he added the sequence he had observed himself in southern France, Italy, and in the Vienna Basin. He consulted Franco Andrea Bonelli (1784–1830) of Turin and Oronzio Gabriele Costa (1787–1867) of Naples, who had assembled large collections of invertebrate fossils. Later in Paris he met and collaborated with Paul Deshayes, an invertebrate paleontologist, who had his large fossil shell collection arranged according to their occurrence in the previously classified sequences of stratified rocks. Lyell and Deshayes found that the rocks stratigraphically lower yielded fewer species bearing close relationship to those in the seas today and that the nearer the formations were to the top of the stratigraphic column, the greater was the percentage of living kinds. Even before Lyell met Deshayes he had decided to set up the four subdivisions, which we still include in the Cenozoic, on the basis of their stratigraphy and on the fossil data.

LYELL'S TYPE LOCALITIES

Type of the first subdivision	*Eocene*	London and Paris Basins
Type of the second subdivision	*Miocene*	Touraine, France
Type of the third subdivision	*Older Pliocene*	Subapennine Strata, Northern Italy
Type of the fourth subdivision	*Newer Pliocene*	Ischia and Val di Noto, Italy

Lyell stated that further work might reveal the necessity of recognizing additional divisions of the Cenozoic. He seemed particularly cognizant of a time interval between the Eocene and the Miocene. When the Cenozoic sequence became better known the Oligocene and Paleocene Epochs were added to the time table.

CENOZOIC EPOCHS CURRENTLY RECOGNIZED

Epoch	Author	Date	Original Evidence	Country
Pleistocene	Lyell	1839	Invertebrates from marine sediments (now recognized on glacial and interglacial Ages)	Italy
Pliocene	Lyell	1833	Invertebrates from marine sediments	Italy
Miocene	Lyell	1833	Invertebrates from marine sediments	France, Italy, and Austria
Oligocene	Beyrich	1854	Marine sediments	Germany
Eocene	Lyell	1833	Invertebrates from marine sediments	France and England
Paleocene	Schimper	1874	Plants and brackish to continental sediments	France

Since the Paleocene was the last of the Cenozoic Epochs to be proposed it was more than 50 years before it was accepted by most geologists and paleontologists. It was usually considered as basal Eocene. Mammalian paleontologists were among the first to advocate its extent and usefulness. In the Rocky Mountain area of the United States a long interval between the last survivals of the dinosaurs, ammonites, and other distinctive Cretaceous forms and the first eohippus has been clearly indicated by the sequence of mammalian faunas. Furthermore, detailed work on the marine fossils and the formations in which they occur has also justified the recognition of another Epoch.

Cenozoic rocks for the most part are less indurated, or consolidated, than those in earlier Periods of geologic time. Excellent sequences occur in the United States. In contrast to the Paleozoic and Mesozoic Periods, the deposits in marine embayments with their abundant invertebrate life are restricted to the margins of the continents, whereas extensive fresh-water formations with evidence of terrestrial life are found toward the interior.

Tertiary sedimentary rocks of the United States have yielded excellent fossil materials in the six paleontologic provinces: *Atlantic coastal plain, Gulf coastal plain, Great Plains, Rocky Mountain, Great Basin,* and *Pacific coast.*

The *Atlantic coastal plain,* which stretches almost continuously from New Jersey to Florida, is the least extensive, though formations referable in age to each of the Tertiary Epochs are represented. Aside from Florida's coastal, land-laid Miocene and Pliocene rocks, with their characteristic mammalian remains, the formations along the Atlantic coast are marine, though Miocene land mammals are infrequently found in the Calvert marine formation.

Marine waters spread much more widely over the *Gulf coastal plain.* Indeed the Gulf coast, much of Mexico, Central America, and the continental borders of the Caribbean sea were inundated at one time or another during the Cenozoic. In the early Tertiary marine waters extended up the Mississippi Valley as far as the confluence of the Ohio and the Mississippi rivers. From this northern most point they gradually retreated in subsequent Epochs to their present position. Continental sediments containing plant or animal fossils were washed in from the adjacent land from time to time or were buried on coastal plains and in structural depressions in places as clues to the contemporaneity of life on land and in the sea.

The *Great Plains* province east of the Rocky Mountain front stretches from the panhandles of Oklahoma and Texas north to Alberta in a blanket of continental sediments, which are almost exclusively middle and late Tertiary in age. They were laid down in stream and flood-plain deposits primarily derived from uplifts in the Rocky Mountains. Some of the later formations are made up partly or, in some cases, entirely of reworked sediments. This area developed into one of the most important centers of evolution and dispersal of open grassland mammals known to the world.

The early Tertiary continental record in North America is contained primarily in the *Rocky Mountain* province, and Paleocene and Eocene formations have been located discontinuously from New Mexico to Saskatchewan. These deposits, in lakes, in stream channels, and on flood plains, were laid down in the old intermontane basins formed by the Laramide disturbances and have yielded the best and most complete record of early Cenozoic land life to be found anywhere in the world. Scattered late Cenozoic faunas and floras have also been discovered in this province, but as in the Great Plains there are no marine formations.

The *Great Basin* of Nevada, southern Idaho, and eastern Oregon also has no marine formations. Though some formations are reputed to be of early Tertiary age, fossils have not yet been found to support this age assignment. Miocene and Pliocene mammals and plants have been found almost throughout the area, clearly indicating much more humidity than prevails there today. Forested areas, with dawn redwoods and other trees, grasslands, and an abundance of mammalian life, are clearly demonstrated by the fossil record.

The *Pacific coast* province and the Gulf coast are somewhat alike in their thick marine sections and interfingering of continental formations. The abundant information now available on Cenozoic marine invertebrates in rocks all the way from Mexico to Washington and even in Alaska has offered a clear picture of the different environments in which the shell-life lived. In 1919 the temperatures of these Pacific waters was determined and the isotherms were plotted with considerable accuracy by comparison of the fossils with the Recent faunas by James Perrin Smith of Stanford University. The marine Eocene faunas show that a tropical-to-warm temperate climate extended from the equator far to the north, quite in contrast to the Pleistocene when at times subboreal temperatures occurred nearly as far south as San Diego. Similar temperature conditions are evident for other parts of the world. The marine formations, with their large fossil assemblages, are unusually thick in many of the old inland troughs where deposition was almost continuous. The marine sequence in the Coast Range of California has been studied and mapped in considerable detail during the past 60 years.

In the Northern Hemisphere the most extensive marine embayments

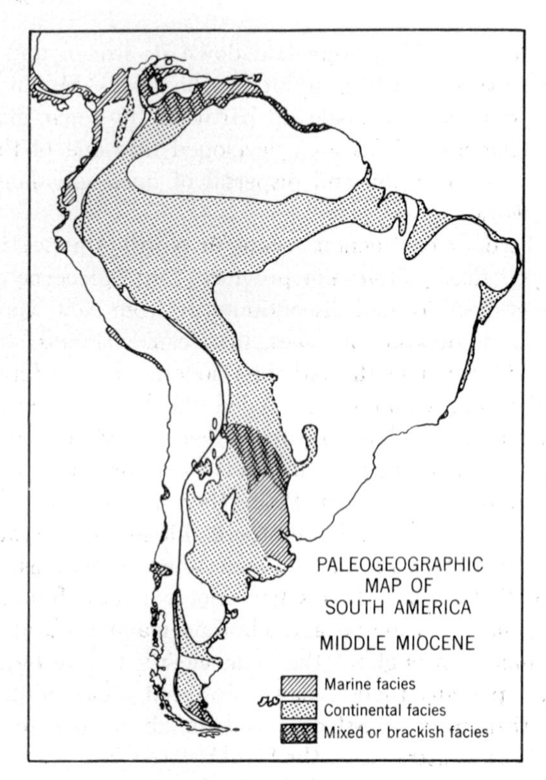

Fig. 155. A map showing the distribution of sedimentary rocks. After Weeks.

have been mapped from Paleocene, Eocene, and Miocene rocks. In Europe Tertiary rocks are widely known in the classic areas of England, Belgium, France and Italy. Many other areas, such as the Vienna Basin, much of central Europe and Russia, have yielded much to our knowledge. Fortunately, the occurrence of both marine and continental formations in some areas has offered an opportunity to establish approximate contemporaneity of life on land with that in the sea. This has been particularly valuable when the fossils have been compared with those from the type sections.

The Tertiary record of land and marine life is excellent from northern Africa and southern Europe, southwestern Asia, India, Burma, and on into the Indo-Pacific. The American Museum's Asiatic expeditions and, more recently, Russian and Chinese paleontologists have disclosed the potential of an outstanding record of terrestrial life in central Mongolia and in other areas in the interior of Asia. Many parts of Africa are also unexplored, but intriguing reports have been written on the

Tertiary mammals of Kenya and southwest Africa. It seems evident that clues to the origins of many families and genera of mammals are awaiting discovery on that continent.

Marine Tertiary deposits lay farther inland than the present shore lines in Tasmania and southern and western Australia, but no Cenozoic marine embayments penetrated so far into the mainlands as in previous Eras. In New Guinea Tertiary formations occur primarily in structural troughs. Continental formations with the fascinating remains of marsupials have recently been found east of Lake Eyre and elsewhere on the continent of Australia, as well as in New Guinea.

South America was isolated from North America some time in the Cretaceous or in the early Paleocene until the early Pleistocene, as clearly indicated by the distinctness of the Tertiary land mammals. Consequently, the terrestrial life in South America evolved quite independently from that on other continents, except for certain chance dispersals which brought some representation (monkeys and rodents) from other regions. The records of these peculiar land animals have been found mostly in Argentina, Brazil, and Colombia, though some continental formations have yielded materials from other areas. Marine embayments did not penetrate so deeply into the continent as they did in the Paleozoic and Mesozoic (Fig. 155).

22 Tertiary Life in the Sea

Agglutinate and calcareous foraminifers abundant;
gastropods, pelecypods, and echinoids,
important groups of larger invertebrates;
teleost fish most abundant marine vertebrates;
origin and evolution of whales,
sirenians, desmostylids, and pinnipeds.

PALEOCENE INVERTEBRATES have been recognized on every continent. For the most part the genera and species were not so widely distributed as in previous Periods. Though genera, such as the nautiloids *Hercoglossa*, the gastropods *Turritella*, and the pelecypods *Venericardia*, as well as some planktonic foraminifers, such as *Globigerina* and *Globotruncana*, seem to have had their species distributed around the globe, identical species have seldom been found in areas so far removed as California and the Caribbean. But areas as close as Copenhagen in Denmark and Mons in Belgium have few species in common. Environmental tolerances evidently account for these differences.

Foraminifers. Many families of Tertiary foraminifers appear first in the Cretaceous, then expand rapidly in the early Cenozoic. The warm tropical-to-subtropical waters in the Paleocene and Eocene swarmed with these microscopic organisms. Agglutinated (arenaceous) and calcareous forms were particularly abundant. In certain rocks the presence of planktonic kinds are clearly indicative of open connections from coast line marine embayments to the open ocean. Neritic, bathyal, and abyssal depths in these inland or near-shore troughs are determined by comparison of the fossils with their nearest living relatives and the environments in which they live.

The abundance of these microorganisms has facilitated Stage and Zone refinement in the classification of marine time-rock units. Sixteen Stages have been recognized in the Cenozoic of California, from the Santa Ana Mountains to the San Francisco Bay area. These form an almost uninterrupted vertical sequence, though not represented in totality in any one section.

For the most part our present knowledge will not support a correlation of Stages and Zones beyond the boundaries of provinces such as the Pacific coast or Gulf coast. Moreover, Stages and Zones are much more readily recognized in rocks of marine origin than in most continental rocks.

Among the early Tertiary larger foraminifers are the benthonic genera *Nummulites* and *Discocyclina.* The nummulites may have been the first fossil foraminifers to be observed in rocks. Egyptians encountered them almost 5000 years ago while constructing their Pyramids of stone made up largely of coin-shaped tests. Species of the genus *Nummulites* were widely distributed in the old Tethys Sea which stretched from northern Africa and southern Europe through southwest Asia to India and into the East Indies. These foraminifers were most abundant in the middle and late Eocene and throughout the Old World characterize the interval from the base of the Paleocene through the early

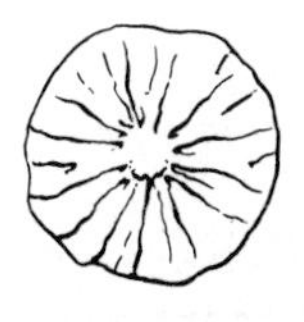

Fig. 156. Eocene foraminiferan, Discocyclina radians, $\times 2\frac{1}{2}$. *Family Discocyclinidae. After Cushman, from* Foraminifera, *Harvard University Press.*

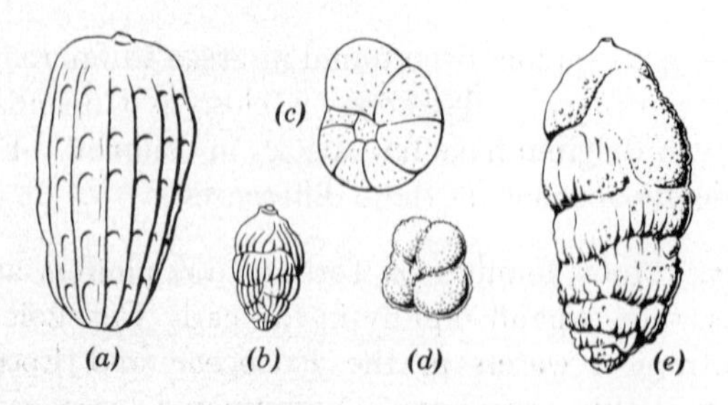

Fig. 157. Foraminifers from the Miocene of California: (a) Siphogenerina reedi, × 25; (b) Uvigerinella obesa, × 21; (e) Uvigerina joaqinensis, × 35. *Family Buliminidae;* (c) Valvulineria californica, × 45. *Family Rotaliidae;* (d) Globigerina bulloides, × 29. *Family Globigerinidae. Redrawn from Kleinpell,* Miocene Stratigraphy of California, *by permission of American Association of Petroleum Geologists,*

Oligocene. In the Western Hemisphere their place was taken largely by the related genus *Operculina* in the Caribbean province, where *Nummulites* were rare and are known only from the late Eocene. Another family, the Discocyclinidae, represented by *Discocylina* and a few other genera with intraseptal canals and marginal cords which show relationships to the nummulites, was distributed around the world in the same latitudes. Wherever discocyclinids are found they are indicative of the Paleocene or the Eocene Epoch (Fig. 156).

Lepidocyclina, of the family Orbitoididae, which lacks a central canal system and is not so closely related to the nummulites, seems to have taken up where *Discocyclina* gave way in the late Eocene. This genus was most abundant in the early Miocene and survived through the late Miocene. Orbitoids experienced world-wide distribution in tropical and warm temperate latitudes. Many other larger foraminifers are also important in the Tertiary record.

Less conspicuous but in far greater numbers are the smaller foraminifers. Included here are the planktonic Globigerinidae. The genus *Globigerina* is recognized readily by its cluster of globular chambers which are dotted with pores. The species are widely distributed in all Cenozoic Epochs. Their abundance is indicated in sediments which solidified from oozes on the floors of oceans, in which some rocks are made up almost entirely of the tests, or shells, of these microscopic organisms. This and a closely related and comparably specialized family, the Globorotaliidae, include the planktonic foraminifers of our present oceans.

Another well-known family is the Buliminidae, which includes numerous genera widely distributed in Cenozoic faunas. One of the most useful genera among these tiny benthonic foraminifers with calcareous tests is *Siphogenerina.* Several Zones in the Miocene of California take their names from its species. Their beautiful tests tend to be elongate, though many youthful specimens still are rather short. The pointed ends of these microscopic foraminifers and the parallel ribs with which they are generally ornamented are suggestive of the forms seen in some tropical fruit. Though species occur in other provinces, many of them were endemic to the Pacific coast and the California embayments in particular. Phyletic lineages are clear in the family Buliminidae. Other genera referable to the family and often having comparably decorative tests are *Uvigerina* and *Bolivina* (Fig. 157).

These examples from the numerous families and genera should give some idea of the countless numbers of tiny organisms with an almost unbelievable array of pleasing shell patterns that inhabited oceanic waters.

Other Invertebrates. Modern methods of preparation with different chemicals have made it possible not only to recover microscopic radiolarians but to reveal much of their ornate hard parts, sometimes even their long delicate spines. These protozoans are pelagic and widely distributed in oceanic waters. They have been recorded from all of the Cenozoic Epochs.

The lacy structures of bryozoans are also frequently encountered, but their smaller species are difficult to recognize from fossil materials. Corals are important, but their greater numbers are found in warm waters.

The extinction of numerous groups of Cretaceous invertebrates and changes in water temperatures evidently opened the way for the evolution of other invertebrates that were present in small numbers in the later Mesozoic faunas. Conspicuous radiations in the larger invertebrates occurred in the gastropods and pelecypods. In sheltered areas on the fine silty bottom of the oceans, where wave action was greatly reduced, species of the gastropod *Turritella* occurred in abundance. Evidently they were gregareous and moved about slowly, carrying their high narrow spiraled shells at an angle. There they fed on the detritus of marine algae or diatoms. Their probable habitat was in the neritic zone some twenty to forty fathoms deep. In some places they were buried in tremendous numbers. Normally the species had short time ranges and have been valuable index fossils within limited areas such as the early Cenozoic inland seas of California and the Gulf coast. These attractive shelled creatures and their descendents formed an important part of the Cenozoic faunas. Like many other invertebrates, the succeeding species of *Turritella* retreated toward the equator as

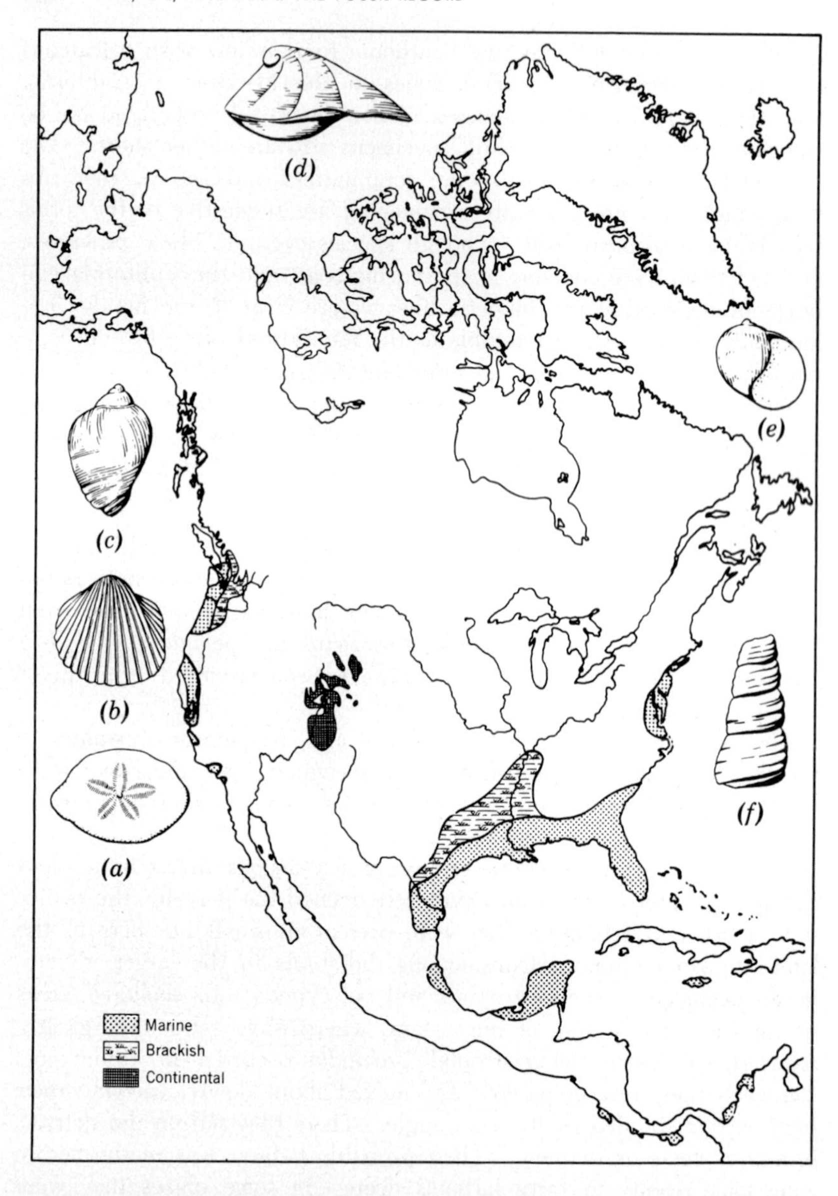

Fig. 158. Marine, brackish, and continental early Eocene sedimentary rocks in North America. After Schuchert. Marine invertebrates illustrated: (a) Eoscutella coosensis, *echinoid;* (b) Loxocardium brenerii, *pelecypod;* (c) Pseudoliva dilleri, *gastropod;* (d) Velates schmidelianus, *gastropod;* (e) *Eocernina hannibali,* gastropod; *(f)* Turritella uvasana, *gastropod.* (a), (b), (c), (e) *after Weaver.* (d) *after Zittel,* (f) *after C. W. Merriam.*

the ocean temperatures cooled. Most of the other gastropods were not so gregarious. Though usually well represented in fossil assemblages, they were more widely scattered than the turritellas.

Pelecypods occur in large numbers. The free-moving giant venericards, with their subequal valves and their flat, radial ribs separated by narrow grooves, are common in early Cenozoic rocks. Sedentary oysters were also abundant. Some reefs are made up largely of thick irregular oyster shells which are quite resistent to abrasion. *Ostrea titan* from the Miocene in California is a giant; some of its shells are at least eighteen inches long. Many other pelecypods are equally abundant. The fan-shaped scallops, or pectens, with their equally spaced ribs and valves of different size, are as conspicuous in the rocks as fossils as they are on the beaches today. They are especially well represented in both Miocene and Pliocene formations. Many are excellent index fossils. Several pelecypods are the burrowing kinds that work their way down into the mud or sand or even into hard rocks on the bottoms of shallow seas. They make their contacts with the water above by long or short siphons. Burrowing clams are usually not so abundant as those with different habits.

After the close of the Mesozoic when ammonites became extinct the nautiloids became rather abundant in some environments. This was markedly true in the Paleocene and less so in the Eocene. After that, they diminished almost to the point of extinction in the Miocene and Pliocene. These were jet-propelled animals. Genera such as *Hercoglossa* and *Cimomia* were widely distributed. Species like those in equatorial western Africa, Trinidad, and Alabama evidently were rather closely related. Hovering near the bottom at neritic depths, these nautiloids were ever on the alert for shrimp or other food of their choice.

Decoratively patterned echinoids were also an important part of the life in the Cenozoic seas. The mouth parts, first described by the Greek philosopher Aristotle, are called Aristotle's lantern because of their superficial resemblance to a lantern. A more complicated jaw can hardly be visualized, since it includes forty hard parts operated by sixty muscles. The regular echinoids, subclass Regularia, so named because of their fivefold radial symmetry, include the spiny sea urchins. Fossil specimens are frequently crushed, unless silt has infiltrated into the hollow interior, and the long heavy spines are usually broken off.

Much more important in the fossil record are the irregular echinoids, subclass Irregularia, or sand dollars and heart urchins. These are much more bilaterally symmetrical than the regular echinoids. The sand dollars, known as clypeasteroids, appear in the late Cretaceous and became abundant in the Cenozoic. They live in large colonies partly

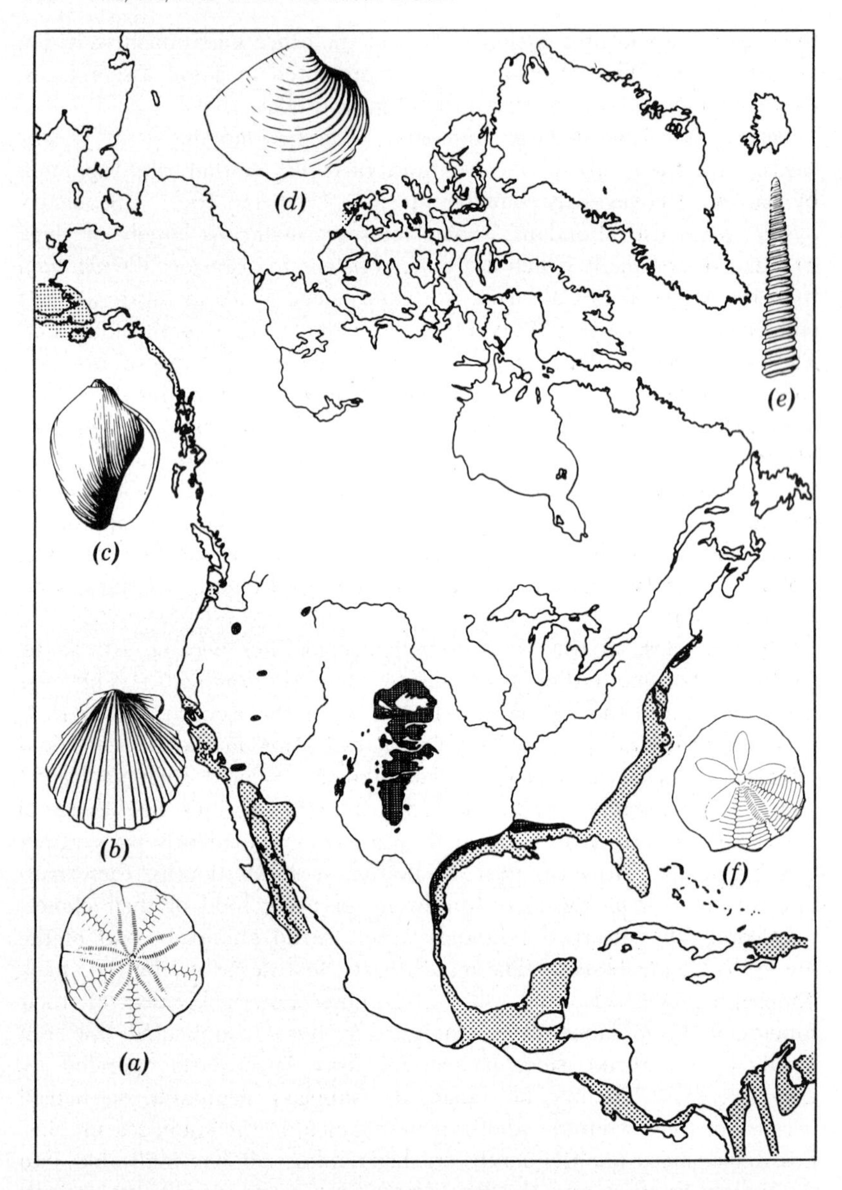

Fig. 159. Marine and continental Miocene sedimentary rocks in North America. After Schuchert. Marine invertebrates illustrated: (a) Dendraster coalingensis, *echinoid, after Kew;* (b) Pecten crassicardo, *pelecypod, after Arnold;* (c) Orthaulax pugnax, *gastropod, after Cooke;* (d) Clementia pertenuis, *pelecypod, after Shimer and Shrock;* (e) Turritella altilira, *gastropod, after C. W. Merriam;* (f) Albertella aberti *echinoid, after Clark and Twitchell.*

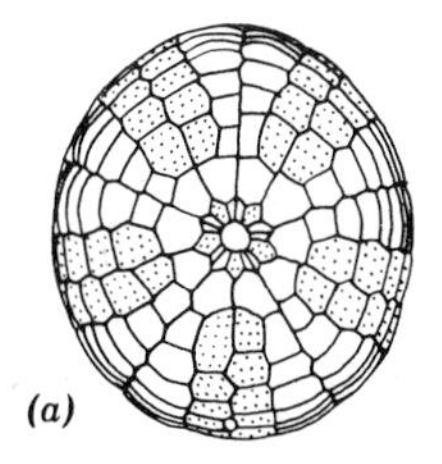

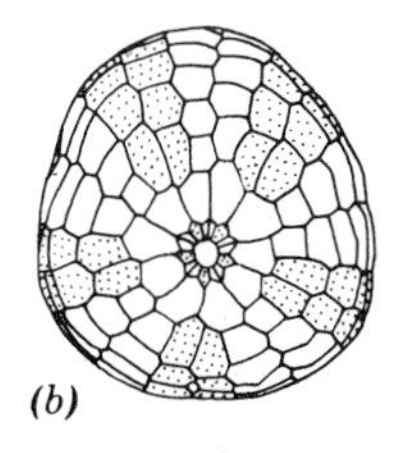

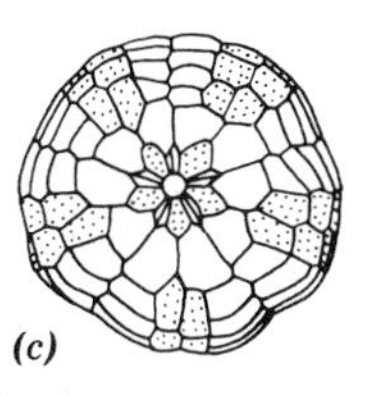

Fig. 160. Sand dollars, or clypeasteroid echinoids, showing pattern of plates on oral (underside) views, from the Pacific coast province. After Durham: (a) Astradapsis anticelli, *Pliocene;* (b) Dendraster gibbsii, *Pliocene;* (c) Vaquerosella merriami, *Miocene.*

buried in the sandy or muddy bottoms of marine waters at both shallow and great depths. They move about on short spines. In the Miocene of the Coast Range in California *Vaquerosella merriami,* a small rounded species, occurs in such numbers that the stratigraphic unit in which they are found has been called the "Button Beds." Two other genera of great importance in the later Cenozoic of western United States are the genera *Astrodapsis* and *Dendraster.* The short geologic range of their species makes them useful in correlation, but phyletic lines are difficult to trace. The species of *Astrodapsis* first appeared in the early Miocene, became abundant in the later Miocene, then died out quickly about the early middle Pliocene. *Dendraster,* though it appeared earlier, then seemed to take over. It was most abundant in the late Pliocene and is still living. Though some genera of sand dollars may be found in regions as far apart as California and India, most species are confined to local provinces (Fig. 160).

Marine Vertebrates. The most abundant marine vertebrates of the Tertiary are the *bony fish* (teleosts) which include tuna, herring, cod, and many others displaying diverse specializations. Sharks also are present throughout the Period; huge triangular serrated teeth nearly 6 inches wide are indicative of some of the giants.

One of the most famous marine fish localities was found near Lompoc, California. Herring and other ocean fish, possibly seeking a place to spawn, were trapped there by the thousands in an embayment. There they perished and were covered with diatomite. Buried with the fish were bodies of oceanic birds such as shearwaters (chapter heading 30) boobies, and murrelets. The birds may have come to feed on the stranded fish, as did the sea lions.

For the most part marine fish are found as individuals in the rocks, though their scales are much more uniformly distributed. Some rather remarkable identifications have been made on these scales.

The most interesting of the open-ocean vertebrates is the whale. The steps through which it evolved from a quadrupedal land-living mammal into one of the most highly specialized pelagic animals is not known. It is quite apparent that it must have evolved rather quickly to take the place of the ichthyosaur and mosasaur of the Mesozoic. Even the earliest ones known in the middle Eocene of Egypt are far removed structurally from other mammals. There is some suggestion of relationship to early carnivores, but this is not at all conclusive. Because of the shape of their teeth the largest of the early whales were at first called zeuglodonts (yolk + teeth) but are now referred to as *archaeocete whales.* The first vertebrae discovered in the late Eocene marine beds in the Gulf coast province were thought to be reptilian because of their enormous size and dense bone structure. Hence the generic name *Basilosaurus.* These Eocene whales, though clearly cetaceans, retained vestiges of the hind limb and pelvis. Even the position of the skull bones is more like that in other mammals instead of being telescoped backward, as in later whales and dolphins, in which many

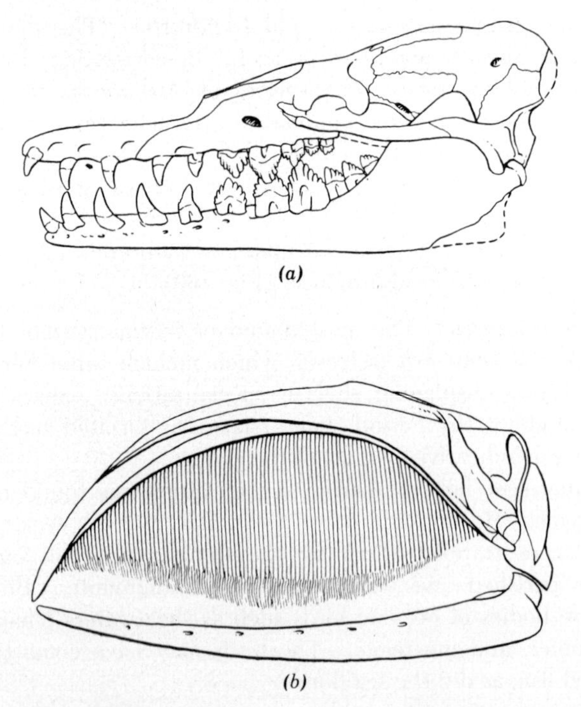

Fig. 161. Whale skulls; (a) *Eocene archaeocete,* Basilosaurus cetoides, $\times \frac{1}{18}$, *after Kellogg.* (b) *Recent baleen whale,* Balaena mysticetus, $\times \frac{1}{84}$.

Fig. 162. A restoration of Desmostylus with occlusal and side views of one of its cheek tooth.

of the bones are overlapping and the nostril openings are far back on the head. It is interesting to note that the ear bones, which are structurally different in the two major groups of later whales (toothed whales and whalebone whales), were also represented in the earliest groups.

Almost as startling as the first appearance of the earliest whales is the sudden influx of modern *whales* and *dolphins* in the late Oligocene and Miocene. There are no good intermediates between these and the archaeocetes. Archaeocetes were greatly reduced in numbers at that time, and the new groups seem to have spread widely. Because of their large size and also because they are frequently found in hard sandstones, whales are difficult to collect and prepare. The living blue whale is the largest of all animals; some individuals have attained a length of 100 feet and a weight of approximately 115 tons (Fig. 161).

Another interesting order of aquatic mammals which inhabits marine waters is the Sirenia. These are the *sea cows* and *dugongs*. The oldest fossils were taken from the Eocene of Egypt and Jamaica. Though apparently more closely related to the dugongs than to the sea cows, the Eocene kinds belong to more primitive families. Though the genera of dugongs were widely distributed in middle latitudes during the Oligocene and the Miocene, they are now distributed in marine, brackish, and sometimes fresh waters of the Eastern Hemisphere. The sea cows, or manatees, have been found only in the Americas.

One of the most peculiar mammals to frequent marine environments, probably near shore, was a creature with hippopotamuslike proportions. These animals were called *Desmostylus* for the arrangement of the bundle of thick enameled pillars in their molars. It is assumed they ate mollusks because of the massive construction of their molars and powerful jaws. *Desmostylus* was widely distributed in neritic waters of the Pacific in the Miocene. A skeleton has been reconstructed from fossils found in Japan, and numerous jaws and isolated teeth have been found in California (Fig. 162).

Appearing at the same time as the modern whales were the pinniped carnivores known to us as *walruses*, *seals*, and *sea lions*. It has not been determined from which group of carnivores they descended, since they share certain characters with several families, such as the otters, dogs, and raccoons. This is another example of terrestrial animals reverting to life in the water. The most extreme specialization seen among these creatures is in the walrus, with his long belligerent-looking canine tusks.

A more recent trend from land to sea is exemplified by the sea otter. No fossil record of these fur-bearing animals is known, but the transition must have taken place as far back as the Pliocene.

23 Tertiary Life on Land

Eohippus

Paleocene Epoch

Archaic mammals as dominant land animals; first tarsiers and lemurs; modern groups of birds; plants trending toward subtropical climates.

THE BEST EVIDENCE of life on land in the Paleocene Epoch is in North America and in the latitudes of the United States, though late Paleocene mammalian faunas from Europe, Asia, and South America are known. The early and middle parts of the Epoch are represented only in North America. Evolution in the mammals indicates that nearly as much time transpired in the Paleocene as in any of the Epochs that followed.

The Cenozoic opened with a trend from temperate climates in the late Cretaceous toward the mild subtropical environments of the Eocene. There were few plant species in common in the latest environments in which dinosaurs lived and in those of the Paleocene where mammals appeared in abundance. There probably was no frost. Seasonal variations were doubtless governed largely by the amount of rainfall. Because of the proximity of the ocean, warmer climates extended farther north along the Pacific coast than into the interior, a condition which seemed to prevail along the continental borders throughout the Northern Hemisphere.

The most marked contrast between the late Cretaceous and the Paleocene faunas, aside from the disappearance of dinosaurs and other groups, was in the land mammals. Three orders, the Multituberculata, Marsupialia, and the Insectivora, continued from the Cretaceous, and nine new orders appeared before the end of the Paleocene. Among these archaic mammals were the first primitive carnivores, tiny lemurs and tarsiers (primates), early squirrellike rodents, condylarths, animals distantly related to the armadillos, and other orders that reached the peak of their evolution and died out in the three succeeding Epochs. Many of these were small mammals, comparable in size to rats and mice, though some were as large as raccoons, and toward the end of the Epoch some, including *Coryphodon,* were as large as tapirs. Many of these first representatives of mammalian orders were closely related, indicating proximity to common ancestral stocks probably of the late Cretaceous and early Paleocene (Fig. 163).

Most of the Paleocene mammals are known from teeth and jaws, though there is enough evidence of body skeletons to make some restorations. Most of the ecologic niches, such as streams, forest floors,

Fig. 163. A Paleocene condylarth, Tetraclaenodon; $\times \frac{1}{14}$. *After Osborn. This five-toed ungulate was near the ancestry of the first horses.*

and trees, seem to have had their occupants. There were carnivores, insectivores, and herbivores. Reptiles were rare in comparison with those in the Mesozoic, but crocodilians and turtles are recorded.

A marked similarity in the late Paleocene mammals of western Europe and North America offers good evidence of a land connection between the two continents at that time. A meager collection from central Mongolia shows some relationships to both North and South America. The South American late Paleocene fauna is so different from that in North America that a land connection to the north may have been replaced by marine waters as early as Cretaceous.

Paleocene rocks were observed in the San Juan Basin of New Mexico by Edward Drinker Cope on October 27, 1874, while he was serving on the Lieutenant W. K. Wheeler Survey. This was the same year in which Wilhelm Philip Schimper proposed the name for the Epoch from the evidence of fossil plants and brackish and continental sedimentary rocks in France. Though Cope, in his hasty survey of the San Juan Basin, recognized the stratigraphic position of dull gray shales (Paleocene) of considerable thickness between Cretaceous and Eocene formations, he found no fossils in them at that time. Not satisfied, Cope sent David Baldwin, one of his collectors, back to the area 7 years later. With more time for prospecting, Baldwin then found the first fossils of American Paleocene mammals. Subsequently, other discoveries were made, mostly in the past 50 years, not only in the famous San Juan Basin but also in Colorado, Utah, Wyoming, and Montana. Specimens from California and a jaw fragment from a deep well in Louisiana have shown that these Paleocene mammals were not restricted to the Rocky Mountain province.

Eocene Epoch

All modern orders of mammals present;
small tarsiers and lemurs still common;
first horses; first artiodactyls;
subtropical forests with heavy rainfall;
little or no frost.

LYELL ORIGINALLY DEFINED the Eocene to include all Tertiary rocks in the London and Paris basins older than the Miocene and younger than the Cretaceous. Even at that time he suggested the possibility of dividing his Eocene into additional Epochs, in which case he suggested that the middle one would become the type of his Eocene.

Eocene continental deposits, though much better known than those in the Paleocene, were laid down for the most part in the same provinces. Early Eocene warm temperate to subtropical forests which persisted in the humid lowlands were more widespread and extended farther north than in the preceding Epoch. There were heavy forests and open country in the Rocky Mountain province of North America. Streams with their wide flood plains flowed into swampy woodlands that were not more than 1500 feet above sea level, and volcanoes in the uplands at times spread volcanic ash over the area.

The presence of heavy forests and limited grassland was reflected in the adaptations of the mammals in the early Eocene of North America. In the trees were primitive opossums, *Peratherium,* tiny multituberculates, *Ectypodus,* the last of this ancient order; numerous insectivores, small lemurs, *Pelycodus,* and tarsiers, *Tetonius.* On the floor of the forest were the first horses, *Hyracotherium,* called eohippus, small tapirlike animals, *Homogalax,* the first of the artiodactyls, *Diacodexis,* other primitive carnivores, creodonts, and rodents, *Paramys* and *Sciuravus.* Several species of these ground-dwelling mammals ranged into the limited grasslands and swamp border areas where there browsed larger forms, such as *Coryphodon, Lambdotherium,* the first small ancestors of the great brontotheres, and rhinoceroses, *Hyrachyus.* Bats, as well adapted for flight as any living species, flitted about in the dark forests or in the twilight over the lakes and open spaces. In these early Eocene faunas the first recognizable representatives of the mammalian orders Perissodactyla (odd-toed ungulates) and Artiodactyla (even-toed ungulates) were *Hyracotherium* and *Diacodexis.* These and other closely related small ungulates were the forerunners of mammalian families destined to become the dominant herbivores of the Cenozoic. Among the perissodactyls were the first members of several families which included the horses, tapirs, rhinoceroses, bronthotheres, and chalicotheres. The latter eventually developed into huge animals with heavy claws on their feet. Before the end of the Eocene many primitive families of artiodactyls, not represented among living mammals, had become well diversified. In these faunas were the first camels smaller than a greyhound.

Ten orders of birds appeared for the first time in North American Eocene faunas, only one of which is now extinct; nine of the twelve families have ranged on to Recent time. There were ducks and geese in the streams and lakes, bitterns, herons, and gallinules along the shores, owls and hawks in the woodlands, and grouselike birds in the grassland. All were generically distinct from their living relatives. The

Fig. 164. Neocathartes, a stilt vulture of the Eocene. After Wetmore. They were 18 inches tall.

most conspicuous of all American Eocene birds was *Diatryma* (Fig. 244), a giant ground bird with greatly reduced wings. This bird evidently was carnivorous, as indicated by its large head and powerful beak. Though it was 7 feet tall, it showed a distant relationship to the rails. It has been suggested that it could have swallowed a small eohippus.

The early Eocene mammals of western Europe were closely related to those of North America. There were at least ten genera common to the two continents. Specimens of eohippus found on both continents were so much alike that some evidently belonged to the same species. Even *Gastornis*, a giant ground bird of Europe was very much like *Diatryma* from Wyoming.

This marked similarity in the land-living animals is suggestive of a broad land connection between the two continents at that time and a rather open interchange of animal life in both directions. Unfortunately, a good early Eocene fauna has not yet been found in Asia. Therefore, it is not known whether this land connection lay across Siberia and Alaska or took the form of a land bridge across the North Atlantic.

The late Paleocene and early Eocene formations in England, Belgium, and northern France show marine beds replaced by lagoonal and estuarine deposits which at times were traversed by stream channels. Farther south, near Paris, there were lakes. As in North America the mammals represented both woodland and grassland types. Among the trees were the sassafras, linden, magnolia, locust, fig, and tulip. Evidently forests like these were dispersed all along the northern border of the Tethys Sea from Burma to England.

The land connections between the Eastern and Western Hemispheres must have been submerged once again in the middle Eocene. In contrast to the closely related early Eocene mammalian genera and species, those in the middle Eocene for the most part represented different genera which had evolved independently in the two regions. The European faunas were well represented and beautifully preserved in the Geiseltal "Braunkohle" and elsewhere. Floras with Indo-Malayan species extended far to the north in western Europe. Mild climates were indicated by palms in the woodlands of England and by giant sea snakes and gavials which swam through the warm waters. The initiation of uplift of the Pyrenees Mountains coincided with an influx of seas over parts of the continent in other areas. A warm climate prevailed. The meager fossil materials from Mongolia and China seem to show closer affinities with European mammals than with those in America. This was also true of the mammals of Burma.

In North America the middle Eocene is best known from the classic Bridger Basin in southwestern Wyoming. The exposures stretch away for miles from old Fort Bridger where Professor O. C. Marsh first prospected in these beds in 1870. His first field work there marks the introduction of a long program of exploration and research in western North America on the most complete sequence of early Tertiary mammalian faunas known anywhere in the world. Many mammals in the sandy shales of the Bridger Basin are the direct descendants of early Eocene genera mentioned on a preceding page. *Orohippus* is another step in the evolution of the horse, most of which took place in North America. Not far away are the Green River shales which contain the most perfectly preserved fresh-water fish skeletons ever found. Here, too, is a large insect fauna faithfully imprinted in the shales. Much of our knowledge of Eocene forests has also come from the Green River formation.

Meanwhile distinct groups of South American mammals on their large insular continent were evolving on their own, as they evidently had done throughout the Paleocene. Some of the earlier groups died out before the end of the Eocene, others continued into the later Cenozoic. As on other continents, crocodilians and turtles were the dominant reptiles, though some snakes and lizards have been found.

In the later Eocene, particularly in North America and in the Old World, there was a trend toward more temperate climates. More mammals became modified to live in grasslands, and some had increased considerably in size, as seen in the brontotheres. Some of the brontotheres showed incipient development of bony protuberances on the cranium. These big animals dispersed widely; related genera occurred in Burma, Mongolia, and North America. North America and Asia evidently were

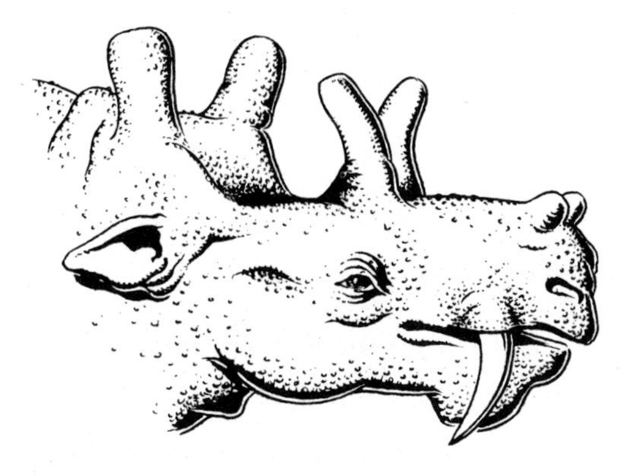

Fig. 165. One of the largest mammals from the late Eocene in the Rocky Mountain province in North America was Eobasileus.

connected also in the late Eocene because some mammalian genera on both continents were closely related. One of the largest mammals in North America was *Eobasileus,* a huge rhinoceroslike creature with sharp tusks and six bony horns on its cranium. These horns must have been covered with skin, as in the giraffe. The evolution of the horse had progressed to the *Epihippus* stage. These little fellows still had four well-formed toes on the front foot, as in eohippus. In the late Eocene the ancestors of many of the modern mammalian families were already recognizable.

One of the most interesting late Eocene faunas is the Qasr-el-Saga found in a fluvio-estuarine deposit not far from the Pyramids of Egypt. Here are the oldest known proboscideans (the order including elephants and mastodonts) called *Moeritherium,* creatures related to the earliest sirenians (dugongs), some of which were found in that area. As mentioned previously, primitive archaeocete whales were also found in those estuarine deposits. More moeritheres and other more strictly land-living forms were discovered in early Oligocene exposures in the Fâyum beds overlying the Qasr-el-Saga.

Another late Eocene assemblage of considerable interest was found in Burma along the present drainage of the Irrawaddy River north of Rangoon. The formation grades from marine rocks into lagoonal sediments and then to continental deposits. Brontotheres, a rhinoceros and primitive tapir related to those in central Mongolia and in North America frequented the stream courses and lagoonal borders. With them were artiodactyls known as anthracotheres. These were the Eocene forerunners of the hippopotamus, though they looked little like a hippopotamus at that time. Arboreal mammals are indicated by two jaw fragments of primates. Until more complete materials are recovered the family relationships of these primates will be in doubt.

Oligocene Epoch

Most modern families of mammals present;
primitive apes and monkeys;
first saber-toothed cats, true cats, and dogs;
oreodonts abundant;
plants trending toward temperate kinds;
mild temperate climates.

All orders of mammals were well differentiated in the Oligocene and most of the families were recognizable. Land connections between Eurasia and North America re-established in the late Eocene continued into the early Oligocene, but there were fewer genera common to both continents than in the early Eocene. Evidently the route of dispersal between the continents lay across the area of Alaska and Siberia. The breaking up of the old Tethys Sea which had separated Eurasia from Africa almost continuously since the Triassic offered avenues of dispersal for the mingling of many groups of mammals between those land masses. South America and Australia remained isolated. A conifer-angiosperm forest spread across central Europe in what is now East Germany and the southern part of the area covered by the Baltic Sea. Resin exuded from the branches and trunks or dropped to the ground from these conifers entangled thousands of insects which were eventually completely enveloped in the sticky substance. The accumulations of soft resin solidified into nodules of rosin and washed into the sea where they were covered with sediments and altered to amber. In these clear matrices the best known fossil insect fauna is preserved.

The best collecting grounds in the world for Oligocene mammals are in the White River badlands of South Dakota, northwestern Nebraska, and northeastern Colorado. The first fossil to come from the Tertiary of the Great Plains province was a jaw fragment of a brontothere picked up in the White River badlands by a fur trader. This came from early

Fig. 166. Early Oligocene brontothere, Brontotherium platyceras; $^1/_{60}$ *natural size. After Osborn. The mythical "thunderhorse" of the Sioux Indians.*

Oligocene beds, probably in South Dakota. Then between 1853 and 1867 geologist and explorer Ferdinand Vandeveer Hayden and others made excellent collections from the badlands. These were submitted to Joseph Leidy whose comprehensive descriptions still form an important part of the literature in vertebrate paleontology.

In Oligocene time the area east of the Rocky Mountains was a vast, nearly level flood plain traversed by meandering rivers and cut by stream channels. Larger trees, hackberries, and sycamores lined the stream borders over most of the flood plain area. Near Florissant, Colorado, there was a temperate upland forest, about 2000 feet above sea level with a redwood dominant. This middle Oligocene flora had many species similar to those in the Green River Eocene of Wyoming. One of the best insect and spider faunas from the American Cenozoic was found in the fine-grained sediments at Florissant. An unusual record was that of a tsetse fly, so dreaded as a vector of sleeping sickness in equatorial Africa today.

The gigantic mammals in the lower beds of the White River badlands were the brontotheres. These huge brutes were much larger than their immediate predecessors in the late Eocene. A thick pair of frontal horns gave them a formidable appearance. Though they were present in great numbers in the early Oligocene, these perissodactyls had reached the acme of their development, and they all died out with startling suddenness before middle Oligocene time. Other Oligocene mammals included medium-sized, slender-limbed rhinoceroses, *Hyracodon,* and camels, *Poebrotherium,* the size of small sheep. Rodents, including the first beavers and primitive pocket mice, were abundant and diversified; rabbits and many other small mammals not known today were also present. There were several kinds of insectivores, including moles and shrews. The most abundant of the medium-sized mammals, the artiodactyls, known as oreodonts, *Merycoidodon*, represented a family that made its first appearance in the late Eocene. Evidently they had browsing and perhaps grazing habits. Though a conspicuous family in the North American middle Tertiary mammalian faunas, they never spread to other continents. The anthracotheres, a family with their earliest records from the Eocene of India and Burma, also reached North America in the early Oligocene. The true carnivores (flesh eaters), among which were dogs, small foxlike animals, true cats, saber-toothed cats, and many other smaller forms, had almost totally replaced the Eocene creodonts. Only one genus of creodonts, *Hyaenodon*, remained, and some of these were as large as grizzly bears. The horses, *Miohippus*, were slenderer but otherwise about the size of sheep. Their low-crowned teeth were indicative of browsing habits, and their feet were reduced to three toes on each foot. In some streams were amphibious rhinoceroses, *Metamynodon*, and in others were peculiar artiodactyls, *Protoceras*, which had short horns on each side of the nostrils and two more back of the eyes. Medium-sized tortoises were everywhere. Loons inhabited the streams, and cormorants watched for fish from the vantage points of dead branches above the water. Hawks and vultures soared above. In the shrubbery were wild turkeys and "American pigs," known as peccaries.

The European Oligocene mammals were as abundant as those of North America. Some genera occurred on both continents, but many genera and some families were restricted to Eurasia. As an example, true pigs had their origin and experienced their evolution in the East-

Fig. 167. Baluchitherium, the largest land mammal known, from the Oligocene of Asia. It was 18 feet high at the shoulders; $^{1}/_{104}$ natural size. After Gregory.

ern Hemisphere. There were some relationships between African and Asian mammalian genera, but for the most part the Asiatic faunas showed closer affinities with the North American assemblages. The American Museum's expeditions to central Mongolia made the amazing discovery of a gigantic perissodactyl of the rhinoceros family. The genus was previously described by C. Forster Cooper on fragmentary materials from Baluchistan. *Baluchitherium,* as it was called, had huge postlike limbs and a hornless skull 4 feet long. It was the largest of all land mammals. Its height of 18 feet at the shoulders permitted it to brouse on tender leaves high in the trees (Fig. 167).

In the famous Fâyum beds near the pyramids of Egypt were the first mastodonts, *Palaeomastodon* and *Phiomia,* which had short tusks in the upper and lower jaws. Though they were much smaller than the later mastodonts and mammoths, the construction of their skulls showed clearly that these early mastodonts had trunks. The Fâyum was a most interesting fauna. It included more of the most primitive proboscideans, *Moeritherium,* which first appeared in the Eocene, large creodonts (primitive carnivores), rodents, hyracoids, which are almost exclusively confined to Africa, anthracotheres, which spread into Africa from India and Burma, and other mammals not known elsewhere. In this stream border and grassland fauna were small primates considerably advanced over those found in the Eocene.

South American mammals, still isolated from the other continents by water barriers, showed remarkable convergences in evolution with unrelated orders in other parts of the world. Small rodents belonging to the families of the American porcupine and chinchilla arrived in South America in the early Oligocene. These rodents were among the first on that continent. A few million years later, near the end of the Oligocene, the oldest known New World monkeys, Cebidae, appeared. Possibly these rodents and monkeys had been evolving somewhere in Central America and by chance drifted across to the shores of South America on floating vegetation.

Miocene Epoch

Rise and rapid evolution of grazing mammals;
apelike creatures; spread and diversification
of mastodonts throughout Northern Hemisphere;
temperate kinds of plants; moderate seasonal climates.

As with most of the other Tertiary Epochs there has been much difference of opinion as to the position of the Oligocene and Miocene boundary. During this questionable interval the mammalian evidence seems to indicate that the Old and New World were again separated. Except for the rodents all but two or three of the mammalian families currently recognized were present in the Miocene. Plants of the warmer Oligocene and early Miocene forests in middle latitudes had shifted farther south and were replaced by plants better suited to slightly cooler and drier environments. The dawn redwoods, *Metasequoia*, were common. Bald cypresses, poplars, alders, madroños, and black and live oaks also gave verdure to the landscapes in western North America.

Uplifts in the Rocky Mountains resulted in the deposition of a blanket of fine sands over much of the northern Great Plains. Some

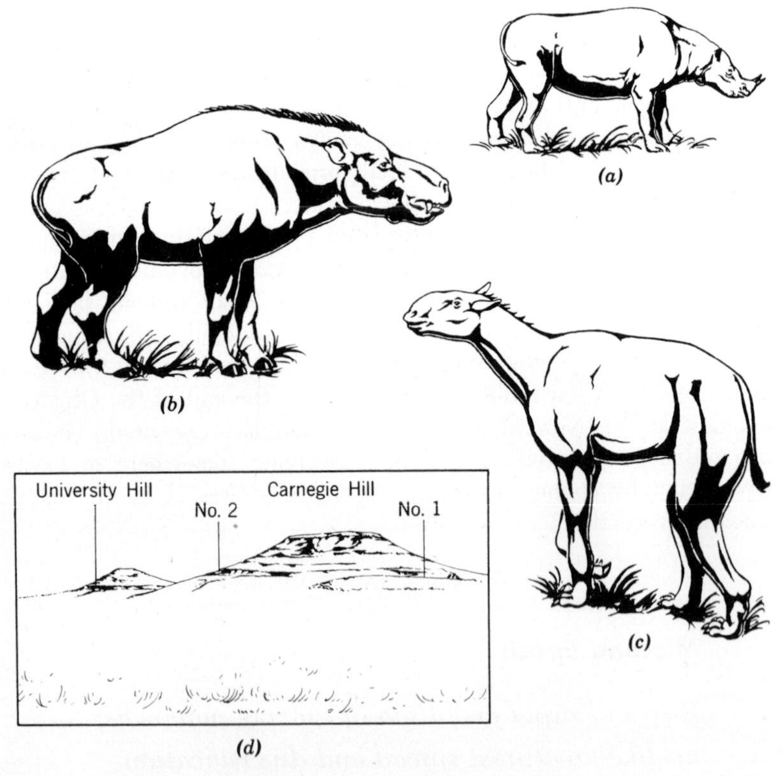

Fig. 168. The Agate Springs Quarry and three kinds of mammals found there; (a) Diceratherium cooki; (b) Daeodon hollandi; (c) Moropus cooki; $^{1}/_{96}$ *natural size; and* (d) *University and Carnegie Hills, showing locations of the quarries.*

of the best fossil mammalian remains from these formations were found on and adjacent to the Cook Ranch on the Niobrara River in Sioux County, Nebraska. The oldest known spear grasses, *Stipidium* (Fig. 215), were discovered there in the early Miocene Harrison formation by Maxim K. Elias.

Two famous fossil localities, the Agate Springs and *Stenomylus* quarries, have yielded skeletons of some remarkable creatures. In the Agate Springs site, located by Captain James H. Cook in 1878, were abundant remains of a medium-sized rhinoceros, *Diceratherium cooki,* an unusual perissodactyl, *Moropus cooki,* with claws instead of hoofs on its feet, and an enormous hoglike animal, *Daeodon hollandi.* The remains of some other mammals were also present, but they were not so common in the quarries. A few miles farther down the river a small exposure was found in which dozens of small camels, *Stenomylus gracilis,* were entombed. Nearly every specimen was a complete skeleton, seemingly lying where the animal had died. It is still a mystery why so many died at that one place.

Many other mammals have been recovered from the richly fossiliferous Harrison formation. Some of the largest oreodonts, *Promerycochoerus,* long, slender-limbed camels, *Oxydactylus,* another of the peculiar horned artiodactyls, *Syndyoceras,* bear dogs, *Daphaenodon,* small foxes, *Pseudocyonidictis,* peccaries, *Perchoerus,* and many others were present. The dominant horses were *Parahippus,* representing the next step in the evolution toward the modern horse. Though their teeth were still low-crowned, the basic pattern for the development of high-crowned teeth was becoming established. The side toes were slightly reduced, and the limbs were considerably longer than those of their Oligocene predecessors. Another interesting animal was the beaver, *Palaeocastor.* It was much smaller than the living beavers and may have had habits like the ground squirrel. Skeletons of some of these rodents were found in peculiar spiral structures, called *Daemonelix,* or in straight tubelike tunnels leading into the spirals. It has been assumed that these were the dens of *Palaeocastor;* other investigators have contended that *Daemonelix* comprised the forms left from plant roots. Both interpretations may be correct.

Near the beginning of middle Miocene time there was another interchange of Old and New World terrestrial mammals which continued in a limited fashion throughout the remainder of the Tertiary. Among the Miocene mammals found on both continents were beavers, cats, dogs, "bear dogs," horses, and mastodonts. There is no evidence that any of these animals moved in mass migrations from one continent to another, as the word "migration," used so frequently in the literature, might imply. In all probability the species of these genera reached such distant places by the normal spread of their populations into adjacent areas over several hundred thousand years, or as long as suitable areas for their survival and reproduction were available for them to spread.

The middle and late Miocene in North America has been typified by species of the equid genus *Merychippus,* which marked the second major step in the development of hypsodonty in grazing horses. The side toes on these more advanced horses were slightly higher from the ground than in *Parahippus.* Other mammals also were diagnostic. Mastodonts and "bear dogs" spread into North America from Eurasia about the beginning of the middle Miocene; then both groups experienced considerable evolution in both hemispheres, especially the mastodonts. Some of the "bear dogs," *Amphicyon,* reached enormous sizes for carnivorous animals. Crania were as much as 2 feet long. Though bears evolved from Oligocene doglike animals, the animals called "bear dogs" represented another group. Camels were abundant. Among the many kinds were those modified for living in a grassland environment; others were browsers along bushy stream courses. Pronghorn antilocaprids also populated the late Miocene grasslands in great numbers.

Though there were some mammalian genera found in both Eurasia and North America, many families were restricted to the continents of their origin. In Eurasia the Miocene Epoch marked the appearance of hyenas, deer, giraffes, and bovines (true antelopes). Horses of the genus *Anchitherium,* with heavy side toes and low-crowned teeth, had spread over from America near the close of the early Miocene and persisted in Eurasia throughout the Epoch before they died out. Rhinoceroses were also abundant. Mastodonts and the peculiar deinotheres (also proboscideans) were the largest mammals. Pigs, tragulines (small artiodactyls), tapirs, dogs, cats, and civets helped make up the faunas. Among the most interesting mammals were Old World monkeys and *Pliopithecus,* gibbonlike creatures. In most areas the European continent was rather heavily forested, and stretches of grassland were limited, except possibly in parts of Russia. There were warm temperate floras near the coast of western Europe. In the European Miocene numerous birds included herons, rails, curlews, ducks, sea eagles, eagles, hawks, owls, partridges, pheasants, hornbills, crows, and sparrows. As in the Oligocene, many showed relationships with African birds.

The initiation of uplift of the Alps in Europe and the Himalayas in Asia possibly was in the Oligocene, and this elevation continued in the following Epochs.

A lake and flood-plain deposit in Mongolia, called Tung Gur, has yielded a large series of shovel-tusked mastodonts, *Platybelodon*, which had small, short, rounded tusks in the cranium and a scoop-shaped mandible with rather short wide tusks. They must have been amphibious creatures, since they probably fed on water plants. These mastodonts, a beaver, *Anchitheriomys*, a horse, *Anchitherium*, a hyenoid dog, *Aelurodon*, and other forms were common to both North America and Europe. But, as in the Miocene faunas of Europe and India, the Tung Gur deposits yielded the remains of bovids, giraffids, cervids (deer), suids (pigs), hyenids, and other mammals that were restricted to the Old World.

A Miocene forest fauna from Kenya Colony in equatorial Africa and another from southwestern Africa have offered us an insight into African middle Tertiary mammals. Many of the genera were restricted to that continent; others have been recorded from Eurasia, and only five were distributed as far as North America. These were a creodont, *Hyaenodon*, a "bear dog," *Amphicyon*, a cat, *Pseudaelurus*, a chalicothere, *Macrotherium*, and a mastodont, *Gomphotherium*. Some of these mammals have been useful in intercontinental correlation, though *Hyaenodon* died out in the late Oligocene in North America. The genera restricted to Africa were mostly rodents and hyracoids (cavies), some artiodactyls, including the multiple-tined *Climacoceras*, and a few primates. Among the primates were small bush babies called *Progalago*, two apelike creatures, *Limnopithecus*, and *Xenopithecus;* and the fascinating *Proconsul*, which has monkey-, ape-, and manlike characters.

After the Patagonian marine invasion of Argentina there was an accumulation of 1500 feet of continental sands, gravels, and volcanic ash in the area of Santa Cruz. Buried at that time were the remains of one of the finest Tertiary mammalian faunas to be found on the South American continent. It remained for the Ameghino brothers to direct the attention of the scientific world to these and other Tertiary faunas in Argentina. Carlos collected and Florentino described many of these extraordinary mammals in the early days. Except for the rodents and New World monkeys, Cebidae, that somehow found their way to that continent in the Oligocene, the mammalian families were strictly South American and had been on that continent since the early Cenozoic. In many of their anatomical features the mammals showed a remarkable convergence with those in other parts of the world. They resembled horses, camels, rhinoceroses, mastodonts, and other northern groups, but they actually belonged to different orders. The marsupials took the place of the carnivores and the insectivores. In the upper Magdalena River valley of Colombia a similar but slightly later fauna has been found more recently by the Geological Survey of Colombia and the University of California.

The oldest known pouched mammal from the Australian region was discovered at Table Cape in Tasmania where part of its skeleton had washed into marine deposits. The best indication of its age was derived from a study of the foraminifers in the same rocks by M. F. Glaessner of the University of Adelaide. *Wynyardia,* as it was called, was found to be related to bush-tailed possums still living in that region. Much of the cranium, parts of the mandible, and some limb bones were recovered, but unfortunately no teeth were preserved. Near the end of the Epoch there were medium-sized forerunners of the huge diprotodonts, Diprotodontidae, and other marsupials in Victoria. More recently part of the jaw of a koala, *Perikoala,* and other marsupials thought to be of Miocene age have been found east of Lake Eyre in South Australia.

Pliocene Epoch

Abundance of mammals reaching their peak in evolution; apelike creatures with manlike characters in their teeth and skeletons; large camels and giraffes; antilocaprids and bovines; hyenoid dogs and hyenas; plants of drier and cooler environments.

IN THE COOLER AND DRIER climates of middle latitudes of the world there was an acceleration in the evolution of many mammals. For the most part they were larger and more specialized than their Miocene predecessors. Many genera became extinct toward the end of the Epoch, and some lingered on in a limited number of species into the early Pleistocene. The faunal interchange between North America and Eurasia was of about the same magnitude in the Pliocene as it was in the middle and late Miocene. The carnivores were most successful in this intercontinental dispersal, though mastodonts, rhinoceroses, horses, and rodents were also involved.

The number of specializations in the mastodonts of the Northern

Fig. 169. Pliocene in the San Joaquin Valley, California, showing the camel, Paracamelus, *the horse* Pliohippus, *and the pronghorn* Sphenophalos.

Hemisphere was amazing. There were long-jawed, short-jawed, break-jawed, spiral-tuskers, shovel-tuskers, and others. Some were adapted to life in the woodlands, some in the streams and shallow lakes, and others on the plains or hills. Camels, horses, pronghorns, rhinoceroses, and the last of the fork-horned creatures, *Synthetoceras,* frequented their preferred habitats. Rhinoceroses after a long and varied history in North America died out, except for a few stragglers, at the close of the Pliocene. Camels were abundant, and some were as large as the

Fig. 170. Synthetoceras tricornatus, *a fork-horned artiodactyl from the early Pliocene of Texas;* $\times \frac{1}{14}$ *natural size.*

Fig. 171. Osteoborus, a hyenoid dog with a short face and powerful jaws for crushing bone; about the size of a coyote. Permission California State Bureau of Mines.

living species. Horses had evolved into several genera from the Miocene species of *Merychippus*. All showed an increase in the height of their cheek teeth, when compared with their Miocene ancestors. *Pliohippus*, which had lost its side toes, was in the line that eventually gave rise to the modern horse. There were saber-toothed cats, true cats, hyenalike dogs, and many mustelids related to wolverines, martens, weasels, badgers and skunks. On the whole the mammals were larger and much more modern in appearance than in earlier Epochs.

Typical of the Eurasian Pliocene land animals were the so-called *hipparion faunas*. These were distributed mostly in middle latitudes from western Europe and northern Africa across southwestern Asia to India, Burma, and China. This was a region of lakes, streams, limited forests, and grasslands inhabited by vast herds of grazing mammals, carnivores, deinotheres, mastodonts, rodents and birds. *Hipparion* was a genus of

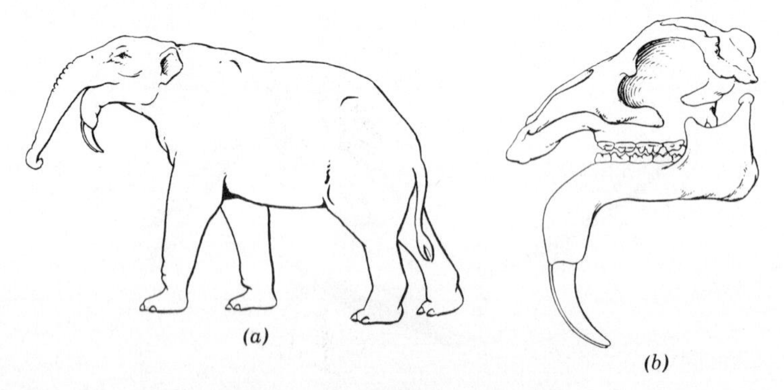

Fig. 172. Deinotherium; mastodontlike deinothere of the European Pliocene; (a) restoration, $^1/_{80}$ natural size; (b) skull, $^1/_{32}$ natural size. After Osborn.

horses with high-crowned cheek teeth and heavy side toes; the first ones had spread into the Old World after they had evolved from one of the American Miocene species of *Merychippus.* Crenulations of the enamel in the surface patterns of their teeth are among the most useful features in recognizing the genus. In contrast to the American Pliocene faunal assemblages of the grasslands, there were many kinds of giraffes, including the giant *Samotherium,* as well as numerous bovine animals. These herbivores were present in enormous herds. Hyenas, dogs, and cats followed the herds, levying on the young and the weak or feeding

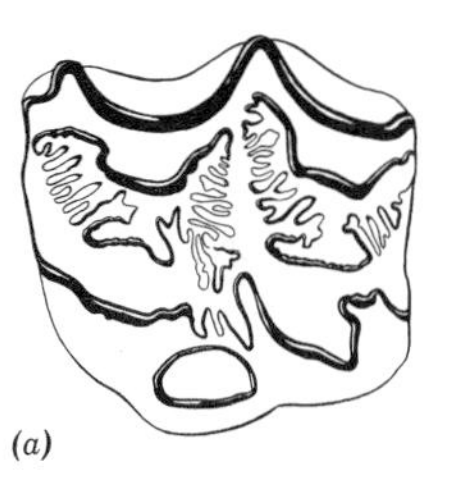

(a)

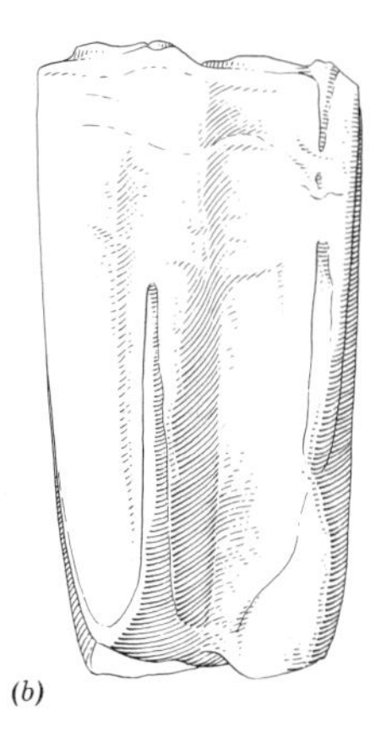

(b)

Fig. 173. An upper molar of a Hipparion *horse from the Pliocene of China:* (a) *worn surface showing crenulations of enamel pattern;* (b) *side view showing height of crown, natural size.*

on the scattered carcasses. In the woodlands and along stream borders were pigs, deer, carnivores, hedgehogs, beavers, porcupines, and other rodents. Another marked difference in the North American and Old World Pliocene mammals was the presence in the latter of cercopithecine monkeys, *Mesopithecus* (Fig. 273), and its relatives. Characters in *Dryopithecus* (Fig. 275) from the numerous jaws and teeth that have come from Europe and the Siwalik formations in India showed affinities with the chimpanzee, the gorilla, and man. The few limb bones that have been found suggest these Pliocene apes lived both on the ground and in the trees.

The north African Pliocene mammals were little different from those in Europe and India. Unfortunately, mammals from the Pliocene of equatorial Africa have not been found. In the heavy forested regions there are few exposures, yet more detailed exploration will undoubtedly reveal much of that important record.

Though South America was still isolated from North America in the Pliocene, the barrier evidently was less formidable, for some animals filtered through in both directions. This dispersal continued in increasing numbers until a good terrestrial pathway became established in the early Pleistocene. From north to south the first in these dispersals were members of the raccoon family and then the mustelids, followed by a peculiar group of horses, *Hippidion*, peccaries, sabertoothed cats, dogs, and mice. Certain ground sloths, with their long evolutionary history in South America, found their way into North America.

An Australian marsupial fauna, apparently of Pliocene age, was found recently in stream channel deposits at Lake Palankarinna east of Lake Eyre in South Australia. The fossils were discovered in channel deposits of an old Cooper's Creek flood plain. There were medium-sized diprotodonts, *Meniscolophus*, kangaroos, *Prionotemnus*, and a bandicoot, *Ischnodon*. Among the other fossils was evidence of turtles, crocodiles, lungfish teeth, and other fish bones. Fragmentary marsupial remains have been taken also in the states of Victoria and Queensland. Diprotodonts and kangaroos related to those in Australia, possibly of Pliocene age, also have been taken in New Guinea. Careful and patient exploration throughout the Australian region will surely produce more of the fascinating history of these interesting mammals.

24 Pleistocene Epoch

Giant bison and California ground squirrel

Dominance of man, and large mammals; mountain building; fluctuating climates; advances and retreats of glacial ice sheets.

THE PLEISTOCENE as outlined here has been estimated at 600,000 years in extent.* It was one of the most interesting of geologic Epochs. In this time man progressed to a position of domination over all other kinds of life. On all continents several groups of mammals that had reached a maximum in the size of certain species became extinct. There were pronounced fluctuations in the north-south distribution of plants, land animals, and marine life. During the colder Ages mammals from tundra habitats, such as musk oxen, woolly mammoth, barren ground caribou, and lemmings, dispersed far to the south. They probably spread back to the north in the warm intervals. The Pleistocene was an Epoch of epeirogenic uplifts and orogenic disturbances in the earth's crust that resulted in a maximum of continental relief and mountain building. There were pronounced fluctuations in the climates from warm temperate to boreal in middle latitudes. Continental glaciers spread down over much of North America and Eurasia, only to subside as the melting fronts retreated back to the north and valley glaciers melted back into the higher mountainous areas. This was repeated four times, with mild interglacial intervals. In addition, there were minor fluctuations, as clearly demonstrated, in the glacial and interglacial Ages. Thus the Pleistocene has become adopted as the "Ice Age."

The name Pleistocene was proposed by Sir Charles Lyell in 1839 to replace his older term Newer Pliocene. This time term was based on the percentage of marine invertebrate species, represented in the upper layers of rocks along the southwestern coast of Italy, on Sicily, and on the island of Ischia, which were recognized as living in the sea at that time. Later it was assumed that Lyell's Pleistocene represented the same time interval as the glacial and interglacial Ages.

Glaciation. Toward the close of the nineteenth century it was recognized that glaciers had been much more extensive in the past. Many of the geologic features in the Alps and the nature of the deposits in the U-shaped glacial valleys clearly demonstrated that glaciers had at one time deposited their till in end moraines far below the present edge of melting ice. Some of these valley glaciers had joined on the plains below to form piedmont glaciers. Many years later it was realized that all of northern Europe had been glaciated. Valley glaciers that occur today in high mountains of equatorial regions have greatly receded, and others are totally gone from where they were in the Pleistocene.

It has been estimated that one third of the earth's surface was covered with ice during one of the glacial Ages. Most of the ice was in the Northern Hemisphere, for nearly all of the high mountains and much

*Cesare Emiliani, on the basis of radiocarbon age determinations of deep-sea core samples of pelagic foraminifers and isolation data from three periodic phenomena, suggests that the Günzian and Nebraskan glaciations started about 280,000 years ago.

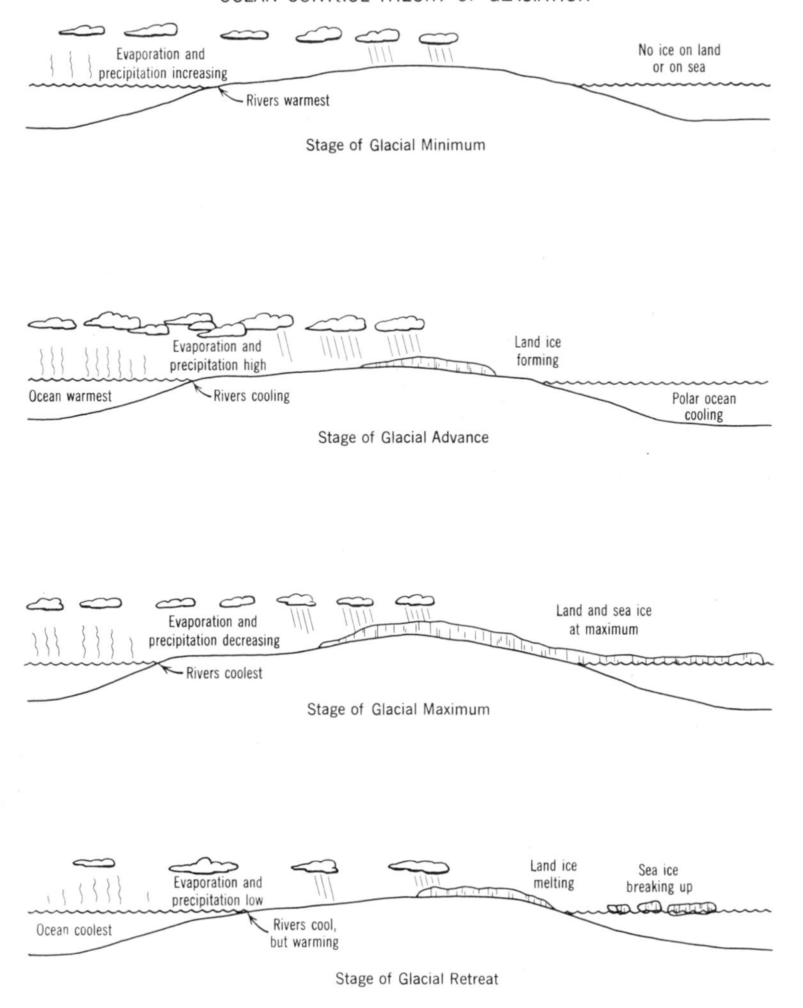

Fig. 174. Sketches of the ocean control theory of glaciation, as proposed by William Lee Stokes.

of the relatively flat interior regions of North America and Eurasia were covered with ice and snow in the process of changing into glacial ice. There were two kinds of glaciers, as there are today: valley glaciers and continental or sheet glaciers. The valley glaciers were restricted to mountainous areas of the world, though termination of the flow ice extended to the lower valleys into lakes or into the sea. The continental glaciers spread out in all directions from inland centers, but the greatest extensions of flow ice was toward lower latitudes.

The possible causes of glaciation are still not adequately understood, but it has been suggested that continental elevation, changes in atmospheric composition, higher evaporation from rising temperatures, topographical and isolation effects, air currents, ocean currents, and variation in solar radiation have been contributing factors.

Situations likely to contribute to such accumulations of ice on the earth's surface were recently proposed by William Lee Stokes in his *ocean control theory.* These are (1) a maximum in the elevation of continental areas and mountain building; (2) alteration in air currents, possibly affected by orogenic disturbances, which carried moisture laden clouds into areas with low temperatures; (3) greater precipitation associated with the lowering of temperatures over the land and evaporation from warm oceanic waters (Fig. 174).

Laurentide Continental Glacier. Continental glaciers are of more interest to the paleontologist because plant and animal remains found in the deposits of the different Ages give an idea of the sequence of life as it was effected by the changes in climate. The ice caps of Greenland and Antarctica are vivid reminders of their gigantic cousins which spread over the northern continents during the Pleistocene.

Nearly one half of North America was covered by the Laurentide glacier. It evidently originated as small valley glaciers in high land areas in eastern and northeastern Canada. As moisture-laden clouds which had formed over warm oceanic waters in lower latitudes came in contact with low temperature conditions over elevated land areas in the north, the resultant snowfall accumulated to considerable thickness. Under the pressure of great thickness the flaky snow recrystalized into loose granular ice, known as névé. Each granule was only a fraction of a millimeter in diameter. After several seasons when the névé was buried to a depth of 100 feet it metamorphosed into a continuous solid containing tiny trapped air bubbles. The constituent grains were then a few millimeters in diameter. This was glacier or flow ice. As the pressure continued to build up from the enormous weight of overlying ice, the basal mobile mass of flow ice began to move slowly down the valleys. On the steeper slopes the rigid outer crust fractured into deep crevasses. Eventually the valley glaciers coalesced with other valley glaciers into piedmont glaciers on the plains below. The piedmont glaciers came in contact with others. Finally this complex formed a great ice cap that flowed out in all directions but mostly to the south and west. It probably reached a thickness of nearly 2 miles in some areas. Surface fractures in the form of deep crevasses may have extended down as much as 200 feet. Toward the east the ice spread out from the land over the ocean as sea ice.

In the meantime pressure from the weight of the glacial ice on land caused the plastic and elastic substratum of rocks below the ice to depress. It has been estimated that the earth's crust in these areas subsided to about a third of the maximum thickness of the ice above.

When glaciation reached a maximum, evaporation and precipitation decreased. By this time sea ice and cold water from the rivers had cooled the surface waters of the oceans, and cold waters in the polar areas had sunk and spread toward lower latitudes.

In a southwesterly direction the line of ablation (evaporation and melting) of the continental glacier was as far south as Missouri and Kansas. The first advance of the ice has been called the Nebraskan Glacial Age. Its outermost ablation line has been difficult to trace be-

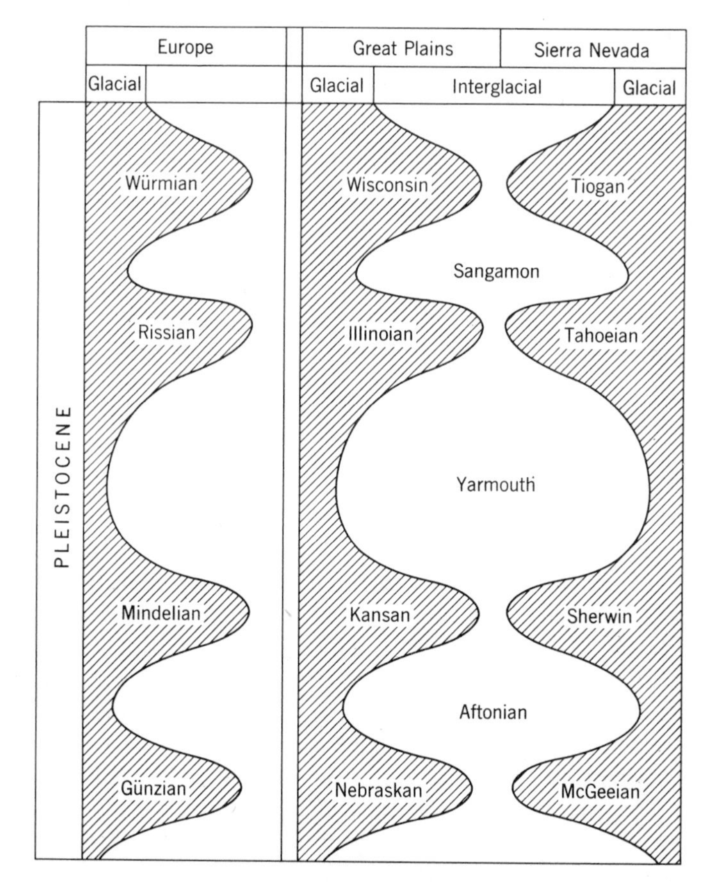

Fig. 175. Diagrammatic sketch of glacial and interglacial Ages.

cause the three subsequent ice sheets, Kansan, Illinoian, and Wisconsin, not only covered the Nebraskan terminal moraines with later glacial till but tended to destroy evidence of the older glaciers. To the west and northwest the Nebraskan probably coalesced with the piedmont glaciers of the Canadian Rockies. Its northern limits have not been determined, but it probably extended northeast as far as Greenland.

At glacial maximum a state of equilibrium was reached when melting and evaporation were equivalent to the rate of accumulation; yet because of a time lag in the cooling of oceanic waters a stage of glacial retreat was retarded for perhaps as much as 40,000 years. This resulted in a reversal in which melting exceeded accumulation. As melting continued, terminal moraines and many other glacial features were left behind as indisputable evidence of glaciation, and at the edge of the ocean huge blocks of ice calved off into the water and floated away as icebergs.

The advance of the Nebraskan ice was then followed by the mild climates of the Aftonian Interglacial interval. This was clearly demonstrated by the nature of the sediments and fossil evidence. For example, a sequence of fossil pollens in gumbotil and stream deposits from five different exposures in Iowa were described by H. G. Lane. He found conifer pollens immediately above the Nebraskan till which indicated a cool temperate climate, grasses which grew in a climate much like that of today, then evidence of a short time in which oaks lived (particularly in eastern Iowa), and conifers which flourished before the advance of the Kansan, or second, glaciation. After the beginning of Aftonian time, when the ice had melted and the oceanic waters again became warm, the process was renewed by another advance of the ice sheet. Similar interglacial Ages, the Yarmouth and the Sangamon, followed the Kansan and Illinoian glacial Ages.

GLACIATION IN EUROPE. Much like the Laurentide was the smaller Scandinavian continental glacier in Europe, taking its name from its point of origin. It produced three or possibly four great advances of ice with interglacial intervals essentially simultaneous with those in North America. The European glaciers spread south over central Europe and east to the steppes of Russia; one lobe crossed the English Channel.

The classic terminology for the four glacial Ages was based on Alpine glaciation. This great mountain system, except for its highest peaks, was covered with glacial ice. Though as many as seven terminologies have been proposed for sequences from different parts of Europe, the one from the Alps has been most widely adopted. From oldest to youngest these are Günzian, Mindelian, Rissian, and Würmian. The interglacial intervals have not been given distinct names as have those in North America.

OTHER GLACIATED REGIONS. There were ice caps and valley glaciers in Siberia, but they were not so extensive as those in Europe. Most of the high mountains of Asia and some in Africa experienced glaciation. In South America there were valley glaciers along the Andes Mountains, and the accumulation was so great in Patagonia that piedmont glaciers formed on the plains below. The great ice cap of Antarctica was much thicker and more extensive than at present. New Zealand, Tasmania, and southeastern Australia also show evidence of Pleistocene glacial ice. Indeed, nearly every country in the world shows some evidence of glaciation or its effect on life in the different regions at that time.

Lowering of Sea Level. The amount of water substance at the earth's surface is nearly constant. Therefore, when snow and ice accumulate on land there must be a corresponding reduction in sea water. This is accomplished by the process of evaporation over the oceans and transportation by winds to points on land where it precipitates in the form of rain or snow.

At times of maximum glaciation, when approximately one third of the earth's surface was covered with ice to depths as great as 3000 feet locally, the level of oceanic waters must have been 150 to 300 feet lower than it is today. Evidence of this has been found in beach terraces along coast lines undisturbed by diastrophism. Terrestrial animals could have crossed from Siberia to Alaska and from the Malayan Peninsula to Java and Sumatra on dry land. The British Isles were similarily populated at times from the mainland of Europe. Melting of the ice with approaching interglacial intervals resulted in water again returning to the sea and a progressive elevation of sea level.

Marine Invertebrates. The Pleistocene marine invertebrate faunas differ only slightly from the living ones. Some species are now extinct and others have modified their distribution patterns. There is evidence of north and south fluctuations of cool- and warm-water marine faunas, but these times of different temperatures in oceanic waters have not yet been satisfactorily correlated with the glacial advances and interglacial intervals. One of the finest sections of marine Pleistocene rocks, nearly 2 miles thick, is in the Ventura Basin in California. Another excellent sequence is in the Los Angeles Basin.

Inland Lakes. It has been suggested that reduction of heat and decreased evaporation during the glacial Ages caused cool-water streams to fill inland-lake areas. One of the best examples was old Lake Bonneville of which the present Great Salt Lake is one of the saline remnants. During its second high-water stage Lake Bonneville covered an area of nearly 20,000 square miles and was almost 1000 feet deeper than the waters of Great Salt Lake today. Subsequent erosion of the outlet to the north through Red Rock Pass reduced the lake level 375 feet. Now the waters have evaporated further until they are only 150 feet at the deepest point. The old high-water levels left clearly marked strand lines along the hills and mountains in the vicinity of Salt Lake, and these are clearly visable to travelers along U. S. Highway 40.

Another lake of similar origin, but not so large, covered much of western Nevada. This was called Lahontan. Winnemucca Lake, Carson Sink, and Pyramid and Walker lakes are remnants of much larger Lahontan. Not only the terrace lines but the monumental residuals of calcareous tufa are indicative of the earlier presence of a large lake in that area.

The ages of the different water levels in these Pleistocene lakes have not yet been correlated with glacial advances, but the time of their maximum extent was probably Wisconsin.

The Great Lakes, on the other hand, were formed in quite a different way. Before glaciation the area was a lowland with rather gentle folds in the underlying Paleozoic rocks. The movement of glacial ice tended to gouge out less resistant rocks in the synclines, in places cutting to considerable depths. During the early existence of lakes in the area the strand lines were much higher than they are today; the water was bounded by higher terrain to the south and glacial ice to the north. After the retreat of the last continental glacier the Great Lakes assumed their present outline.

Land-Inhabiting Vertebrates

At the end of the Pliocene and the beginning of the Pleistocene there were increasing crustal disturbances that culminated in mountain making and glaciation. This resulted in profound changes in the faunas and floras as the older groups died out and the modern ones appeared. Because of the elevation of land in certain areas and because of a lowering of the level in marine waters land connections were established between all of the continental land masses except Australia.

EARLY PLEISTOCENE

There was a profound interchange in the mammals of North America and Eurasia. Zebralike horses which had evolved from *Pliohippus* found their way into the Old World. Dispersing in the same direction were the first camels to reach that part of the world. On the other hand, deer, bears, modern beavers, and other mammals came to North America across an Alaskan-Siberian land connection. At the same time cattle, *Bos*, as well as other bovines and antelope forms appeared for the first time in the Eurasiatic faunas. Mammoths, hippopotamuses, and numerous other genera not present in the Pliocene were also represented. On the whole, the African faunas showed many genera in common with those of Asia and Europe, and many Pliocene genera of northern latitudes also survived in Africa, although they died out elsewhere. In North America some of the conspicuous mammals, apparently confined to the continent, were the hyenoid dogs, *Borophagus*, saber-toothed cats, *Ischyrosmilus*, giant camels, *Gigantocamelus*, small, slender, three-toed horses, *Nannippus phlaegon*, and some pronghorns, *Sphenophalos* and *Ceratomeryx*. Early Pleistocene faunas in the Old World are usually referred to as belonging to the Villafranchian Age; in North America they are called Blancan.

North and South America were again connected after being separated for 70 million years or more. Horses, camels, deer, bears, cats, saber-toothed cats, dogs, mustelids, mastodonts, tapirs, and numerous small rodents dispersed into the southern continent; and more ground sloths, armadillos, chlamytheres, glyptodonts, capabaras, and porcupines spread into North America. The distinctive South American orders, such as Notoungulata and Litopterna which played such an important role among the Tertiary mammalian faunas on that continent, evidently could not compete with the invaders from the north and consequently became extinct during the Pleistocene.

MIDDLE PLEISTOCENE

North America. One of the greatest uplifts in the earth's crust followed the first glacial Age. Mammoths and true horses came into North America from Eurasia but were blocked from South America by a heavy rain forest across parts of Panama and northwestern Colombia. Among the mammals there were only a few holdovers of the peculiar South American genera from Pliocene lines of descent.

One of the best middle Pleistocene mammalian faunas has been taken from the Irvington gravel pits in the San Francisco Bay area. The remains of these animals were buried in a deltaic flood plain by meandering streams that flowed from the south. Some of the first Columbian mammoths, *Mammuthus*, bathed in the streams while their distant relatives, the American mastodonts, *Mammut*, crashed their way through dense willow thickets. Small turtles toppled off dead logs into the protection of deep pools. Deer, *Odocoileus*, browsed in the shrubbery while peccaries rooted about in the sticks and leaves. On the grasslands were herds of the first true horses, *Equus*, large camels, *Camelops*, and smaller, slender-limbed camels, *Tanupolama*, long-tined pronghorns, *Tetrameryx*, and occasional *Euceratherium* which resembled musk oxen. Ground squirrels raced into their burrows as coyotes and the huge dire wolves following the herds of grazing mammals came too close for comfort. An ever present menace was the saber-toothed cat, *Dinobastis*. Gopher mounds belonging to a species of the genus *Thomomys* speckled the ground where these animals took advantage of the loose soil mantle, and pack rats, *Neotoma*, built their stick nests in the dense thickets. Deer mice, pocket mice, and meadow voles had runways in the areas most suitable for their purposes of daily life. Present also were ground sloths, *Megalonyx*, and huge short-faced bears, *Tremarctotherium*, that showed a marked resemblance at first glance but had nothing in common but shaggy coats and similar shapes. Flocks of Canadian geese sought swampy grassland to nibble the fresh green blades and chatter at each other. This interesting fauna represented a major segment of middle Pleistocene life in the middle latitudes of North America, though faunas of the same age differed slightly in composition in different provinces. For example, the giant beaver, *Castoroides ohioticus*, the giant ground sloth, *Megatherium americanum*, plus some other North American genera never reached the Pacific coast province.

Europe. The European Pleistocene faunas are better documented as a whole than those of North America, and in many respects they are much more accurately correlated with the glacial Ages. On that continent also is much of the early history of man.

A fauna possibly an Age older than the Irvington in the European sequence was the Mauer, First Interglacial Mindel-Riss, fauna found in an old bed of the Neckar River near Heidelberg, Germany. In 1907 the scientific world was stimulated by the discovery of a massive primitive human mandible in the Mauer sands deep below the surface of the ground. Though excavations at Mauer have been watched carefully since that time, no additional human remains have been uncovered.

This primitive man, known as *Homo heidelbergensis*, may have hunted in this woodland area where there was an abundance of mammals. Some he may have killed for food. In the swampy ground areas were water voles, *Arvicola greeni*, large beavers, *Trogontherium boisvilletti*, moose, *Libralces reynoldsi*, giant deer, *Megaloceros mosbachensis*, and pigs, *Sus scofra*. Frequenting the streams were hippopotamuses and true beavers, *Castor fiber.* The woods were populated with bears, *Ursus deningeri* and *U. avernensis,* dogs, *Canis mosbachensis,* ancestors of the domestic cats, *Felis catus,* cave lions, *Felis leo spelaeus,* leopards, *Felis pardus,* and lynx, *Lynx issiodorensis.* Lurking about were hyenas, *Hyaena avernensis,* and the last of the machairodont saber-toothed cats in Europe, *Machairodus latidens,* and possibly *Dinobastis.* More conspicuous were the elephant, *Loxodonta antiqua,* and the mammoth, *Mammuthus trogontherii,* a descendent of an early Peking man, *Pithecanthropus pekinensis.* These men had mandibles rhinoceros, *Dicerorhinus etruscus.* Zebralike horses, *Equus stenonis,* were being replaced by the more progressive true horse, *Equus mosbachensis.* Deeper in the woodland were the stags, *Cervus benindei,* the roe deers, *Capreolus capreolus,* and the Mauer bison, *Bison schoetenstacki.*

This fauna contrasts with the cold tundra and the open grassland steppe faunas that spread over Europe with the cold tundra and warm temperate climates. There are only a few genera in common with those of similar age in North America and none of the species is the same.

China. In parts of China at this time the climate was milder and somewhat cooler than that of today. This was the environment of Peking man, *Pithecanthropus pekinensis.* These men had mandibles much like the Heidelberg man but less massive in structure. The remains of Peking man were found in a northward facing cave called Chou-k'ou-tien into which he had brought animals killed for food in the surrounding area. There around the hearth which had been kept aglow with redbud wood were early Palaeolithic stone implements, bones of edible mammals, and skull parts of Peking man with the bottom of the brain case broken open. It has been suggested that the occupants of this cave may have been cannibalizing their neighboring tribes.

The mammal remains about this ancient hearth included the elephant, *Loxodonta namadica,* Merck's rhinoceros, *Dicerorhinus merckii,* sika deer, *Pseudaxis grayi,* thick-jawed deer, *Megaloceros pachyosteus,* Teilhard's Asiatic buffalo, *Bubalus teilhardi,* cave bear, *Ursus spelaeus,* Young's cat, *Felis youngi,* hyena, *Hyaena sinensis,* and mole mice, *Miospalax fontanieri.* This was a temperate fauna without hippopotamuses and anthropoid apes so characteristic of contemporary faunas in the cool tropical environments in Java. The only species in common with faunas of a similar age south of the Himalayan Mountains was the widely distributed elephant, *Loxodonta namadica.*

Java. This brings us to a consideration of the Trinil fauna of Java, made famous by the Dutch army surgeon Eugène Dubois and his discovery of *Pithecanthropus erectus*, the ape man of Java. Cool pluvial conditions and the connection of Java to the mainland of Asia brought into that area certain continental species that mingled with the local fauna.

Peking man and the ape man of Java were closely related, as their generic names indicate, yet they lived in quite different surroundings. *Pithecanthropus erectus* lived in a heavy forested area of monsoon rains where shaggy orangs, *Pongo pygmaeus*, gibbons, *Hylobates*, and Old World monkeys watched his activities. Hippopotamuses swam about with great buoyancy in the deep streams. In the forests and limited grassland were horses, *Equus namadicus*, antelopes, *Duboisia kroesenii*, Lydekker stags, *Cervus lydekkeri*, stegodonts, *Stegodon trigonocephalus*, and the wide-ranging Asiatic elephant, *Loxodonta namadica*.

Middle Pleistocene mammalian assemblages on other continents are, for the most part, less clearly differentiated than those in the later part of the Epoch where mammalian remains are much more abundant in the fossil record.

South Africa. As on other continents, Pleistocene fossil remains have been found almost throughout Africa, but in 1924 the discovery of a young skull with both human and apelike features in a remnant of travertine cave breccia near Taungs focused the attention of paleontologists and anthropologists on the Union of South Africa. Foremost among these scientists were Raymond A. Dart and Robert Broom, who closely followed the operations of the commercial companies that were exploiting these travertines for lime. Within a few years their untiring efforts were rewarded. The discoveries of the remains of fossil men at Sterkfontein, Kromdraai, Swartzkrans, Makapan, and elsewhere shed considerable new light on the story of early man.

From these deposits came good fossil records of the local mammalian faunas that were prevalent, though some of the largest mammals, such as elephants, are extremely rare in the cave deposits. The faunal composition differs in the numerous localities, but a few species are common to all the assemblages. At present the oldest faunas are recognized by the presence of relatively more extinct genera and species that reflect affinities with those of the Pliocene of Europe and Asia. Among these are the saber-toothed cats, *Machairodus* and *Megantereon*, a hyena, *Hyaenictis*, a chalichothere, *Metaschizotherium*, a heavy-bodied giraffe, *Siuatherium*, and a hipparionlike horse, *Stylohipparion*.

Common occupants of those ancient caverns were packs of baboons that came in at night for protection and shelter. There were several kinds; one, *Dinopithecus*, was a giant, perhaps as big as a gorilla.

Saber-toothed cats and other large cats related to the lion and leopard must have made life unpleasant for the baboons and other mammals that frequented the caves and neighboring territory. Possibly the big cats dragged parts of the carcasses of large herbivores, such as rhinoceroses, giraffes, wart hogs, and bovines (cattlelike forms and true antelopes), into grottoes and caves for their young. Bones and teeth of smaller mammals, such as hyenas, serval cats, herpestines, jackals, foxes, bat-eared foxes, wolves, aardvarks, dassies, golden moles, hedgehogs, porcupines, and many others, including smaller rodents and bats, were scattered all through the cave and fissure breccias.

The age relationships of these South African mammalian faunas with those of Europe and Asia are difficult to determine because of the distance and consequent lack of species in common between the two regions. Some effort has been made to establish contemporaneity of pluvial conditions in South Africa with the glacial advances in Europe, and this may prove to be a successful method of recognizing synchrony of events between the two continents, but the mammalian sequence still shows promise of being the most successful.

H. B. S. Cooke, R. F. Ewer, and other vertebrate paleontologists in South Africa have recognized that something like 40 percent of the Sterkfontein and Makapan faunas are represented by extinct species. On the basis of similar studies, the Swartzkrans and Kromdraai* faunas are thought to be slightly younger. The Taungs fauna has been considered by some authorities as about the equivalent of Sterkfontein. Probably some of the oldest South African cave and fissure faunas are early Pleistocene in age, but such refinement in correlation is difficult to attain at this time. The best representation of the later faunas in South Africa occur in gravel deposits of the Vaal River.

East Africa. Some of the best clues to the ages of early man in Africa and the mammals with which he was associated come from the stratified deposits of Olduvai Gorge in eastern Africa. There, abundant and varied mammalian faunas occur in superposition where they are associated with human stone cultures that range from primitive pebble tools in the lower levels to advanced hand axes in the upper units. Among the earlier mammals in the Olduvai Gorge the giraffe, *Sivatherium*, the chalicothere *Metaschizotherium*, a mammoth, and the peculiar proboscidean, *Deinotherium* (Fig. 172), though associated with move advanced forms, are suggestive of similar ones in the Pliocene and early Pleistocene of northern Africa and Eurasia. Mammal-

*The Kromdraai mammalian fauna comes from a neighboring deposit and not from the site at which the ape man skull was found.

ian remains from the higher levels are move advanced and more like the living ones in Africa.

LATE PLEISTOCENE

North America. The most complete Pleistocene vertebrate fauna known was found in the Rancho La Brea tar pits. Sites of earlier excavations can now be seen in Hancock Park along Wilshire Boulevard in the heart of Los Angeles. For the average person it is difficult to realize that these old tar seeps were located in the center of a chaparral-covered landscape, with grasslands, pines, cypress, junipers, live oaks, manzanitas, blue elderberries, and hackberries. Even more obscure is the realization that long before civilized man moved into Los Angeles Valley it was populated with many huge mammals no longer living anywhere.

The tar pools were formed by black, viscous, petroliferous ooze that seeped up through fractures in the earth's crust from the underlying Miocene source rocks. At times quite liquid and again with a hardened asphalt crust, these pools were unique as death traps. Rain from the first autumnal storms vanished almost as fast as it fell into the dry soil, but on the surface of the tar pools a film of water remained. This attracted herds of grazing mammals such as the true horses, American camels, *Camelops hesternus*, pronghorns, *Antilocapra americana*, small La Brea pronghorns, *Breameryx minor*, and bisons, *Bison antiquus*. Gathering around these pools to quench their thirst, these herbivores were charged by giant cats, *Felis atrox*, by agile mountain lions, *Felis concolor*, or by the huge short-faced bears, *Tremarctotherium californicum*. In their haste to escape, the herbivores sometimes sprang into the viscous pools, with the big carnivores intent on the kill in close pursuit. Both were trapped. The more they struggled to free themselves, the deeper they sank into the tar. Larger beasts, such as Columbian mammoths, *Mammuthus columbi*, American mastodons, *Mammut americanum*, Harlan ground sloths, *Paramylodon harlani*, Jefferson ground sloths, *Megalonyx jeffersoni*, and tapirs, *Tapirus haysii*, at other times broke through the hardened asphalt crusts which formed over the surface. Once their feet became entangled in the quagmire, they were never able to extricate themselves. Saber-toothed cats, *Smilodon californicus*, and the giant dire wolves, *Canis dirus*, came in to feed on the partly exposed carcasses and in turn slipped into the tar. Giant condors, *Teratornis merriami* and other condors and vultures, as well as long-legged eagles, *Breagyps clarki*, with the same intent, met a similar fate. Flocks of California turkeys, *Parapavo californicus*, and

California quail, *Lophortyx californica,* common in brushy thickets, probably got into the tar by chasing insects or other edibles. Seemingly, no mammal, bird, reptile, amphibian, or insect species, large or small, escaped these death traps. Mammals ranging in size from bears and wolves to tiny shrews and harvest mice, as well as all kinds of birds from eagles and ravens to chickadees and goldfinches were there in great numbers. Though the number of specimens representing different species varied considerably, over 1000 skulls of the saber-toothed cat and at least 100,000 bird bones have been taken from the pits.

The Rancho La Brea fauna may represent most of later Pleistocene time. Because of the movement of the tar and the trampling of the larger mammals, all of the fossil remains are mixed. The numerous mounted skeletons in museums throughout the world have been made up of the bones of different individuals.

Europe. Neanderthal man, *Homo neanderthalensis,* with his Mousterian stone industry, flourished in Europe during the last interglacial and the early part of the final glacial Age (Würmian). During the interglacial he spent much of his time in the open, resorting to overhanging cliffs or shallow grottos as temporary shelters. The forests, meadows, and streams were populated with an abundance of animal life. Hippopotamuses in the streams were characteristic of the milder interglacial climates. Of the largest mammals there were the straight tusked elephant, *Loxodonta antiqua,* an early mammoth, *Mammuthus trogontherii,* and Merck's rhinoceros, *Dicerorhinus merckii.* Also in the forested areas were the stocky forest horse, *Equus robustus,* the uroch

Fig. 176. Woolly rhinoceros, Coelodonta antiquitatus. *After Osborn.*

Fig. 177. European cave bear, Ursus spelaeus. After Abel and Kyrle.

cattle, *Bos primigenius*, the moose, *Alces alces*, the giant deer, *Megaloceros giganteus*, and the stag, *Cervus elaphus*. From the steppes came the Przewalski horse, *Equus caballus przewalskii*, the wisent, *Bison bonasus*, the hyena, *Crocuta spelaea*, the European badger, *Meles taxus*, the suslik, *Citellus rufescens*, the steppe hen, and many others. In the mountains were the ibex, *Capra ibex*, the chamois, *Rupicapra rupicapra*, the marmot, *Marmota bobac*, and the alpine wolf, *Cuon fossilis*. In the caves were the cave leopard, *Felis spelaeus*, and the cave bear, *Ursus spelaeus*.

With the advance of ice fields from the great centers of Scandinavia and the Alps, the climates changed. Some mammals so typical of the interglacial interval became extinct, others adjusted themselves to the cold and survived with the tundra fauna that came down from the north. There may have been some seasonal migrations by which some elements of the colder fauna reached their southernmost distribution in winter and went back north in summer. The most conspicuous animals from the tundras were the woolly mammoth, *Mammuthus primigenius*, and the woolly rhinoceros, *Coelodonta antiquitatus*. With their shaggy hair and dense under fur, they were well protected against rigorous weather. Also typical of the far north were the musk oxen, *Ovibos moschatus*, the Arctic hare, *Lepus variabilis*, the banded lemming, *Lemmus torquatus*, and the Arctic fox, *Alopex lagopus*. Tundra and tundra border plants, such as poplars, willows, dwarf birch, spruce, and fir, also spread into central Europe. These replaced the temperate interglacial floras until the next retreat of the ice caps. Neanderthal man used the skins of reindeer to protect himself from the bitter cold, and he sought the deeper caves for shelter where he sometimes encountered the great cave bears, *Ursus spelaeus*, the cave lion, *Felis leo spelaeus*, or the cave leopard, *Felis pardus spelaeus*.

Toward the end of the last Ice Age, Neanderthal man was superseded by the Cro-Magñon race. These men were taller and in many ways more advanced than their predecessors in western Europe. Their immediate ancestry and their center of origin has not been determined, but they are referable to the same species as modern man, *Homo sapiens*. The fauna and flora that surrounded Cro-Magñon was little different from that seen by Neanderthal.

Java. A marked similarity to the neanderthaloids has been observed in the skulls of Solo man, *Homo soloensis*, from Java. The stratigraphic sequence along the Solo River suggests that these fossils of man and other mammals may also be of the third glacial Age. Some mammals contemporary with *Pithecanthropus* of the second glacial Age apparently were missing, and others that had been so abundant, such as the Lydekker stag, *Cervus lydekkeri*, were rare. Stegodonts, elephants, and hippopotamuses were more specialized than those of the older Trinil fauna, and pigs, cattle, water buffaloes, and rhinoceroses were much more like those living today. Most abundant were the little axis deer, *Axis axis*. The Ngandong, as this fauna is called, is an equatorial assemblage and, though far from complete, shows a marked contrast with those in far away Europe and America.

Fig. 178. (a) The living red kangaroo, Megaleia rufa; *(b) extinct short-faced kangaroo,* Procoptodon goliah; *drawn to scale.*

Australia. Australia is not far from Java and southeastern Asia geographically, but faunistically these areas have been distinct for at least 100 million years. Unfortunately, the best fossil record from the Australasian region of New Guinea, Australia, and Tasmania is from the late Pleistocene. Marsupials (pouched mammals) in their varied specializations are dominant. Monotremes (egg layers) of greater interest than the marsupials are even less numerous than the murid rodents that evidently came in fortuitously sometime in the Pliocene. The first Australian aborigines appear in the record before some of the large marsupials died out. A slightly later invasion of aborigines evidently is responsible for bringing in the dingos.

As on other continents, large mammals were abundant in the Pleistocene of Australia. *Diprotodon optatus* was the giant of them all. These huge quadrupedal herbivores were distributed all over Australia, but none has yet been discovered in New Guinea or Tasmania. In the "outback" of South Australia, at Lake Callabonna, hundreds of these big marsupials bogged down in the unctuous mud flats around a big mound spring. During periods of drought they must have come in from all directions to quench their thirst. With them were giant wombats, *Phascolonus gigas,* big ground birds, *Genyornis newtoni,* and probably many other kinds of animals.

Equally spectacular was the variety of large and small kangaroo remains, found eroded out of flood-plain and stream-channel sands in the famous Darling Downs of Queensland. Big ones such as *Procoptodon goliah* were short-faced, and stockily built. There were many other genera, now extinct, showing different specializations. Still others were closely related to the living genera and species. There must have

Fig. 179. **Euowenia grata, *a diprotodont marsupial from the Pleistocene of Australia. These mammals were a little larger than a tapir.***

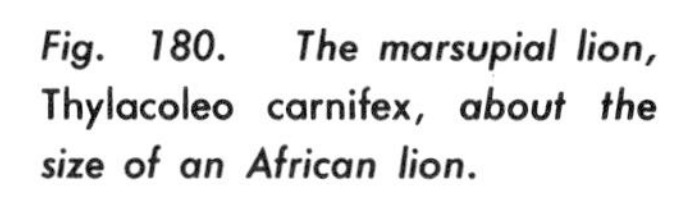
Fig. 180. The marsupial lion, Thylacoleo carnifex, *about the size of an African lion.*

been thousands of these kangaroos when the Pleistocene climate of Australia was more favorable than it is today. Among the other marsupials there were the genera *Nototherium* and *Euowenia,* related to but smaller than *Diprotodon,* that ranged into favorable habitats throughout the continent. Perhaps the most spectacular of all was the so-called marsupial lion, *Thylacoleo carnifex.* Its head was nearly as large as that of the lion, but its shearing cheek teeth were much larger and more specialized. The canines were vestigial, and only the median incisors were well developed in the front of the mouth. Only recently skeletal bones have been collected by Norman B. Tindale and Paul F. Lawson in South Australia, the forms of which should give us an idea of their functions in this interesting creature.

Cave and fissure deposits such as those at the Wellington Caves in New South Wales have yielded most of the remains of the smaller marsupials and rodents, though bones and jaws of the larger mammals are also found there.

Madagascar. It is possible that the island of Madagascar has been separated from the African continent throughout the age of mammals, if not longer. Unfortunately, only Pleistocene (possibly Subrecent) and Recent mammals are known from the island, but when the seven orders and nine families are reviewed it is apparent that the smaller ones could have got there on floating vegetation, by swimming, or by the agency of man. If there had been a land connection at any time during the Cenozoic, some of the strictly terrestrial mammals, so plentiful on the mainland, surely would have spread to Madagascar.

Twelve genera of fossil lemurs have been found. Though some are Recent, others are extinct. There are numerous specializations among them. *Megaladapis,* large forms secondarily adapted for living on the ground, were the size of a gorilla. The ancestors of the Madagascar lemurs probably reached that island in the early Cenozoic. The families represented there are not known elsewhere in the world. On this island also were the largest known ground birds, *Aepyornis* and *Mullerornis,* whose eggs were of gallon capacity or even larger. Nothing is known of the ancestors of these birds.

South America. By Pleistocene time many mammals had infiltrated from North and Central America into South America. Some of the older South American groups survived until the Pleistocene when they became extinct.

The most widely known of all South American formations and its vertebrate fauna is the Pampean of Argentina. It was in consolidated loess (wind-blown sediments) in this formation that a skeleton of the largest ground sloths, *Megatherium americanum,* was first found and sent to Spain. It was described by Cuvier in 1797. Ground sloths, having reached the climax in their evolution, were abundant and diversified. They included *Glossotherium robustum, Mylodon darwini, Lestodon armatus* and others. The armored glyptodonts, after a long history in South America, also reached their maximum in size. These were fantastic creatures with a short deep head, a body covered with thick bony plates, short legs, and heavy feet. The most conspicuous differences between them was in the tail. In *Glyptodon clavipes* the tail sheath was composed of overlapping bony rings, each with heavy short spines. Another *Daedicurus clavicaudatus* had a long clublike tail with hornlike spines in a cluster at its end. Armadillos ranged in size from tiny fellows like *Chlamyphorus truncatus,* less than 6 inches long, to large Recent ones like *Priodontes gigas,* more than 3 feet long. The chlamythere, *Chlamytherium,* had a body five feet long. Two of the most peculiar Pleistocene mammals were first found by Charles Darwin, on his cruise around the world on the H.M.S. *Beagle,* when he went ashore in Argentina. One was a huge skull of rhinoceros proportions but with long, strong, curved upper teeth. It was called *Toxodon platensis.* In the years that followed complete skeletons were found. The other specimen that attracted Darwin's attention was part of a skeleton of *Macrauchenia patachonica,* which he thought was related to the camel because of the construction of its neck vertebrae. Actually, these animals were the end products of a long and prolific evolutionary history of peculiar South American ungulate mammals. A smaller animal about the size of a pig, *Typotherium protum,* superficially looked something like a huge rodent but belonged to one of the suborders of mammals, Typotheria, peculiar to South America.

The rodents showed a mingling of the groups related to porcupines and chinchillas and the early Pleistocene arrivals from North and Cen-

Fig. 181. Large Pleistocene mammals of South America: (a) *Toxodont,* Toxodon platensis, (b) *macrauchenid,* Macrauchenia patachonica, (c) *ground sloth,* Megatherium americanum, (d) *glyptodont,* Daedicurus clavicaudatus. *After Scott.*

tral America. Other mammalian invaders from the north were horses, llamas, deer, mastodonts, dogs, cats, saber-toothed cats, bears, and numerous mustelids. Most experienced a rapid evolution in a relatively short time. The exact time of man's arrival in South America has not been determined, but it was either late Pleistocene or Subrecent.

25

Foraminifera and Oil

Foraminifera (Lat. *foramen*, opening; *ferre*, to bear) is an order of animals, usually microscopic, referable to the phylum Protozoa. Oil geologists frequently refer to them as "bugs" or "forams," expressions which are almost universal in the petroleum industry. It is impossible to estimate the tremendous numbers of these organisms that existed in the oceans throughout much of known geologic time. It is equally difficult to conceive of the role they played, especially as a source of food for other life in the sea.

The foraminiferal organism, though single-celled at maturity, has only one nucleus (mononucleate), but at some stage in its life history it may temporarily have several nuclei. In either case the nuclei are surrounded by cytoplasm, which is the protoplasm outside the nucleus in a cell. Foraminifers belong to the class of protozoans having "false feet" (pseu-

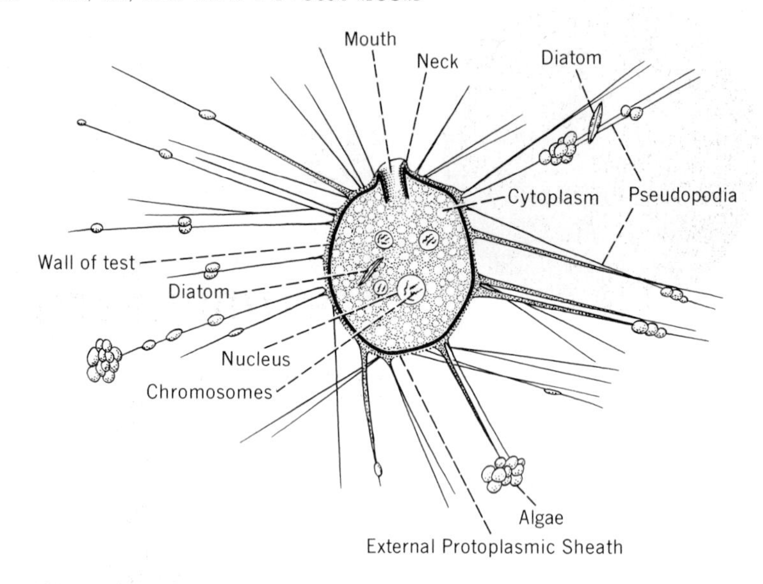

Fig. 182. A living foraminiferan, Allogromia laticollaris, *showing its different parts and captured diatoms. Prepared under the direction of Zack Arnold.*

dopodia) as their locomotor structures. They differ from higher organisms in that there is no well-defined division of labor between the body units, though the nucleus and cytoplasm are involved in different functions at the time of reproduction (Fig. 182).

In the adults the species range in size from about 60 microns (six hundredths of a millimeter) to nearly 10 inches long. Among the smallest is *Lagena*, which first appeared in the Jurassic; its species are still living in the oceans. Those recorded as 10 microns in size are probably immature. One of the largest is *Loftusia*, which is 7 or 8 inches long. The smaller foraminifers, by far the greatest in numbers, are less than 1 millimeter in diameter. These microorganisms occur at different depths in marine waters from polar areas to the warm oceans in tropical regions. They are so numerous in some places that the beach sands are made up almost entirely of tests which have washed ashore.

The Tests. The hard parts are usually called *tests* (Lat. *testa*, shell, pot). Some are perforated with tiny openings called pores; others are imperforate. The mistaken notion is still prevalent that the name "Foraminifera" was proposed because of these surface openings. Actually, the name had reference to the large opening now known as the *aperture* found in each chamber and in the earlier formed chambers occurring as openings between the internal chambers. These chamber connections were once thought to correspond to the siphuncle of cephalopod mollusks. Consequently, Lamarck, d'Orbigny, and other early students placed the Foraminifera and Cephalopoda in the same group.

The tests may be straight, bulbous, or coiled in one or more planes. In plan of growth, or arrangement of chambers, foraminifers assume almost every conceivable shape in some of the most aestheically pleasing patterns imaginable. Several lineages which differ in their shell-wall material and structure may exhibit much the same plan of growth; yet each lineage within the order characteristically follows one or another form in its chamber arrangement. Genera which change their plan of growth within an individual life span are termed *multiferea* and have been thought to reveal thereby intermediate relationships between ancestral and descendant stocks.

The most primitive of all foraminifers, the family Allogromiidae, include living genera with or without chitinous parts. Other families have tests of agglutinate material, lime, or even silica. Many foraminifers change the composition of their tests during the growth of the individual from the young to the adult stages.

The apertures, or large openings, of the tests, through which food substances are taken in to be assimilated by the protoplasm, may also vary greatly in shape from one lineage to another. In this respect, too, they offer many phyletically diagnostic features. These openings may also change considerably in different growth stages. In fact, there may be one or many apertures. Finally the chambers themselves tend to vary in form from group to group, and so does the surface ornamentation which often becomes delicately intricate. But the composition and structure of the test wall, as secreted by the soft parts, remains the most deep-seated trait inherent in the various lineages of the groups.

CHITINOUS TESTS. Chitin is a sort of horny substance similar to the covers (elytra) of beetle wings. The oldest known foraminifers from Cambrian rocks apparently were covered with thin transparent chitin without minute openings or pores. But throughout the fossil record such material tends to be replaced by other minerals of secondary origin.

AGGLUTINATED (ARENACEOUS) TESTS. Agglutinated tests are composed of a thin chitinous base with foreign particles cemented to the surface in a chitinous, ferruginous, siliceous, or calcareous complex. These form efficient protective coverings for the body protoplasm. Particles solidly cemented into the test include sand grains, flakes of mica, sponge spicules, parts of tests of other foraminifers, or other kinds of minute objects. By some means, still not known, some species are very selective in the particles used in agglutination, others are not. Still other forms tend to lose the agglutinated particles, as the individuals become mature, to

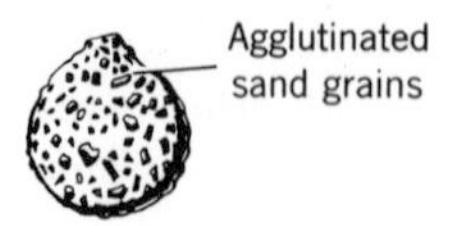

Fig. 183. Agglutinated test, Saccammina sphaerica, *× 14. After Cushman. Redrawn from* Foraminifera, *Harvard University Press. Recent. Family Saccamminidae.*

the extent that the earlier coatings of such extraneous materials are limited to the early growth stages of the individual. This is frequently exemplified in adults with calcareous or siliceous tests (Fig. 183).

SILICEOUS TESTS. Tests composed essentially of silica, a hard white colorless mineral, are relatively rare in foraminifers. Consequently, aside from agglutinated forms with siliceous cement, silica occurs only in the family Silicinidae. Marine formations containing considerable volcanic ash are often rich in other fossil microorganisms with siliceous tests, such as diatoms, silico-flagellates, and radiolarians. The first foraminifers with essentially siliceous tests are recorded from the Jurassic.

CALCAREOUS TESTS. Calcareous tests are composed of calcium carbonate, a limy compound usually in the form of calcite or aragonite. As already stated, foraminifers with agglutinated tests in a more youthful stage of development sometimes develop a calcareous test in old age. This change may occur in those with a ferrous or calcareous agglutination matrix. The calcareous tests may be either imperforate or perforate. In imperforate tests, which characterize some five or six families, calcareous crystals are oriented lengthwise to the surface of the test in a sort of felting. But among calcareous foraminifers the perforate tests are by far the most abundant. In these the pores, which vary considerably in size and even in the pattern of their surface arrangement, simply develop between the calcite crystals. The crystals in this kind of test are oriented with the long axis at right angles to the surface.

Environments. Most foraminifers live in the oceans, though some prefer brackish waters, and a few are fresh-water inhabitants. The greatest number of oceanic kinds are *benthonic,* or bottom-dwellers. The majority of the marine species occur on the floor of the ocean in waters of medium depth, *neritic,* between the level of low tide and 100 fathoms, or, *bathyal,* i.e., on the continental slope beyond the edge of the shelf. Though some species with large numbers of individuals occur at *abyssal* depths. Benthonic forms, such as *Robulus, Eponides, Bolivina,* and many others, move slowly about on their pseudopodia or "false feet." The netlike pseudopodia, which extend out of the perforations or apertures of the tests, are utilized also in gathering in food particles.

Fig. 184. Bathyal foraminiferan, Bolivina bramletti, × 150. *After Kleinpell,* Miocene Stratigraphy. *Permission of American Association of Petroleum Geologists. Miocene. Family Buliminidae.*

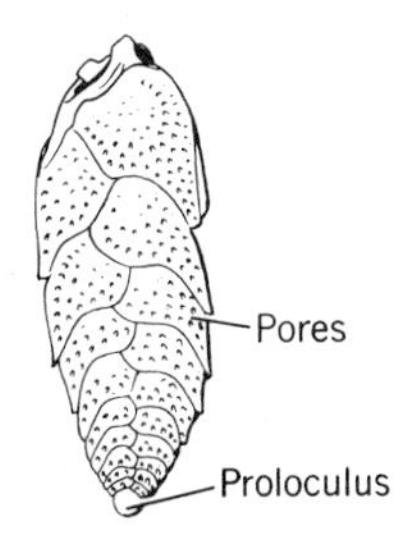

Other kinds, such as *Dendronina,* living on the ocean floor, attach themselves to various objects. They are called *sessile.* Sessile foraminifers are distributed as young while in a mobile state. The adults often assume a dendritic or branching form or reflect in other aspects of a specialized morphology their change to an attached mode of life.

Still other foraminifers live in a *pelagic* environment in the waters of the oceans above the benthos (bottom). Of these the *planktonic* genera, such as *Globigerina,* have no means of locomotion of their own. Their globular chambers partly filled with air permit them to float, and thus they are dispersed by ocean currents or by waves whipped up by the wind. Still other planktonic species, such as *Hastigerinella digistata,* are not pelagic but float about on the surface in the restricted vicinity of coral reefs. No *nektonic* foraminifers, those capable of moving by their own means through open oceanic waters, are known.

Many of the benthonic and sessile genera and species have local environmental preferences. This can be detected in the fossil assemblages as well as with the living organisms. Bottom conditions, then, are governing environmental factors. Rocky, sandy, and muddy surfaces, together with depth, temperature, wave action, currents, presence of other organisms, and many other conditions, determine the habitats of these bottom-living foraminifers. On the whole, temperature seems to be the most important among these controlling factors. Some of these conditions also affect the distribution of planktonic forms.

Fig. 185. Sessile foraminiferan, Dendronina aborescens, × 14. *After Cushman. Redrawn from* Foraminifera, *Harvard University Press. Recent. Family Hyperamminidae.*

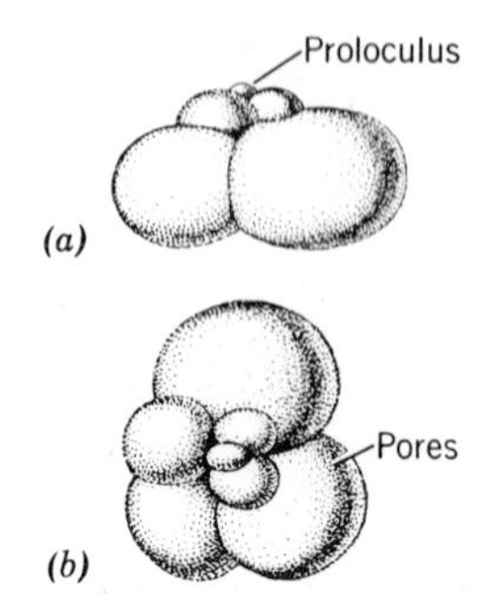

Fig. 186. Planktonic foraminiferan, **Globigerina bulloides:** (a) *Side view,* (b) *top view;* × 60. *Miocene. Family Globigerinidae.*

It is impossible to estimate the almost infinite numbers of foraminifers that inhabited the oceans during geologic time (or even today). During the late Cretaceous, when the ocean surface of the world was close to a maximum for all of geologic history, the planktonic pelagic foraminifers first appeared. Most of the world's natural chalk is composed largely of the tests of *Gümbelina* (Fig. 148*b*), an extinct genus which flourished in the oceanic plankton of those times. It has been said that 30 per cent of the present ocean floor is covered by a sticky calcareous mud or ooze made up of the tests of foraminifers, most of which is composed of species of the planktonic families Globigerinidae and Globorotaliidae. When one stops to consider that these microorganisms have a tremendous rate of reproduction after they have reached maturity in a life span of one to two years and that the lifeless tests are settling continuously through the water, the figure does not seem unreasonable.

As climatic conditions changed throughout geologic time, many of the families and genera of foraminifers have dispersed to faunal provinces with more favorable environments. Microfaunas much like those in the Eocene and Oligocene of Europe are now found in Indo-Pacific and Australian waters. These are the remnants of early Tertiary microfaunas of the old Tethys Sea that are thought to have spread as early as the Miocene into the provinces they now occupy. Other foraminiferal faunas that were widespread, like those that existed off the coast of Florida and in the Gulf coast province as late as Oligocene time, have either become extinct or have become preserved in a modified form in the Pacific where related kinds are still living in Philippine waters. There are still other relationships between the faunas of the Australian seas and those of the West Indian waters that reflect past dispersals not yet known. Today, as they have been in the geologic past, foraminiferal faunas are well marked in their regional and provincial distribution.

Six foraminiferal provinces, Antarctic, West Indian, Indo-Pacific, East African, Mediterranean, and Arctic, are recognized today. Though they are not so diverse as those based on the larger living marine invertebrates, their number has increased steadily since the late Cretaceous when foraminiferal faunas were essentially cosmopolitan.

Classification and Geologic History

Since the early classification of Foraminifera by Alcide Dessalines d'Orbigny in 1826, numerous attempts have been made to improve and to clarify the relationships of the families. As in other groups of

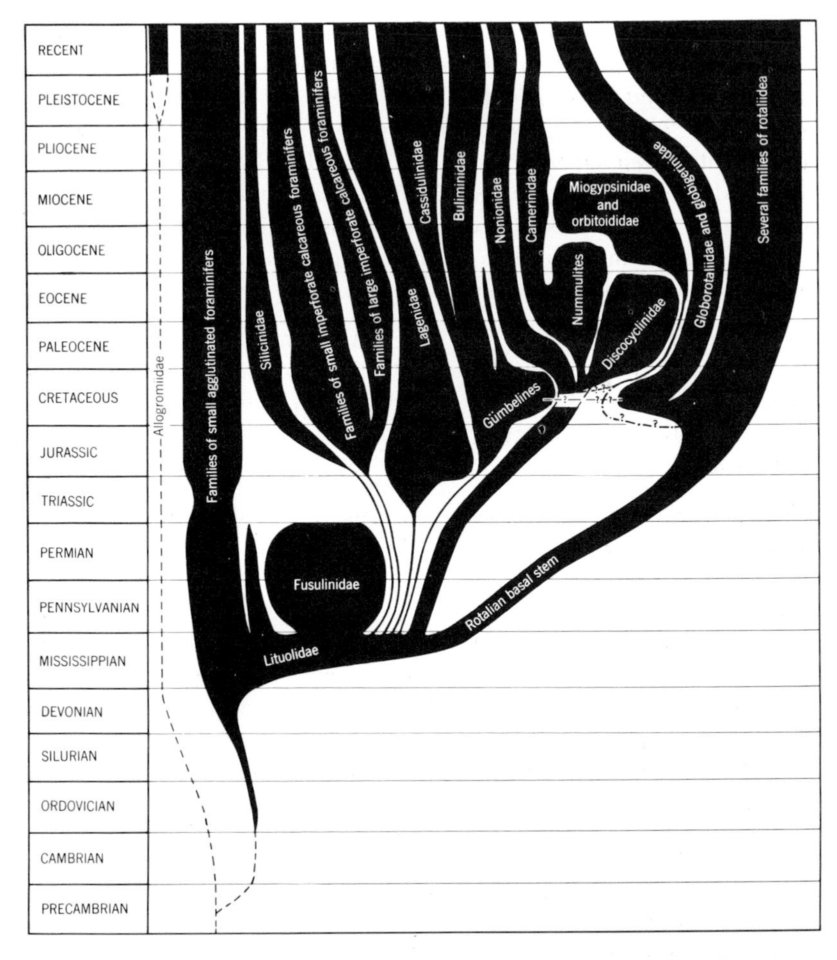

Fig. 187. A simplified phylogeny outlined by Robert M. Kleinpell.

Fig. 188. Pseudastrorhiza silurica, × 25. *After Cushman. Redrawn from* Foraminifera, *Harvard University Press.*

organisms, there is much convergent and parallel evolution in the families, thus making it difficult to recognize phyletic trends in some lines of descent. There is, however, good evidence that many families are derived from others, and this is also true among genera and species. Nevertheless, the present classification of the foraminifers is as satisfactory as that of any group of organisms of comparable magnitude and superior to many because of the tremendous effort devoted to it by countless workers. Fifty families were recognized by Joseph A. Cushman in the fourth edition of his excellent book, *Foraminifers, Their Classification and Economic Use,*" in 1948. M. F. Glaessner, in his useful book *The Principles of Micropaleontology,* divided the known families into seven superfamilies: Astrorhizoidea, Lituoloidea, Endothyroidea, Milioloidea, Lagenoidea, Buliminoidea, and Rotalioidea. It is not intended here to present a complete classification of the higher categories of the Foraminifera but to introduce the student to some of the more important families or groups of families (Fig. 187).

Allogromiidae. The most primitive of the families, the Allogromiidae could probably be traced well back into Pre-Cambrian time if fossil evidence were available. But these protozoans were soft-bodied forms, or possessed only a thin organic wall, and have left no trace of their predecessors.

Small Agglutinated Families. Though foraminifers have been recorded from both the Cambrian and Pre-Cambrian, only a limited number from the Cambrian of England have been confirmed. Records of supposedly "advanced" foraminiferal genera and families from the early and middle Paleozoic have been based on recognition of plan of growth alone, without regard to primary test-wall material and structure. Most of the Paleozoic families had agglutinated tests, a stage in evolution beyond those without hard parts or those with thin chitinized tests, yet still not "advanced" in the sense of those dominant in the late Mesozoic, the Cenozoic, or in the seas today (Figs. 188, 190).

Fig. 189. Raibosammina mica, × 17½. *After Cushman. Redrawn from* Foraminifera, *Harvard University Press. Ordovician. Family Saccamminidae. Small agglutinated foraminiferan.*

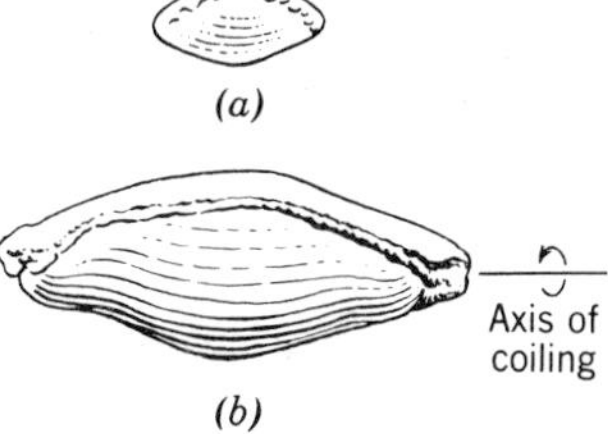

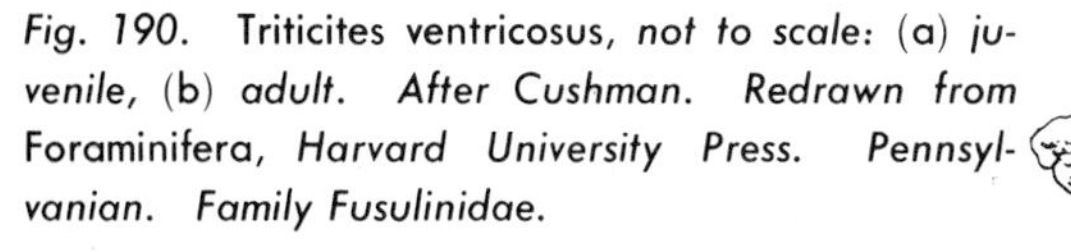
Fig. 190. Triticites ventricosus, *not to scale:* (a) *juvenile,* (b) *adult. After Cushman. Redrawn from* Foraminifera, *Harvard University Press. Pennsylvanian. Family Fusulinidae.*

The families Astrorhizidae, Rhizamminidae, and Saccamminidae, which appear early in the Paleozoic, could well have been derived directly from an earlier allogromid stock. Other small-sized but more complex agglutinate families (Reophacidae, Ammodiscidae, Trochamminidae, Lituolidae) had also appeared by the early or middle Paleozoic. Several of the most ancient and ancestral genera referable to these families are still living. The Textulariidae and their derivatives developed later in the Paleozoic. A few large agglutinated genera, such as *Orbitolina, Loftusia,* and *Dictyoconus,* that belong to three other families did not evolve until the Cretaceous.

Fusulinidae. One of the foraminiferan families most useful in correlation is the Fusulinidae. The first representatives appeared toward the end of the Mississippian, and the genera and species increased tremendously during the Pennsylvanian and Permian. More than 600 species and about 50 genera have been described. Strangely enough, the family died out suddenly at the end of the Permian (Fig. 190).

Fusulinids were benthonic foraminifers with granularly calcareous test-wall material at maturity. The tests resembled grains of wheat in size and shape. Others were as much as 2 inches long. The internal chambers were coiled planispirally about an axis which showed an evolutionary tendency toward elongation, thus eventually producing a spindle or dirigible-shaped test. Most of the diagnostic characters are internal, and the specimens must be carefully sectioned for identification. Since fusulinids were widely distributed and evolved rapidly, they have been very useful to the petroleum geologist in stratigraphic correlation. In some localities they make up most of the rocks in which they occur. They were particularly abundant in the clear, shallow water of the inland seaways in the present mid-continental and in southwestern United States.

This remarkable radiation of the fusulinids stemmed from the genus *Endothyra,* which were fairly small, agglutinated, coiled descendants of the family Lituolidae. It was also the first genus of foraminifers to become abundant. After the large fusulinids became extinct at the end

Fig. 191. **Endothyra media,** × **20.** *After Cushman. Redrawn from* **Foraminifera,** *Harvard University Press. Pennsylvanian. Family Lituolidae.*

of the Permian, foraminifers once more were exceedingly rare. This situation prevailed throughout the Triassic. In the meantime the small imperforate and perforate calcareous foraminifers made their appearance.

Imperforate Calcareous Families. Imperforate calcareous foraminifers appear in the Carboniferous. Seemingly, they are highly stenotopic (greatly restricted ranges within certain depths) in their environmental tolerances. They are, therefore, very spotty in distribution, both geologically and geographically. In fact, they have seldom been abundant except in local concentrations, though the genera have evolved in diverse directions. It is by no means certain that they all came from a common ancestry, and not more than two or three lineages of major imperforate groups seem ever to have evolved.

The families Ophthalminiidae and Miliolidae are composed mostly of small forms with their tests coiled about a continually changing axis. These little foraminifers seem to thrive in clear current-sheltered waters, often on rocky bottoms and at warm temperatures. Even so, the miliolid genus *Pyrgo* appears to have become adapted to cold water. The Peneroplidae and Alveolinellidae are families now living in tropical waters and are relicts of the Indo-Pacific that occupied large tracts of the ocean bottom during the Eocene and Miocene. In their evolution the Alveolinellidae paralleled the Loftusiidae and the Fusulinidae of earlier Periods and Epochs (Fig. 192).

Lagenidae. The lagenids were, and still are, a highly variable and morphologically plastic family, though they have uniformly perforate calcareous tests and possess a most distinctive radiate form of the aperture. During their geologic range from the Mississippian to the Recent they neither diversified nor multiplied notably until the Jurassic. Then they reached the peak of their evolution in an efflorescence, which actually carried on into the Cretaceous (Fig. 148*a*), but by that time they were one of many other newly rising groups of foraminifers. Lagenid species from Cenozoic rocks are utilized in correlation every day by paleontologists and stratigraphers.

Fig. 192. **Pyrgo murrhina,** × **20.** *After Cushman. Redrawn from* **Foraminifera,** *Harvard University Press. Recent. Family Miliolidae. Small imperforate calcareous foraminiferan.*

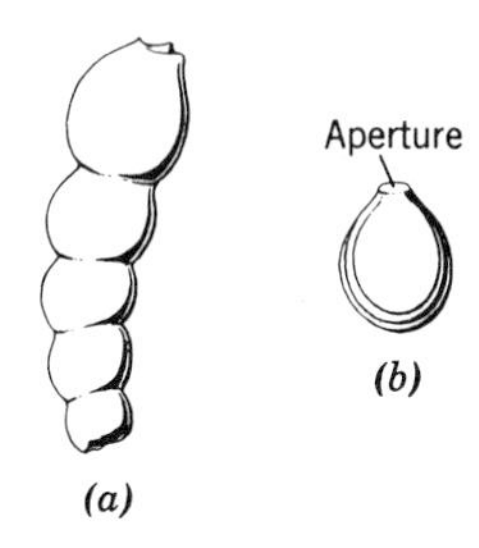

Fig. 193. (a) Lagena marginata, × 45. (b) Dentalina obliqua, × 20. *After Kleinpell.* Miocene Stratigraphy. *Permission of American Association of Petroleum Geologists. Miocene. Family Lagenidae.*

In these benthonic foraminifers the tests have frequently uncoiled into straight or elongate forms, whereas some are compressed into the shape of an inverted v. In abundance, today and apparently throughout their geologic history, they characterize the medium-depth forms on ocean bottoms. The coiled ones, especially, are more abundant in the shallower neritic waters, and the rectilinear kinds are found in the deeper, bathyal environments on the continental slope (Figs. 148a, 193).

Buliminidae. The buliminids have elongate or cylindrical calcareous and finely perforate tests. In the tests an asymmetrical coil of chambers has been extended into a high spire. These are multiserial in *Buliminella,* triserial in *Bulimina,* biserial in *Virgulina,* uniserial in *Rectobolivina,* and even fan-shaped in certain end genera such as *Pavonina.* The buliminids, too, are benthonic, or bottom-dwellers. The family ranges from Jurassic to Recent. Many genera and species were common in the late Cretaceous chalk formations. Species of the triserial *Uvigerina,* the tropical uniserial *Siphogenerina* and the biserial *Bolivina* have been used extensively in zonal terminology within the early and middle Cenozoic, particularly in the Pacific coast province of North America (Figs. 157a,b,e, 194).

Rotaliidae. Historically, the first genera of this family to be scientifically defined were *Rotalia* and *Discorbis.* They were described by Lamarck as early as 1804. Subsequently, as apparent similarities in the

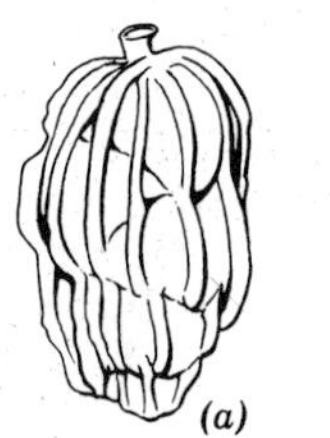

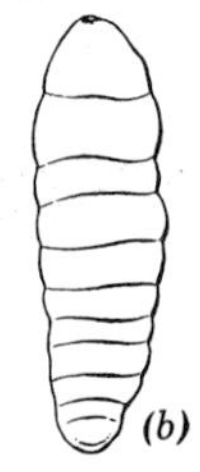

Fig. 194. (a) Uvigerina gallowayi, × 37. (b) Siphogenerina hughsei, × 21. *After Kleinpell.* Miocene Stratigraphy. *Permission of American Association of Petroleum Geologists. Miocene. Family Buliminidae.*

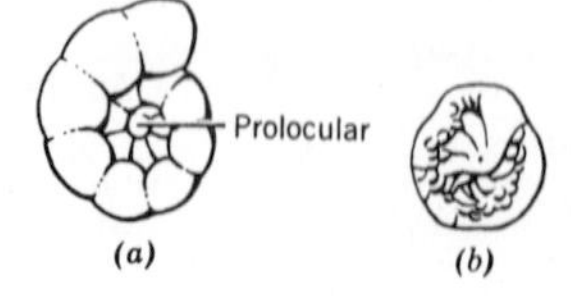

Fig. 195. (a) Discorbis vesicularis, × 10. (b) Rotalia trochidiformis, × 10. *After Cushman. Redrawn from* Foraminifera, *Harvard University Press. Eocene. Family Rotaliidae.*

tests of other related genera were observed, the family came to include a multitude of subfamilies and genera from the late Paleozoic to the Recent. The Rotaliidae included benthonic, trochoid (top-shaped), and asymmetrically coiled forms. Many of the supposedly component lineages have been recognized as having long independent histories and have been raised to family rank. These include the families Anomalinidae, Calcarinidae, and Amphisteginidae. Martin F. Glaessner has proposed the superfamily Rotalioidea for all of these related families. The rotalids are the most numerous and most diverse of all of the groups of Foraminifera and one which has continued to evolve and diversify to this very day (Figs. 157c, 195).

Globigerinidae. This is one of the specialized planktonic families of small foraminifers which have perforate calcareous test-wall material similar to that in their rotalid progenitors. In perfectly preserved specimens long delicate spines extend out from the interpore spaces, frequently serving as shields for the pseudopodia. The more advanced genera possess several globular chambers which make it possible for them to float in pelagic environments (Fig. 157d).

The Globigerinidae and the related family Globorotaliidae make their first appearance in the Cretaceous. Specimens are found in most faunal samples taken from Cretaceous and Cenozoic marine rocks, and both families are still abundant in the oceans. Because of their wide dispersal they are sometimes used in crude intercontinental correlations. The globigerina ooze in our oceans is made up primarily of the tests of these foraminifers.

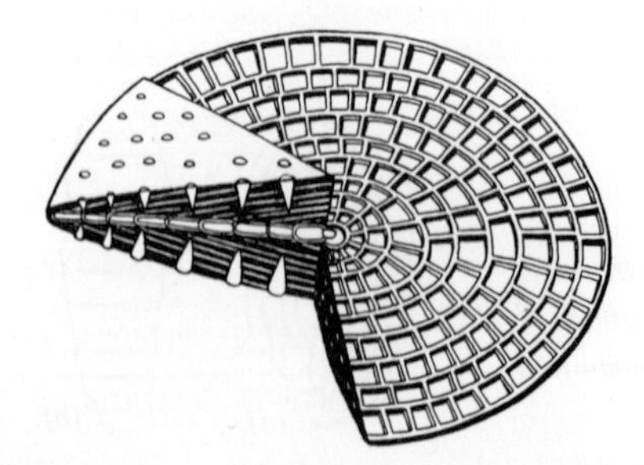

Fig. 196. Cycloclypeus, showing internal structures. After R. Wedekind. Family Camerinidae.

Fig. 197. Operculina, showing internal structures. After R. Wedekind. Family Camerinidae.

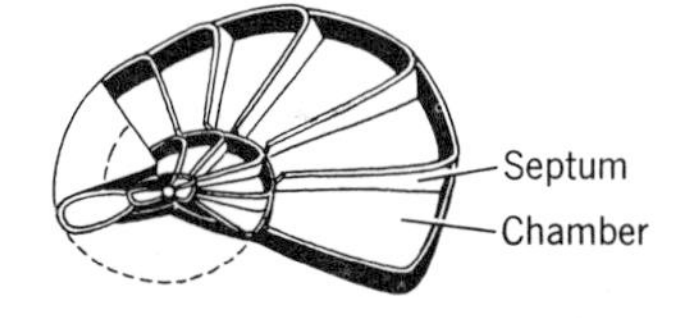

Camerinidae. These are the foraminifers frequently called "nummulites." The generic name *Nummulites* was proposed by Lamarck in 1801. The name was used for a long time before it was disclosed to be a synonym of *Camerina* Bruguiere, 1792. However, the International Commission of Zoological Nomenclature has finally restored the generic name *Nummulites* for *Camerina.*

Nummulites are large benthonic forms with planispiral and perforate calcareous tests. The family includes two or three subfamilies and twelve to fourteen or more genera. Most of the genera are lens- or coin-shaped. The more advanced kinds have a complex of internal chambers or chamberlets and an intricate canal system without septae. Though the family, *sensu stricto,* ranged from the early Cretaceous to Recent time, they were most abundant and diagnostic in the Paleocene, Eocene, and early Oligocene when they reached the size of an American half dollar or an Australian penny (Figs. 196, 197).

Nummulites were dominant in the old Tethys sea which extended from southern Europe and northern Africa across the middle east to India and the East Indies. To a great extent the Egyptian Pyramids are built of blocks of calcareous fine-grained sandstone and limestones that are largely composed of nummulites.

The typically planispiraled Nonionidae and the Camerinidae evidently arose from the same ancestry in the Jurassic, and the Heterosteginidae evolved from the Camerinidae in the Eocene.

Discosyclinidae. The genera in this family are in many ways much like the nummulites with their intraseptal canal system and marginal cord; but, in other features they are more like the orbitoids in their embryonal apparatus, internal pillars, and arrangement of equatorial and lateral chambers. All three families have large lenticular-to-discoidal tests. In the discocyclinids the fairly large equatorial chambers are rectangular in horizontal sections.

Fig. 198. Plectofrondicularia miocenica, × 21. *After Kleinpell.* Miocene Stratigraphy. *Permission of American Association of Petroleum Geologists. Miocene. Family Heterohelicidae.*

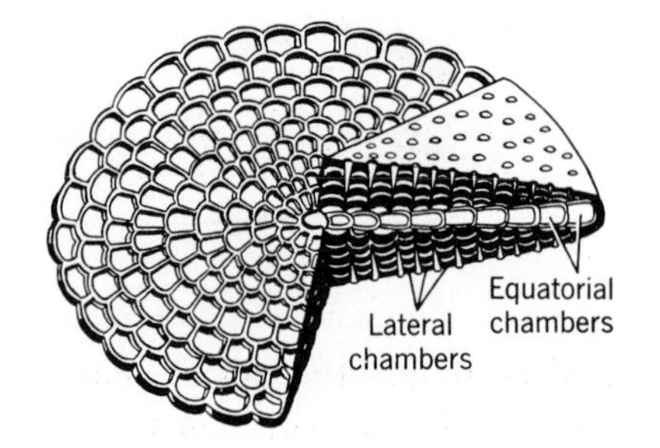

Fig. 199. Lepidocyclina, showing internal structures. After R. Wedekind. Family Orbitoididae.

The family made its first appearance in the late Cretaceous when many other groups with comparable extensive Tertiary histories first appear in the fossil record. *Discocyclina*, a common Paleocene and Eocene genus, was distributed around the world. It occurred in tropical America, New Zealand, the South Seas, the East Indies, India, southern Europe, and northern Africa.

Orbitoididae. The orbitoids are also benthonic. They, too, are large foraminifers, superficially shaped much like the nummulites and the discocyclinids. But they differ from members of those families in the horizontal sections of their equatorial chambers which are hexagonal to lozenge-shaped and in addition have no marginal cord. *Lepidocyclina*, which ranged from the middle Eocene to the late Miocene, was especially numerous and diverse. Furthermore, it was widespread in Oligocene and Miocene time. Its various subgenera, many of which are restricted stratigraphically, have been very useful in correlating within the Tertiary faunal provinces. For example, they have been found in Gulf coast and Caribbean formations, in European and African formations, and in formations of the Indo-Pacific from India and Western Australia to Japan, New Zealand, the East Indies, and the Pacific Islands (Fig. 199).

Miogypsina, another genus of large foraminifers superficially similar to *Lepidocyclina*, formerly grouped with the orbitoids, is now placed in a separate family, Miogypsinidae. It is also widespread, though tropically stenotopic. These have been useful as index fossils in late Oligocene and Miocene formations.

Cassidulinidae. Perhaps the strangest of the small foraminiferal families is the Cassidulinidae which consists principally of two genera, *Cassidulina* and *Ehrenbergina*. Like the family Heterohelicidae, which includes the genera *Plectofrondicularia*, *Nodogenerina*, and *Zeauvigerina*, the origins and affinities of the cassidulinids are still disputed by foraminiferologists (Fig. 198).

In their wall material, biserial chamber arrangement, and comma-shaped aperture the Cassidulinidae show close relationships to the biserial Buliminidae. Nevertheless, the cassidulinid plan of growth uniquely coils symmetrically in the plane of its biserial axis (Fig. 200).

The Cassidulinidae are probably most significant in paleoecology because of their cold temperature optimum and their wide environmental tolerance. They are common and even dominant in most Recent foraminiferal faunas. Usually they inhabit cold marine waters. This may be at shallow depths in polar areas or in tropical deeps. Some casulinids, though, are present in small numbers in shallow tropical seas.

The genus *Cassidulina* made its first appearance in the late Cretaceous, and though it never diversified into numerous species it reached its greatest expansion much later. Its increase kept pace with the widespread disappearance of large tropical foraminifers and the progressive cooling of world-wide climates at the end of the Tertiary and in the Pleistocene.

Economic Use. Foraminifers are of considerable economic importance in stratigraphic correlation of marine sedimentary rocks, particularly in petroleum exploration. The microfossils are used not only from the surface outcrops of the rocks but also from well core samples. This work has been conducted in enormous proportions since 1920. All of the major oil companies have large laboratories in many parts of the world, or employ consulting specialists in this field, for the preparation and identification of foraminifers. Temporary quarters are sometimes set up on promising new oil fields.

Fossil foraminifers are utilized in local and provincial correlation for several reasons.

1. They are abundant in many kinds of marine sedimentary rocks, especially in those of fine texture often so barren of larger fossils.
2. Most sources of oil and petroleum reservoirs are associated with rocks of marine origin.
3. Because of their size, samples of entire faunas can be taken from well cores in the subsurface rocks.
4. The abundance of fossil foraminiferal tests is also such that the

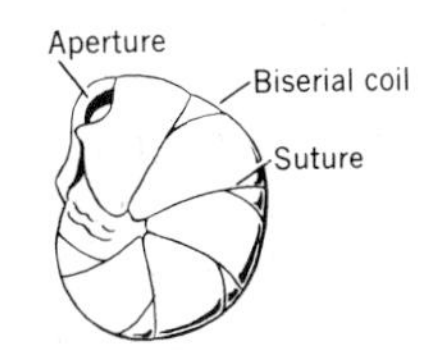

Fig. 200. Cassidulina panzana, × 45. *After Kleinpell.* Miocene Stratigraphy. *Permission of American Association of Petroleum Geologists. Miocene. Family Cassidulinidae.*

continuity of fossiliferous beds in the stratigraphic sections, both locally and along the strike within geographic provinces, can be determined.

Foraminifers serve, then, as readily available evidence in keeping the stratigrapher constantly informed, through intermittent core sampling, of the position in the stratigraphic section while a well is being drilled into the rocks of the earth's crust. Species with short vertical ranges are the most useful in these controls. Many species with long vertical ranges or those infrequently found are sometimes disregarded in the immediate application of well-control data. But in the more thorough industrial laboratories entire faunal samples are mounted on microscope slides. Experience has shown that the interpretive significance of particular paleontologic data is often enhanced, or in other ways modified, in the light of subsequently accumulated data.

For a local area detailed checklists and eventually vertical range charts can be drawn up by using typical specimens of the genera and species locally represented. In an oil field in which the vertical section has been established with fossils and other related information the ages of the rock units can often be recognized as soon as the foraminifers are cleared from the rock matrix and placed under the microscope.

Thousands of dollars may be saved and many more thousands of potential capital realized by information on the stratigraphic position of a well core. Structural trends in the rock units in which folds or faults are usually involved may also be revealed or clarified by the correct identification of a critical sample of a faunal assemblage. At other sites where the reservoir rocks may thin out, lose porosity, be overlapped, or otherwise pinch out, the location of stratigraphic traps for the accumulation of oil are important. A discovery of these subsurface reservoirs may be materially aided by a critical foraminiferal correlation. Likewise, a paleontological evaluation of a foraminiferal assemblage from an area in which the faunal sequence is known may also be helpful. The known conditions necessary for the accumulation of petroleum are

1. Source rocks with an abundance of organic remains accumulated under conditions in which the supply of free oxygen was limited, in which anaerobic bacteria were active, and in which other destructive agents such as scavenger or general oxidation did not operate.
2. Basins of almost continual subsidence with accumulating sediments in which biochemical and thermochemical processes would permit the generation of gas and oil.
3. Reservoir or natural storage rocks of sufficient porosity and permeability in which gas and oil could be retained in commercially adequate quantities.

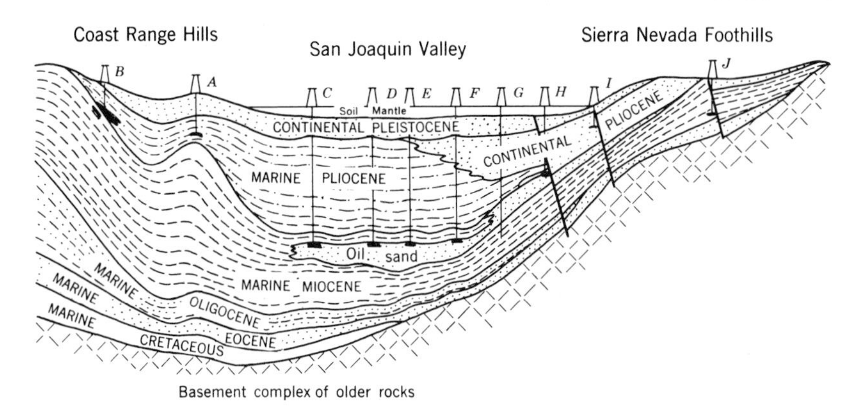

Fig. 201. A generalized section across the San Joaquin Valley, California, showing structure of formations and positions of oil reservoirs. Modified from Hill and Eckis, after Hoots, California Bureau of Mines, Bulletin 118, p. 266.

4. Structural and/or stratigraphic traps overlain by impervious rocks to prevent the escape of gas and oil.

An ideal example of the development of oil fields has been illustrated for central California by Mason Hill and Rollin Eckis (Fig. 201). A situation of this kind is interpretable by rock outcrops, subsurface sequences, well cores, and geophysical data. Oil seeps discovered near the town of McKittrick and elsewhere along the west end of the San Joaquin Valley clearly indicated the presence of oil in the underlying rocks. Eventually well **A** was drilled into an anticline which was clearly a structural trap. Though all such anticlines do not yield oil, this one fortunately did. Other wells were crowded onto the structure as exploration for other producing fields continued. Another fortunate "wildcat" well brought in still another field at site **B.**

In the meantime control fossils were being worked up from the surface exposures in the hills to the west. At first gastropods, pelecypods, and echinoids were used, since they were readily collected in the surface exposures and were better known. Unfortunately, few fossil specimens of this size were retrievable through the drills and least of all in a core barrel of the crude sort employed in the old cable tool operations. However, with the invention of the modern core barrel by Elliott, soon after the close of the World War I, attention of petroleum geologists and paleontologists shifted to microfossils.

As wells **C, D, E,** and **F** were put down, the foraminifers from the core samples were compared with the charts of fossils made from the control sections in the hills and from the samples of the other wells.

This threw considerable light on the stratigraphic problems. The stratigrapher watching the progress of the drilling operations was kept informed of the stratigraphic position by the paleontologist in the laboratory. With samples from the other wells, the depths of the recognizable stratigraphic horizons could be compared and structural trends determined from well site to well site. This offered information to determine within rather narrow limits when the drill was likely to strike the different strata and approximately at what depth the well would "come in," if indeed there were a reservoir where it was to be expected.

It was less difficult to locate other producing wells in the subsurface structures of the middle valley after the first wells had been successfully completed. When operations were extended farther east it was decided to put a well down at site **G.** No fossils were found in the Pleistocene rocks, but a mammoth tooth, *Mammuthus columbi,* taken from an outcrop at the edge of the valley, demonstrated the age of these particular beds and showed that they were continental in origin. In the upper part of the underlying early Pleistocene sands and clays a small microtine mouse, *Mimomys primus,* appeared in the well core. This not only confirmed the age and nature of beds, but showed that sediments of continental origin had been laid down along the eastern margin of the basin, whereas marine sediments and fossils were deposited to the west.

Farther down in the characteristic marine shales neritic, benthonic, and planktonic foraminifers of the early Pliocene were disclosed. As drilling proceeded, other foraminiferan samples showed that the late Miocene oil sands were encountered at a much higher level than in wells **C, D, E,** and **F.** Although well **G** was structurally higher, no oil reservoir was encountered below the foraminiferal zone in which it had been found in the central valley wells. Since no older productive horizon was readily predictable from surface data, drilling operations were stopped and the well was deserted as a "dry hole" or "duster." Eventually wells **H, I,** and **J,** still farther east, struck oil reservoirs in fault-trap accumulations.

Training in Related Fields of Study. A broad basic training in the biological sciences and in the earth sciences is essential for a paleontologist. This is as necessary for those who expect to enter economic work as for those who expect to follow an academic career. As stated previously, a paleontologist in the field may encounter a combination of problems any one of which may hold the key to a final solution.

Structural geology in areas of diastrophic disturbances in the earth's crust usually is of the utmost importance. Formations may be folded and faulted out of their original positions. Knowledge of the structures and positions that these rocks may assume under stress, together with paleontologic and petrographic information of the different units, can often reveal in great detail the historical geology of the area under study.

Stratigraphy in its revelation of the content, the origin, the order, and the position of stratified rocks holds many clues, even for the deciphering of several aspects of geologic structure.

Understanding of the *historical geology*, both locally and regionally, should be the basis from which to launch an investigation. If the field man has a broad regional or provincial picture well in mind, local details will eventually fall into their correct order.

A paleontologist, and especially those concerned with foraminifers and the larger invertebrates, must know something about *marine ecology* as it exists in oceanic environments today. Changes in environmental or depositional conditions, when a formation is traced laterally for a few hundred yards or farther, may seemingly alter the age by different faunal compositions in the samples, yet a trained marine ecologist can recognize the differences between ecological and temporal changes in his faunas. Without some knowledge of marine ecology the paleontologist would be of little aid in directing attention to neritic, bathyal, benthonic, planktonic, and other kinds of marine organisms. This can be extremely important in correlation.

Finally, the investigator will be greatly aided if he understands the principles of *taxonomy and classification* of organisms. These are our means of communication in expressing what we know about the evolution and the phyletic relationships of organisms. Unfortunately, many stratigraphers are too lax in their efforts to understand and apply these principles. If they were not worthy of serious consideration, surely the system would not have been retained since 1758 when it was introduced by Linnaeus.

Micropaleontology. The term micropaleontology is frequently applied to the study of foraminifers. In this respect it is misleading because it also refers, and correctly so, to the study of any microscopic organisms or even to the microscopic parts of megascopic plants and animals. Plants included in the study of micropaleontology are bacteria, algae, some flagellates (Flagellates are classified as Protozoa by most zoologists and Thallophyta by most botanists), and diatoms. Even fossil pollens, in a strict sense the study of *palynology*, are also included. In the animal kingdom any number of tiny parts have been considered under the term micropaleontology. In addition to foraminifers, there are radiolarians, chitinozoan and calpionellid protozoans, ostracod crustaceans, sponge spicules, and conodonts and scolecodonts with toothlike structures from animals of unknown or at least disputed relationships. Other microscopic forms are echinoid spines, holothurian spicules, bryozoans, graptolites, and many minute parts of other animals. To a scientist who is accustomed to arranging organisms in accordance with their relationships the term micropaleontology is misleading and confusing. For these reasons the name **Foraminiferology** is proposed for the study of Foraminifera. The other microscopic organisms as well as microscopic parts of the higher plants and animals can be referred to their respective fields.

26

Cephalopods and Structures in Ammonites

PERHAPS THE MOST spectacular class of mollusks is Cephalopoda. It includes the ancient nautiloids, various intermediate groups, and the living pearly nautilus, subclass Nautiloidea; the ammonites, subclass Ammonoidea; and the squids, cuttlefish, octopuses, and extinct belemnites, subclass Coleoidea.

General Considerations. Cephalopods possess many features indicative of their high development. A series of tentacles bearing hooks and suckerlike disks protrude anteriorly from around the mouth, and below is the funnel-shaped hyponome. The eyes situated laterally on the head are nearly equivalent in complexity to those in the vertebrates, and an advanced hearing structure is organized. Though cephalopods can move slowly about on the bottom, they are also able to move swiftly through the water by a force created in a jetlike expulsion of water from the gill cavity through the hyponome. The extinct belemnites, like the living squids and octopuses, had an ink sac that could be emptied into the lower intestine and hyponome from which they expelled the black fluid to form a smoke screen to aid them in escaping their enemies. The shells of the nautiloids and ammonites are divided into numerous internal chambers or camerae. The members of the first two subclasses are thought to have had four gills, as in the pearly nautilus, and are therefore said to be tetrabranchiate. On the other hand, the extinct belemnites, as in the living squids, cuttlefish, octopuses, and the like, are believed to have been two-gilled, or dibranchiate. The sexes in cephalopods are separate, and the eggs are heavily yoked. Moreover, the young do not pass through a trochophore (free-swimming) stage in their development, as is so frequently the case in other groups of invertebrates.

Evolution in the Class. The earliest cephalopods were small curved- or straight-shelled nautiloids. The first and most primitive ones appeared in the late Cambrian. Then they evolved rapidly into eight orders in the Ordovician, apparently from the curved-shelled kinds. They were present in great numbers at that time. These early nautiloids evolved so rapidly that by the middle of the Ordovician *Endoceras proteiforme*, a giant straight-shelled species, reached a length of 13 feet and had an aperture 8 to 10 inches in diameter. One of the orders, the Nautilida, which had coiled shells, eventually evolved into the living *Nautilus*. The ammonites arose from one of the nautiloid groups in the early Devonian. The ancestry of the squids, cuttlefish, octopuses, and belemnites probably came from a similar ancestry sometime in the later Paleozoic.

Characters of Subclasses. NAUTILOIDEA. Shell straight or coiled; sutures as simple lines; noncalcareous protoconch; with umbilical perforation under shell mantel; living *Nautilus* with four gills. Most abundant in the Ordovician and Silurian.

AMMONOIDEA. Shell usually coiled, sometimes uncoiled or assuming peculiar shapes in some genera; sutures with marked folds or extremely complex undulations; calcareous protoconch; no umbilical perforation; assumed to have had four gills. Most abundant in the Jurassic.

COLEOIDEA. Soft external covering, with or without internal calcalcareous shell; no series of internal gas filled chambers; with ink sac connected to lower intestine. Probably most abundant during the Jurassic.

AMMONOIDEA. One of the most widely known groups of cephalopods are the ammonites. The coiled shells of these fossil invertebrates are so abundant in the marine Mesozoic formations of central Europe that Cro-Magñon men used them with the bones, jaws, and teeth of late Pleistocene mammals in the construction of some of their sepulchres. Ammonites have proven to be the most useful Mesozoic fossils in zonal correlations, as clearly demonstrated by William Smith, Albert Oppel, and many later paleontologists. Bernhard Kummel of Harvard University has stated that approximately 1800 ammonite genera are known.

The Ammonoidea are currently recognized as a subclass of the class Cephalopoda. The more primitive groups were not abundant in the Devonian, Mississippian, and Pennsylvanian, but with the change in climatic conditions in the Permian they evolved into numerous diversified families. Ammonites nearly died out in the late Permian, recovered in and then barely survived the Triassic, and finally became extinct at the end of the Cretaceous.

Only seven genera are known from the late Permian, but marine fossil-bearing formations of that age are not well known. Only one of the Permian genera ranged through into the earliest Triassic. Favorable conditions resulted in such rapid evolution that 128 genera or more appeared before the middle Triassic. The earliest Triassic genera of the family Ophiceratidae are extremely diversified in the shape of their shells. These, in turn, gave rise to the numerous families of the middle and late Triassic. On the whole, some 370 genera are known from Triassic formations, more than twice the number currently known from all Paleozoic rocks. Many Permian and Triassic genera have converged so closely in their characters that relationships between them have been, at times, incorrectly assumed.

From 140 genera in the late Triassic, ammonites declined rapidly during three Stages until only five genera and five species remained as the Period came to a close. Of these, only one stock carried through to the early Jurassic. Here again ammonite history repeated itself, as the only surviving genus quickly recovered from near extinction and evolved into the greatest number of genera existing in any one Period during their long history. Unfortunately, the interfamilial relationships of the Jurassic ammonites have been extremely difficult to interpret.

Ammonites reached a climax in their specializations and in the size of individuals in the Cretaceous. The number of genera, though greatly reduced from those in the Jurassic, still made up the most conspicuous elements of the marine invertebrate faunas. As in previous Periods, the shells in nearly all of the genera were planispirally coiled in a horizontal plane, others assumed various shapes. *Turrilites* was vertically spiraled; *Ancyloceras* was boat-shaped; *Baculites* was long and straight,

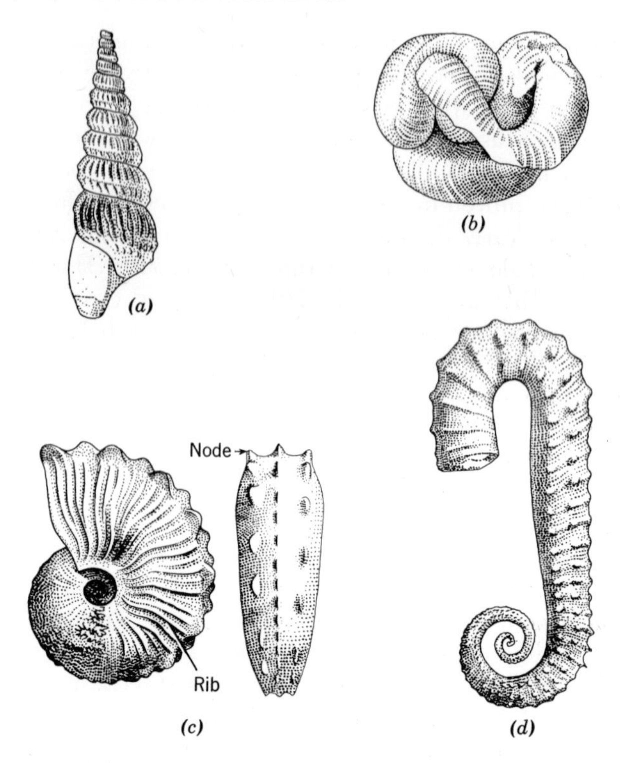

Fig. 202. Some shell shapes in ammonites: (a) Turrilites, *vertically spiralled;* (b) Nipponites, *coiled into wormlike knot;* (c) Taramelliceras, *planispiralled;* (d) Ancyloceras, *boat-shaped.*

except for an early coil; and *Nipponites* was coiled into a wormlike knot. Apparently coiling and uncoiling took place in different groups of both nautiloids and ammonites at different times. Descriptive terms have been given to the various kinds of shell shapes (Fig. 202).

Among the largest of the smooth-shelled ammonites were individuals of the genus *Pachydiscus suppenradensis*, which were supposed to be 7 feet or more in diameter (Fig. 149), whereas some of the smallest ammonites were no larger than the end of a man's finger. Nearly all of the Cretaceous kinds possessed extremely complicated sutures, but some experienced a reversal in their evolutionary trend in developing more simplified ceratitelike or even goniatitelike sutures.

The number of ammonite genera and species are so great and their characters so complex that specialists still tend to concentrate their attention on the taxonomic groups most readily available. Phylogenies showing the relationships of the orders have been drawn through the combined efforts of several authors in the "Treatise on Invertebrate Paleontology, Part 4, Mollusca 4," edited by Raymond C. Moore.

Terminology for Shell Features. The different structural features in each group of organisms must be named in detail to offer a terminology for descriptions. These names are similar, or sometimes the same, in closely related organisms but may be quite different in distantly related groups. Anyone studying the ammonites or the nautiloids must be familiar with the terms used for those subclasses. The features to which these names apply are utilized by paleontologists in classification.

The *conch,* or shell, is divided into two parts, the outer *living chamber* and the *phragmocone,* which consists of all the internal chambers back to the apex. In orientation it should be borne in mind that the ventral area is the outer side of the curved surface and the dorsal area is the inner side of the whorl, not visable in a complete shell except at the aperture, or opening, of the shell. These are called *venter* and *dorsum,* terms sometimes confusing to a beginning student, since it is

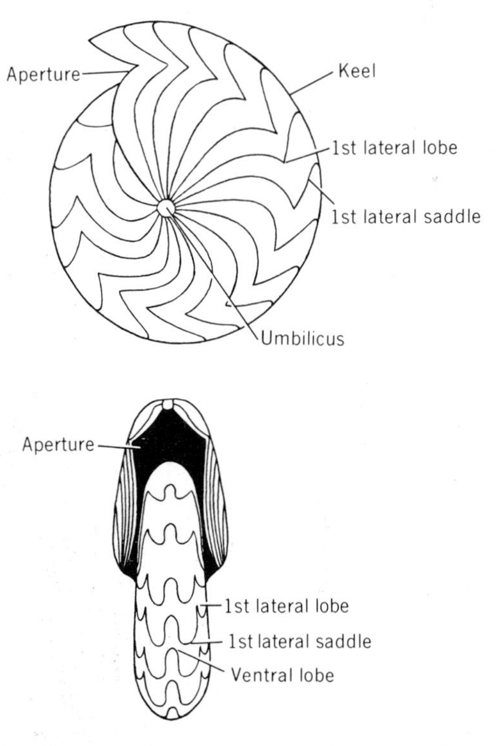

Fig. 203. Ammonite showing ceratite sutures and other structures.

believed by cephalopod specialists that a reverse coiling has taken place in some groups in which the orientation of the ventral and dorsal areas would be reversed (Fig. 205).

The surface sculpture of ammonite shells has been used extensively and successfully in describing genera and species. Of these the *keel*, or keels, denotes the outermost edges of the whorls on the extreme ventral surface from the aperture to the apex of the central whorl. These may be visualized in comparing the shape and position of a keel on a boat. They may assume different shapes or be inconspicuous. Normally at right angles to the keel are the transverse *ribs*, which are elevated surface features, frequently parallel, running from the keel across the ventral surface of the shell to the *umbilical seam* or inner side of the whorl. The ribs may be sharp, rounded, or slightly elevated. Sometimes they are continuous across the ventral surface, or they may fade out in either direction. Frequently they divide into two branches before they reach the keel or keels. Many genera and species display a series of *nodes* or *tubercles;* though situated on the ribs they form a line around the lateral ventral surface parallel to the keels. In some ammonites the nodes or tubercles occur as short sharp crests or spines. Such protuberances may appear in two or more rows or may occur on the

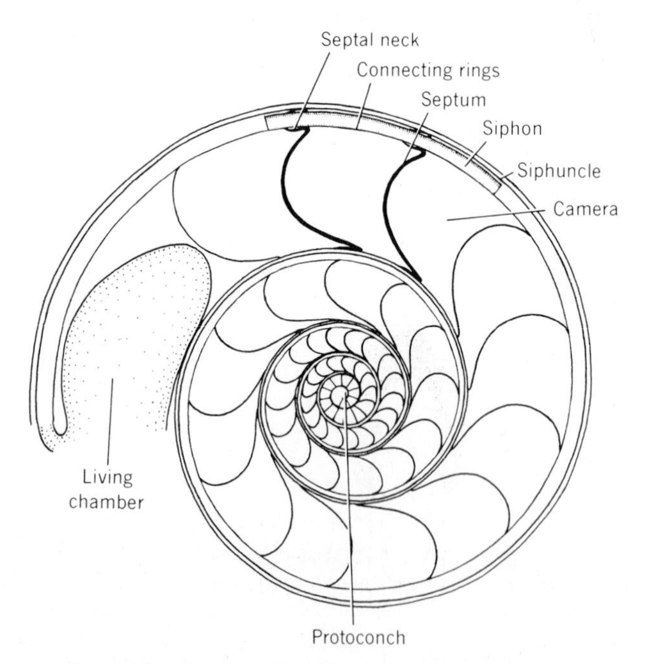

Fig. 204. Median longitudinal section of ammonite showing internal structures.

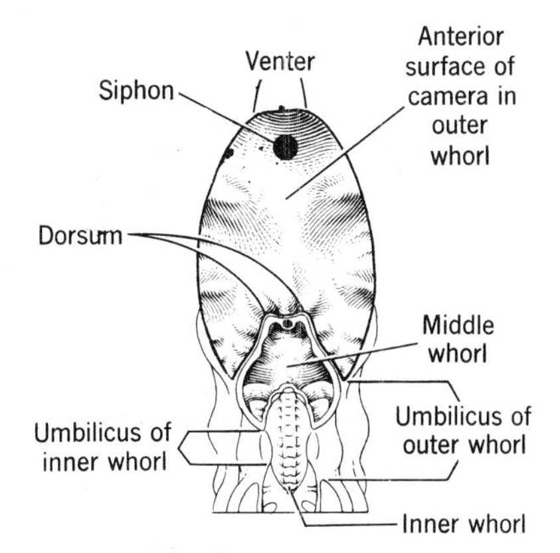

Fig. 205. Transverse section of ammonite showing positions of whorls and internal structures.

keels. These ornate surface features have arrested the attention of innumerable budding paleontologists who eventually got much deeper into the subject of ammonite structures and history (Fig. 202c).

The terminal openings, or *apertures*, are of varied shapes. They may be round, oval, elliptical, or strongly compressed laterally. Here, too, is the *hyponomic sinus* so frequently referred to, if fortunate enough to be preserved in the fossil.

Newly hatched ammonites secreted a calcareous sac around the body called the *protoconch*. As the animal grew, the margin of the shell was built forward. The body of the animal also moved forward and periodically a transverse *septum* was added to separate the head and body in the living chamber from the area it occupied previously. These internal chambers, known as *camerae*, progressively increased in size from the inner to the outer whorls and are confined by the septal walls.

Another structure of considerable importance is the *siphuncle* that extends from the oral end of the visceral hump back through (toward apex of shell) all of the camerae to the first chamber but not into the protoconch. The siphuncle is composed of short, tubular *septal necks* that extend out from and are part of the septum, together with the calcite *connecting rings* that join the septal necks from one septal perforation to the next. The construction of the siphuncle is quite complex and has offered useful characters in classification. Within the siphuncle tube is the *siphon;* apparently it contained fleshy protoplasm and blood vessels. Evidently there was some connection between the siphon and the gas-filled camerae, if an analysis can be drawn between the living *Nautilus* and the ammonites. The camerae must have functioned hydrostatically for submerging and emerging, thus allowing the animals to disperse widely in oceanic waters (Fig. 204).

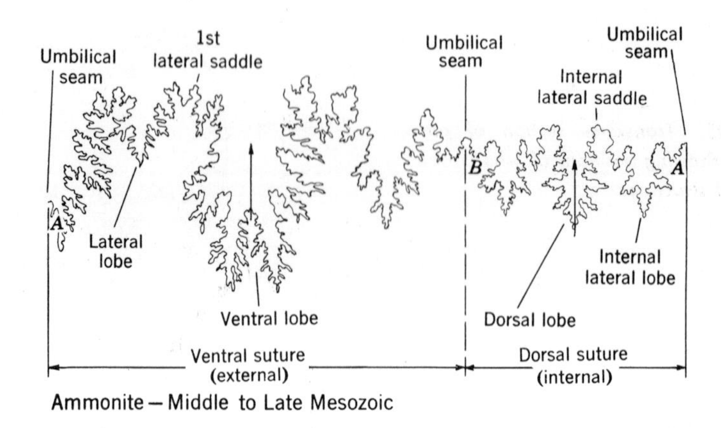

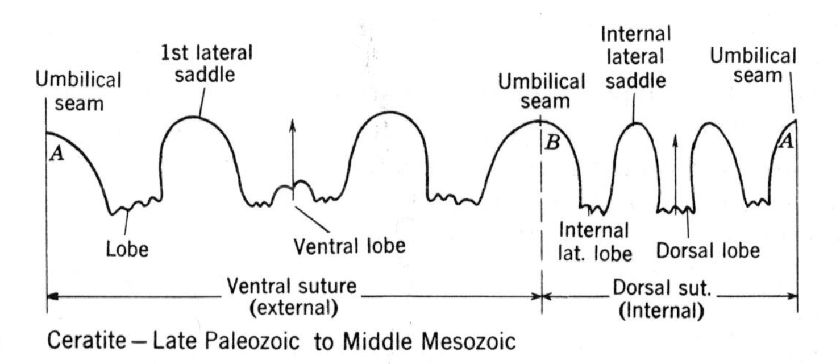

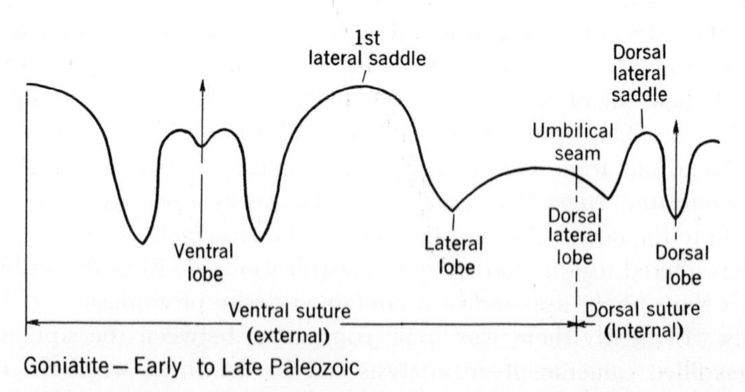

Fig. 206. Diagrammatic sketches showing three kinds of ammonite sutures.

When the thin outer surface of the shell is removed or broken away *suture* lines disclosing the contact of the septa with the internal surface of the outer shell are revealed. Sutures are gently curved or nearly straight in nautiloids; on the other hand, the sutures in most ammonites are conspicuously convulate. If the undulations of the sutures are convex away from the aperture, they are called *lobes;* when they are convex toward the aperture they are termed *saddles* Figs. 203, 206).

Three kinds of sutures are recognized in the ammonites. *Goniatite sutures* have rounded saddles and rather angulate lateral lobes. These appeared in Devonian forms and prevailed as the dominant kinds until the Permian. *Ceratite sutures* are more complex. In these the saddles are rounded and the lobes are crenulate. The first of these sutures evolved in the Mississippian but did not survive beyond the Triassic. The third kind, *ammonite sutures,* is by far the most complex. Both the saddles and the lobes are minutely crenulated into dendritic patterns. The most complicated ammonite sutures occur in Cretaceous genera, though the earliest ones were present in the Permian.

The wide dispersal of these fascinating free-swimming creatures, together with the numerous features in the shells which exemplify their rapid evolution, has made ammonites most useful in late Paleozoic and Mesozoic intercontinental correlations.

27 Highlights in the Evolution of Plants

Psilophyton, Asteroxylon, Horneaphyton

The importance of plants to man and to the existence of all other organisms can hardly be overestimated. Through the process of photosynthesis, carbon dioxide and water in the presence of the green pigment known as chlorophyll produce carbohydrates, fats, and proteins that are stored in the tissues of the plants. This has supplied food for animals, either directly as in the case of herbivores or indirectly for carnivores. In addition, wood has been used not only to keep us warm but to construct most of our homes and to supply us with many essential products. There are also many other plant products too numerous to mention. Indeed, it would be a bleak miserable world if there were no plants. Life as we know it would be impossible without the food, the shelter, and the beauty of plants.

The major divisions in the classification of plants is based primarily on a combination of characters. These are the nature and selection of leaf and stem, vascular anatomy, and position of sporangia or seeds. Because of the lack of discovery in the fossil record there are few recognizable intergrades between the major groups as they are now recognized.

The history of plants evidently extends back 2 billion years or more. The first were the simplest kinds of aquatic organisms which gradually evolved into more complicated forms. By Cambrian time some plants had gained a foothold on land, but it was not until Devonian that we have evidence of their spreading over much of the land as primordial forests. After that their evolution was much more rapid.

Earliest Plants. It is not known which is the oldest record of fossil plants because Pre-Cambrian rocks in which they occur cannot yet be correlated accurately from one region to another. Even in the same area it is usually impossible to determine whether the rocks are Archaeozoic or Proterozoic unless their ages have been determined by radioactive minerals. Pre-Cambrian limestones, graphites, and iron must have formed in part from primitive plants. An iron-forming bacteria such as

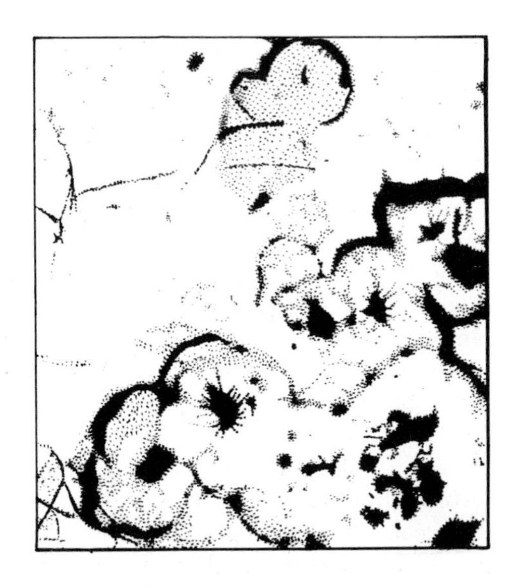

Fig. 207. Colonies of blue-green Algae from the base of the Gunflint Formation, Ontario, Canada, thought to be about 2 billion years old. After Tyler and Barghoorn.

Leptothrix may have formed iron by utilizing the energy of chemical oxidization in an organic synthesis. Limestones and graphite were probably formed from primitive algae.

The oldest known identifiable plants are blue-green algae, fungi, and possibly a flagellate from the black flints of Ontario, Canada, which are thought to be 2 billion years old. The process of photosynthesis evidently had been developed at that time. If these Pre-Cambrian algae were like the living blue-green algae, as their structure suggests, they must have reproduced asexually. Among other primitive characters they probably had no nuclei, with the protoplasm not differentiated but probably much the same throughout the cell. With the introduction of the green pigment chlorophyll and the ability of these early plants to produce food substances by photosynthesis, a major step in the evolution of plants and of all life had been attained.

Some of the known Proterozoic fossils are algal reefs described by Carroll H. Fenton and Mildred A. Fenton from Montana. They refer these algae to the genus *Collenia*. In certain localities, in which they occur in moldlike colonies, concentrates of these microscopic plants are abundant. These calcium carbonate precipitates are arranged in globular or hemispheric masses with concentric laminations, but no plant structures have been observed. Similar structures as well as bits of carbon with ratios characteristic of organic concentrations are known from other Pre-Cambrian rocks.

Primitive Land Plants. The next highlight in the evolution of plants was the development of plants that could live out of water. This important event probably came about in the Proterozoic. The most widely held theory is that certain kinds of highly adaptive marine algae established themselves in lagoonal and estuarine waters and through natural selection over a period of several million years gradually adapted themselves to living out of water. Eventually the more progressive kinds spread out onto the moist soils adjacent to the water. The first, as yet unknown, land plants were evidently related to the Psilopsida or to some primitive plants closely related to them. Plants much like these hypothetical ancestors have been found in Silurian and Devonian rocks, and fragmentary stems and pollen indicate that they were also present in the earlier Periods of the Paleozoic. In the simplicity of their structures the most primitive psilopsids are intermediate in their stage of evolution between marine algae and the more progressive land plants. It is possible however that land plants as we know them may have evolved from different groups of Pre-Cambrian algae.

In contrast to the algae that depended on the media of water to retain necessary moisture in their tissues, the Psilopsida possessed a superficial layer of cells, the epidermis, which protected the more delicate inner tissues and prevented the loss of water so vital to the life of plants. These primitive land plants also had a simple thin vascular system to conduct water containing dissolved mineral substances up through the xylem tissues of the underground rhizomes and stems to the branches, where photosynthesis took place. The foods were then conducted back into the cells of the plant through the phloem tissues. Some stems had pithy centers, but in others the pith was greatly reduced or wanting. The annular elongate, or tracheid, cells of the xylem supported the plants in their upright positions. Perhaps the tallest plants were less than 6 feet high and most of them were much smaller.

The genera *Rhynia* and *Horneaphyton*, so well represented in the Rhynie middle Devonian cherts of Scotland, are the simplest of the Psilopsida. They bear terminal sporangia at the ends of their erect cylindrical stems. The stems are naked and infrequently branched dichotomously and are seldom more than 18 inches high. *Asteroxylon* is a more advanced plant than *Horneaphyton*. *Psilophyton* is from the Devonian of Canada.

Several other genera of primitive land plants have been described, and the subphylum seems to have been widely distributed about the world. It is the consensus that the Paleozoic psilopsids arose from aquatic algae or algaelike plants. Two genera are still living in temperate and tropical regions, but the group as a whole must have diminished greatly in numbers after it set the stage for the evolution of the more advanced plants that eventually came to blanket the land with verdant splendor.

OLDEST KNOWN LAND PLANTS. During the past 20 years paleobotanists in different parts of the world have directed attention to the presence of spores and bits of woody tissues in Silurian, Ordovician, and Cambrian rocks. The results of these investigations are highly suggestive of plants even more advanced than the Psilopsida. Direct evidence of more progressive plants was disclosed by William Henry Lang and Isabel C. Cookson in 1935 from specimens recovered from middle Silurian rocks in Australia. The most advanced plant was named *Baragwanathia longifolia*. This fossil is much more complex than any of the Silurian or Devonian psilopsids.

Though the characters of *Baragwanathia* (Fig. 105) reveal a somewhat intermediate position between the two subphyla, it is clearly more closely related to the Lycopsida than to the Psilopsida. The relatively wide stems (5 millimeters) are thickly clothed with long (4 millimeters) narrow, soft-leaved structures with traces of veins; the leaves are arranged spirally around the stems. Sporangia occur on the stems and branches at the bases of the leaves.

Even more impressive is the 1953 report from eastern Siberia of small middle Cambrian lycopsids by the Russian paleobotanist, Afrikan Nikolaevich Kryschtofowitch. These oldest known land plants, *Aldanophyton antiquissimum,* are represented by four impressions of shoots. Though not complete, one specimen is over 80 millimeters long and another is 13 millimeters wide. The stems are thickly covered with thin, delicate leaf structures, some of which are 9 millimeters long. Vascular bundles have been traced through the stems, but there is no evidence of terminal sporangia as in the psilopsids. These most interesting fossil plants, together with the different kinds of spores that have been reported repeatedly from Cambrian rocks, indicate a diversified, though primitive, land flora in the Cambrian. Such evidence inevitably leads to the conclusion that land plants must have originated from marine plants not later than the Proterozoic.

OLDEST FORESTS. Though the early Devonian floras were dominated by simple herbaceous psilopsids that resembled certain kinds of rushes, they gave way to more advanced plants in the later part of the Period. These more progressive plants made up the oldest known forests, which were restricted to moist lowland areas. They were the forerunners of the great Carboniferous forests that succeeded them in the Mississippian and Pennsylvanian. In these early forests were the first scale trees, *Protolepidodendron,* ferns, *Archaeopteris,* seed ferns, *Eospermatopteris,* some of which were tall and palmlike, primitive sphenopsids, *Spheno-*

Fig. 208. Trees of the earliest Devonian forests: scale tree, Protolepidodendron, *and seed fern,* Eospermatopteris. *Modified from C. R. Knight.*

phyllum, and other plants, *Palaeopitys*, which had cordaiten and pteridosperm structures in the stems. They provided moist shelter and protection from the sun for spiders and other air-breathing arthropods, as well as cool shady places for the first amphibians that ventured onto land. The plants had well-developed roots, stronger stems, leaves, and more adequate means of dispersing their spores and seeds. Some of the trees with pithy trunks were 30 to 40 feet high. One of the most significant advancements in plant development was the seed habit, which had already been attained at this time. Devonian fossil plant localities have shown that mild climates extended as far north as Spitzbergen and south to Antartica well within our present polar regions (Fig. 208).

The seed is essentially an undeveloped plant composed of an embryo, endosperm (food-storage tissue), and seed coat. The embryo has four important parts: *cotyledons*, or seed leaves, capable of absorption, storage, and digestion of food from the food storage tissues; *epicotyl*, the part above the cotyledons which grows into the primitive stem; *hypocotyl*, the stem below the cotyledons; and the *radicle*, which develops the primary root. The outer coats of seeds reduce the possibility of evaporation and in other ways protect the embryo prior to germination. Evolution of the seed habit made it possible for plant embryos to survive in a quiescent stage during winters and other unfavorable climatic conditions.

MISSISSIPPIAN AND PENNSYLVANIAN SWAMP FORESTS. The larger plants so characteristic of the later Devonian increased in numbers and kinds during the Mississippian. Their habitats were somewhat restricted by marine waters that still inundated most of the inland troughs and geosynclines, but numerous fossils have been found, particularly in eastern North America and in Europe. The Mississippian plants were evolving as the local and provincial environments fluctuated. Those in the early part of the Period were still much like the Devonian plants, and the late Mississippian species were remarkably close to those in Pennsylvanian formations. One of the most interesting plants appearing for the first time was the coniferlike *Cordaites*.

No record of fossil plants has been more impressive than that of the great swamp forests of the Pennsylvanian. The tremendous abundance of plant life, especially in the old Appalachian geosyncline, has been demonstrated time and again by the great thicknesses of the coal beds and by the presence of well-preserved plants in adjacent shales. Conditions were ideal for the preservation of these plants in a moist, mild, humid climate where the trunks, branches, leaves, and fruiting bodies accumulated in almost unbelievable numbers, later to be covered by sediments washed in from the land areas nearby. Most of the plants

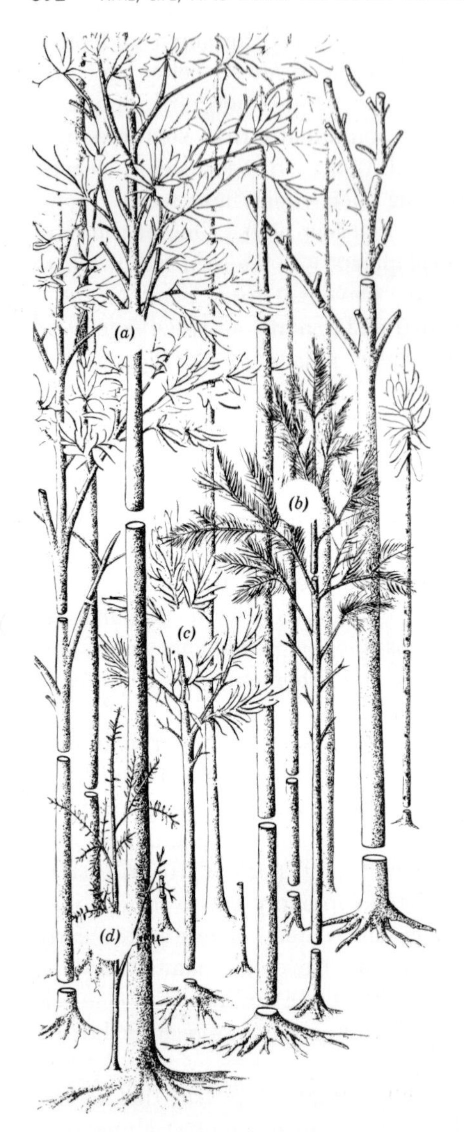

Fig. 209. Coniferlike plants: Cordaites (a, b, c) *showing centers of trunks, tree habits, and leaf forms; and* (d) *a primitive conifer,* Dicranophyllum. *From the Carboniferous of France. After Grand'Eury.*

were fast-growing, with shallow root systems, in soft ground or mud. Trees must have toppled over frequently in layer after layer of vegetative materials. Pollens, like microspores, were produced in such tremendous numbers to insure reproduction that the dense stifling air in some seasons must have appeared as a yellow smog settling among the trees and over the stagnant water. In places they accumulated in such thicknesses that yellowish cannel coals were made up almost exclusively of spores.

The most abundant and spectacular plants were the scale trees known as *Lepidodendron* and *Sigillaria* (chapter heading 15). They were called scale trees because the leaf scars on the bark bore a superficial resemblance to scales on the skins of certain reptiles. Indeed, it has been said that some of the early coal miners thought the flattened fossil stems and trunks were the remains of enormous snakes. Root structures of the scale trees were called *Stigmaria*, and at first they were thought to represent a different genus of plants. Trunks have been found preserved in the coal in an upright position with the roots imbedded in the underlying shales, showing that coal formed where the plants grew. Trunks and branches had large pithy centers within a woody cylinder, which in turn was coated with a corklike bark.

The trunks of the great lepidodendrons 4 to 6 feet in diameter tapered upward to an estimated 100 feet. The upper one third was forked repeatedly into stubby branches that were covered with slender straplike leaves. As the older leaves were shed they left a spiraled pattern of roughly diamond-shaped scars. Spores were borne in cone-like structures at the ends of the branches.

Sigillaria was as large as *Lepidodendron* but quite different in shape. There was little or no branching at the top where bladelike leaves much longer than those in *Lepidodendron* spread out from the stem. In *Sigillaria* the rounded leaf scars on the trunks were arranged in vertical rows (Fig. 117).

Scouring rushes, called *Calamites*, were much smaller than the huge scale trees. The largest trunks were about 1 foot in diameter and the height was approximately 30 to 35 feet. They evidently grew in dense thickets in the wettest habitats. Like the living horsetails, *Equisetum*, these Pennsylvanian plants are readily recognized by their jointed and vertically ribbed stems. A bractlike whorl grew out of each of the joint segments, and these in turn at regular intervals had narrow leaves borne in whorls. These small leaf whorls were at first mistaken for flowers and named *Annularia* (Fig. 118b).

Sphenophyllum, another sphenopsid plant with a climbing or vine-like habit, persisted from the Devonian into the Permian. The units in the leaf whorls were wider than in *Calamites* and were arranged in multiples of threes, usually in numbers of six or nine.

Much of the forest floor was covered with seed ferns, but true ferns were also represented. Fossil leaves have been beautifully preserved. Well-preserved specimens of seed ferns have been found in small concretions and nodules of sandstone at Mazon Creek, Illinois. Some of the most common genera are *Neuropteris*, *Pecopteris*, and *Alethopteris*. The true ferns are variable, but when examined in detail have been found to be forerunners of primitive modern fern families.

One of the most interesting orders of Paleozoic fossil ferns are the Coenopteridales. They are tree-fernlike plants with a central stem encircled by leaf-stalks and adventitious roots. The structure of the fern axis is unusual in its possession of secondary wood. The reproductive parts (sporangia) are relatively large, club-shaped, pear-shaped or round, and may occur singly or clustered at the end of the leaf; others occur on the lateral margins of the leaf where the tip of the leaf has been lost by reduction. These gametophytes (plants with sex organs) apparently are much like the other true ferns in their early stages of development but their four synangial (encased in thick layer cells, or case) sporangia, or variously united sporangia, are in a compound structure that opens at maturity. This is suggestive of an early evolutionary stage in the origin of seeds.

We have seen that there were suggestions of coniferlike plants in the Devonian and that the genus *Cordaites* appeared in the Mississippian, but in the Pennsylvanian the conifers became firmly established. *Cordaites*, with its well-developed softwood trunk, long bladelike leaves, and seeds born in racemes instead of cones, were abundant in the drier habitats along the emerging Appalachian geosyncline. They were tall slender trees, possibly 100 feet high. Some cycadophytes also made their first appearance but were not prominent. Pennsylvanian climates were relatively drier in the area now west of the Mississippi River, and this condition was reflected in the kinds of plants known from that province. There were more seed ferns, and, of more importance, there were true conifers represented by the genera *Walchia* and *Lebachia* (Fig. 123). These conifers bore their seeds in small cones much like the modern kinds. Ginkgos, *Dichophyllum* (Fig. 123), also found in the late Pennsylvanian at Garnett, Kansas, show that the next highlight in the evolution of plants so clearly exemplified in the later Permian had already started in some areas in the Pennsylvanian.

Elevation of Continents and Changes in Floras. Floras of the Devonian until the early Permian were more or less cosmopolitan in their distribution in moist mild climates over much of the land surface of the earth. There was a trend toward plants with small cutinized leaves, stronger woody parts, and better reproductive systems. These upland plants expanded and spread as the continents were increasingly elevated in the later Permian. There were new genera of seed ferns, and ginkgos became important floral components. Even more conspicuous were the conifers, *Walchia*, *Lebachia*, and *Voltzia*. The scale trees and calamites, so dominant in the Pennsylvanian swamp floras, were restricted to wet swampy areas. Similar changes were reflected in the vertebrate animals and in the arthropods. This Period marked one of the most profound alterations in the plants and animals of the world.

Diastrophic disturbances in the earth's crust resulted in the earliest outlines of the great Tethys geosyncline across southern Eurasia and a somewhat similar depression between North and South America. This culminated in the evolution of different floras in the Northern and Southern Hemispheres. The southern flora, which spread around the globe from New Zealand and Australia to peninsular India, thence to Madagascar, southern Africa, and southern South America, has been called the Gondwana or *Glossopteris* flora.

The elongate leaves of *Glossopteris* are usually elliptical or ovate, and the apices are obtuse, but they taper gradually toward the base where they merge into the petiole, or stem. There a narrow flange that is not veined continues along the edge of the strengthened center. The thin lamina of the leaf shows anastomosing venation. The venation near the midrib in some leaves is longitudinal and distinct from that in the lateral laminae but tends to fan out at the apex.

Fossil leaves of *Glossopteris* and the related *Gangamopteris* have been found abundantly in late Pennsylvanian and Permian floras. The French botanist Adolphe Théodore Brongniart (1801–1876) described the first species in 1828. Since then, many other species of *Glossopteris* have been recorded, but the exact systematic relationships of these plants remained in doubt for nearly 125 years. They were called

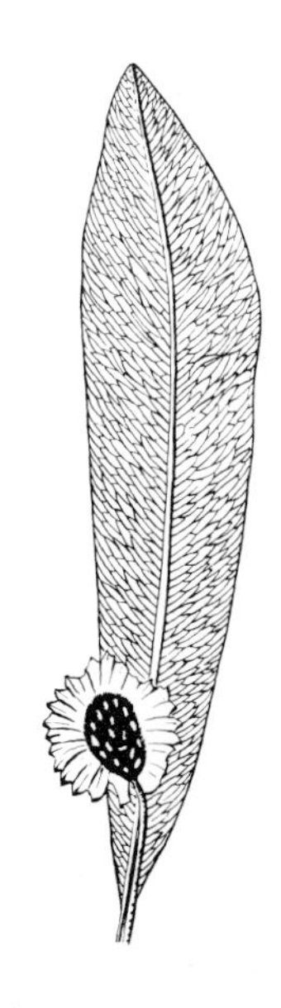

Fig. 210. Glossopteris leaf with fructifications attached. Redrawn from Plumstead.

"tongue-leafed ferns," and were generally thought to be related to the seed ferns. Much light has been thrown on the problem by the recent discoveries of Stephanus F. le Roux, an enthusiastic amateur collector of Vereeniging, South Africa. This keen observer and skilled technician, at the instigation of Edna P. Plumstead of the University of the Witwatersrand, found fructifications attached to *Glossopteris* leaves that were taken from the pale buff, silty clays and from the deep red clays of an early Permian Ecca formation.

The fruiting capsules are attached to the midlines of the leaves by short pedicles. The capsule encloses a head that bears a flat peripheral wing, which is usually fluted and striated and has a dentate margin. Imbedded in the tissue of the head are tiny (1 to 2 millimeters) oval sacs with central hard cores, apparently seeds. This information, disclosed by Mrs. Plumstead, represents one of the important paleobotanical discoveries of our time. It now seems possible, as some paleobotanists suggest, that these fascinating plants may represent another subclass of the class Gymnospermae, equivalent in rank to the seed ferns, conifers, and ginkgos. More recently similar fructifications have been found on *Gangamopteris* leaves (Fig. 210).

Many other plants also occur in the *Glossopteris* floras. Among them is a horsetail, *Shizoneura*, clearly different from the northern *Cala-*

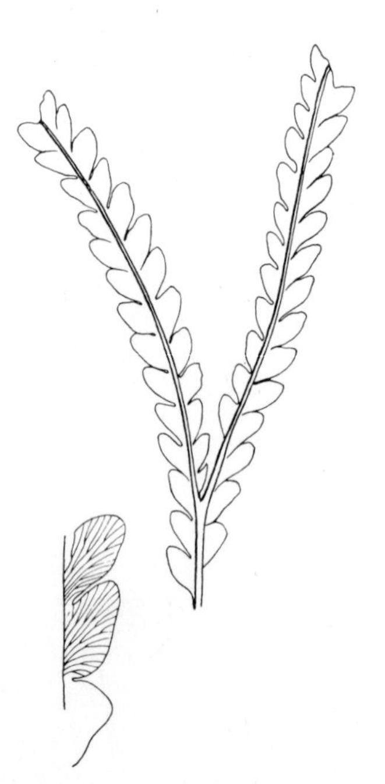

Fig. 211. Part of a frond and leaflets of Dicroidium odontopteroides, a seed fern from the Triassic of Australia. After Feistmantel.

mites; true seed ferns, such as *Sphenopteris* and *Dicroidium (-Thinnfeldia),* as well as such conifers as *Walkomia* and *Paranocladus,* are clearly distinct from the genera in the Northern Hemisphere. Later the *Dicroidium* floras of the Triassic replaced the *Glossopteris* floras of the Permian of the Southern Hemisphere (Fig. 211).

Era of Ginkgos, Conifers, and Cycadeoids. For the most part, the early Triassic was drier and poorly suited to the growth of forests. Later in the Period a greater number and variety of plants occupied more favorable environments. The great change brought about by the uplift of continents in the Permian effected the beginning of a profound evolution in plants. The scale trees *Lepidodendron* and *Sigillaria,* some seed ferns, together with a small number of tropical ferns, lingered on for a short time in favorable habitats in the Triassic. Large horsetails still survived, but more progressive seed-producing plants, such as the conifers, ginkgos, and cycadeoids, were the dominant trees.

At the end of the Permian and the beginning of the Triassic the cycadophyte plants which first appeared in the late Pennsylvanian evidently gave rise to a group of trees with short barrel-shaped trunks, forked stems, and palmlike leaves. These were the cycadeoids that spread into the lowland areas while the true cycads, which evidently existed at the same time, may have preferred higher altitudes. The cycadeoids continued to expand in the Triassic and became one of the most abundant plants in the Jurassic and early Cretaceous. The ginkgos, typified by their small leaves with parallel venation, followed a similar evolutionary and distributional pattern (Figs. 40, 150).

The conifers also experienced considerable evolution and diversification during and after the Permian. They tended to spread down into lower altitudes. Included among them were the ancestors of the cedars, redwoods, pines, and other conifers. Perhaps the most spectacular of all Mesozoic conifers was the enormous *Araucarioxylon* found in the Petrified Forest of Arizona (Fig. 136). Some of these late Triassic giants with trunks 3 to 4 feet in diameter must have towered nearly 120 feet above the ground, though most of the logs measured were from 60 to 100 feet long. Unfortunately, their leaves are not certainly known. Their nearest living relatives are the monkey puzzle pines, *Araucaria,* now confined chiefly to the Southern Hemisphere. Lyman H. Daugherty has recorded thirty-eight species of plants from the Chinle formation.

Conifers continued in abundance until the early Cretaceous when they gave way to the more progressive flowering plants, the Angiospermae. Ferns also were abundant in the Mesozoic floras.

Origin and Spread of Flowering Plants. Another important highlight is the origin of angiosperms, the flowering plants. It is still not known exactly when this took place, but they appeared in abundance in the middle Cretaceous and by the end of the Period had spread to every continent. However, they were in existence long before that. Fossil pollens, wood fragments, and some leaves are recorded from both the Jurassic and the Triassic. Further substantial support to their early Mesozoic existence is given by Roland W. Brown of the United States Geological Survey in his recent description of fossil palmlike leaves, *Sanmiguelia lewisi*, from the middle Triassic of Colorado. These beautifully preserved leaves, with delicate parallel venation between the many accordian-pleated ribs, confirm the presence of a monocotyledonous plant at that time. Unfortunately, no evidence of fruiting bodies was found with the large (fifteen-inch), simple elliptical leaves. Nevertheless, their relationships to the other classes of seed-bearing plants seem clear (Fig. 212).

This Triassic palmlike plant, together with the profound changes in the insect faunas and the meager evidence of other angiosperms in the Jurassic and the Triassic, strongly supports the belief of Daniel I. Axelrod and other paleobotanists that the first representatives of the class appeared in the Permian. As previously stated, the Permian was a Period of epeirogenic continental uplifts that resulted in cooler and drier climates. Our fossil records show that many groups of organisms

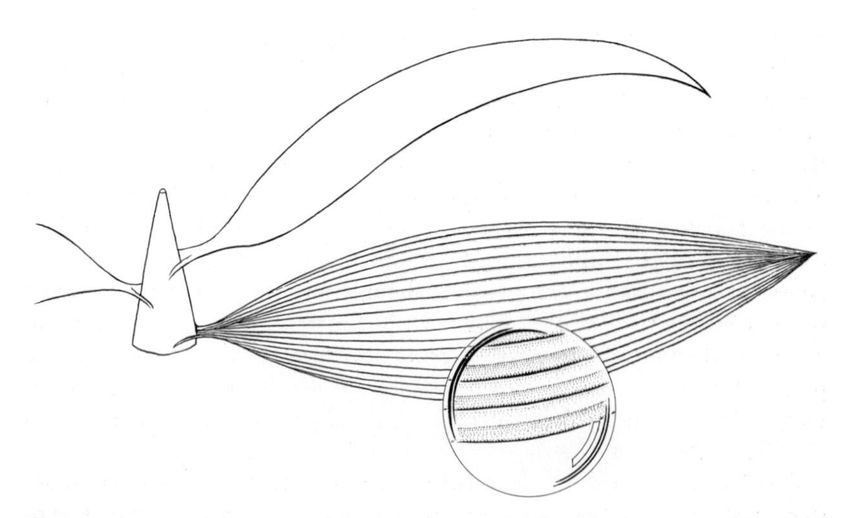

Fig. 212. The oldest known angiosperm, a palmlike monocotyledon, Sanmiguelia lewesi. *After R. W. Brown.*

Fig. 213. Early Cretaceous angiosperms from the Patapsco formation in Maryland: (a) Cissites, *grapelike leaves;* (b) Sapindopsis, *soapberrylike leaves. After Berry.*

experienced rapid and divergent evolution, or extinction, at that time, as they adapted themselves to the new environments. Some paleobotanists have maintained that primitive flowering plants experienced their early evolution in the upland areas. There the leaves and other parts were buried in the sediments of intermontane basins that for the most part have been destroyed by subsequent uplifts and erosion. During the Mesozoic these more progressive plants gradually became adapted to the lowland environments in which by middle Cretaceous time they had largely replaced the cycadeoids, the ginkgos, and the conifers (Fig. 213).

All major groups of modern plants were well established at the end of the Mesozoic. The last great transformation in plant communities took place in the Cretaceous. As in the Permian, the floras in the Northern Hemisphere again differed from those in the Southern Hemisphere.

Greater specialization and diversification in their vascular systems and reproductive structures, as well as their greater potential to evolve, enabled the angiosperms to replace the conifers as the dominant group of land plants. They became adapted to all kinds of environments. There were tropical-to-subtropical floras well equipped to live in warm humid areas of heavy rainfall. Others were adapted to temperate zones where they developed better root systems, shorter stems, and larger flowers. In the drier areas there was a reduction in leaf volume and other modifications to inhibit evaporation. The deciduous trees developed systems of seasonal growth and rest to carry them through seasonal climates. There are amazing adaptations and specializations in almost endless kinds in the flowering plants that have spread over the land surfaces of the globe. Angiosperms make up 80 to 90 percent of the modern plants.

Sequences and Shifting of Cenozoic Land Floras. Climatic zones on land as expressed by the vegetation shifted toward the equator as the climates changed gradually during the Tertiary from warm and moist to cool and arid conditions. This coincided with rather continuous orogenies which elevated the continents higher above sea level. An example of this change in floras at a given latitude has been outlined by Ralph W. Chaney and his students at the University of California. Their field work over nearly two score years in the John Day Basin of Oregon has disclosed a sequence from the tropical and subtropical forests in the Eocene to the semidesert environments today. This research and similar evidence from other provinces has confirmed the earlier conclusions of J. P. Smith (1919) when he plotted the position of isotherms from Lower California to the Gulf of Alaska for each of the Tertiary Epochs. Smith's evidence was derived from the marine invertebrate faunas which were checked by him with the latitudinal positions of the Tertiary land floras.

Mild and warm moist climates with tropical and subtropical floras spread into middle latitudes in the Eocene. Most floras were forests stimulated into luxurious growth by heavy rainfall. Grasslands were restricted, and there was little or no frost. The forests in central Oregon at that time were similar to rain forests now living in central and northern South America. There were figs, laurels, legumes, oaks, palms, tree ferns, and numerous vines. The leaves were relatively large, thick, and frequently pinnate, i.e., with pointed tips. Rainfall has been estimated at 80 inches annually. At that time dawn redwoods *(Metasequoia)*, pines, spruce, birch, maples, elms, cottonwoods, walnuts, willows, and sedges were living on Saint Lawrence Island, halfway between Alaska and Siberia, and at many other places in Alaska and Greenland. Floras like these gradually retreated toward the equa-

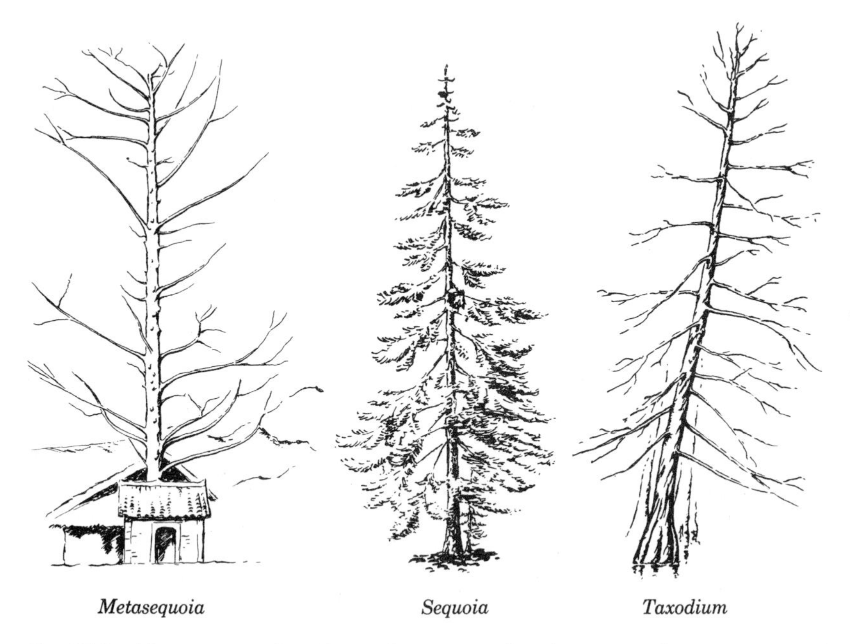

Fig. 214. Three genera of conifers well represented in the Tertiary floras. Dawn redwood, Metasequoia, *an old tree. After Chaney. Redwood,* Sequoia, *a young tree; and bald cypress,* Taxodium, *a young tree.*

tor as rainfall decreased and temperatures lowered in the Epochs that followed (Fig. 214).

The Miocene floras in central Oregon have indicated a temperate climate in which tropical conditions prevailed some 30 million years before. Dawn redwoods were dominant in lowlands along well-watered drainageways, and bald cypresses were also represented. A mild temperate climate is indicated by black oaks, evergreen oaks, white oaks, elms, maples, cottonwoods, beech, birch, and alders. Grasslands were parklike openings in areas previously blanketed by trees. The rainfall probably did not exceed 50 inches.

In the middle Miocene the first Cascadian disturbances began in western North America and have continued at various intervals since that time. Pliocene floras from central Oregon are rare, but those that are known indicate a somewhat drier climate than that of the Miocene. Eventually the Cascade mountains formed a rain shadow; thus moisture-laden clouds did not reach the interior. Today, central Oregon is a semidesert in which willows, cottonwoods, and dogwoods are confined to the stream banks. Back from the streams sagebrush and other plants occupy drier habitats. The annual rainfall is about 15 inches over the lowlands.

Latitudinal shifts in Cenozoic floras similar to that in the John Day Basin have occurred throughout the world and have culminated in the present distributional patterns of our land plants.

FOSSIL GRASSES IN THE GREAT PLAINS. The early Miocene evidently marked the beginning of plains conditions in the interior middle latitudes of North America. In 1935 Maxim K. Elias reported the discovery of siliceous grass seed husks and the evidence of other herbaceous plants in the widespread sandy flood-plain deposits in the Great Plains province. The most abundant of these were spear grasses referable to the genus *Stipidium*. Spear grasses evidently became widespread in the Great Plains province shortly after their first appearance because their siliceous husks have been found in abundance in the sandy stream channel deposits of Miocene and Pliocene age (Fig. 215).

Hackberries, cottonwoods, and poplars grew along the stream banks. Thus, apparently by intermittent uplifts in the mountains to the west, an environment was established over the Great Plains that became one of the most important centers of dispersal and evolution for plains mammals in the world. A certain amount of silica and the dry, harsh condition of grass in the late summer and autumn had its effect in the wearing down of teeth, but this was probably secondary in its effect to the abrasiveness of sand grains lodged in between the stems and blades of grass. A combination of conditions existing in this environment was so effective in wearing down the teeth of grazing mammals that natural selection tended to favor those species that acquired slightly higher crowned teeth and other features that strengthened the molars. Just how such changes arose as heritable modifications and resulted in the progressive development of higher crowned teeth, or hypsodonty, is not clear from the fossil record. Such information must be obtained from studies of living mammals.

FOSSIL AND LIVING DAWN REDWOODS. One of the most interesting botanical discoveries in recent years was announced in a report that decid-

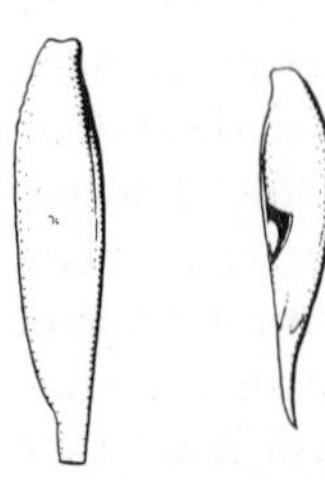

Fig. 215. Siliceous husks of spear grass, Stipidium schereri, *$\times 7\frac{1}{2}$, from the early Miocene Harrison formation in western Nebraska. Redrawn from Elias.*

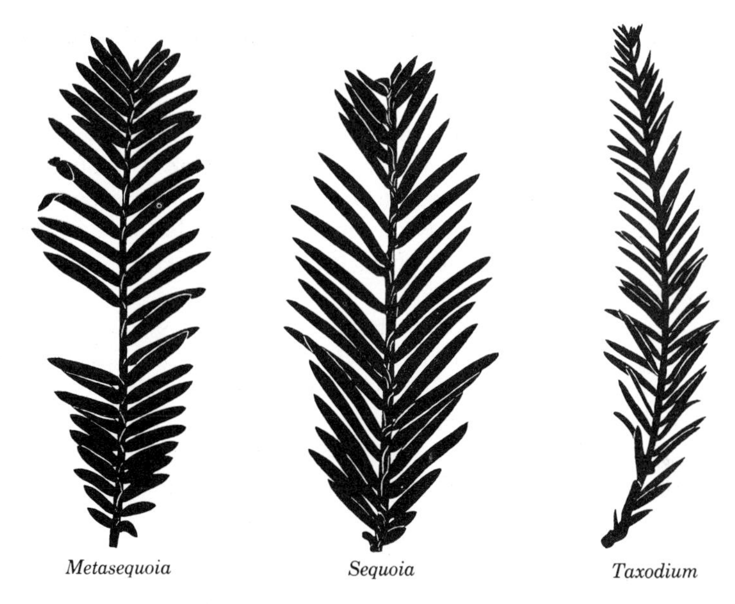

Fig. 216. Branchlets of the dawn redwood, Metasequoia, *redwood,* Sequoia, *and bald cypress,* Taxodium.

uous redwood-like trees were living in central China. A forester, T. Wang of the Chinese National Government, saw the first of these unknown trees near the small village of Mo-tao-chi in the Province of Szechwan. When specimens were taken to the National Central University they were immediately recognized as a new and strange kind of conifer evidently most closely related to the redwoods. It so happened that this was to be one of those most unusual cases in which a fossil genus had been described before the living representatives were found.

John S. Newberry described some Cretaceous and Tertiary specimens of *Metasequoia* from North America as early as 1863. Apparently recognizing their deciduous habits, he referred them to the bald cypress genus, *Taxodium.* He realized that they were not redwoods, *Sequoia,* and evidently thought they might belong to a distinct genus because he listed some of the characters now known to be diagnostic of *Metasequoia.* Later specimens from other Cenozoic floras in the Northern Hemisphere were also called redwoods or bald cypresses, though some of the *Metasequoia* characters were recognized. Then in 1941 a Japanese paleobotanist, Shigeru Miki, described the genus *Metasequoia,* basing it on a fossil species previously called *Sequoia japonica* by the

Swiss paleobotanist Oswald Heer. The significance of the research by Miki was not realized until the Chinese discovery was announced. Paleobotanists then re-examined their collections and commenced to recognize *Metasequoia* in considerable numbers, especially in Cretaceous and early-to-middle Cenozoic floras. In 1951 Ralph W. Chaney revised the fossils of all three genera in western North America. These trees had been common not only in middle latitudes but also far to the north within the Arctic Circle where they were associated with other deciduous trees (Fig. 216).

Metasequoia is a deciduous conifer, not an evergreen. In other words it sheds its branchlets in the winter like the elms, maples, and many other trees we know so well in middle latitudes. The cone scales, needles, and twigs are opposite and not spirally attached or alternate as in the redwood. Among other distinctive features the cones are borne at the ends of naked stalks, not on the needle-bearing twigs.

28 Highlights in the Evolution of Vertebrates

American opossum

Vertebrate animals tend to arrest our attention more than any other major group of organisms. Other groups may be of greater economic importance, but even the most enthusiastic invertebrate paleontologist will thrill to the discovery of a horse tooth in the rocks where he is taking samples. A collector of aquatic insects will freeze in his tracks when a trout darts from the shelter of a large boulder, and a herd of pronghorn antelopes will be watched by nearly anyone until they speedily fade away in the distance. Not only their size and actions, but their methods of reproduction and the care of their young, the modifications in their anatomical structures, their physiology, and other functions attract our interest. This is understandable because, whatever else we may attribute to man, he, too, is a vertebrate animal.

Primitive Vertebrate Animals. The geologic history of the vertebrates is long, even longer than our present records show. The oldest known fossil vertebrate remains show certain specializations that must have required many millions of years in their evolution, but how much longer than the middle Ordovician we cannot accurately predict.

The oldest known fossil vertebrates occur in rocks of middle Ordovician age in Colorado, South Dakota, Wyoming, and Michigan. These fossils, though very incomplete, have been identified as fragments of primitive jawless fishlike vertebrates called ostracoderms. Part of a head shield and hundreds of scalelike elements have been recovered. The first ostracoderm fragments from Colorado were named *Astrapsis desiderata* and *Eriptychius americanus* by Charles D. Walcott. Fortunately, ostracoderms have also been discovered in late Silurian and Devonian rocks of Greenland and Spitzbergen. These specimens are so perfectly preserved that their relationships to the living semiparasitic lampreys and hagfish, known as cyclostomes, have been clearly established by Erik A Stensiö and his associates (Fig. 217).

The extinct ostracoderms were found chiefly in fresh- and brackish-water deposits. Some such as *Hemicephalaspis* had large head shields and flat bodies covered with bony plates. These creatures evidently groveled about on the bottoms of pools and brackish embayments sucking organic matter into their jawless mouths. Other armored kinds, *Pteraspis* and *Pterolepis*, had bodies shaped more like those in most of the true fish. *Pterolepis* even had a mouth somewhat suggestive of that seen in the swift-swimming teleost or bony fish. These more agile ostracoderms probably fed on minute organisms in the water. One primitive fishlike specimen, *Jaymoytius kerwoodi*, from the Silurian in England, apparently had a thin skin and no armor like the primitive lancelet chordates, but it may not have been an ostracoderm.

The ostracoderm's most outstanding feature was the absence of jaws. The cranium was a single ossified unit underlying the dermal shield and composed of true bone with a surface layer of dentine, but there was no trace of a bony body skeleton. Some, but not all, had structures like paired fins. There were indications of a notochord and two semicircular canals in the inner ear region. Instead of gills some of the ostracoderms had as many as ten pairs of gill sacs or pouches that were used in feeding and breathing. Few of these most primitive vertebrates were longer than 12 inches.

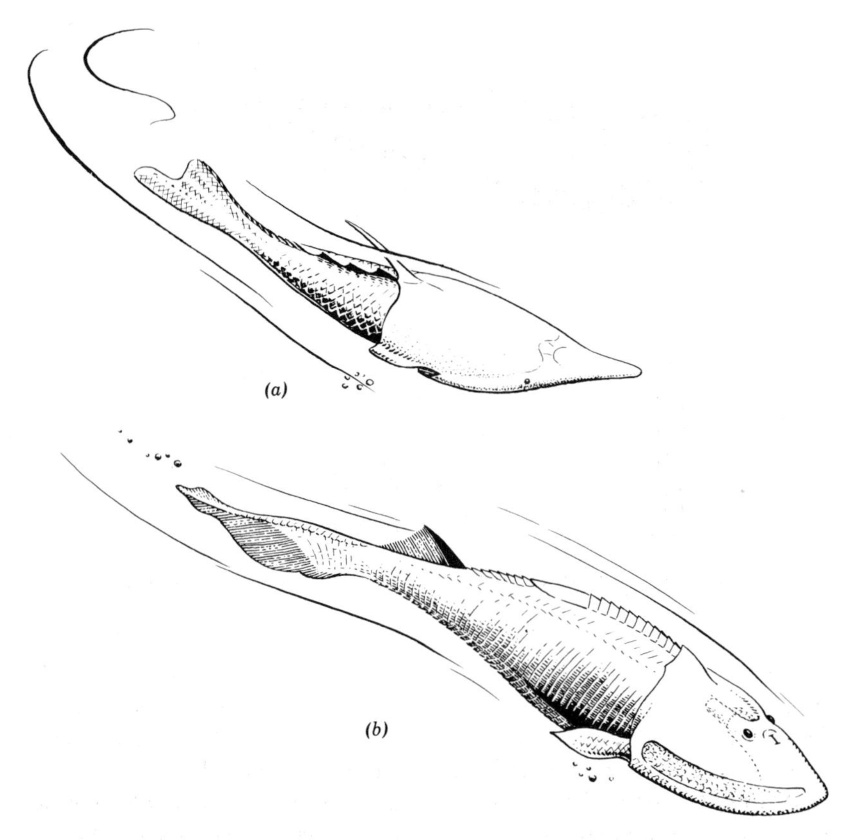

Fig. 217. Devonian ostracoderms: (a) Pteraspis, *about 2.5 inches long;* (b) Hemicephalaspis, *about* 8 *inches long. After Colbert.*

Their methods of reproduction can only be inferred from the habits of the living cyclostomes. In these the sexes are separate, and the eggs are fertilized by the male sperm shortly after the female releases them over a depression on rocky bottoms or in riffles of streams. The eggs are soon covered with silt. Since the adults die after mating, they offer no parental protection, but the thousands of young that later emerge from the silt insure survival of enough to perpetuate the species.

Appearance of Fishes With Jaws. Evidence of the next major step in the evolution of vertebrates also occurs in late Silurian and Devonian rocks. No less than six orders of fishes in the class Placoderma show various specializations in a great radiation of fishes that was taking place at that time. All placoderms, or plate-skinned fish, had three important characters in common: (1) bones in their internal and external

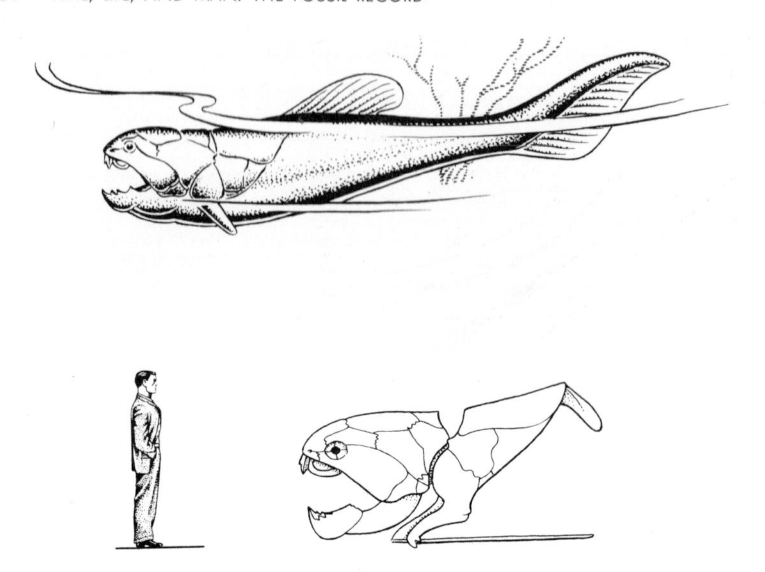

Fig. 218. Joint-necked arthrodie, Dinichthys, *about 30 feet long.*

skeletons; (2) paired pectoral and pelvic fins usually present; and (3) upper and lower jaws of a primitive type. The most spectacular of these orders were the arthrodires (joint-necked fish), the acanthodians (spine-finned fish), and the antiarchs (bottom-dwelling fish).

ARTHRODIRES. Arthrodires were the largest and most formidable scourgers of the epicontinental seaways. Some species of the genera *Dinichthys* (Fig. 218) and *Titanichthys* were 30 feet or more in length. The head and neck and the front of the body (thoracic region) were covered with heavy bony plates connected by a joint on each side of the neck; thus they have been called joint-necked fish. The back part of the body was covered with skin and scattered dermal ossicles. In the mouth were peculiar bony jaws with sharp, toothlike points and shearing edges. As these monstrous arthrodires moved in hot pursuit of other fish, the elevation of the cranium on the hinged joints permitted the mouth to open wider as the lower jaw remained fixed or moved slightly downward. Not all arthrodires were so large. Many small kinds first appeared in the late Silurian and early Devonian in fresh and brackish waters. The well-known genus *Coccosteus* in the Old Red Sandstone of England and Scotland included some species 2 feet long.

ACANTHODIANS. In many respects the acanthodians or spine-finned fish were more interesting. These small minnow-sized fish were completely armored with diamond-shaped bony scales. The many small bony plates in the skull were arranged in definite patterns. Though there were five well-developed gill arches, they were more primitive than in later fish. Each of the arches bore gill rakers. In both the upper and lower jaws there were one to three separate centers of ossification. The toothed lower jaw was not articulated with the cranium through the hyomandibular bones, as in more advanced fish. Pectoral, pelvic, and dorsal fins, as well as numerous pairs of smaller fins along each side of the belly, were present, and each fin was strengthened by a sharp anterior spine. Among the most primitive jawed vertebrates these little spine-finned fish were nearer to an ancestral position for the later groups than the more specialized arthrodires, though both made their first appearance in the late Silurian (Fig. 219).

ANTIARCHS. The antiarchs apparently competed with the ostracoderms and other bottom-dwelling fish in Devonian waters. They appeared in abundance in the middle Devonian and prevailed for the remainder of the Period; no trace of them has been found in the Mississippian. They were no larger than most ostracoderms, but many structures, such as jaws, paired fins, and the like, have indicated that they may have arisen from a primitive arthrodire stock. The head and much of the body were covered with hard dermal plates (Fig. 220).

These and other little-known orders of placoderms, or plate-skinned fish, predominately Devonian in age, indicate a diversified evolution into various specializations. Fishes intermediate between them and

Fig. 219. Spiny acanthodian, Climatius, *about 4 inches long. After Colbert.*

the ostracoderms have not been found, so that their immediate ancestry is not clearly understood. Nevertheless, in general they do represent an intermediate stage between the most primitive vertebrates known and the more advanced fishes.

SHARKS. Sharks also appear for the first time in the Devonian (Fig. 110). They, too, have a long and diversified history and are still living, mostly in marine waters. No less than five major groups of these dreaded creatures are known to have existed at one time or another. Their skeletons are cartilaginous, and though the ear has three semicircular canals, as in all higher vertebrates, there are no external auditory openings for a more distinct reception of sound. The sexes are separate and fertilization is internal. In most sharks the female retains the fertilized eggs in the oviduct, where they hatch; later the active young are born. Sharks of the genus *Mustelus* have developed a placentalike modification in the oviduct to nourish the embryos when the food supply in the yolk sac has been depleted.

BONY FISH. The fourth and greatest class of fishes are the Osteichthyes or bony fish. They are divided into two subclasses: the Actinopterygii, ray-finned fish, and the air-breathing fish Choanichthyes. In contrast to the other classes, the bony fish have a well-ossified internal skeleton (except in the sturgeons and spoonbills), lower jaws connected to the cranium through the hyoid arch, and swim bladders or lung structures. Both of the subclasses appeared in the Devonian, which shows that the oldest and most primitive bony fish must have extended back into the Silurian.

RAY-FINNED FISH. The first ray-finned fish appear in the middle Devonian. The oldest order, known as Palaeoniscoidea, is well exemplified by the genus *Cheirolepis* (Fig. 111). Palaeoniscoids were rare at first, but as the ostracoderms and placoderms faded out these more advanced fishes expanded. They are most abundant in the Mississippian and

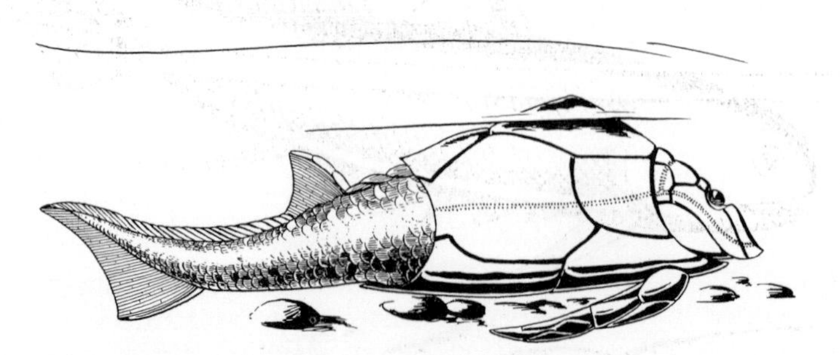

Fig. 220. Plate-skinned antiarch, Pterichthyoides, *about 6 inches long. After Traquair.*

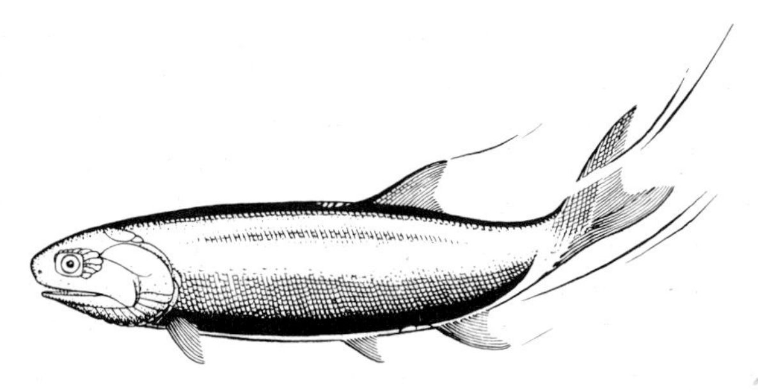

Fig. 221. Palaeoniscus, a primitive ray-finned fish, about 9 inches long. After Traquair.

the Pennsylvanian. *Cheirolepis* is nearly 12 inches long and is covered with thick, shiny ganoid scales. The fins are composed of nearly parallel rays from which is derived the name rayfin. The palaeoniscoids like the contemporary air-breathing fish possess lung structures, but these primitive lungs are replaced by air bladders in the later ray-fins. In the Triassic and Jurassic the sturgeon and garlike fishes, obviously derived from a paleoniscoid ancestry, come into the fossil record. Finally, the most successful fishes of all time, the teleosts, or true bony fish, appear in the early Jurassic, then expand tremendously in the later Jurassic and Cretaceous. One of the giants is the 15-foot *Hypsodon* of the Cretaceous seas. Most of the teleosts are covered with thin bony scales, though some like the catfish have a smooth skin. No group of water-living vertebrates has attained the extraordinary expansion and adaptive radiation of these fishes. They include most of the living fishes, such as the trout, perch, bass, tuna, pike, and eel.

Most fishes are oviparous; they lay small transparent eggs in the water. Since the eggs have no protective covering to retain their essential moist condition, the water is a necessary element for their protection. Some are deposited in depressions or nests on the bottom; others may float, remain in suspension at certain depths, or be attached to water plants. Many fishes, however, are ovoviviparous; i.e., they retain the eggs internally until they are hatched. Usually, but not always, the eggs or young are reproduced in great numbers. Some fishes, such as the trout, spawn several hundred to a few thousand eggs, and the cod is said to extrude several million eggs. Insect larvae and other predators levy heavily upon the eggs and young fishes. It is only by means of their tremendous reproductive potential that most kinds of fishes are able to survive. Parental protection for the most part is extremely limited or nonexistent.

AIR-BREATHING FISH. As we have seen, vertebrate life was abundant in water, particularly in fresh water, during the Devonian Period. All the classes of fishes were represented at that time. They had specialized in many directions and had already occupied most of the environments available to them. But even more momentous were the critical evolutionary changes which took place in stagnant swamps and pools, where the waters were low in oxygen and subject to periodic evaporation. These environments were occupied by a superclass of bony fishes called Choanichthyes, or air-breathing fish with internal nostrils (choanae). Since they possessed lung structures, they became proficient in gulping their oxygen directly from the air when they rose to the surface. There were three groups: the Rhipidistia, the Coelacanthini, and the Dipnoi. Only the coelacanths, represented by the deep-sea *Latimeria* off the coast of Africa, and the dipnoans, or true lungfish, of Australia, Africa, and South America have survived.

The most conspicuous features differentiating the Choanichthyes from the other bony fish are their two dorsal fins and their rhombic scales of cosmoid structure. Since the rhipidistians and coelacanths are more closely related to each other than either is to the dipnoans, they are usually classified as the order Crossopterygii (lobe-finned fish), and the lungfish are recognized as representing the Dipnoi, another order.

Lungfish are still living. The genus *Protopterus* is found in the rivers of Africa; *Lepidosiren*, the smallest and most specialized, occurs in South America; and *Epiceratodus* lives in the streams of Eastern Australia. *Epiceratodus* looks quite like the genus *Ceratodus*, which was widely distributed over the world in the Triassic. Lungfish first appear in the middle Devonian, where they soon became numerous. The earliest, as typified by the genus *Dipterus*, are much more like the crossopterygians than the living genera. Nevertheless, *Dipterus* shows a reduction and decalcification of the internal skeleton and an increase in small bony plates in the skull. Furthermore, the brain case is not well ossified, and the teeth are reduced and lost from the margins of the jaws. These and other features clearly remove these interesting fishes from an ancestral position to the tetrapods, or four-footed, land-living vertebrates.

The coelacanths have long been recognized as a specialized side branch of the crossopterygians. In the Devonian they were found closely associated with the earliest lungfish and rhipidistians, but they soon spread into oceanic waters. The fossil forms retained many degenerate characters which showed a rather close relationship to the rhipidistians, but the bones in their lobed fins, were reduced, the fin rays tended to increase, and the lung was calcified. For nearly a century scientists thought that the last coelacanths died out toward the end of the Mesozoic, when so many of the larger vertebrates became extinct. Then in 1939 a South African fisherman cast his nets deep into the ocean and brought up in his haul a fish entirely unknown to ichthyologists. This unusual fish was about 5 feet long; it was covered with deep blue scales and had two dorsal fins. Unfortunately, before a scientist could reach the scene the specimen had deteriorated to such a degree that only the skin could be saved. Its relationships were soon recognized, and the name *Latimeria chalumunae* for this coelacanth flashed around the world. This was like bringing a fossil back to life. More recently several more were taken off the coast of Madagascar, and details of their anatomy and habits have enriched our knowledge of these interesting fish which have become adapted to live in certain oceanic environments.

As stated previously, the rhipidistians are in the direct line of ancestry of the amphibians. It is interesting to note that one of their genera has been recorded from early Devonian rocks, and no other bony fish are known until middle Devonian time. But it must be logically assumed that all bony fish must have arisen from a common ancestry, possibly in the late Silurian. Though the coelacanths, their closest relatives, eventually dispersed into oceanic and inland embayments of salty water, the rhipidistians flourished in poorly drained swamps and pools where they were the predominant predators. Natural selection favored those with sharp pointed teeth along the margins of their jaws, better developed lungs, well-ossified skulls, and more strongly developed bones in their pectoral and pelvic fins. Waters with little oxygen were no problem to the rhipidistians, for they could come to the surface and even crawl out along the shores where they breathed their oxygen directly from the air. When the stagnant waters dried up these air-breathing fish could make their way out of water to adjacent pools or possibly venture onto land to capture cockroaches under the ferns and around logs.

Two of the best examples of amphibious fishes are *Osteolepis* and *Eusthenopteron*. The ancestral position of these rhipidistians is clearly indicated in several characters. The bones in the craniums and jaws are well ossified, and the pattern of the bones is much like that in the earliest amphibians. The origins of the skull bones are inherited from their earlier fish ancestors. Some originated as skin plates and are called *dermal bones,* whereas others are referred to as *replacement bones;* that is, the bone is replaced cartilage. Not only the location and shape of the teeth are much the same, but in both the rhipidistians and the

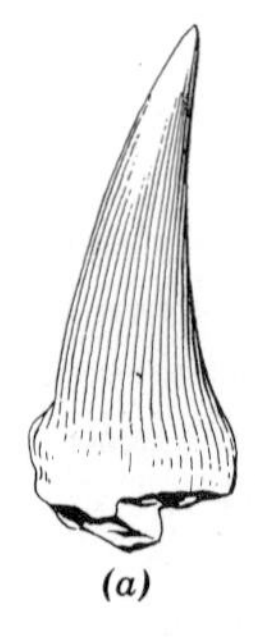

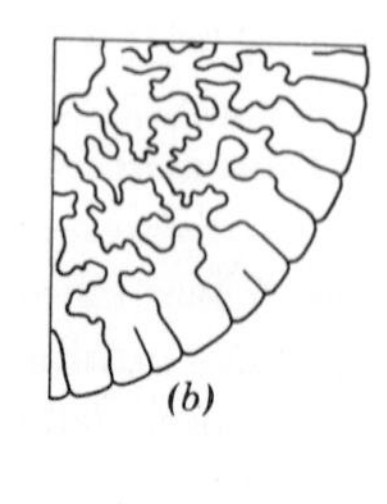

Fig. 222. (a) Labyrinthodont tooth, × 2, (b) quarter section to show labyrinthine pattern, × 22. After R. Owen.

labyrinthodont amphibians there are complicated infoldings of enamel which give them a labyrinthine pattern in sectioned teeth. The well-ossified components of the rhipidistian vertebrae foreshadow the structures that appear in their land-living descendants. In addition, the vertebral column of the tail is straight, and though it is much like that in the crossopterygians it could have readily evolved into an amphibian tail. The pectoral and pelvic lobed fins have bones corresponding to the humerus, radius, and ulna in the forelimbs and the femur, tibia, and fibula in the hindlimbs of terrestrial vertebrates. Even the smaller bones located more distally could have been easily modified into foot bones. Here, then, is one of the best examples of the evolution from one class of animals to another. This was a momentous event in the evolution of vertebrate animals because the more advanced could live on land and eventually adapt themselves to occupy all of its varied environments.

Emergence of Four-Footed Animals. The continuous expansion during the Devonian of land plants which retained moisture about them had set the stage for the conquest of land by water-living vertebrates. Amphibious animals with moist skins could crawl out of the water into these shaded environments without losing much of their moisture by dehydration. The air-breathing rhipidistians had already evolved a long way in that direction. Even before the Period closed the transition had been made, for there were primitive amphibians in the late Devonian of Greenland.

The earliest labyrinthodonts, *Ichthyostega* (Fig. 112) and *Ichthyostegopsis*, still retained numerous rhipidistian features. The rather flattened skulls were about 6 inches long but of solid construction, and the eyes located farther back on the skull were larger than those of their predecessors. Though finrays of a fishlike tail were retained, the pectoral and pelvic girdles had evolved limbs with the five-digited feet of terrestrial four-footed animals (Fig. 223). Amphibians had arrived and at least partial existence on land was possible. Their eggs still unprotected from dehydration probably were laid in water, and the young went through most of their metamorphosis obtaining their oxygen from the water by means of fishlike gills.

Amphibians evolved rapidly during the Mississippian and Pennsylvanian, filling many ecologic niches in the humid swamp-forest environments. Though they were quite diversified, the labyrinthodonts with their heavy, armored skulls were predominant. The group best adapted to terrestrial life was the rhachitomes. Their limbs were very strong for vertebrates of that time, and there were bony nodules in their skin which in part formed a protective armor. These sluggish creatures

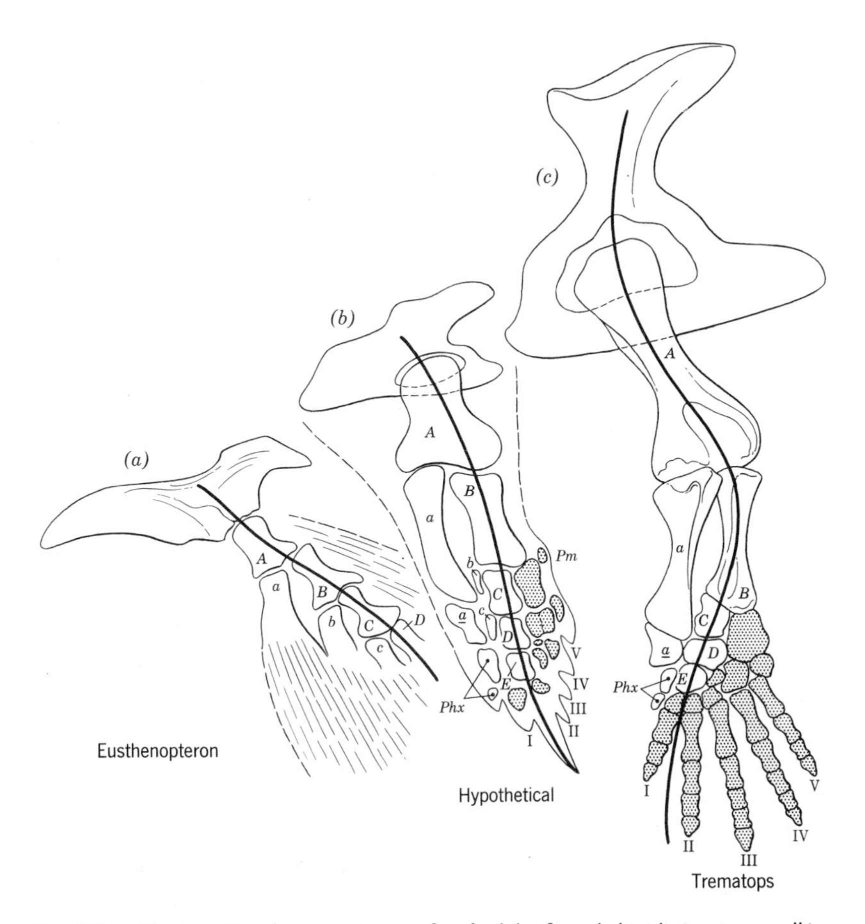

Fig. 223. The transition from a swimming fin of a lobe-finned rhipidistian to a walking foot of an amphibian, as suggested by Gregory and Raven. Phx, prehallux (the hallux is the big toe).

reached their maximum size in the early Permian species of *Eryops* (Fig. 127), some of which were 6 feet long.

The largest Pennsylvanian amphibians were the 15-foot embolomeres, well exemplified by *Eogyrinus* (Fig. 120). In this genus the limbs were reduced, the head was rather deep and narrow, and the long flexible body and tail was adapted to pursuit in swamp waters of fish and other amphibians.

Among the last labyrinthodonts in North America were the Triassic stereospondyls called *Eupelor* (Fig. 137*b*) from the Painted Desert of Arizona. These also were water-living forms with large flat heads, reduced limbs, and short bodies and tails. *Buettneria* from the North American Triassic was one of the largest. It was about 9 feet long. Some of the genera of stereospondyls were the only amphibians known to invade marine waters. The stereospondyls were the last survivors of the impressive labyrinthodonts. The latest questionable record is from the Jurassic of Australia. They evidently could not compete with the different reptilian groups which had reverted to aquatic habits.

In contrast to the labyrinthodonts are the frogs and toads which had no solid, bony-roofed skulls, no tails, but long hopping hind legs. Their history can be traced back to the Jurassic, but beyond that the record is sketchy, though enough of it is known to indicate that they probably descended from a Pennsylvanian labyrinthodont ancestry.

The waters of the great Mississippian and Pennsylvanian coal swamps were well populated with other very unusual amphibians referable to the subclass Lepospondyli. Among them were the flat, angulate-skulled bottom-dwellers known as diplocaulids. The last ones, *Diplocaulus* (Fig. 129), died out in the Permian. Some of the coal-swamp lepospondyls in the order Microsauria retained rather well-formed limbs; others were eel-like amphibians, as in *Sauropleura* (Fig. 121), in which the limbs were lost or only vestiges of them remained. In many respects the most peculiar were snakelike amphibians, *Ophiderpeton* (Fig. 120), with no trace of limbs. Some of this genus were 2 feet long. The modern newts, salamanders, and blind, wormlike caecilians may have arisen from a microsaur ancestry as far back as the Mississippian. It has been suggested that the lepospondyls may have descended from a different rhipidistian ancestry than the other amphibians.

Reptiles Evolve from Amphibians. Though the first reptiles are recorded from middle (Fig. 122) and late Pennsylvanian rocks (Fig. 123), their remains are rare in that Period. Nevertheless, somewhere in the early part of the Pennsylvanian a group of amphibians, well adapted to life on land, gradually evolved into reptiles. In this process there were some rather profound changes. There is no direct proof from the fossil record, but we can readily hypothesize the conditions under which it came about.

These transitional forms possibly deposited their eggs in shallow, sandy-bottomed pools, and there the water may have evaporated, leaving only limited moisture around the eggs. Natural selection favored those eggs that tended to have a more resistant covering and a more liberal supply of yolklike substance to nourish a developing embryo. Eventually an amniote egg evolved from the amphibian kind. Instead of returning to the water to lay their eggs, these more advanced vertebrates gradually altered their habits and deposited their eggs in warm moist sands away from the water's edge.

The eggs were probably more like the rather soft-shelled eggs of the turtle found along sandy beaches and stream banks today. After the eggs were fertilized in the body, the parent female laid the clutch in moist sand where the warmth of the sun carried them through to incubation. As the embryo developed, a large yolk sac formed for its nourishment. An amnionic sac filled with fluid enveloped the embryo for its protection, and another sac, the allantois, developed for the reception of waste materials. Meanwhile, the outer shell and its inner lining, the chorion, were so constructed that oxygen could enter and carbon dioxide escape without dehydrating the internal content of the egg.

Many years ago a deposit of fossil vertebrates found on West Coffee Creek near the town of Seymour, Texas, by Charles H. Sternberg was destined to throw much light on the transition between amphibians and reptiles. The animals were named *Seymouria baylorensis* (Fig. 128). This has proved to be one of the most important discoveries in vertebrate paleontology, for here are animals with characters so intermediate between the labyrinthodont amphibians and the reptiles that their systematic allocation is still in doubt. No one knows whether they laid eggs on land or in the water. But it is interesting to note that vertebrate animals must have been depositing their eggs on land at that time because Alfred S. Romer recently described a fossil egg, taken from beds of early Permian age in the Red Beds of Texas, of the kind one would expect from such a transitional stage. Nevertheless, the *Seymouria* from West Coffee Creek cannot be the ancestral reptile, since reptiles had appeared before early Permian time. It obviously represents a primitive group that continued with little modification for several million years after the first reptiles evolved from the labyrinthodonts in the Pennsylvanian or earlier, a phenomenon not uncommon in the evolution of life.

RADIATION OF REPTILES. With the perfection of the amniote egg and the development of external coverings by which the animals could retain moisture within their bodies, reptiles were much better equipped than amphibians to occupy more diverse environments on land. After their appearance in the Pennsylvanian they evolved rapidly into different groups. In the Permian reptiles outnumbered amphibians. The late Pennsylvanian and the Permian were times of climatic extremes. Epeirogenic uplifts in continents throughout the world, coupled with orogenic disturbances, resulted in aridity in some areas and even extensive glaciation in the Southern Hemisphere. This was reflected by changes in the reptiles to meet the new environmental conditions. Many groups continued their conquest of the wide expanses of land areas; others reverted to life in the water where there was an abundant food supply. Some of the herbivorous Permian land reptiles were as large as rhinoceroses, and the carnivorous forms, though averaging smaller in size, were equally abundant.

In the Mesozoic some of the dinosaurs evolved into the largest land animals known, and the different groups adaptively radiated into diverse specializations. Many of the Mesozoic reptilian orders were already distinct in the Triassic. These include turtles, ichthyosaurs, plesiosaurs, rhynchocephalians, crocodilians, phytosaurs, and the two orders of dinosaurs.

POSTORBITAL OPENINGS IN REPTILE SKULLS. Though characters in the skeletons are useful in recognizing the different groups of reptiles, the positions and numbers of the temporal openings between the bones back of the eyes are basic for beginning students. These openings, or fenestrations, are associated with the attachment areas for the powerful temporal muscles used in closing the jaws. These modifications are utilized even further in the large dinosaurs by reducing the weight of the head while retaining a strong structure.

In his book on the evolution of vertebrates, Edwin H. Colbert has illustrated skulls with or without openings. The cotylosaur, with no openings, is *anapsid;* the ichthyosaur with one opening high on the skull and bordered below by the postfrontal *(pf)* and supretemporal *(st)* bones is *parapsid;* the plesiosaur also with one opening high on the skull but bordered below by the postorbital *(po)* and squamosal *(sq)* bones is *euryapsid;* the pelycosaur with one opening on the side of the skull is *synapsid;* and the eosuchian with two openings back of the eye is *diapsid* (Fig. 224).

Water Reptiles. Water and land-living reptiles were about equal in numbers. Perhaps the first truly aquatic reptiles were the small mesosaurs (Fig. 130). Some were about 36 inches long, but most were

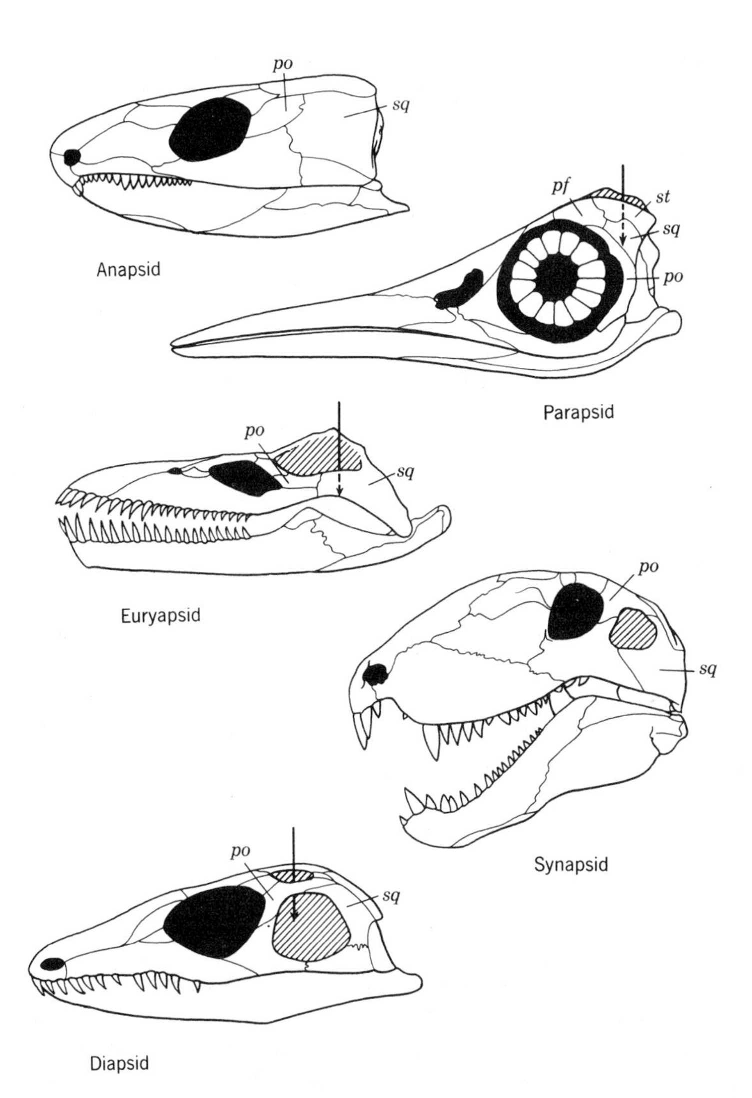

Fig. 224. The openings in reptile skulls and the positions of adjacent bones: Anapsid, Captorhinus, *a cotylosaur; parapsid,* Ichthyosaurus, *an ichthyosaur; euryapsid,* Muraenosaurus, *a plesiosaur; synapsid,* Dimetrodon, *sail reptile; diapsid,* Youngina, *an eosuchian.* pf, *postfrontal;* po, *postorbital;* sq, *squamosal;* st, *supratemporal; not drawn to scale. After Colbert.*

smaller. Their long flexible bodies and tails and paddlelike limbs were well adapted to rapid propulsion through the water. The skull was elongate and narrow, and the jaws were lined with over 300 needlelike teeth. These active fish-catchers have been found in the late Pennsylvanian of both South Africa and South America.

The first true turtles, or chelonians, appear in the Triassic, though there is some suggestion of an ancestral form in the Permian. With their bony carapace and plastron for protection, they have been one of the most successful groups of reptiles. Different kinds, from tortoises in the deserts to the sea turtles in oceanic waters, are adapted to many environments. Their skulls are a modified anapsid type, fitted with a horny beak. Among the chelonians, the tortoises have developed highly arched carapaces and live on dry land or even in deserts; the terrapins have a very flat shell and live in swamps and rivers; and the true turtles with a rather intermediate shell may be found in rivers and in the oceans. Some of the marine turtles are gigantic. The great Cretaceous marine turtle *Archelon* is 12 feet long. An additional primitive group failed to develop a hard shell; this was the soft-shelled, or trionychid "turtle."

The most sharklike and dolphinlike of the reptiles were the ichthyosaurs. No reptiles were better adapted for life in the water. They existed throughout the Mesozoic, but were most abundant in the Jurassic. Some Triassic forms in inland seas of western North America were 30 feet long, but most were much smaller. Since they were strictly aquatic animals, the females gave birth to young. The skeleton of a female was found in Germany with the remains of five unborn young in the body cavity and two just outside it but near the pelvic region. The first skeleton of an ichthyosaur was found by a little girl, Mary Anning, on the southern coast of England in 1811.

The spectacular plesiosaurs (chapter heading 20) form another group of marine reptiles. They are recognized by their euryapsid skulls and long paddles instead of feet. The paddles may have as many as seventeen bones in the longest fingers. The first plesiosaurs appeared in the Triassic, and in the Jurassic the suborder diverged into two contrasting groups that continued on through the Cretaceous. The species of one group evolved short necks and long heads. An extreme of this specialization is seen in a gigantic skull from Queensland, Australia, called *Kronosaurus*. It is 9 feet 6 inches long and is the largest known reptile skull. In marked contrast, the other group developed long necks and short heads. This group culminated in a species, *Thalassomedon haningtoni*, from Colorado, described by S. P. Welles, that was some 40 feet in length and had a neck 18 feet long. All plesiosaurs were equipped with sharp interlocking teeth well adapted for catching fish.

Since these sharp-pointed teeth were useless for chewing, the plesiosaurs used another method of breaking down their food for digestion. The fish were swallowed whole and were ground to bits by the stomach stones, or gastroliths, in the gizzard.

Rhynchocephalians have diapsid skulls in which the lower jaw is firmly hinged to the cranium. Some of these reptiles are quite crocodilian in appearance, particularly in the Mesozoic; others have heads and beaks somewhat like the turtle. The only living one is the tuatera, *Sphenodon*, of New Zealand. It is strictly terrestrial in habits, but its ancestors may have been marine animals.

The crocodiles, gavials, and alligators, with nostril openings at the end of the snout, are the largest living reptiles. They have had a long and varied history since the Jurassic. Though most of them were rather amphibious in habits, some invaded marine waters. *Protosuchus* from the early Jurassic of northern Arizona is an ideal ancestor for all the later crocodilians.

The Triassic phytosaurs (chapter heading 18) show a marked similarity to the crocodilians, but they are much more closely related to the dinosaurs. Their habits must have been much like those of the crocodiles. All phytosaurs have an armor of bony plates, and some developed horns along the neck and the edge of the back. They differ from crocodiles in having the nostril openings near the middle of the head not far from the eyes. They differ also in many other features. They are known only from the Triassic.

Lizards and snakes are descended from a diapsid ancestry and are grouped together in the superorder Squamata. Since they have lost the lower temporal arch, the hinge for the lower jaw has become loosely attached to the skull and a wider gape has developed. Lizards first appear in the Jurassic and snakes in the Cretaceous. Both groups had aquatic kinds. The most remarkable, with aquatic adaptations, were the mosasaurs which were closely related to the living varanid lizards. Some were nearly 30 feet long. Evidently these large fish-catching mosasaurs were confined to the late Cretaceous. They were abundant in the Niobrara Sea in middle North America.

Flying Reptiles. One of the most spectacular groups of reptiles is the order Pterosauria, which includes the pterodactyls with their vestigial tails and the rhamphorhynchids with fan-shaped membranes at the end of a long tail. Their forearms are modified into wings by greatly elongated phalanges of the fourth digit. Even in the specimens from Solnhofen, Bavaria, were indications of soft parts are so faithfully preserved, there are no traces of scales or feathers. Wide thin membranes sustained these reptiles in flight. Though the pterosaurs and birds are descended from similar reptilian ancestry, they are no more closely related to each other than either is to the dinosaurs.

Pterosaurs range in size from the Jurassic species of *Pterodactylus* (Fig. 141), no larger than a sparrow, to the great *Pteranodon* of the Cretaceous with a wingspread of 27 feet. The head in *Pterodactylus* is narrow and elongate. The eyes are well developed, and the teeth in both jaws are directed forward. *Pteranodon* has no teeth but possesses a long sharp beak and an equally long posterior extension of the skull back from the eyes. These winged reptiles are well adapted for soaring over water in search of fish, much in the manner of oceanic birds, but they are not so well equipped as modern birds to direct their flight. It is doubtful that they could get about on land with two or even all four of their limbs without considerable effort. Much is still to be learned about their flight habits and their methods of taking off and landing. Possibly they had resting habits like bats and were able to cling to stony cliffs and trees with the sharp claws on their feet and wings. Flying reptiles appear first in the early Jurassic. Their remains are most abundant in rocks of that Period but diminish in numbers in the Cretaceous and disappear before the Cenozoic when the more progressive birds took their place.

Land Reptiles. The most conspicuous terrestrial land animals were the dinosaurs, though some of these saurians developed amphibious habits. But the dinosaurs, lizards, snakes, and other later land-living forms were preceded by enormous numbers of more primitive terrestrial reptiles, especially in the Permian. There were plant eaters, flesh eaters, and others which must have fed on insects. The two earliest orders found in Pennsylvanian rocks were the anapsid Cotylosauria and the synapsid Pelycosauria. The early cotylosaurs, with their solid roofed skulls, were closely related to *Seymouria* (Fig. 128), the animal with many amphibian characters. Evidently all of the reptilian orders stemmed out of the cotylosaurs and the pelycosaurs.

More interesting to us were the synapsid pelycosaurs, for they were the ancestral group of mammallike reptiles which in turn gave rise to mammals. Pelycosaurs were abundant in the Permian deltaic flood plains of Texas. The genera *Dimetrodon, Edaphosaurus* (Figs. 125, 126), and *Casea* were characteristic of the order.

The mammallike reptiles of the order Therapsida, so numerous in the middle and late Permian, Triassic, and early Jurassic of South Africa, seem to have descended from carnivorous pelycosaurs. They, too, were greatly diversified. Some were large herbivorous creatures, but most of the smaller ones were carnivorous. All of them had some mammallike

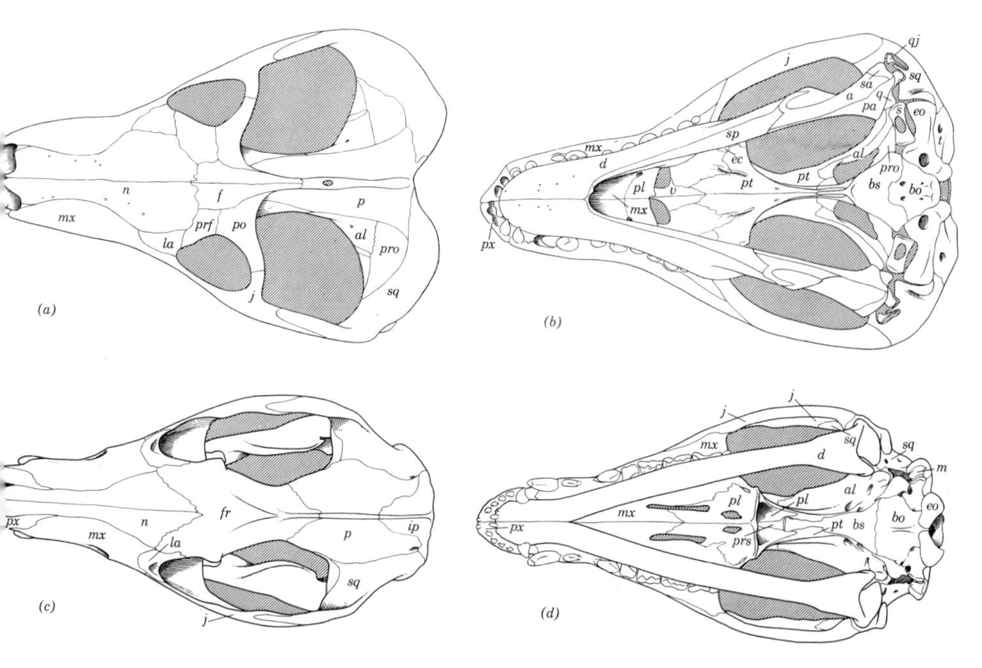

Fig. 225. The mammallike reptile skull of Thrinaxodon: (a) *dorsal view,* (b) *ventral view; compared with the American opossum skull,* Didelphis, (c) *dorsal view,* (d) *ventral view (see also Fig. 138).* a, *angle;* al, *alisphenoid;* bo, *basioccipital;* bs, *basisphenoid;* d, *dentary;* ec, *ectopterygoid;* eo, *exoccipital;* f, *frontal;* ip, *interparietal;* j, *jugal;* la, *lacrimal;* m, *mastoid;* mx, *maxillary;* n, *nasal;* p, *parietal;* pa, *prearticular;* pl, *palatine;* prf, *postfrontal;* pro, *proötic;* pt, *pterygoid;* px, *premaxillary;* q, *quadrate;* qj, *quadratojugal;* s, *stapes;* sa, *surangular;* sp, *splenial;* sq, *squamosal;* sx, *septomaxillary;* v, *vomer.*

characters, but some were much closer to the mammals than others. In fact several ictidosaurs were very close indeed to the earliest mammals.

There is abundant evidence that these extremely advanced reptiles are the ancestors of mammals. The intergradation is so complete that it is frequently difficult to tell whether one is dealing with a mammal or a reptile. The limbs had shifted from a wide lateral stance to a more direct position under the body; the digital formula (number of phalangeal bones in the toes) of 2-3-4-5-3 in the reptiles had changed or was changing to 2-3-3-3-3, so typical of mammals. Bones of the cranium and mandible were reducing to the number and to the positions seen in mammals. Most of these creatures probably had a coat of hair and primitive milk glands and may have partly broken their food down by

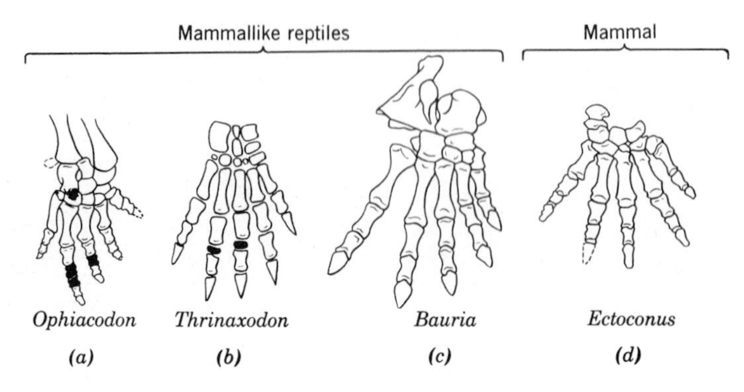

Fig. 226. Three feet of mammallike reptiles and one of an early Paleocene mammal showing different stages of reduction of the phalanges. (a, b, d), *front feet and* (c) *hind foot. After W. K. Gregory.*

mastication instead of swallowing it whole, as the other reptiles did. Indeed the mammallike reptiles were much more closely related to the entire class of mammals than to most of the orders of reptiles (Fig. 226).

The egg-laying platypus and echidna, or monotremes, in Australia are quite like the mammallike reptiles. Unfortunately, essentially nothing is known of their fossil record. If nothing but their fossil skeletons were available to us for study, they would most certainly be called specialized reptiles. Many characters in their soft anatomy and skeletons also are reptilian, particularly the pectoral girdle; and most of the characters retained in the skeletons of monotremes are possessed by mammallike reptiles. These interesting animals are the platypus, *Ornithorhynchus*, and the echidna anteaters of the genus *Tachyglossus*. They are clearly more primitive than any known mammal, living or extinct. Both lay soft-shelled eggs which have a liberal supply of yolk. The fertilized eggs are incubated in a short time from the warmth of the mother's body. Milk glands are distributed as a sheet of lacteal tissue, or as a multiple of small saclike glands, under the skin of the abdomen. The milk is exuded through numerous porous milk ducts, and the young obtain their nourishment by licking the milk from the hair or skin. Some authorities have referred the monotremes to the mammals and others to the reptiles (Fig. 227).

Evolution of the Mammalian Ear and Mandible. There were many interesting evolutionary changes in all parts of the skeleton, from fish to mammals, but the most extraordinary of all were the changes that culminated in the mammalian ear and the mandible, or lower jaw.

These evolutionary stages are traced through four stages: the fish, the amphibian, the mammallike reptile, and the mammal (Fig. 228).

It has been observed that there were as many as ten gill arches in the ostracoderms, or jawless fish. Critical research by Professor D. M. S. Watson at the University of London on the acanthodian placoderm, *Acanthodes,* together with studies on the embryology of living fishes and the arrangement of the cranial nerves, indicates that some of the ostracoderm gill arches were lost in the evolution toward higher fish. Evidently the third arch was modified into a pair of upper (palato-quadrate) and lower jaws (Meckel's cartilages). Back of this was the hyomandibular arch of which the single upper bone (hyomandibular) in *Acanthodes* had no contact with the jaws, as in later fish, but its upper end rested against the brain case near the inner ear. Since the water in which these primitive fishes lived was a good auditory medium, sound could be transmitted readily through the skull to the inner ear and then to the brain. The space in front of the hyoid arch was still a gill-slit opening, not a spiracle as in the sharks. The mandible was composed of two or three cartilage bones supported by a long dermal bone.

Fig. 227. The duck-billed monotreme, or platypus, Ornithorhynchus anatinus. *From Braznor's* Mammals of Victoria. *After Browning.*

As the connections between the jaws and the brain case tightened in the crossopterygian fishes, the hyomandibular of the old gill arch, which had functioned as a support for the upper and lower jaws, was no longer needed for that purpose. It gradually shifted into the old spiracular canal, where in the amphibians it assumed the new function of transmitting sound vibrations between the tympanic membrane, or ear drum covering the outer opening of the spiracle, and the inner ear. At that time the hyomandibular (-stapes) developed the stapedial foramen near its inner end. The stapedial artery and nerve passed through this foramen. The lower end of the spiracular canal developed into the eustachian tube that connected with the mouth. In the meantime during the evolution of the fishes each mandible had developed seven to ten distinct bones.

The next stage in this fascinating evolutionary sequence has been demonstrated in the pelycosaurs (early mammallike reptiles) and in the therapsids (advanced mammallike reptiles). In these tetrapods the bones at the posterior end of the mandibles reduced in size as the dentary bones became larger. Concurrently the outer end of the stapes shifted from its location high on the cranium wall, as seen in amphibians, to a lower position opposite the small quadrate bone of the cranium and the reduced articular of the mandible. Finally, when the most advanced therapsids and early mammals established a new articulation surface between the dentary of the lower jaw and the squamosal of the cranium to facilitate mastication, the quadrate and articular became enclosed in the middle ear. There they formed a series of small auditory ossicles (hyomandibular-stapes, quadrate-incus, articular-malleus) to amplify and transmit air-borne sound vibrations from the tympanic membrane of the outer ear to the inner ear. The bones of the middle ear have also been called hammer (malleus), anvil (incus), and stirrup (stapes) in the mammalian ear.

In their evolution it can be seen that these bones in the beginning functioned as part of the respiratory system in the gill arches; later they gradually transformed as supports and parts of the jaws operating in the digestive system; and finally, when they were no longer needed for that function, they were taken over by the auditory system to amplify and transmit sounds to the nervous system.

Mammals. It is quite clear that mammals arose from reptiles, but, as in other transitions of this kind, it is difficult for the systematist to draw a precise distinction between the two classes. Usually we think of mammals as possessing hair, milk glands, a high body temperature (warm blood), a single bone in each lower jaw (dentary), three small bones in

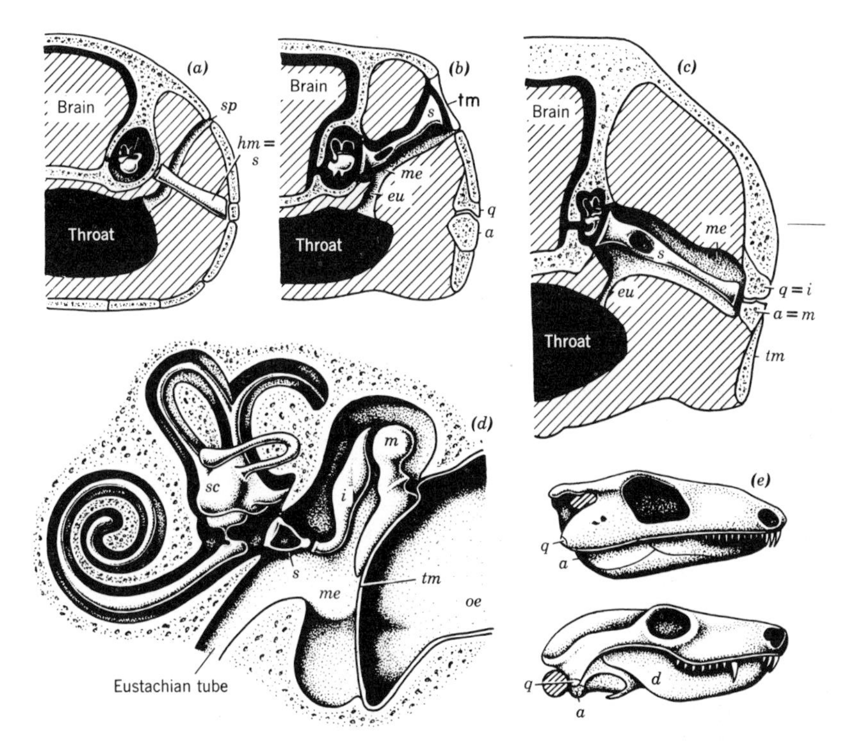

Fig. 228. A series of stages in the evolution of the ear apparatus: (a) *Cross section of half a fish skull in the ear region. The ear structures consist only of deep-lying sacs and semicircular canals,* sc. (b) *An amphibian. The hyomandibular bone of the fish,* hm, *is pressed into service as a sound transmitter, the stapes,* s; *the first gill slit, the spiracle* sp, *becomes the Eustachian tube* eu *and the middle-ear cavity* me; *the outer end of the spiracle is closed by the tympanic membrane* tm. (c) *A mammallike reptile. The stapes passes close to two skull bones (*q, *quadrate,* a, *articular) which form the jaw joint.* (d) *Man (the ear region only on a larger scale). The two jaw-joint bones have been pressed into service as accessory ossicles, the malleus* m *and the incus* i. (e) *A primitive land mammal and a mammallike reptile show the relation of the eardrum to the jaw joint. At first in a notch high on the side of the skull (the otic notch) occupying the place of the fish spiracle; it then shifts in mammallike reptiles to the jaw region. In mammals the jaw is formed of one bone only (*d, *the dentary), and the bones of the jaw-joint region are freed to act as accessory hearing organs;* oe, *tube of outer ear. After Romer with permission of the University of Chicago Press.*

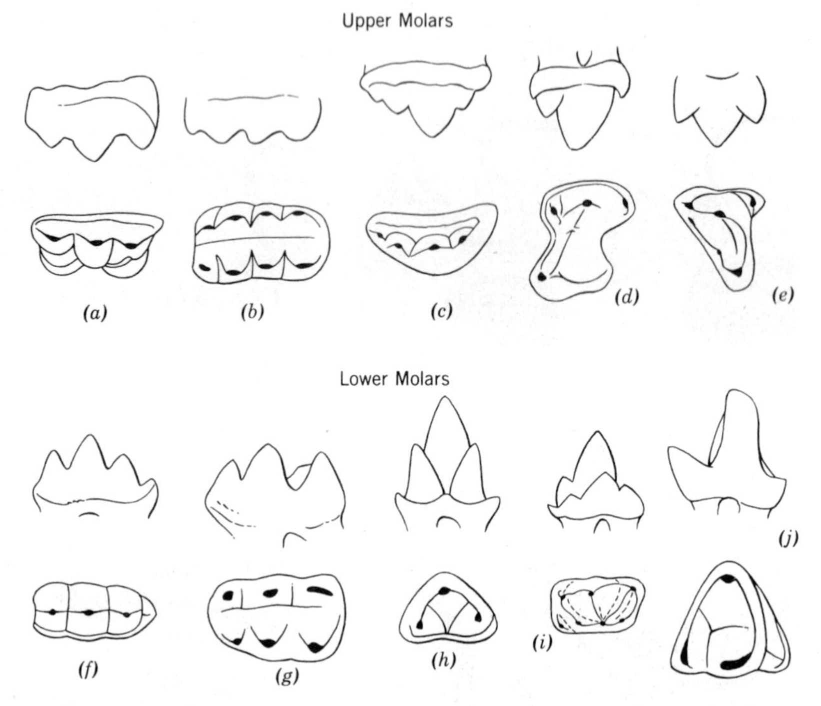

Fig. 229. Upper and lower molars of Jurassic mammals showing occlusal (wear surface) and side views: (a, f) *triconodont* (Priacodon), (b, g) *multituberculate* (Ctenacodon), (c, h) *symmetrodont,* (Eurylambda), (d, i) *docodont* (Docodon), (e, j) *pantothere* (Dryolestes). *After Simpson.*

the middle ear, and a simplified pectoral girdle. Nevertheless, all or nearly all of these characters may have been possessed by one or another groups of the most advanced mammallike reptiles. This is to be expected in the evolutionary patterns and has been demonstrated time and again.

The oldest known true mammals occur in the middle Jurassic of England, though some isolated teeth from the late Triassic or early Jurassic of central Europe may have already reached the mammalian stage of development. Five Jurassic orders are known from teeth and jaws (Figs. 140, 141 and 229). They are the multituberculates, triconodonts, symmetrodonts, docodonts, and pantotheres. The construction of the teeth indicates that the pantotheres were closer to the marsupials and the placental mammals than were the other orders. The multituberculates are known to have continued until early Eocene, and the triconodonts extend into early Cretaceous.

Appearing for the first time in the later Cretaceous were the marsupials (pouched mammals) and the insectivores (earliest placental mammals). Though these Mesozoic mammals are, for the most part, small inconspicuous creatures in a world dominated by huge reptiles, they are the forerunners of another tremendous radiation of vertebrate animals. Twenty-six additional orders of placental animals appear in the Cenozoic, all of which probably have representatives somewhere in the Eocene and more than half in the Paleocene. They include such well-known orders as the bats, primates, edentates (armadillos, etc.), rodents, carnivores (flesh eaters), whales, sea cows, mastodonts and elephants, perissodactyls (rhinoceroses, horses, etc.), and artiodactyls (pigs, camels, deer, cattle, etc.).

Marsupials. Marsupials are the most primitive of the living mammals, if we exclude the egg-laying monotremes. Some of the most widely known examples of these pouched mammals are the opossums, the kangaroos, and the koalas (Fig. 230).

Opossumlike mammals are known from the Cretaceous of North America, where their teeth and jaws have given some idea of the early an-

Fig. 230. Gray kangaroo, Macropus major, a doe, with a joey in the pouch.

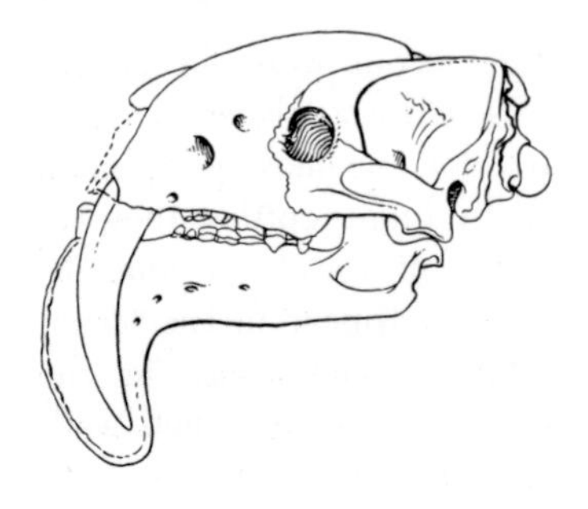

Fig. 231. The skull of a South American sabre-toothed marsupial, Thylacosmilus atrox. *It was about the size of a mountain lion. The skull displays a remarkable convergence with those in sabre-toothed cats.*

cestry of the order. It is not known when marsupials first reached Australia, but it must have been as early as Cretaceous, for they have evolved into extremely diversified groups on that island continent. Cretaceous insectivores and other early Cenozoic placental mammals apparently did not have access to Australia when they were so well represented in the Northern Hemisphere.

In South America flesh- and insect-eating marsupials assumed the role of the true carnivores (Fig. 231) and rodents, since the first of the rodents did not arrive there until much later. Though mammals that carry their young in pouches are known to nearly every youngster, the unique method of reproduction and care of the young in these fascinating animals is still a mystery to many people.

The ova, or eggs, in mammals are modified amniote eggs without yolk and are the means of nourishment for the developing embryo. After the eggs pass down from the ovaries into the oviducts, they are fertilized by sperm that work their way up those passages. From this stage on the method of nourishing the embryos in marsupials is quite different from that in the placental mammals. The period of gestation, or time between fertilization of the egg and birth, in the pouched mammals is short: between 12 and 13 days in the American opossum and about forty days in kangaroos. It is not known how the embryos are nourished during this interval.

Thus the young are born in a tiny embryonic state but work their way into the pouch by a swimming motion of the forelimbs. A pathway of dampened hair is prepared by the mother by licking the area with her tongue. Upon reaching the pouch the successful ones find nipples and for some time remain attached. The young are then fed by specialized, contracting muscles of the mother which force the milk into their mouths. Under the protection and warmth of the pouch they continue to develop and grow until they can move about and eventually are weaned. Actually this is a very efficient system of reproduction and protection of the young. The failure of most marsupials to compete successfully with the placental mammals, wherever they occur together, probably has been due to their inferior mental capacity and not to their method of reproduction.

Placental Mammals. The greater number of mammals familiar to the average person develop their young by means of allantoic placenta. The first of these are the small shrewlike insectivores of the Cretaceous. Many teeth and parts of jaws have been found in North America, and the skulls and jaws recovered by the American Museum Central Asiatic expeditions have given us important information concerning these fascinating little mammals. The placental mammals, such as men, rodents, elephants, horses, dogs, and camels, are the dominant terrestrial vertebrates of the Cenozoic. Others, such as monkeys and squirrels, are arboreal, bats are well adapted to fly, and whales, seals, and related groups have invaded the oceans.

In these mammals the fertilized eggs, or ova, soon adhere to the inner lining of the uterus. There the foetal membranes form placental structures through which the embryos are nourished. Waste is carried away by the blood systems of the embryo and of the mother, but the blood of the parent and embryo does not mix. Oxygen and nutriment are brought into the placenta in the blood stream of the mother; there they pass through intervening tissue to the adjacent embryonic placenta and to the blood system of the embryo. The blood system of the embryo then conducts these materials through the umbilical cord to the developing embryo. In turn, carbon dioxide and waste are transported from the embryo back to the mother's blood. system where they can be eliminated.

The placental method of reproduction probably first appeared in the Cretaceous, though evolution in that direction, as well as the marsupial method, may have been well underway in Jurassic pantotheres.

In some placentals the young are *precocial,* or born as active individuals well coated with hair, their eyes open, and within a few hours are capable of moving actively. Good examples are colts, calves, and young jackrabbits. *Altricial* young, on the other hand, are born in a more or less inactive state, frequently without a protective coat of hair and in need of postnatal care for a much longer period. These may be exemplified by human babies, most young rodents and baby cottontail rabbits.

Tyrannosaurus rex

29 Dinosaurs

THE WORD DINOSAUR, meaning terrible lizard, is awe-inspiring. Mounted skeletons and restorations of these beasts in numerous museums have justified in the minds of most people all that the name implies. Indeed, some of the plant-eating amphibious dinosaurs were the largest land animals that ever lived, but the most ferocious were the ever-present flesh eaters such as the powerful *Tyrannosaurus* that reached its peak in the Cretaceous. All dinosaurs were not great ponderous creatures, for some were no larger than barnyard fowls. No fossils have more appeal to the public than the dinosaurs. They are amazing animals, startling in their various specializations, the dominant reptiles of the Mesozoic.

Quite contrary to the concept of Sir Richard Owen, who proposed the name dinosaur over a hundred years ago, there are two orders of these great reptiles, which are as distinct in their relationships as are the crocodiles and the flying pterosaurs. Critical studies by many paleontologists have revealed that the two orders of dinosaurs (Saurischia and Ornithischia), the phytosaurs, crocodiles, pterosaurs, and primitive birds probably arose from a more primitive late Permian or early Triassic diapsid ancestry. All except the birds are included in the Diapsida, though some herpetologists prefer other names for the subclass.

Dinosaurian Ancestry

The early ancestry of the dinosaurs and related reptilian orders began in the late Permian when certain reptiles known as eosuchians developed two openings in the cranium back of the eye. Two of these genera were *Youngina* and *Youngoides* of South Africa. They were small reptiles with elongate bodies and slender limbs. There were sharp teeth on their palates and at the edges of their jaws.

From this ancestry there appeared in the early Triassic an order of more advanced diapsid reptiles about the size of lizards, the thecodonts, with another opening in front of the eye. Active little flesh eaters in this group, such as *Euparkeria* from South Africa, and *Ornithosuchus* and *Saltoposuchus* from Europe, showed a reduction in the forelimbs, a triradiate pelvis, strong hind limbs, rows of small bony plates down their backs, and a long tail for balance in their bipedal gait. Their teeth were

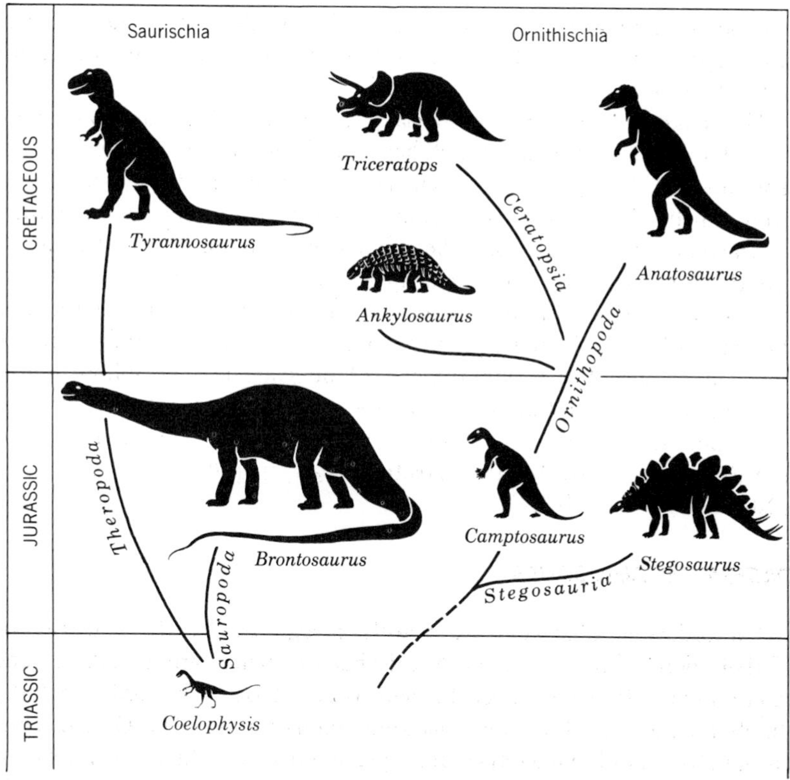

Fig. 232. A simplified phylogeny of the dinosaurs.

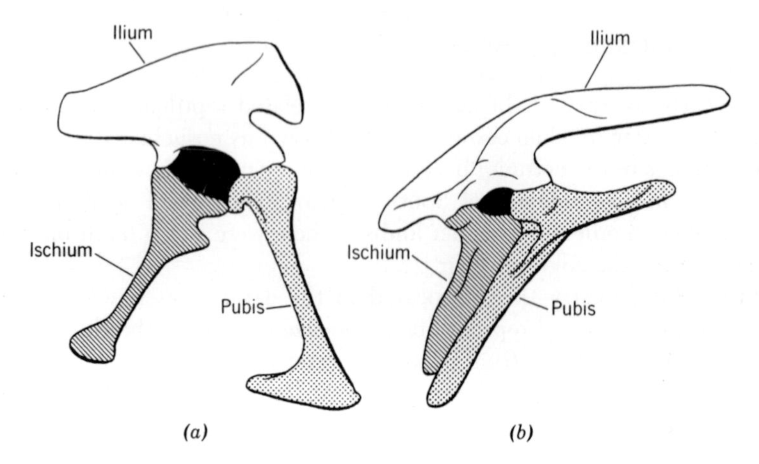

Fig. 233. Pelvic bones of the two orders of dinosaurs: (a) *Saurischia,* (b) *Ornithischia.*

confined to the edges of their jaws, as in all later diapsid reptiles. It was from reptiles like these that both orders of the earliest dinosaurs arose in the late Triassic.

The separation of primitive dinosaurs from some of the most advanced thecodonts frequently poses a difficult problem for the systematic paleontologist. In other words, when can a genus in these lineages of evolution be classified as belonging to one of the orders of dinosaurs? Characters ascribed to both thecodonts and dinosaurs are frequently found in the remains of one specimen. Then it becomes a matter of opinion as to which of the features are most important in determining its relationships. The most primitive dinosaurs tend to differ from thecodonts in having (1) no skin plates, (2) no clavicle, (3) hollow bones, (4) a perforated acetabulum (the socket joint where the femur attaches to the pelvis), (5) a more birdlike hind foot with three prominent digits, and (6) the femur shifted to a position nearly parallel with the anteroposterior axis of the body.

ORDERS OF DINOSAURS

The orders of dinosaurs are readily recognized by the construction of their pelvic bones. In the Saurischia the bones are triradiate, and in lateral view they have been likened to milkstools. The order includes the flesh eaters and the enormous amphibious forms. The Ornithischia, on the other hand, have tetraradiate pelvic bones in which the pubis has shifted to a backward position. This construction is similar to that in birds, but it is now clear that birds did not ascend directly from these dinosaurs (Fig. 233).

Order. Saurischia (reptile + pelvis). The great reign of dinosaurs started in the middle and late Triassic, with small carnivorous saurischians which were not far removed from their thecodont ancestry. About three dozen genera of these saurischians have been recorded from Triassic formations and are referable to two suborders, Theropoda and Sauropoda, of the order Saurischia.

Suborder. Theropoda (beast + foot) **or Carnivorous Dinosaurs.** These flesh-eating saurischians included the most ferocious brutes of the dinosaur world. They evolved into numerous families, but all of them retained their bipedal means of locomotion. Some were huge predators, as exemplified by *Allosaurus* of the Jurassic and *Tyrannosaurus* of the Cretaceous, whereas others, such as *Podokesaurus* of the late Triassic and *Compsognathus* of the Jurassic, were no larger than turkeys (Fig. 234).

The genus *Coelophysis*, as described by Edwin H. Colbert from a series of skeletons found in one late Triassic deposit in New Mexico, has given us an excellent sample of one of the early theropods, or carnivorous dinosaurs. These lightly constructed, hollow-boned, slender-bodied reptiles were about 8 feet in length. Their long tails, strong hind limbs, and greatly reduced forelimbs were indicative of better balanced bipedal posture than in their thecodont ancestors. There were sharp claws on their feet, and their short maneuverable forelimbs were well adapted for holding their prey. Sharp serrate teeth were also indicative of their carnivorous habits. They must have been the most successful land-living predators of their time.

Coelurus of the Jurassic, frequently called *Ornitholestes*, was another small agile dinosaur not greatly different from the Triassic *Coelophysis*, though the lateral digits in the forefoot were more reduced, leaving three elongate fingers with claws. *Ornithomimus* of the Cretaceous carried these specializations much further. Its proportions and size were much like those of the ostrich. These Cretaceous fleet-footed dinosaurs had three long fingers in the hand, with one somewhat opposable, as in the human hand. Moreover, they had lost all of their teeth and had developed a small, lightly built, birdlike skull. There has been much speculation about the food habits of *Ornithomimus*. Some authors have maintained that they fed on the eggs of other dinosaurs. Some suggestion of this, though not conclusive, was the discovery in Mongolia of an ostrichlike dinosaur skull near a nest of dinosaur eggs. Other paleontologists have suggested that *Ornithomimus* and related kinds fed on fruits of the flowering plants which had spread over the world in Cretaceous time. Perhaps the genera and species of these ostrichlike dinosaurs had different food preferences.

Even more fascinating were the smallest primitive dinosaurs of the Triassic and Jurassic. They were preceded by other small, little-known Triassic forms such as *Procompsognathus.* These little creatures probably searched about fallen trees and under low-lying foliage for insects and small reptiles. In a sense, they were miniature editions of the rather distantly related huge carnosaurs which have attracted world-wide attention.

The big-game hunters of the Jurassic were the active and powerful *Megalosaurus* and *Allosaurus.* Some of these huge beasts, bulky as rhinoceroses, were 35 feet long from the end of the nose to the tip of the tail. Their skulls were large in proportion to their bodies and were equipped with jaws 3 feet long. There were three strong claws on their front feet; and they had three-toed, birdlike hind feet, in which a fourth digit behind functioned at times as a prop. These big carnosaurs are thought to have been capable of killing the largest herbivores that frequented the same environments.

More powerful still was *Tyrannosaurus* and related genera of the Cretaceous. *Tyrannosaurus* was 47 feet in length and has been estimated to have weighed from 8 to 10 tons. In contrast to *Allosaurus,* the claws in the handlike forefeet were reduced to two. The huge head,

Fig. 234. A contrast in the size of carnivorous dinosaurs is exemplified in the Jurassic Compsognathus, *about the size of a rooster, and the Cretaceous* Tyrannosaurus, *with its head from 12 to 15 feet above the ground.*

at least 4 feet long, with some teeth 6 inches long, was 12 to 15 feet above the ground. These were the most formidable vertebrate animals of their time (Fig. 234).

Suborder. Sauropoda (reptile + foot) **or Amphibious Dinosaurs.** These, the largest of all dinosaurs, are known from about 3 dozen specimens in Triassic formations. They attained their maximum evolution and diversification in the Jurassic and diminished greatly in numbers during the Cretaceous.

Though the thecodont reptiles had evolved toward species adapted to run on their hind legs while holding their bodies in a somewhat upright and forward posture, some of the first dinosaurs were tending to increase in size and revert to walking on all four feet. A fine specimen of the 8-foot *Yaleosaurus* was taken from old mudstones in the Connecticut Valley, and more advanced *Plateosaurus* skeletons, 20 feet in length, have been found in Germany. The teeth in these early dinosaurs were basically a carnivorous type but were becoming blunt-pointed or spatulate for eating plants. These reptiles were the Triassic forerunners of the sauropods which in the next Period evolved into the largest four-footed animals ever to inhabit the earth.

The swamps and bayous of the Jurassic were the rendezvous of these immense amphibious dinosaurs. There they waded and floundered about, feeding on soft water plants. Others frequented stream courses where they spent part of their time along the banks or on sand bars lazily warming their huge bodies in the sun. These enormous reptiles had long tails, long necks, and small heads. Their nostrils were located on the tops of their heads so that they could breathe without exposing much of themselves above the surface. While on land their huge bodies were supported by four postlike legs and short, rounded, stubbytoed feet. Usually there was one large claw on each front foot and three on each hind foot, perhaps reminiscent of their thecodont ancestry.

Names such as *Diplodocus* (double + beam), *Brontosaurus* (thunder + reptile), and *Brachiosaurus* (arm + reptile) have become familiar to everyone (Fig. 145) at all interested in dinosaurs because complete skeletons of these animals have been mounted. *Diplodocus* and *Brontosaurus* were much alike, though they differed in many details and in their proportions. *Diplodicus* was slenderer, its tail was longer and more whiplike, and its head was thought to be smaller than that of its contemporary *Brontosaurus*. One specimen of *Diplodicus* was at least 87 feet long, and *Brontosaurus* ranged between 70 and 80 feet from end to end. These huge dinosaurs must have weighed 30 tons or more.

Brachiosaurus, on the other hand, was even more remarkable. Its body and tail were comparatively short, but its body was unusually bulky. Some of the ribs were 8 feet long. It has been estimated that adults may have weighed as much as 50 tons. The enormous size of these dinosaurs can be visualized when it is realized that one of the largest African elephants weighed about 6.5 tons. Another outstanding feature of *Brachiosaurus* was its forelimbs, which were much longer than the hind ones and gave the reptile a giraffelike posture. Its high shoulders and long neck equipped it well for wading and foraging in deep waters.

The sauropods must have been slow-moving dinosaurs that spent most of their time in the water. Other than their enormous size, they had no apparent means of defense. The young ones at times were probably killed by carnosaurs.

Order. Ornithischia (bird + pelvis). All four suborders of these dinosaurs had birdlike pelvi and were plant eaters. The teeth were specialized, but in some forms they were degenerate. In most of the genera the teeth were lost in the front of the mouth, and a birdlike beak developed. Many of them had flattened hooflike structures instead of claws on their toes. The order was not clearly differentiated in the Triassic but expanded in the Jurasssic and became abundant and greatly diversified in the Cretaceous. It has been divided into four major groups: duck-billed, plated, armored, and horned dinosaurs.

Suborder. Ornithopoda (bird + foot) **or Duck-billed Dinosaurs.** Irrespective of the name, the feet of these dinosaurs were not so birdlike as those of the flesh-eating kinds. Duckbill, though, is quite an appropriate name, since all kinds had a somewhat flattened ducklike beak. In many of the genera there were more than a thousand leaflike teeth in the jaws. These dinosaurs were unarmored and relatively thin skinned, and at least some of them were covered with thin lizardlike scales. They never became completely bipedal, though they could stand upright on their hind legs, with the tail on the ground forming a tripod support. Most of the duckbills must have been excellent swimmers. Their feet were well adapted for paddling, though their vertically flattened tails may have helped to propel them through the water. It is thought by some students of dinosaurs that there were special breathing devices in many genera by which an extra supply of oxygen could be retained in long passageways in the bony structures of the skull.

Camptosaurus was a genus of rather generalized Jurassic duckbills. The head was not so specialized as those of the Cretaceous genera. They were rather small dinosaurs, not exceeding 8 feet. Although some of the species survived as late as the Cretaceous, both the duck-billed and the horned dinosaurs must have evolved from an ancestor much like *Camptosaurus*.

The first discovery and description of dinosaur remains were teeth found by Gideon Mantell and his wife in the early Cretaceous rocks of England. When the teeth were directed to the attention of the French paleontologist Georges Cuvier he identified them at first as rhinoceros, and later, when bones were discovered, as hippopotamus. Mantell was not convinced because his studies indicated that the teeth were reptilian. Therefore, in 1825 he named them *Iguanodon*. Later Richard Owen of the British Museum proposed the name Dinosauria for these and other large reptilian remains. At that time no one had the faintest notion that there were two distinct orders of gigantic land reptiles.

The duck-billed *Iguanodon* was abundantly represented in Europe. Seventeen fine skeletons were found in a coal mine at Bernissart, Belgium. A group has been mounted in l'Institut Royal des Sciences

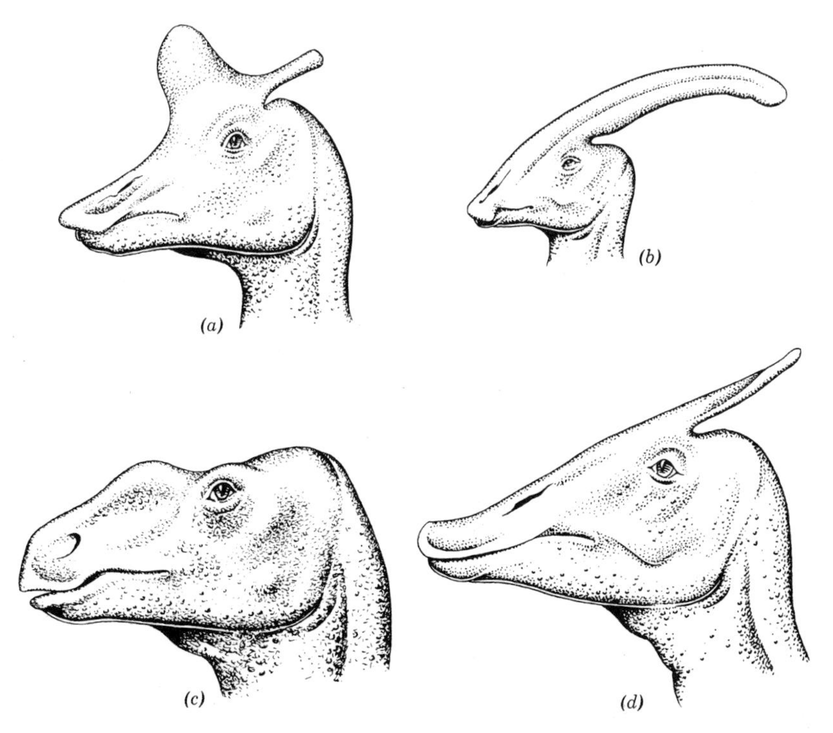

Fig. 235. Four kinds of duck-billed dinosaurs: (a) Lambeosaurus, (b) Parasaurolophus, (c) Gryposaurus, (d) Saurolophus; $^{1}/_{20}$ *natural size.*

naturelles of Brussels. Associated with *Iguanodon* in the early Cretaceous were small dinosaurs called *Hypsilophodon* (Fig. 150). William Elgin Swinton, Curator of Reptiles at the British Museum, has directed attention to several structures in the bones of *Hypsilophodon* which he believes are indicative of arboreal habits. After all, these little dinosaurs, perhaps a yard in length, were not too large for getting about in the trees.

One of the most common Cretaceous duckbills in North America was *Anatosaurus* (Fig. 151*b*). These dinosaurs were three to four times the size of the Jurassic *Camptosaurus,* and their mouths were expanded into a broad ducklike bill somewhat like the shoveler in our wild ducks. The well-known fossil collector Charles M. Sternberg discovered two most unusual specimens. In fact, they were fossils of what had been dinosaur mummies. The skin in these remarkable fossils had been tightly shrunken around the bones, showing that the carcasses had dehydrated as mummies before they were quickly buried in fine sandy clay sediments. The fossilized impressions of the original skin showed that it had been covered with nonoverlapping (lizardlike) scales. The scales were somewhat thicker on the head than on the body and tail. Also, there were impressions of a web between the toes. These dinosaurs were among the best swimmers in the Cretaceous swamps (Fig. 235).

The duck-billed dinosaurs evolved into numerous genera with crested and other peculiar bony expansions of the nasal and premaxillary bones. These structures housed long and sometimes quite tortuous passages from the nostrils to the pharynx. The exact functions of these passages are not clearly understood. The most bizarre of the duck-billed dinosaurs was the "bow-horned" *Parasaurolophus.* In this unusual dinosaur a long, curved bony structure extended back from the top of the head. *Corythosaurus* (Fig. 151*a*) and *Lambeosaurus* had quite different hooded structures cresting their heads. Equally amazing were the so-called "bonehead" dinosaurs of the North American Cretaceous. They, too,

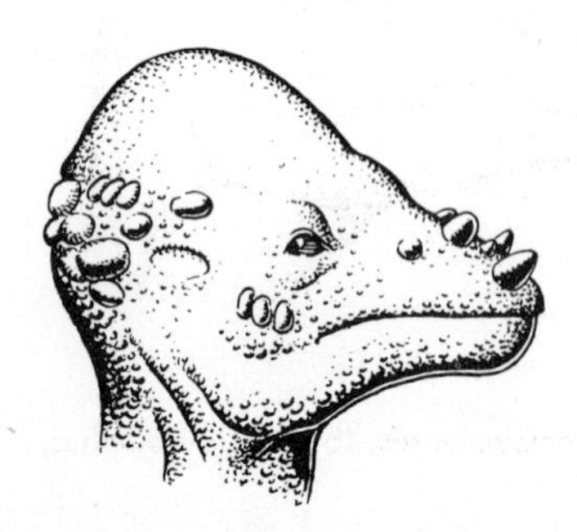

Fig. 236. The "bonehead" dinosaur, Pachycephalosaurus; 1/20 natural size. After Colbert.

Fig. 237. The Jurassic plated dinosaur, Stegosaurus, *was 20 feet long and had a small head and heavy plated armor.*

belonged to the suborder with the duckbills; *Pachycephalosaurus* was at least 6 feet tall. The top of its head was formed into a dome of solid bone about 9 inches thick. Its appearance was not particularly enhanced by wartlike bony rugosities and short spikes at the back of the head and across the nose. The dentition was greatly reduced (Fig. 236).

Suborder. Stegosauria (cover + reptile) **or Plated Dinosaurs.** These spectacular dinosaurs, typified by the genus *Stegosaurus,* which was some 20 feet in length, lent a weird aspect to the land life of the Jurassic. Among their most conspicuous features were the large alternating bony plates arranged in two rows along each side of the spinal column from the neck to the tail. The small head, short neck, high arched hips, and tail equipped with long thick spines contributed to their awesome appearance. Thus natural selection had provided these slow-moving reptiles with a means of defense against the even larger flesh eaters such as *Allosaurus.*

The enormous bony plates in *Stegosaurus* reached their maximum size above the base of the tail where they were as much as 2 feet high, 2.5 feet long, and 4 inches thick at the base. In life, evidently, these bony plates were covered with a horny substance that would have increased their over-all dimensions. The tail spines, 2 feet long and about 5 to 6 inches in diameter at their bases, must have been rather discouraging to enemies approaching from the rear. The head was small and the brain was tiny for such a large animal. The brain probably weighed no more than 2.5 ounces, but a large nervous ganglion in the sacral region was 10 to 20 times larger than the brain. The body was supported by post-like limbs and rounded feet with hoofs on the toes (Fig. 237).

Plated dinosaurs made their first appearance in the early Jurassic. Other genera survived as late as early Cretaceous.

Suborder. Ankylosauria (curved + reptile) **or Armored Dinosaurs.** The armored dinosaurs took up where the plated dinosaurs left off in the early Cretaceous. They were the dominant armored land reptiles until the end of the Period. The name *Ankylosaurus* was derived from the strongly curved ribs in that genus, but the armor in *Ankylosaurus* and in the other genera was much more characteristic of the suborder.

In the genus *Palaeoscincus* the body was rather flat and broad and was supported by short limbs. The back was covered with large and small bony plates showing a marked resemblance to the bony scutes that cover the backs of armadillos. *Palaeoscincus* was further protected by sharp bony structures at the lower edges of the bony plates. These formed a row of spikes that extended along each side of the body from the shoulders back to and along the sides of the tail. At the end of the tail was a flat structure of dense bone probably used in defense.

In many respects the armored dinosaurs were like the armadillos and glyptodonts among the mammals and the "hornytoads" among the lizards. Once a *Palaeoscincus* tucked his legs under and flattened out on the ground he was quite invulnerable. This suborder of dinosaurs has been called the reptilian armored tanks of the Mesozoic.

Suborder. Ceratopsia (horn + face) **or Horned Dinosaurs.** None of the plant-eating dinosaurs was better equipped for defense than the Ceratopsians. Though they were relatively small when compared with some of the dinosaurian giants, their immense heads with sharp horns and wide bony shields over their necks made them formidable opponents for any of the flesh eaters. Their four comparatively short but sturdy legs enabled them to pivot quickly in rhinoceros fashion to confront an attacking adversary. When the first of these huge horns was found in eastern Colorado in 1887 Professor Marsh naturally assumed that it had come from Pleistocene beds and that it belonged to a giant bison, since nothing like it had ever been seen in the large reptiles. Two years later, however, a skull with horns was discovered in Wyoming and its dinosaurian affinities were recognized (Fig. 152).

The ceratopsians were the last of the suborders to evolve from a late Jurassic or early Cretaceous group of duck-billed dinosaurs. One of the most primitive, a small dinosaur named *Psittacosaurus,* discovered in a Cretaceous formation in the heart of Mongolia, reflected duckbill relationships in its bipedal posture and in other characters, but its deep narrow head and parrotlike beak were clearly ceratopsian features. It had no horns and the bony shield over the neck had not evolved (Fig. 238e).

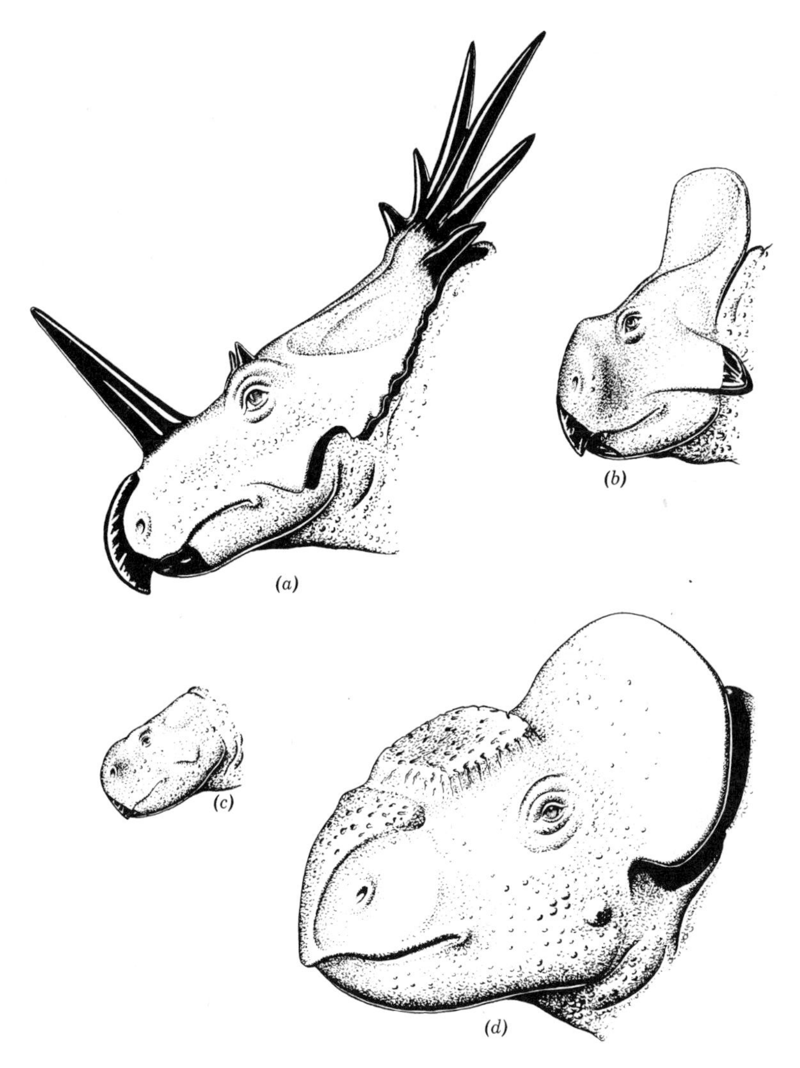

Fig. 238. Ceratopsian dinosaurs: (a) Styracosaurus, (b) Protoceratops, (c) Psittacosaurus, (d) Pachyrhinosaurus; *not to scale.*

Even more fascinating were the dinosaurs and their eggs, so abundant in the late Cretaceous red sandstones of the sun-parched desert around the Flaming Cliffs in Mongolia. They were called *Protoceratops.* These small dinosaurs were 6 feet in length or longer, much larger than *Psittacosaurus,* and though there were no horns on the head the bony neck shield was well developed. These active four-footed reptiles were abundant in central Mongolia. Only the time to collect and facilities to transport them limited the number of specimens that Walter Granger and his American Museum party could bring out. The collection included a fine series showing every stage of development from young, apparently freshly hatched, to adults. The last skeleton was discovered and collected by Peter C. Kaisen just before the last expedition left the field (Fig. 238b).

At first no one would believe George Olsen when he discovered the first dinosaur nest in July 1923, though Walter Granger had found evidence of an egg the year before. However, it was finally concluded that these elongate specimens, 3 inches in circumference and 7 inches long, could hardly be anything but *Protoceratops* eggs (Fig. 153). Some of the eggs were smaller. When other clutches were found with as many as 15 eggs arranged neatly in scooped out depressions in the sandstone their identification was fully confirmed. Most of the shells were fractured but some were perfect. Two eggs broken in two showed the delicate bones of embryos. All of the fossil eggs apparently did not belong to the protoceratopsians, for they were found in different sizes, shapes, thicknesses of shell, and surface patterns. This locality must have been ideal for a long time for the nesting habits of dinosaurs and for the incubation of their eggs, since the eggs located highest in the section were something like 200 feet statigraphically above those in the lowest nest. At this nesting site the eggs received enough moisture from the sand to offset dehydration, yet enough warmth from the sun to incubate. Evidently it was a semiarid area with low desert shrubs. The nesting sites were made up of fine wind-blown sands that piled up at the north end of a lake.

The largest of the horned dinosaurs were spectacular elements in the late Cretaceous landscapes of North America. At that time they had evolved into a dozen or more genera with conspicuous differences in the position and length of the horns and in the construction of the neck shields (Fig. 238a). The teeth were disposed in batteries much like those in the duck-billed dinosaurs, though they were fewer in number. Perhaps the unique feature of the ceratopsians was the turtlelike beak formed from the two predentary bones.

30 Birds

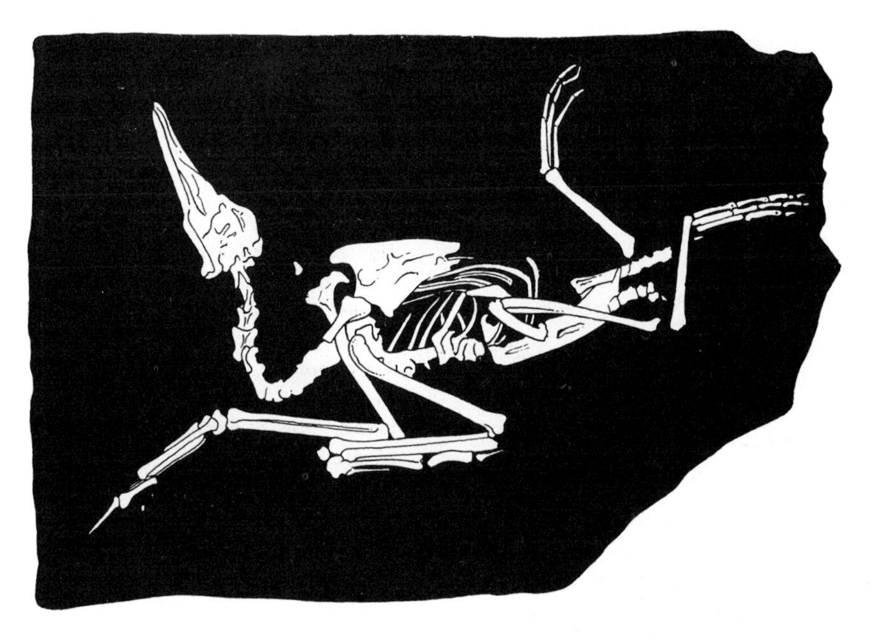

Lompoc shearwater

AMONG THE MOST BEAUTIFUL creatures we see about us today are the birds. Whether it is the kookaburra in the old gum tree, a flock of geese flying south in wedge formation, the brilliant quetzal in a tropical cloud forest, or the bluebird in the backyard, they all have their charm. Birds are readily differentiated from all other kinds of vertebrates in having a covering of feathers to insulate the body. Yet feathers in many ways are quite like the scales in reptiles. However, instead of simple scales, feathers have barbs and barbules that make up the fluffy plumage. Like mammals, they have warm blood, a regulated body temperature, and a four-chambered heart, but they differ from mammals in having a right instead of a left aortic arch. Actually,

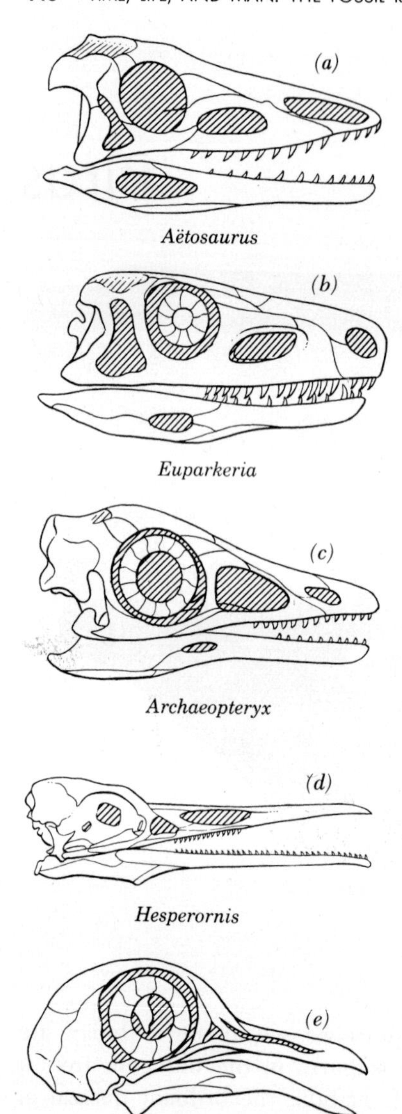

Fig. 239. Possible stages in the evolution of the bird skull: (a) *Triassic diapsid reptile,* Aëtosaurus; (b) *Triassic diapsid reptile,* Euparkeria; (c) *Jurassic bird,* Archaeopteryx; (d) *Cretaceous bird,* Hesperornis; (e) *Recent pigeon,* Columba; *not drawn to scale. After Simpson and others.*

though, they are so closely related to reptiles that some authorities have thought seriously of placing them in a class Sauropsida with the reptiles (Fig. 239).

There are bird lovers in every country of the world. Consequently, birds have been so carefully studied that no living group of animals of their magnitude is better known. But this is not true of their fossil record. For the most part bird fossils are rare, a possible exception being the tar pits of Pleistocene age in California, where thousands of bones have been collected. If bones are to be preserved as fossils, they must be buried quickly before bacterial decay and weathering can destroy them. The very structure of the bird makes this unlikely. Most of the hard parts of animals that became fossils were buried in water-borne sediments in inland seaways or troughs, lagoons, lakes, or streams. If the body of a bird fell into any of these waters, it would tend to float for some time because of its feathered covering and because the body cavity in most birds is filled with air sacs that penetrate even into the bones, especially those in the legs, wings, neck vertebrae, and head. Such specimens would have been subjected to the ravages of fish, reptiles, and other kinds of hungry creatures. Mostly detached water-soaked parts sink to the bottom, and these are what we usually find. The end of a humerus, part of a limb bone, or a neck vertebra may be

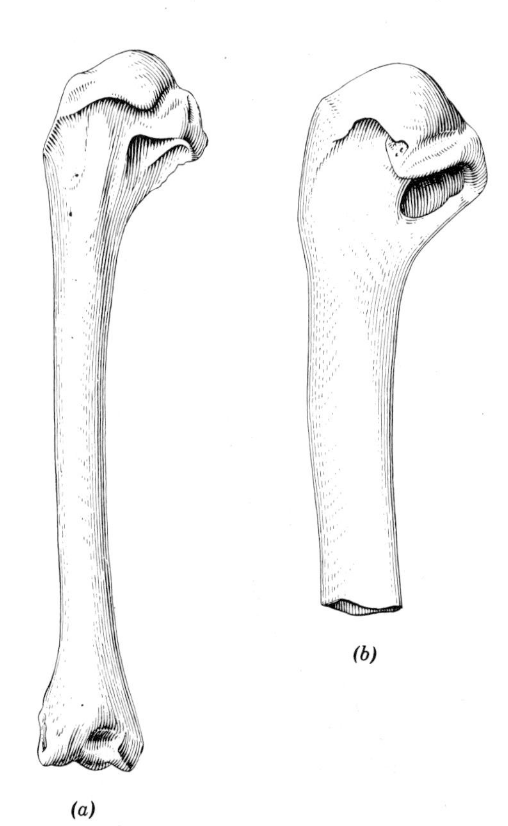

Fig. 240. Structural features on the heads of two bird humeri showing some of the characters used by paleo-ornithologists in identifying bones: (a) Miortyx, *quail,* × 1; (b) Dendrochen, *related to tree ducks and geese,* × 2. *From the early Miocene of South Dakota. After A. H. Miller.*

all that remains of a complete skeleton. Skulls are seldom found. The head seems to be a delicacy in the tastes of many carnivorous animals and is the first part devoured (Fig. 240).

Nevertheless, the paleoornithologist has found that small structural features on the bones offer excellent evidence for identification. The different mannerisms so clearly displayed by birds in flight, their method of swimming, or the way they get about on the surface of the ground are reflected in muscle attachments or in the shape and orientation of the bone. The ends of the humerus, radius, and ulna in the wing and the tibiotarsus and tarsometatarsus in the leg are particularly diagnostic. Other more basic features are characteristic of the orders and families. Many fossil bird types are represented by ends of bones, frequently all that is known of an extinct genus or of a family. There are exceptions, such as the discovery at Lompoc, California, of about twenty specimens referable to six species of Miocene marine birds. These are mostly impressions of partial skeletons in diatomaceous deposits. Fissure and lake deposits in France also have yielded large series of beautifully preserved bird bones, but in these localities the bones of the different species are mixed. In a recent publication, "A Check List of the Fossil and Prehistoric Birds of North America and the West Indies," Alexander Wetmore records 248 extinct species. The greatest number by far in Tertiary rocks are those of diving, swimming, and wading birds. Next come the falcons, vultures, hawks, and related groups, which are followed by the gallinaceous birds, grouse, quail, turkeys, and the like. The perching birds, such as crows, shrikes, and sparrows, are poorly represented, though they must have been much more abundant than the record indicates.

One of the arguments cited by those who were opposed to Darwin's theory of evolution was the origin of birds. It was thought that living birds were so distinct from other vertebrate animals that surely nothing other than special creation could explain their existence. At that time comparatively nothing was known of the fossil record of birds, partly because paleontology was still in its early growth, and few fossils were known. But this interpretation did not prevail. Just 12 years after Darwin's book on the origin of species appeared, a fossil imprint of a feather was discovered by a stone worker in the Solnhofen slates in the Altmuhl Valley of Bavaria. Though the evidence was meager, it was quite clear that some animal with feathers had lived as early as the late Jurassic. Christian Erich Herman von Meyer (1801–1869) called it *Archaeopteryx* (Gr. *archae*, ancient; *pteryx*, wing) *lithographica*. Fortunately, a year later, in 1861, the presence of a feathered creature in those gray lithographic slates was fully confirmed. The second dis-

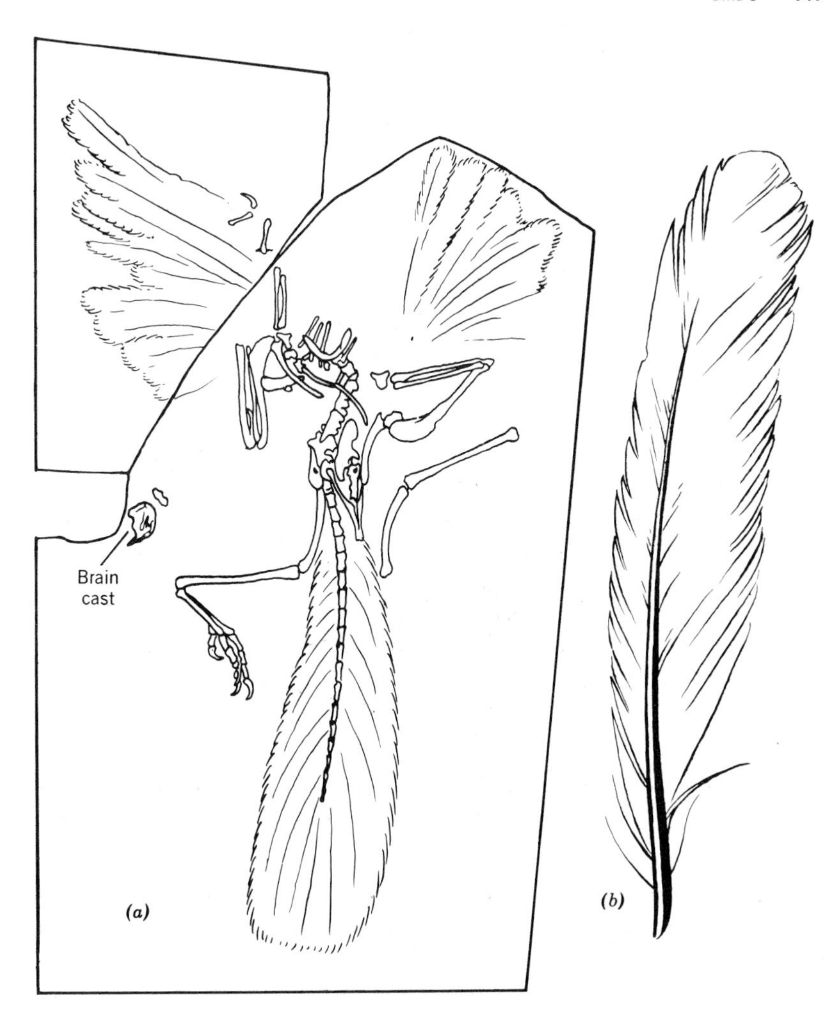

Fig. 241. Archaeopteryx lithographica *H. von Meyer:* (a) *specimen partly consumed by some predaceous animal, cast of brain on the left side of the stone slab,* $^2/_3$ *natural size, specimen in the British Museum (Natural History);* (b) *the first fossil feather imprint of* Archaeopteryx *discovered, about natural size. After de Beer. Permission of the British Museum (Natural History).*

covery was made near Pappenheim in the same valley. Evidently the specimen had been partly consumed and disarranged by some predaceous animal before it was covered with fine-grained sediments. Parts of the body, one leg, and the head and neck were missing. The specimen showed clear impressions of the feathers around the body, along

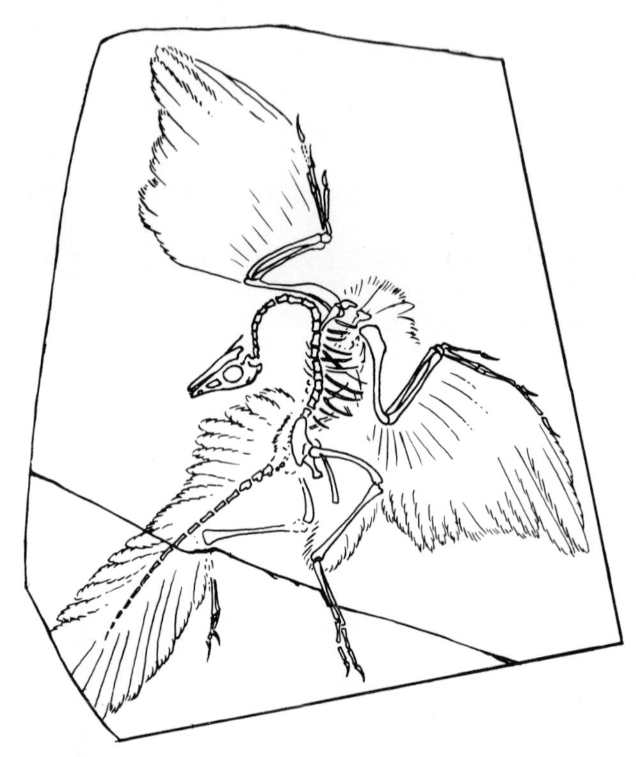

Fig. 242. The second skeleton of Archaeopteryx, in the Berlin Museum; 2/3 natural size. After de Beer. Permission of the British Museum (Natural History).

its long tail, and on its legs, but characters in the bones were primarily reptilian. This then, was the answer to Darwin's critics. Sir Richard Owen of the British Museum insisted that it was a bird, though Johann Andreas Wagner (1797–1861) of Munich referred to it as a winged reptile. For many years British scientists recognized that a small part detached from the specimen and on the same slab of slate was an endocranial cast with a few bone fragments and part of the maxillary containing a few teeth. The endocranial, or brain, cast was not described accurately until 1926 when Tilly Edinger of Harvard University went over it in detail (Fig. 241).

A third specimen found near Eichstätt in 1877 was in a much better state of preservation. This one, usually seen in restorations, was called *Archaeornis siemensi* by Wilhelm Barnim Dames (1843–1898). It had the head attached to the end of the neck vertebrae which were curved posteriorly over the back. As in the other specimen, the feathers were clearly imprinted in the gray shale (Fig. 242).

Except for size, no one has clearly demonstrated that *Archaeornis* differs appreciably from *Archaeopteryx*, to which it is so frequently referred. The one with a head is in Berlin, and the other is in the British Museum. No more have been found.

These specimens are ideal intermediates between birds and reptiles. If there had been no imprints of the feathers, they surely would have been called reptiles.

1. There are teeth in sockets in their jaws.
2. There were two small openings in the skull back of the eye and one in front, as in many diapsid reptiles.
3. The bones are dense, but neither hollow nor pneumatic, as in flying birds.
4. The centra of the vertebrae are concave at both ends (amphicoelus), not jockey-saddle-shaped as in birds.
5. The tail is made up of a long series of vertebrae (twenty or twenty-one) with feathers arranged on both sides down to the tip, not a short series of vertebrae terminated by a pygostyle (a specialized keellike terminal bone made up of several fused vertebrae for the attachment of feathers in a fan-shaped tail used for direction in flight.
6. There were belly ribs, as in some reptiles, instead of the heavy-keeled sternum of later birds.
7. Furthermore, the wing or front limb showed three distinct fingers with terminal claws. These and other features in the skeletons disclosed marked similarities to the diapsid reptiles (with two temporal openings) which include the phytosaurs, crocodiles, pterosaurs, and dinosaurs. The evidence indicates that birds were not derived directly from the dinosaurs, as suggested, but that they descended from Triassic thecodont reptiles much like *Euparkeria*.

Though there is a predominance of these reptilian characters in the skeleton of *Archaeopteryx*, certain features in the skull seem clearly to foreshadow structures that appeared later in the true birds. Even so, the presence of feathers is the most convincing evidence for placing *Archaeopteryx* in the class Aves.

In 1881 O. C. Marsh described as bird tracks some fossil foot imprints from the Jurassic beds at Como Bluff, Wyoming. These four-toed impressions showed evidence of the three front toes having been connected by webs. They may be amphibian tracks.

Cretaceous Birds from Kansas. Even before the third Jurassic skeleton was found at Eichstätt, Marsh announced his discoveries of toothed birds in the Cretaceous Niobrara Chalks in Kansas. The most sensational find was the almost complete skeleton of a large flightless diving

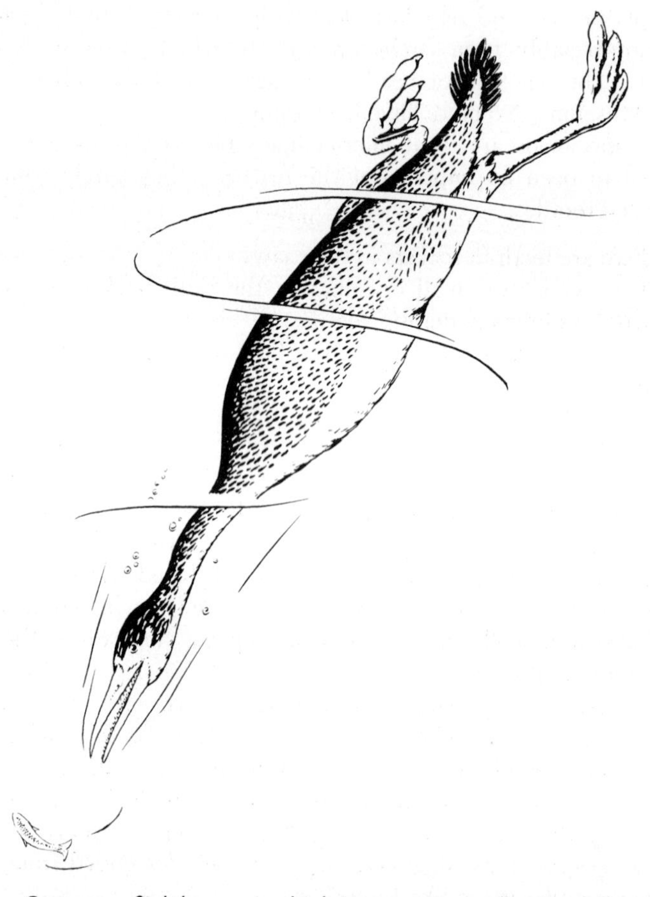

Fig. 243. Cretaceous flightless marine bird, Hesperornis regalis, *about the size of a loon. After Marsh.*

bird, *Hesperornis regalis* Marsh. Superficially it looked like a loon but was larger than the living bird. It retained teeth that were imbedded in grooves, but none was present in the premaxillary, a fact which offered some indication of the development of a birdlike beak. The wings were vestigial and the sternum rather smooth and rounded. This showed a remarkable specialization even at that early time in the evolution of birds. This toothed bird had already lost its ability to fly. Though the tail was short, there was no rudderlike pygostyle at its end, as in most modern birds. Another interesting feature in *Hesperornis* is the remarkable convergence in the construction of its mandible with that of the modified lizards known as mosasaurs, which were competing as fish catchers with *Hesperornis* in the Niobrara Sea. These are adaptive modifications which frequently result in convergence (Fig. 243).

Another famous bird from the Niobrara chalks was called *Ichthyornis dispar* Marsh. Unlike those of *Hesperornis*, its well-developed wings and deeply keeled sternum for the attachment of powerful flight muscles were indicative of its ability to sustain long flight. Its primitive pygostyle also showed that it had already acquired the means of better directing its course. It was about the size of a tern. A pair of mandibles and a maxillary fragment, found by Marsh's collectors, were closely associated with the skeleton and were thought at first to belong to a young mosasaur. Marsh then concluded that they were the toothed jaws of *Ichthyornis*. For 79 years it was listed in all text books and in numerous scientific papers as a toothed bird showing remarkable reptilian relationships. Finally, in 1951, Joseph Tracy Gregory of Yale University carefully reviewed all of the evidence and demonstrated that Marsh's original idea was correct. The jaws in fact did belong to a young mosasaur and the jaws of *Ichthyornis* are not known.

There is other fragmentary evidence of Cretaceous birds in North America. Though not referable to the orders of living birds, they are heronlike and gooselike. In Europe there are four genera recorded from the Cretaceous. *Elopteryx* is suggestive of early relationships between the cormorants and the pelicans; *Scaniornis* and *Parascaniornis* possess features reflected in the cranes and flamingos; and *Gallornis* is questionably placed in the duck family. On the whole, the evidence suggests that Cretaceous birds were already showing diverse specializations. Like the mammals and the insects, they evidently were responding to the opportunities of a multiple of new environments brought about by alterations of life in the water and by the spread and diversification of the flowering plants on land. From that Period on birds assumed an ever-increasing role in the animal life of the world.

Paleocene and Eocene Birds. Though mammals are now well documented from the Paleocene, especially in North America, very little is known of the bird life which must have been fairly abundant. The only known Paleocene birds come from the Hornerstown marl, a locality in New Jersey that was worked enthusiastically by Edward Drinker Cope and Othniel Charles Marsh while they were still friends. At that time the marls were thought to be Cretaceous. Marsh described three genera and seven species, and except for one species proposed 45 years later by R. W. Shufeldt this is all we know of Paleocene birds. There are rails, primitive woodcocklike waders, and early cormorantlike fishing birds. All belong to families that are still living.

In the Eocene we get a much clearer picture of bird life, which had progressed a long way in its evolution by that time. Today eleven orders are known from fossils in North America and Europe; all but one, Diatrymidae, the giant ground-birds related to the rails, have living representatives. Of the twenty-eight families of early Cenozoic birds only twelve are extinct. Two of the living genera, one a rail *(Rallus)* in Europe and the other a crane *(Grus)* in Wyoming, extended as far back as the Eocene. There is no representation of living species in those Eocene faunas.

If a bird lover of today were to be placed suddenly back among warm temperate-to-tropical lakes, streams, and forests of the Eocene, he would be hard pressed to recognize the families of the feathered creatures about him. There he would see swimming and diving birds ancestrally close to the ducks and the geese. Perched on the bough of a dead tree would be another much like a cormorant or a boobie. As the wide-eyed observer moved along the water's edge, there would be birds resembling loons, tropic-birds, gulls, pelicans, cranes, herons, flamingos, curlews, killdeers, sandpipers, gallinules, rails, and many others, all after fish or other kinds of life in the water, along the muddy shores, and among the reeds. Soaring overhead would be vulturelike birds (Fig. 164) on the sharp lookout for dead mammals or any other carcass. One of them, the stilt vulture, so named because his long slender legs carried his head some 18 inches above the mud, seemed to prefer dead fish. Back in the forest would be hawks, owls, and something that looked like a woodpecker. Out in the semi-open grassland there would appear an

Fig. 244. Diatryma steini, large flightless ground bird of the Eocene, about 7 feet tall.

entirely different bird. It would stand 7 feet high, but most conspicuous would be its powerful head and beak. This huge ground bird could have been *Diatryma* or *Gastornis* with vestigial wings incapable of flight (Fig. 244). Less apparent would be birds somewhat like grouse or a perching bird not unlike a shrike.

There was also a pelicanlike bird in the late Eocene of Egypt. Two fossils, one that shows relationships to the rails and the other to the cranes, have been found in the Eocene of Asia. Elsewhere, Eocene birds have not been recovered, though they must have been on all continents. On the whole, though, the early Cenozoic birds appear to have been much more like living kinds than did the mammals, of which there were no living genera in the Eocene and many less living families.

Oligocene Birds. The trend toward modern birds was exemplified even more in the Oligocene during which ten living genera appeared in Europe and five in North America.

OLIGOCENE GENERA NOW LIVING

Europe		North America
Puffinus, puffins	*Vanellus*, lapwings	*Phalacrocorax*, cormorants
Pelicanus, pelicans	*Totanus*, yellow legs	*Buteo*, large hawk
Phalacrocorax, cormorants	*Bubo*, horned owls	*Meleagris*, turkeys
Phoenicopterus, flamingos	*Asio*, long-eared owls	*Grus*, cranes
Anas, large river ducks	*Passer*, weaver finches	*Charadrius*, plovers

One of the best fossil bird and mammal localities in Europe is in the famous fissure deposits near Quercy in south central France. The fissure developed during Eocene time in Jurassic marine limestones. It is 3 to 6 meters wide and about 35 meters long. Layer after layer of scattered bones, most of which are remarkably complete and beautifully preserved, are coated with lime phosphate. This fissure accumulation is widely known as the Phosphorites of Quercy.

Carnivorous mammals and owls were probably primarily responsible for this great assemblage of vertebrate fossils, since these animals are known to carry their captured prey into dens and shelters of this kind. The bone levels here range from middle Eocene well into the Oligocene when the greatest accumulation took place.

The bird remains are helpful in an interpretation of the climate and the nature of the surrounding country. The structure and relationships of representatives of the different families and genera indicate a warm temperate-to-subtropical countryside with local forests and open grassland. Some are related to birds now living in warm temperate and tropical America, and others show affinities to birds in Africa,

India, and southeastern Asia. These fissure deposits are markedly different from most other fossil bird assemblages, for no diving, swimming, or wading birds occur.

Relationships to African birds were still strong in the early Miocene or late Oligocene lacustrine (lake) beds at Saint-Gerand-le-Puy in central France. There were pelicans, ibis, marabous, flamingos, ducks, cormorants, grebes, gulls, sandgrouse, couroucous, and secretary birds.

Miocene and Pliocene Oceanic Birds. After some restriction in the Oligocene, the Miocene marine waters covered much of the present area of California. This, of course, has been clearly reflected in the nature of the sediments as well as in the marine invertebrate and vertebrate fossils. Oceanic birds glided over these marine troughs or perched along their rocky shores ready to plunge into the surf for fish. Loye Holmes Miller, the dean of Pacific coast paleoornithologists, has given us excellent descriptions of oceanic birds found in diatomite near Lompoc, and there are numerous other scattered localities. Albatrosses in their effortless flight were followed by petrels and shearwaters as they ranged far and wide to satisfy their daily wants. Though the species were different, many genera were like those living today. One of the shore-loving boobies belonged to *Miosula*, a distinct genus known only

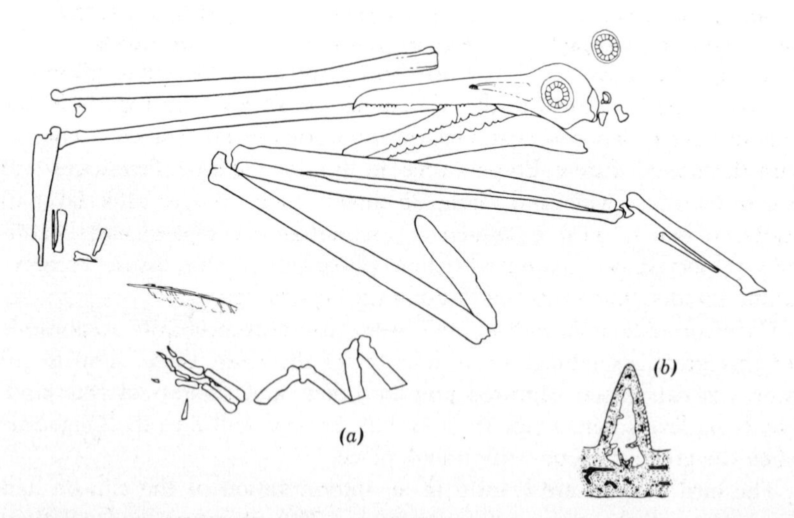

Fig. 245 Outline of part of the crushed skeleton of a gigantic fish-catching bird, Osteodontornis orri, from the Miocene of California; its wingspread was 14 to 16 feet: (a) skull, sclerotic bones of eye, and feather partly restored; (b) toothlike bony projections on jaws, drawn to show internal bony structure and serrate edge; × ½. Redrawn from photographs. After H. Howard.

Fig. 246. A Pliocene flightless auk, Mancella diegense, about the size of a mallard duck. Prepared under the direction of L. H. Miller.

as fossil. The wing bones demonstrated that this bird possessed less ability to fly than the modern boobies. There were other boobies much more like the living ones. The single imprint of the leg and foot bones of an aukletlike bird from Lompoc has been called *Cerorhinca dubia* Miller because of its dubious relationships.

The most spectacular of these oceanic birds is a gigantic fish catcher, *Osteodontornis orri,* recently described by Hildegard Howard. Its crushed skeleton was found on two slabs of sandy flagstone in a quarry near Santa Maria, California. Impressions of some feathers also occur on the sandy shale. The bill is not much longer than that of a pelican, but the head and bill are much larger than in the living bird. Its legs are short, but its enormous wingspread is estimated to have been somewhere between 14 and 16 feet. Even more interesting is the presence of bony toothlike structures along the edges of the jaws. These are variable in size, but it is thought that there were as many as nineteen in the lower jaw and probably more in the upper jaw. The toothlike processes are composed of spongy bone that is structurally continuous with the bone in the jaws. This huge bird belongs to Odontopterygiiformes, an extinct order that shares a combination of characters with two orders of oceanic birds. Some of these features occur in the shearwaters and fulmars and others are seen in the pelicans (Fig. 245).

There are scattered finds of oceanic birds in Miocene beds in other parts of the world, especially in Maryland and in Europe. They are less known from Pliocene rocks, but an albatross, a gannet, two boobies, a gull, and an auk have been described recently from the Bone Valley formation in Florida. One of the most interesting Pliocene species is *Mancella diegense,* a flightless auk from California. One wing bone of another species of this genus was found in 1901 while a tunnel was being put through a hill on Third Street in Los Angeles. It was the first fossil bird recorded from California. Even at that early date in Pacific coast paleoornithology it was correctly predicted that the auk from Third Street was incapable of flight, but, like the penguins, had wings adapted to help propel it through the water. Subsequent discoveries of all the skeletal bones have sustained that interpretation of form and function from a single bone. Loye H. Miller believes that the occurrence of these birds indicates the presence of insular breeding grounds, probably as archipelagos, in southern California at that time.

Early Miocene Birds on the Great Plains. An insight into the bird life of the Great Plains province of central United States in early Miocene time has been recovered from the Harrison and Sheep Creek formations in western Nebraska and from the Flint Hill quarry (Fig. 240) in South Dakota. This was an extensive flood plain at that time, an area traversed by streams and dotted with lakes. The birds when visualized as inhabiting environments much like their closest living relatives suggest the presence of a stream bordered by forests, brush land, and open grassland farther back from the stream.

The hawks, eagles, kites, falcons, and owls were widely distributed, perhaps with related genera in both tropical America and in the Old World. There was a primitive anserine of a distinct family, Paranyrocidae, showing affinities with both ducks and geese. A small duck related to the blue-winged teal must also have been widely distributed. The only wading bird now known from this province in the early Miocene was like the oyster catcher. Among the birds inhabiting the more open brush and grasslands was a prairie chicken, a grouselike bird, and a quail. More typical of tropical regions were tree ducks and a parrot, and a chachalaca reminds one of Mexico and tropical America. Rather unexpected, though, were two kinds of Old World vultures. Of the eighteen genera known, eight are still living and there were no modern species.

Tertiary Perching Birds. It has long been assumed that the passerine, or perching birds (nuthatches, crows, thrushes, sparrows, larks, etc.), accomplished most of their evolution much later than the other orders of birds. This is possibly true, but it should be noted that these birds, so abundant today, have habits that might readily account for their rarity in the fossil record. They usually inhabit forests, brushland, meadows, or plains country, where their bones are less likely to become entombed in sedimentary deposits. Nevertheless, in the Miocene deposits at Sansan in southern France bones of many crows, hornbills, and sparrowlike birds have been found with those of eagles, hawks, owls, pheasants, partridges, herons, rails, curlews, and ducks. The oldest passerines, *Laurillardia,* are shrikelike birds, and *Palaegithalus,* a titmouselike form, from the Eocene of Europe. Of the twenty records older than Pleistocene, only one is from the Pliocene of Asia, two are from the Oligocene and Pliocene in North America, and the remainder are from Europe. Of the twenty-one now known, only five represent extinct genera.

Penguins. The penguins, a peculiar yet most fascinating order of birds, have attracted human attention universally perhaps because of their walking upright and because of their almost human antics. They seem to have been restricted throughout their existence to the Southern Hemisphere. They are, of course, common in Antarctica and may be found in the water or on floating ice in the South Pacific and South Atlantic. Some species occur on the shores of South America, South Africa, and Australia. The Galápagos penguin is the only species as far north as the equator (Fig. 247).

Their relationships to other birds are not clearly understood, though it seems that they may have had a common ancestry with the oceanic albatrosses, shearwaters, and fulmurs, possibly in the late Cretaceous or early Cenozoic. Even in the earliest known Oligocene and Miocene species they seem to have been quite incapable of flight. There are

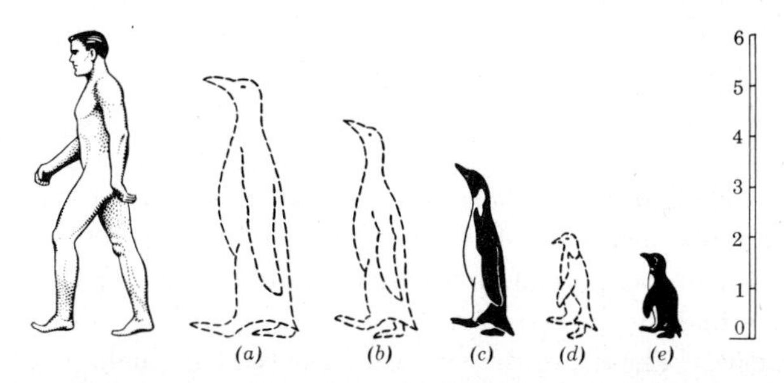

Fig. 247. The sizes of extinct and living penquins: (a) Pachydyptes *or* Anthropornis, *Miocene;* (b) Palaeeudyptes, *Miocene;* (c) Aptenodytes, *Recent;* (d) Palaeospheniscus *Miocene;* (e) Spheniscus, *Recent. After Simpson.*

three theories as to how they became adapted so admirably to their life in the water.

1. In evolving from primitively protoavian types, they became aquatic.
2. At first their ancestors could fly, they then became terrestrial and incapable of flight and finally developed strictly aquatic habits.
3. From flying land birds they altered to oceanic flying kinds and finally from diving into the water for food they eventually assumed an exclusively submarine kind of flight in which their wings changed to function as flippers.

Fossil remains of several genera are known. They occur on the Island of Seymour and in South America. A few fragments have been discovered in Australia and New Zealand. In 1933 George Gaylord Simpson and his field party recovered the first fossil skeleton from Miocene rocks in Argentina. Marples has described other fossils from the Oligocene of New Zealand. The largest extinct penguin is estimated to have been about 5 feet tall.

Large Ground Birds. The most gigantic of all the feathered creatures are some of the ground birds. At first it was thought that all of them must be closely related, but later detailed studies disclosed that they belong to not less than seven orders. It seems then that many groups of birds have quite independently lost the power of flight which their predecessors had attained well back in the Mesozoic. Unfortunately, the fossil record is still too incomplete to demonstrate these evolutionary trends. As the different groups evolved strictly terrestrial habits, their increase in size and rather ferocious nature when cornered tended to protect them from their enemies.

DIATRYMAS. One of the most formidable creatures to be reckoned with in early Eocene times was *Diatryma* (Fig. 244) and his relatives. These large ground birds, with greatly reduced wings, stalked about with their massive heads and jaws about 7 feet above the ground. The construction of the head and jaws suggests that they may have fed on other animals. The order to which they belong has been placed in proximity to that of the cranes and rails, early genera of which were already in existence in the Eocene. *Diatryma* had three prominent toes in front and a small one behind. Their remains have been found in North America and Europe, and another closely related genus, *Gastornis,* occurs in Europe in beds of the same age.

PHOROHACIDS. These South American birds bear such a marked resemblance to the diatrymas that they were at first thought to belong to the same family. Some phorohacids had heads as large as that of a horse. Their bones occur in Oligocene, Miocene, and Pliocene rocks in Argentina. A careful survey of their characters seems to reveal an even closer relationship to the cranes and rails than in *Diatryma*. Numerous genera are now known.

OSTRICHES. Ostriches are living today in a wild state in Africa and Arabia. Though their fossil record is not well known, there is evidence that some of their forerunners existed in the Pliocene and Pleistocene of Eurasia. Because of their well-padded, two-toed feet they are capable of getting about quite rapidly over stony and sandy deserts. They are the largest living birds.

RHEAS. The rheas of the Argentine are called ostriches by the local people, but they are not at all closely related to the ostriches, as their three-toed feet will quickly show. Actually they belong in an order of their own. They have now been found in South America in formations as old as Pliocene.

EMUS AND CASSOWARIES. This great order of ground birds is restricted in its distribution to Australia and New Guinea. The wings are even more reduced than those of the ostrich and rhea. The emus enter prominently into the legends of the Australian aborigines. Evidently these birds have a long history on the continent "down under." Parts of three skeletons and other bones were found in old moundspring deposits by the South Australian Museum Expedition of 1893. There the extinct genus and species *Genyornis newtoni* Stirling and Zietz was found as a contemporary of the giant marsupial *Diprotodon*. The first bone of these large birds was recovered in 1836 in the Wellington Caves of New South Wales by the Sir Thomas Mitchell Expedition to the interior.

MOAS. In 1839 John Rule showed up in London with part of the shaft of a femur which he thought was a "relic of bird-life rare and original." This historic bone, picked up somewhere in New Zealand, had both ends missing, but the surface features were well preserved. It was only 6 inches long but was 5.5 inches in circumference. After some persistent persuasion on the part of Rule, Professor Richard Owen of the Royal College of Surgeons examined the bone critically. With his unexcelled knowledge in anatomy as a background, Owen not only confirmed Rule's opinion that it belonged to a large bird but stated that the bone had belonged to a bird heavier and more sluggish than an ostrich. It was named *Dinornis nova-Zealandiae* by Owen. In the meantime in New Zealand the Reverend William Colenso, explorer and naturalist, had come much to the same conclusion, but because of the native Moari legends he assumed erroneously that the birds could fly (Fig. 248).

Little did these early investigators realize the abundance of the remains or the number of genera and species of these gigantic ground birds that would eventually turn up on the island. The bones were found buried in stream beds, bogged down in swamps, covered with drifting sand, and in aboriginal camp sites. Abundant evidence has been found to show that the early natives killed the moas for food and cooked them in earthen pits 5 feet in diameter. Egg shells 10 inches in diameter and 24 inches in circumference were found at some of the camp sites and buried in sand. The birds ranged from turkey size to others nearly 12 feet tall.

KIWIS AND FLIGHTLESS RAILS. Bones of flightless rails and of the kiwi also occur at some of the sites. The Kiwis still living in New Zealand are small almost wingless birds, with a long bill and tiny eyes. They are recorded also in the Pleistocene of New Zealand.

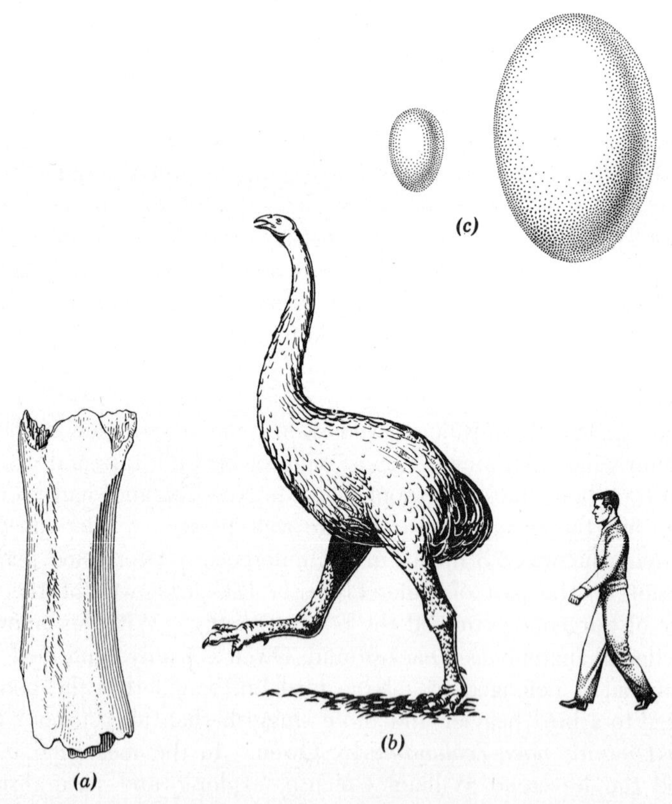

Fig. 248. (a) Part of the shaft of a femur, the first bone of a moa brought to London by Dr. John Rule in 1839; (b) a restoration of one of the moas, Dinornis, *the largest known ground birds; (c) a hen's egg, compared with egg of a moa.*

Fig. 249. Didus, the dodo, a flightless ground bird about the size of a swan, killed off by Dutch settlers on Mauritius, in the Indian Ocean southeast of Madagascar.

ELEPHANT BIRDS. French expeditions to the island of Madagascar have uncovered another order of birds of Subrecent to Pleistocene age. Two genera, *Aepyornis* and *Mullerornis*, have been described. Most of them were about the size of ostriches, but others must have been larger, for eggs said to be of 2-gallon capacity have been found. Some specimens from the early Oligocene also have been referred to this order.

THE DODO AND THE GREAT AUK. The ugly dodos were abundant for some reason on Mauritius Island in the western Indian Ocean. They were about the size of swans, had large hooked beaks, wings too small for flight, and short taillike feathers on their backs. Extraordinary as it may seem, they were related to the pigeons. Arrival of Portuguese explorers in 1505 and later the Dutch in 1598 with their dogs, cats, and pigs resulted in the end of dodo history. By 1691 the dodos were gone, and strangely enough there is no skin in any museum today (Fig. 249).

A similar fate befell the great auks in the area of Iceland in the North Atlantic. The last one was said to have been killed on the island of Eldeu on June 4, 1844. The great auk was a huge bird about the size of a goose, and there is evidence of its having been as far south as Saint Lawrence Bay on the east coast of North America.

Pleistocene Birds. By far the greatest number of fossil birds have been found in Pleistocene deposits throughout the world. The most famous are the tar pits of the Rancho La Brea now in the heart of Los Angeles, California. This was an open woodland in which a great variety of birds was associated with a large mammalian fauna. Many of the bird species found in the bitumen deposits are still living. Among the more spectacular extinct kinds were the giant condor *Teratornis merriami* and the California turkey *Parapavo californicus.* More than 100,000 bones representing at least 125 species have been recovered from the pits.

From a bituminous deposit near the small town of McKittrick along the west side of the San Joaquin Valley, also in California, another impressive collection has been made. Not only birds such as roadrunners and others indicative of an arid desert environment but an abundance of shore and water varieties showed that the old tar pool was on a migratory route. Evidently the birds on their way south in the autumn or flying north in the spring at dusk mistook the gleam of the oily surface for a body of water. Once they alighted they were never able to extricate themselves.

Many other Pleistocene bird bones occur in caves, fissures, lake beds, stream sediments, and in any number of places.

31 Evolution of Horses

Hyracotherium *and* Equus

HORSES ARE AMONG our most graceful and beautiful mammals. They have been treasured and loved by man for centuries, but their ancestry extends much farther back into geologic time to the beginning of the Eocene some 60 million years ago. At that time little eohippus, the dawn horse, looked little like horses as we know them. When in 1839 part of a skull was found in the early Eocene London Clays at a place called Studd Hill in Kent, England, even the most eminent paleontologists of the day little suspected that it belonged to the horse family Equidae. In fact Sir Richard Owen named its genus *Hyraco-*

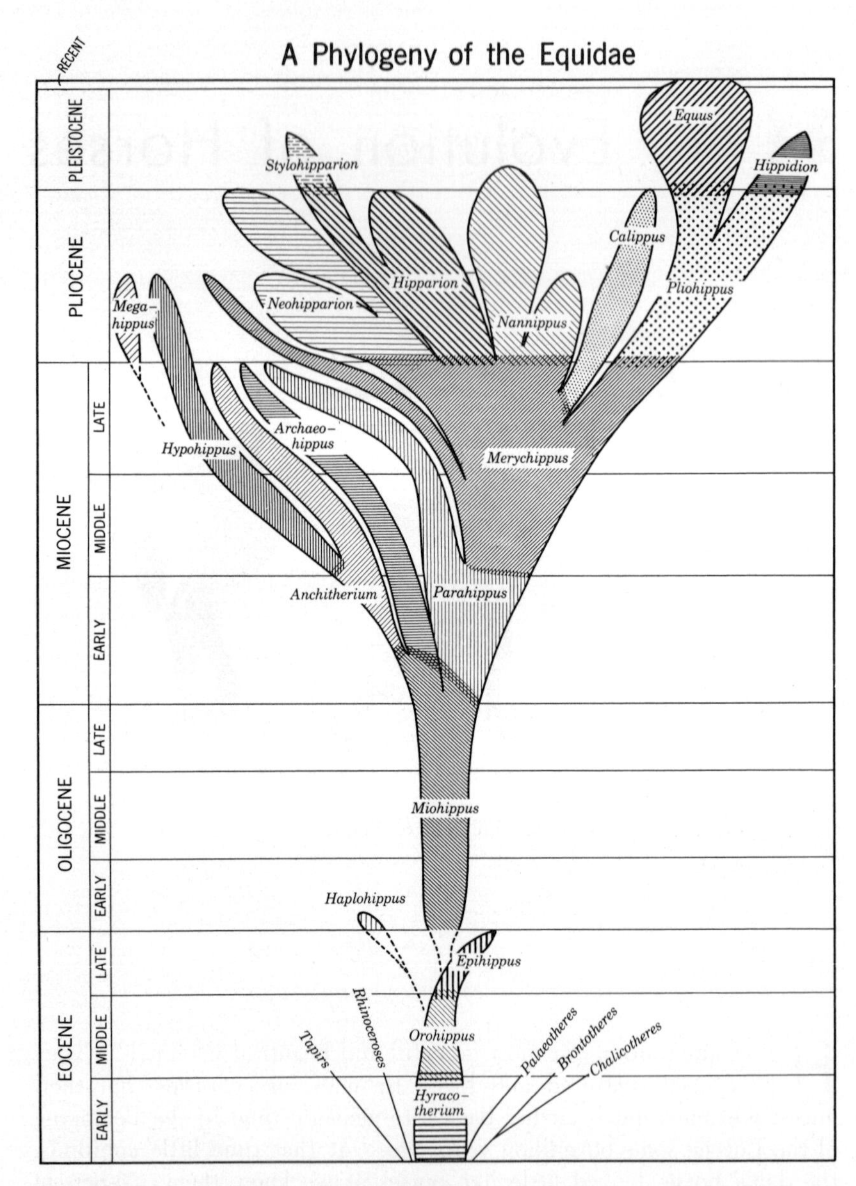

Fig. 250. A revised phylogeny of the Equidae. A phylogeny is an author's interpretation of the evolution and relationships of a group of organisms through time.

therium (hyraxlike + beast) when he compared it with conies, pigs, and rodents. The ancestry of the earliest horses has not been traced back into the Paleocene, but there is considerable evidence to show that they came out of the Condylarthra, a primitive order of five-toed extinct ungulates.

Horses belong to the mammalian order Perissodactyla or odd-toed ungulates, and eohippus is very close to a common ancestry with the other superfamilies of the order, including the gigantic horned brontotheres, the chalicotheres, with huge claws on their feet, the tapirs, and the rhinoceroses. Horses have given us the best evolutionary sequence known in the fossil records because it is so complete. As in other family phylogenies, there are many divergent as well as parallel lineages which finally died out, though each of these and the line of ascent leading to the living horses with their related species can be traced back through essentially uninterrupted sequences to an eohippus ancestry. The record is now so complete in the sequence of Cenozoic rocks and the intergradation between the species and genera so well blended that the lines drawn between them for purposes of classification must be arbitrary. The greater part of horse evolution occurs in North America where abundant remains have been found in stream-channel and flood-plain deposits. The Great Plains province is the greatest center of evolution and radiation of these interesting animals. The evidence shows that from time to time some of them spread into the Old World, only to die out and subsequently be replaced by more progressive kinds which also came from North America. The various genera and species are well established as index fossils for the Epochs and Ages of the Cenozoic. Even an isolated cheek-tooth offers enough evidence to determine the age of the formation in which it occurs.

Earliest Horses. Species of *Hyracotherium,* or eohippus, as they were called, were common in North America and western Europe in the early part of the Eocene. Evidently Europe and North America were connected at that time by a rather wide land bridge because there were many of the same genera of mammals found in both regions and even some species of eohippus were common to both continents.

These first horses were small animals, most of them about the size of a cat or a fox terrier, though some were twice that size. The arched flexible back, long tail, and four functional toes on the front feet and three behind showed almost no resemblance to the modern horse. It was the digit corresponding to our little finger that was vestigial in the front foot and the equivalents of our big and little toes that had been lost in the hind feet. Though there were little hoofs on the terminal phalanges, most of the weight was born on cushioned pads back of the hoof and below the fetlock (Fig. 251).

The face was not elongate and the eyes were situated near the middle of the head. The brain was that of an archaic mammal; actually it looked more like that of an opossum than of one of our horses. Some of the most diagnostic characters were found in the brachyodont, or low-crowned cheek teeth: molars and premolars. There were, of course, three typical molars in each of the upper and lower jaws, but the premolars still reflected their condylarthran affinities. None was molariform, though the last ones were tending in that direction. These little browsing animals, then, were ancestrial to all later horses.

Toward the beginning of middle Eocene time North America and Europe were again separated by a water barrier, and the land animals in both regions evolved into different groups. In Europe the eohippuses gave rise to a family called Palaeotheriidae that seemed quite like the first horses, but none was really a horse. It was in North America that the Eocene and Oligocene horses evolved.

All Eocene horses were rather conservative and seemed to evolve slowly. There was practically no difference in their sizes from the beginning to the end of the Epoch, and the feet with four toes in front and three behind changed only in minor details. In the middle

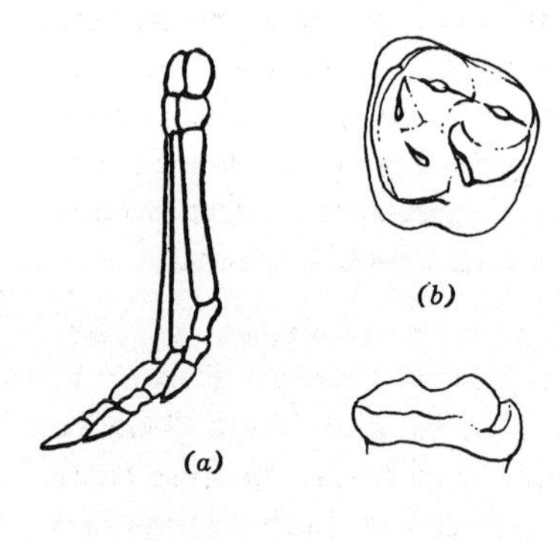

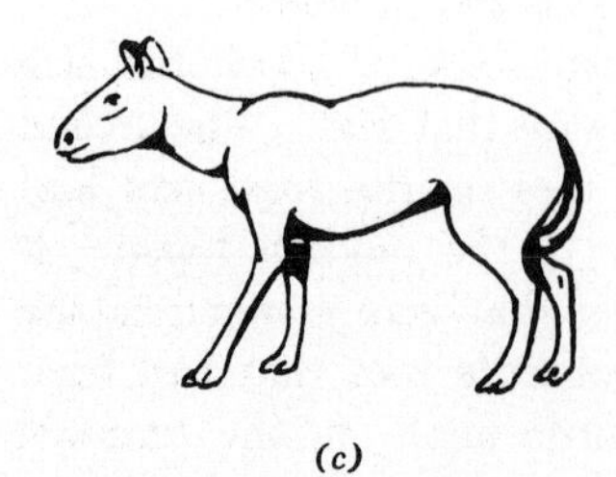

Fig. 251. Hyracotherium, early Eocene: (a) front foot, $\times \frac{1}{3}$; (b) crown surface and side views of an upper molar, $\times 2$; (c) restoration, $\times \frac{1}{15}$.

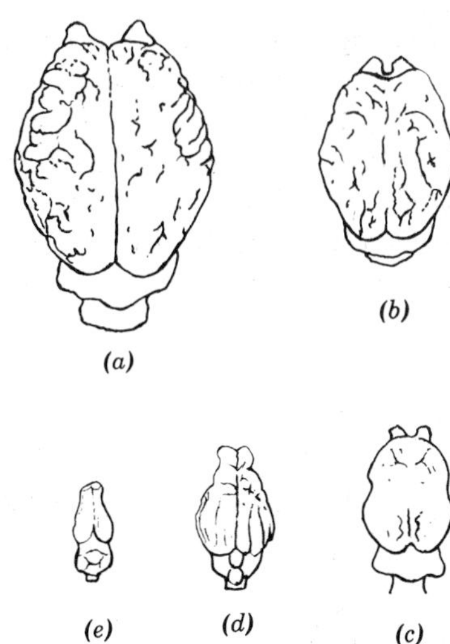

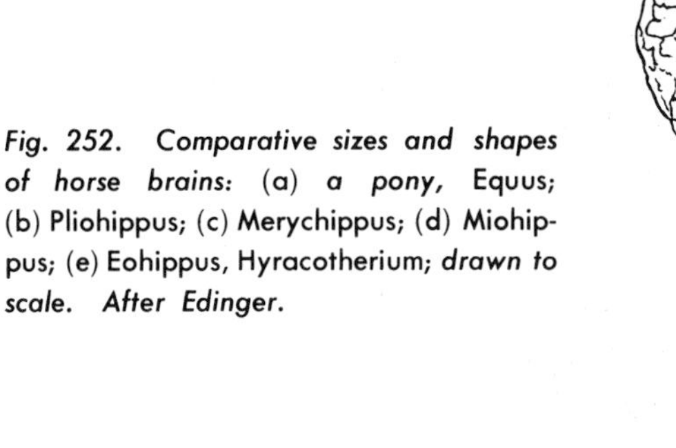

Fig. 252. Comparative sizes and shapes of horse brains: (a) *a pony,* Equus; (b) Pliohippus; (c) Merychippus; (d) Miohippus; (e) Eohippus, Hyracotherium; *drawn to scale. After Edinger.*

Eocene genus *Orohippus* the last, or fourth, premolars had become molariform. This was carried a step further in *Epihippus* from the late Eocene when both the third and fourth premolars came to look like molars (the student should recall that molars are not replacement teeth but the premolars are, since they develop in the jaws below the milk, or baby, teeth and eventually replace them in more mature mammals). There were other detailed changes in the teeth that showed evolutionary trends toward the more progressive Oligocene horses.

First Three-Toed Horses. The Oligocene horse, *Miohippus*, was about the size of a sheep, though some were smaller and others were larger. This little horse, with its slender body and slightly arched back, looked a little more like horses as we know them today. There were only three toes on each foot. The central metapodials, long bones in the fore and hind feet, had developed into cannon bones, and the cushioned pads were greatly reduced, since more of the weight was born on the terminal hoofed phalanges. This was the *first three-toed horse.* Being fleeter of foot, it was better equipped to escape the saber-toothed cats, primitive cats, and early dogs that had made their appearance at that time.

The head of *Miohippus* was more horselike, for the eyes lay a little farther back than in its Eocene predecessors. Its more advanced status was also reflected in the brain that showed expansion and convolutions in the cerebral hemispheres. This demonstrated that this little horse was well on its way to becoming an alert intelligent animal. In addition, another premolar had become molariform. This gave it six grinding and crushing brachyodont cheek teeth in each jaw. No additional premolars came to look like molars in any of the later horses, though the teeth changed in other ways. The slightly elevated crests or lophs, already apparent in the teeth of *Epihippus*, were perfected to a greater degree. Some individuals showed an incipient tendency to connect the oblique transverse lophs in the upper cheek teeth, a feature that became extremely important in later horses (Fig. 253).

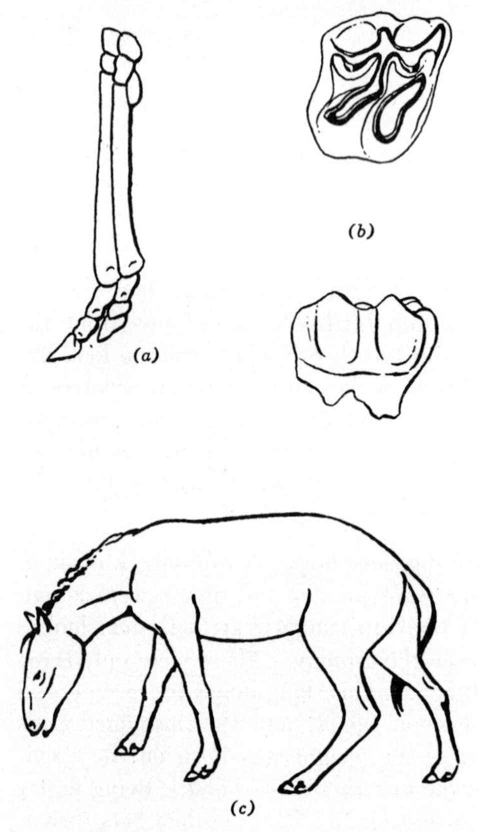

Fig. 253. Miohippus, early Oligocene to early Miocene: (a) front foot, $\times\ ^1/_3$*; (b) crown surface and side views of an upper molar,* $\times$ *2; (c) restoration,* $\times\ ^1/_{15}$.

Though the species of *Miohippus* are abundant in North America, none has been found on other continents. Skeletons, skulls, and teeth are common in the White River badlands of South Dakota and in beds of the same ages or slightly later in Colorado, Wyoming, Nebraska, and Oregon. The most advanced species range into the early Miocene.

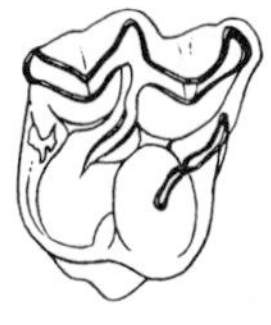
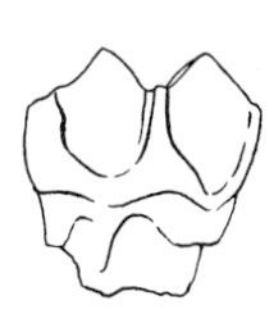

Fig. 254. Anchitherium, *Miocene, crown surface and side views of an upper premolar;* × 1.

Other Horses With Low-Crowned Teeth. *Miohippus* continued into the early Miocene where different species gave rise to other genera, two of which retained low-crowned teeth. One of these, *Anchitherium*, dispersed into Eurasia where it was never abundant and died out at the end of Miocene. In North America *Hypohippus*, a larger and heavier group of horses, emerged from an *Anchitherium* ancestry at the beginning of the middle Miocene. The teeth in this so-called forest horse were wider and much larger than in its predecessors. In addition, the crests or lophs were much sharper, but the teeth were still low-crowned, with only the roots extending into the bone of the jaws. This horse walked more on the flat of its feet than those in the line leading toward modern horses. The side toes were relatively heavy and still carried some of the weight. *Hypohippus* even reached Eurasia where it survived for a short time in the Pliocene as it did in North America.

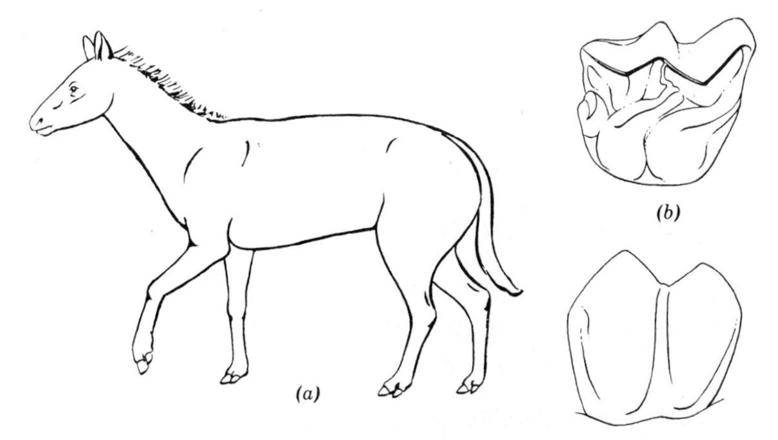

Fig. 254. Hypohippus, *middle Miocene to early Pliocene:* (a) *restoration,* × $^{1}/_{30}$; (b) *crown surface and side views of an upper molar,* × 1.

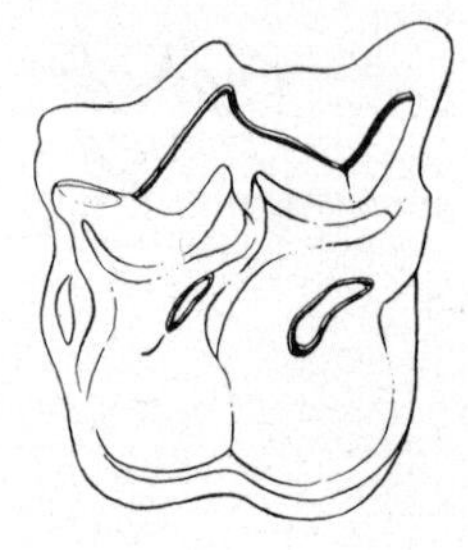

Fig. 256. **Megahippus, *early Pliocene, crown surface and side views of an upper premolar;* × 1.**

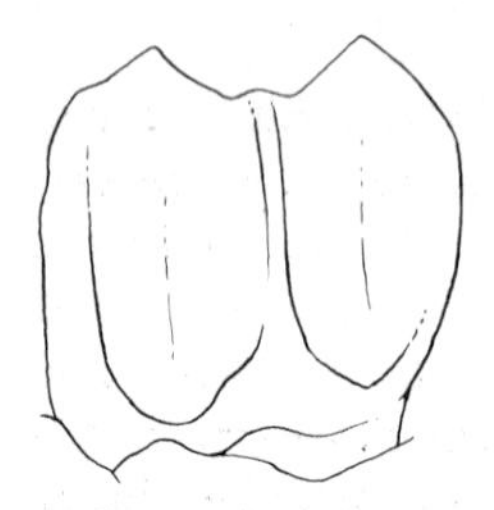

An even larger and more specialized genus, *Megahippus*, with peculiar enlarged cropping teeth, branched out of the *Hypohippus* line in the late Miocene. No other horse was like it, yet very little has been learned about these unusual beasts (Fig. 256).

Archaeohippus represented another group of small, brachyodont, slender-limbed horses that evolved from *Miohippus* in the early Miocene. It was relatively rare compared to the numbers in most other genera and was confined to the North American Miocene. The teeth were even smaller than in most species of *Miohippus*, but as in *Hypohippus* the crests across the cheek teeth were sharper. Evidently it was a fast runner.

Expansion of Grasslands and the Origin of Grazing Horses. A third genus that arose from *Miohippus* in the early Miocene was *Parahippus*. At that time uplifts in the Rocky Mountains resulted in the deposition of a blanket of fine sands and silts in stream-channel and flood-plain deposits over the Great Plains province to the east. In these sedimentary rocks Maxim K. Elias found the first fossil husks of spear grass seeds. This expansion of grasslands between the wooded stream courses was soon reflected in the structures and specialization of the mammals.

The species of *Parahippus* showed a tendency to strengthen their teeth with cement, where it filled the depressions and inflections. In addition more complicated enamel patterns were developed in the cheek teeth. This evidently was a natural selection response in their occupation of open grasslands and was an incipient stage toward hypsodonty or higher crowns in the molars and premolars. These animals, with their slender, well-developed cannon bones and reduced splint bones, bearing the side toes that barely touched the ground, looked much more like later horses. Only a vestige of the cushioned pad remained at the back of the foot (Fig. 257).

The head was lengthened, and the eyes were farther back from the nose than *Miohippus*. In the skull a narrow bridge of bone from the side of the brain case connected with another extending up from the zygomatic arch at the side of the skull to form an osseous orbital ring for the eye. Insofar as we know, an orbital ring was not found in any of the brachyodont horses. In the species of *Parahippus* selection was favoring those with better brains, straighter backs, feet and legs better adapted for running in the open, teeth with their crowns re-enforced by additional connections of the lophs, and those which were lengthening the crowns of their grinding teeth in the jaws. These responses in natural selection to a change in the environment, as indicated in *Parahippus*, were some of the most important in the evolution of the horse

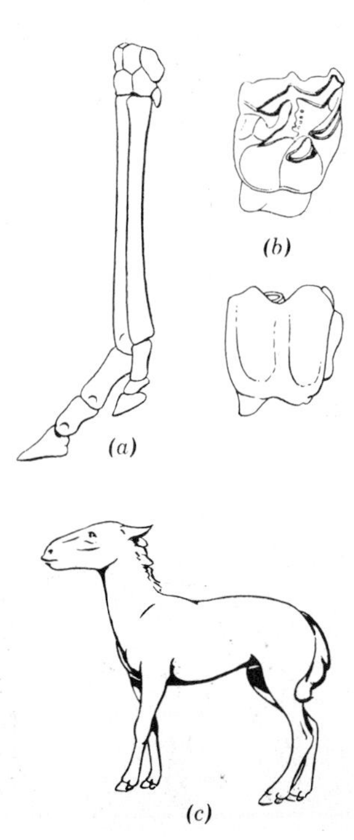

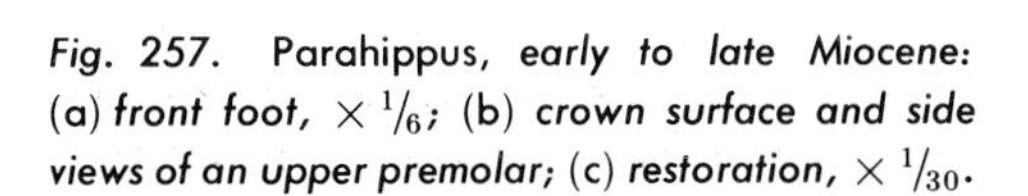

Fig. 257. Parahippus, early to late Miocene: (a) front foot, $\times\ ^1/_6$; (b) crown surface and side views of an upper premolar; (c) restoration, $\times\ ^1/_{30}$.

and were the first stages in the progressive evolution of hypsodont horses which were destined to dominate the horse world throughout the remainder of geologic time.

As the grasslands spread more widely over the North American continent, species of *Parahippus* gradually gave way to a more progressive group of species which had descended from them. The more advanced horses were called *Merychippus.* When they first appeared near the beginning of the middle Miocene they looked very much like *Parahippus,* some of which were still living, but before the Epoch closed most of the species of *Merychippus* had evolved rapidly and far beyond their earlier Miocene ancestors, though a few rare conservative species remained essentially unchanged.

Merychippus species were the first true hypsodont horses. In brachyodont browsing horses the entire crown of the tooth was exposed, with only the roots extending into the alveolar sockets in the jaws; but in hypsodont grazing horses much of the lengthened crown was embedded in jawbones to afford them a stronger anchorage. As the animals grew older and the crowns of their teeth wore down, the ends of the nerves

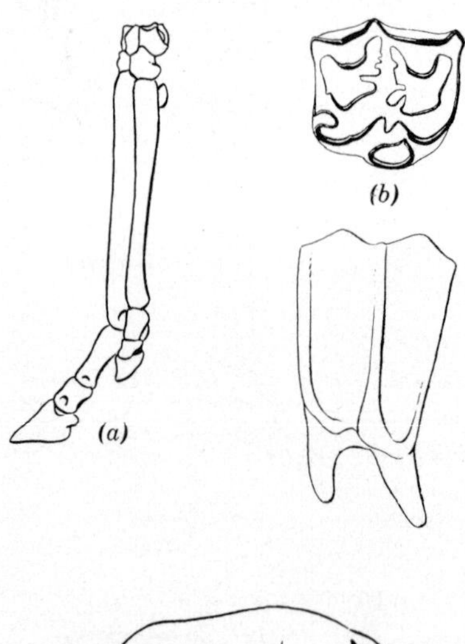

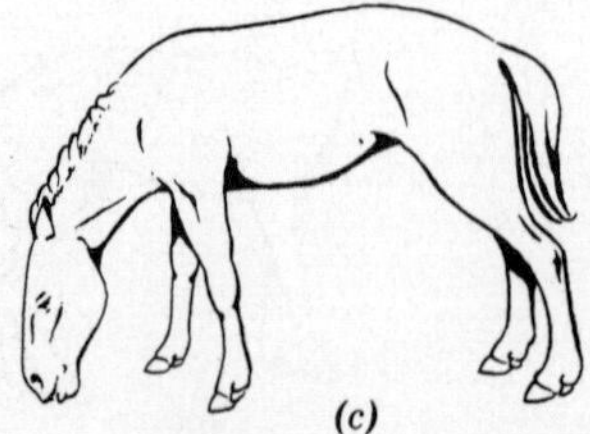

Fig. 258. Merychippus, middle Miocene to early Pliocene: (a) front foot, $\times \frac{1}{6}$; (b) crown surface and side views of an upper premolar, $\times 1$; (c) restoration, $\times \frac{1}{30}$.

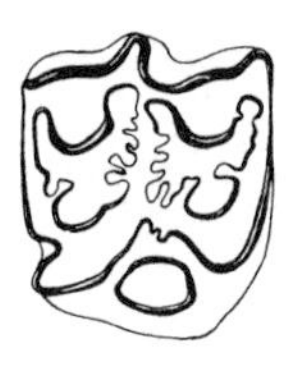

Fig. 259. Hipparion, early Pliocene to Pleistocene, the crown surface of a well worn molar, × 1. Compare the change in pattern, due primarily to wear, with that in Fig. 173.

retreated toward the base of the teeth while supplementary dentine was laid down. Thus horses with teeth like these were well equipped to compete with other herbivorous animals in an environment of siliceous grasses in which grains of sand lodging between the stems and blades were effective abrasive agents (Fig. 258).

The herds of *Merychippus* far outnumbered any of the other kinds of horses of their time. *Merychippus* carried the specializations initiated in *Parahippus* much further, and by the end of the Miocene they had become differentiated into several groups, foreshadowing in their skulls, teeth, and limbs the five Pliocene genera that were to follow.

Great Radiation of Grazing Horses. Wide stretches of grasslands in plains and steppe environments were well established in North America and Eurasia. There was less rainfall in the broad temperate regions than in the previous Periods of the Cenozoic. Evolution of the horses had kept pace with these changes, but there was keen competition among their different genera and species.

The first of the Pliocene genera to be given a generic name was *Hipparion* (Fig. 173) in 1832. This was based on teeth with extremely complicated enamel patterns found near Eppelsheim, Germany. Since then, remains of these three-toed, narrow-nosed horses have been found abundantly in Pliocene rocks all through middle and southern Europe, northern Africa, the Middle East, India, and China, where they were the dominant grazing mammals. They inherited from their *Merychippus* ancestors an isolated protocone or pillarlike structure on the inner side of their upper cheek teeth. *Hipparion* represented the second great invasion of horses into Eurasia from North America. The oldest and most primitive species were in North America, but they did not persist in the New World where other hipparion-like genera prevailed. The Old World species evolved into several groups, under certain conditions displaying marked convergences with species of similar genera in America. A genus called *Stylohipparion* survived as late as the Pleistocene in Africa.

Related genera of the North America Pliocene, with isolated protoconal pillars in their grinding teeth were slender-limbed, three-toed species of the genera *Nannippus* and *Neohipparion*. *Nannippus* evolved

Fig. 260. Nannippus, *early Pliocene to early Pleistocene, crown surface and side view of an upper premolar;* × 1.

from horses of medium size in the early Pliocene to much smaller ones in the early Pleistocene. This was not a reversal in evolution because an increase in size is not always necessary for survival. Other modifications in these little horses were progressive and successful, since as a race they outlived the larger neohipparions that coexisted with them for something like 11 million years in the same environments. Though the cheek teeth in *Nannippus* decreased in the circumference of their worn surface, they increased in crown height. These slender-limbed little horses were almost antelopelike in proportions and must have equaled the antelopes in alertness and speed (Fig. 260).

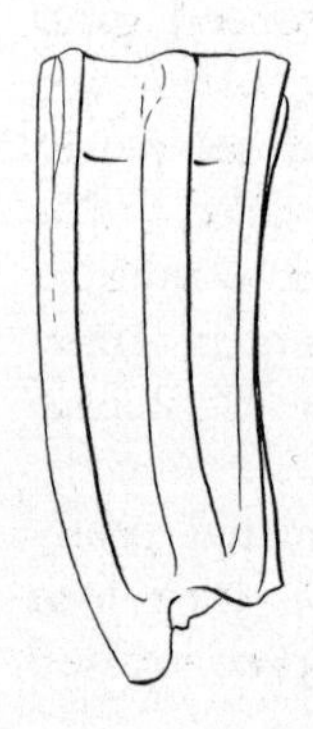

Fig. 261. Neohipparion, *Pliocene, crown surface of an upper premolar;* × 1/6.

The species of *Neohipparion* were larger and tended to evolve flatter and more elongate protocones in their teeth than in the other hipparion-like horses. They were as abundant as *Nannippus* and equally adapted to life on the open plains, but our fossil records show that they became extinct at the end of the Pliocene (Fig. 261).

Another genus that evolved from larger to smaller species was *Calippus*. It was present in the late Miocene and continued through the early Pliocene. No skulls or complete skeletons have been described, but the teeth and feet of these fascinating tiny horses are well known. They had delicate limbs with three toes. They are typified by the simplicity in the enamel patterns on their cheek teeth and the connection of the inner protocone pillar to the front part of the tooth. To pony fanciers they would have been much cuter than eohippus because here was a graceful little horse of gazelle proportions (Fig. 262).

Fig. 262. Calippus, *late Miocene to Pliocene, crown surface of an upper premolar;* $\times \frac{1}{6}$.

Ancestral One-Toed Horses. In the evolution of the horse we have recognized many divergent lineages that tended to reduce in numbers, then die out, but one line continued, eventually to culminate in the horse, onager, ass, and zebra that still survive. This line started with *Hyracotherium* and continued through *Orohippus*, *Epihippus*, *Miohippus*, *Parahippus*, and the *Merychippus* species in which the protocone remained attached to the anterior loph of the tooth. At the beginning of the Pliocene an advanced species of *Merychippus* gave rise to the first species of *Pliohippus* in which the side toes were missing. Only the two slender splint bones remained at the back on each side of the cannon bones. These were the *first one-toed horses.* That these horses were readily adaptable to the changing environments in the Pliocene is clearly indicated in the numerous remains found as fossils. Most of them increased in size, though some of the species were little if any larger than the most advanced species of *Merychippus*. The head and body was quite horselike, though some skulls possessed rather deep pockets in the surface bone in front of the eyes; in others these depressions were shallow or absent. The brain differed only in minor details from those of modern horses. One of the most distinctive features was a strong curvature in the upper molariform teeth. The enamel patterns were simple and the protocones, except in a few exceptional samples, were connected to the main body of the tooth. As in other hypsodont Pliocene horses, the distal ends of the cannon bones were so keeled that lateral movement in the foot was negligible (Fig. 263).

One of the more primitive groups of *Pliohippus* evidently ranged into

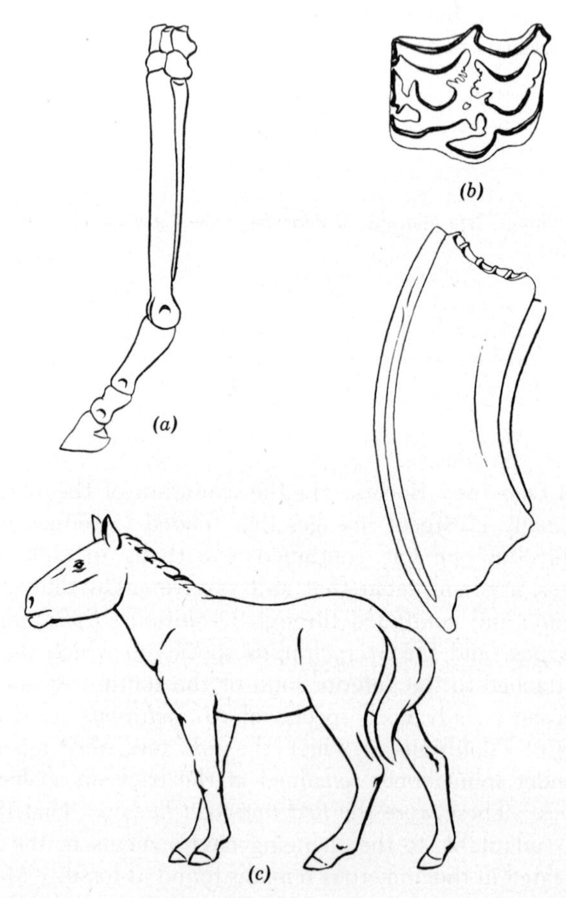

Fig. 263. Pliohippus, Pliocene: (a) front foot; (b) crown surface and side views of an upper premolar, $\times 1$*; (c) restoration,* $\times \frac{1}{30}$.

Central America, where they continued to evolve while retaining many of their primitive features. Then, near the end of the Pliocene, when North America and South America were again connected by land after nearly 74 million years of separation by water barriers, this Central American stock dispersed into the southern continent where they are recognized in the Pleistocene as *Hippidion* and related genera. The hippidions were rather stocky horses with peculiar long nasal bones narrowly connected to the frontal bones posteriorly. The teeth, though larger, were still much like those in *Pliohippus.*

Origin and Dispersal of Equus. In North America other species of *Pliohippus* came to look more like *Equus,* the genus that contains the zebras, asses, onagers, and true horses. The first species of *Equus* in the Pleistocene possessed many zebralike characters in their skulls and teeth. These animals not only spread over North America, but some of them reached Eurasia, as the third great dispersal to that large land mass, where they superseded the hipparions. Certain species moved across the Panamanian land connection into South America either at the same time or shortly after the hippidions. Though the zebralike horses were abundant in North America during the First Advance and Retreat of the Great Ice Sheets, they died out or became greatly reduced in numbers before or during the next glacial advance (Fig. 264).

In the meantime, the zebralike stocks that had spread into the Old World were quite successful. Over the large land masses of Eurasia and Africa they continued to evolve. Typical zebras, asses, onagers, and true horses appeared, and by the time of the Second Advance of the Ice Sheets some of the true horses (caballine horses) reversed the long-established trend by spreading back into North America with the mammoths and some other mammals that had evolved in the Old World. By the time their populations had dispersed as far south as Panama a heavy rain forest blocked their entrance into South America. Though these prehistoric true horses were here when the early Indians arrived, they did not survive but died out seemingly with startling suddenness. The South American zebralike horses and hippidions met the same fate, apparently about the same time.

Zebras, Asses, Onagers, and Wild Horses. In some respects the most spectacular of the living Equidae are the zebras. Their striped coats readily differentiate them from the other equids. Several species have been named but not fully confirmed. The three outstanding ones are the mountain zebra, *Equus zebra,* Grevy's zebra, *Equus grevyi,* and the quagga zebras, including Burchell's and Grant's zebras, *Equus quagga.* All wild zebras are now confined to eastern and southern Africa.

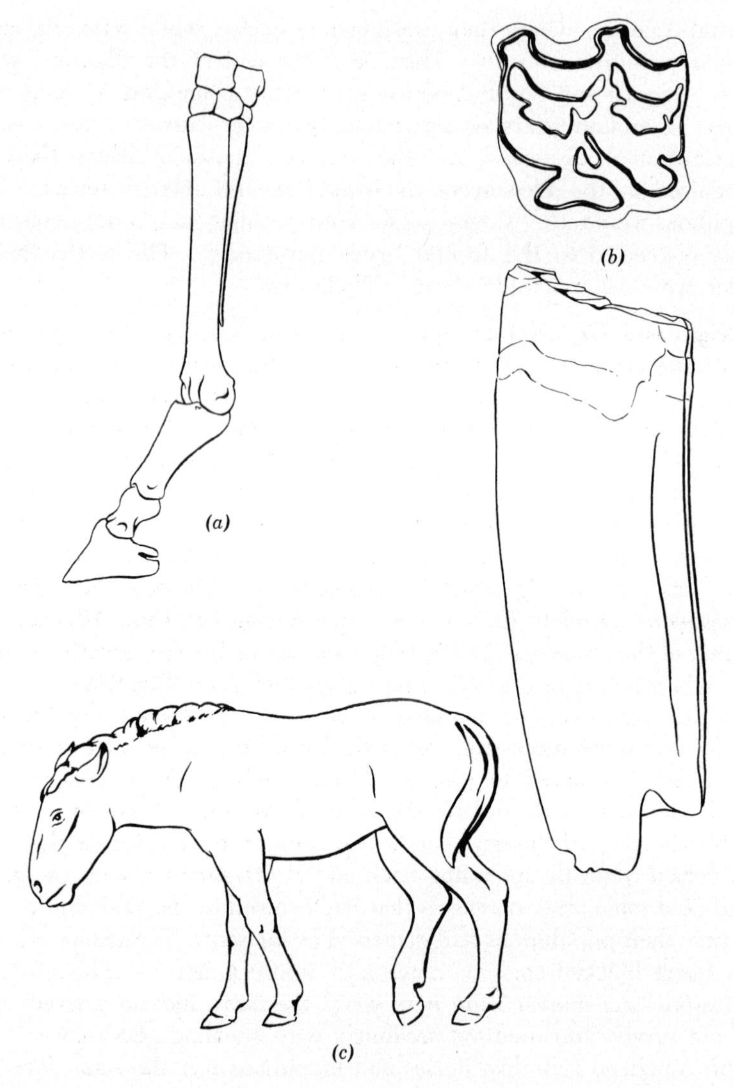

Fig. 264. Equus, *early Pleistocene to Recent:* (a) *front foot;* (b) *crown surface and side views of upper premolar;* (c) *restoration,* $\times \frac{1}{30}$.

The few wild asses, *Equus asinus*, that are still living are also restricted to Africa. The species has previously had a much wider range throughout northeastern Africa, but man in his relentless conquest of everything about him has greatly reduced the wild stocks, either by killing them or by permitting his domestic donkeys to escape as feral animals and mix with the wild populations. The color of the wild animals is ashy gray, and there are dark stripes across the shoulders and down the back.

Onagers, *Equus hemionus,* occur in western Mongolia and adjacent areas. Their former range includes most of steppes and deserts of central Asia and as far to the Southwest as Palestine. They are known by a half dozen or more common names. Onagers are smart-looking animals with ears a little longer than those of the horse and a tail like that on a mule. They have already been stamped as one of the vanishing mammals of the Old World.

We know of two races of true wild horses, though there were possibly more even as late as early historic time. Both of the known races are referred to the species *Equus caballus* Linnaeus. The European wild horses ranged over most of central and southern Europe but are now so mixed with modern horses that they can be said to no longer exist. The other is the Asiatic race known as Przewalski's horse or the Mongolian tarpan. Francis Harper states that it is doubtful that these interesting animals now live in a truly wild state anywhere, though some can be seen in zoos and parks in Europe and elsewhere. They are short, yellowish bay horses with dark erect manes and tails of the same color. Their very light muzzles and lack of forelocks are among their most distinctive features.

Domestication of Horses. Prehistoric man killed horses for meat but evidently did not utilize them as domestic animals. Insofar as we know, the domestication of horses started a little before 2000 B.C. Where this took place and how it came about has not been determined. Once started, it was practiced widely over Europe, Asia, and northern Africa. The slender fleet Barbs and Arabian horses were developed in Arabia, Egypt, and farther west in northern Africa. From central and perhaps northern Europe came stockier horses, with heavy hairy coats and long manes, which were crossed with other breeds as man attempted to improve his horses for different purposes.

After horses had become extinct in North and South America at the end of the Pleistocene, they were again introduced by the Spanish Conquistadors as early as 1519. According to Angel Cabrera, these were Andalusian horses from Spain, frequently called jennets or *jinetas.* Eventually some of them escaped and ran wild as feral horses. Part of the heritage of the plains mustangs came from these horses. In Argentina they are known as *criollos.* The different breeds of domestic horses have been the result of selected breeding over many centuries. Prior to the machine age the horse was mans' most treasured and

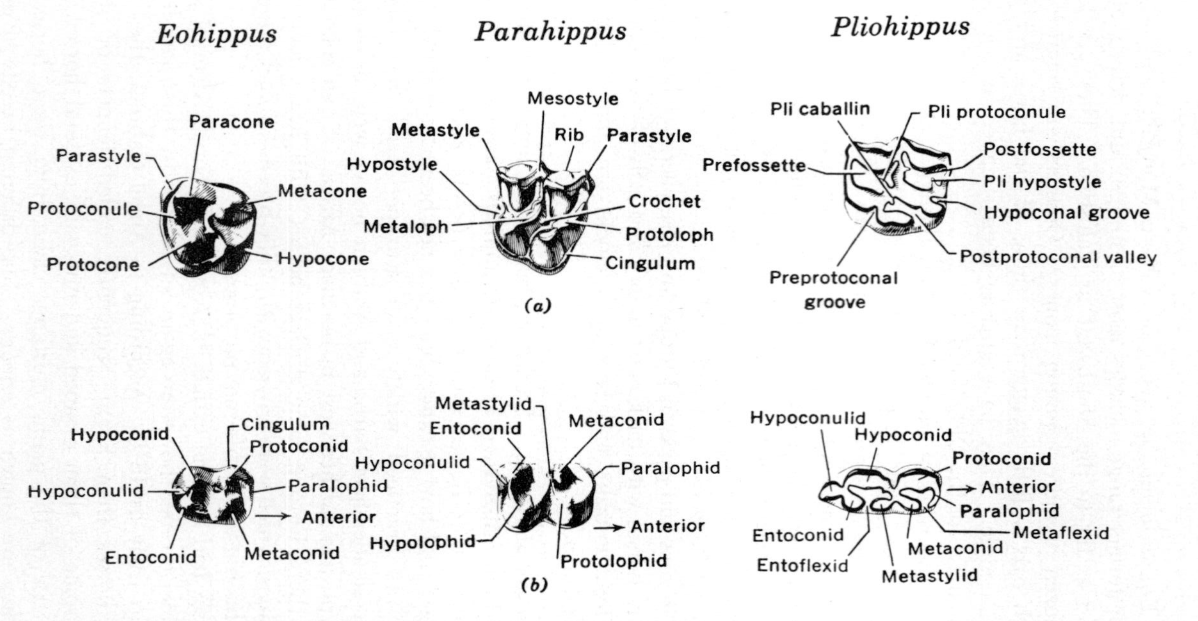

Fig. 265. Dental terminology of surface patterns in horse cheek teeth (molars and premolars): (a) *upper and* (b) *lower cheek teeth; not drawn to scale.*

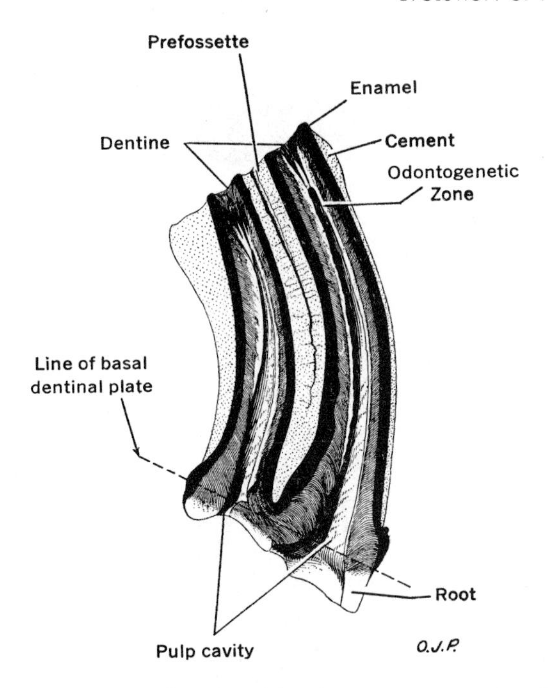

Fig. 266. A transverse vertical section through an upper premolar of a hypsodont horse tooth showing the depth of the prefossette, the pulp cavity where the nerves and blood vessels enter the tooth, and the arrangement of the enamel, dentine, and cement. In the odontogenetic zone the odontoblast cells form secondary dentine to fill the pulp cavity as the tooth wears down and the nerve retreats toward the base of the tooth. The line of the basal dentinal plate marks the lower extent of the enamel where the root becomes fused to the crown of the tooth. Only part of the root is shown.

useful domestic animal and will always serve useful purposes in his activities.

Dental Terminology. Cheek teeth in horses have become so useful in the correlation of Cenozoic rocks that a detailed terminology for them has become well-established in the literature. It has been included here as a source of reference. It is based on the Cope-Osborn theory of dental nomenclature in which it has been demonstrated that most mammalian molar teeth have been derived from a tritubercular (three-cusped) upper tooth and a tuberculo-sectorial (shearing-cusped) lower tooth (Fig. 265).

A tooth has a crown and roots. The crowns of molars, or of molariform premolars, are used in masticating food, and the roots anchor the teeth in the jaws. The crown may be brachyodont (low-crowned) or

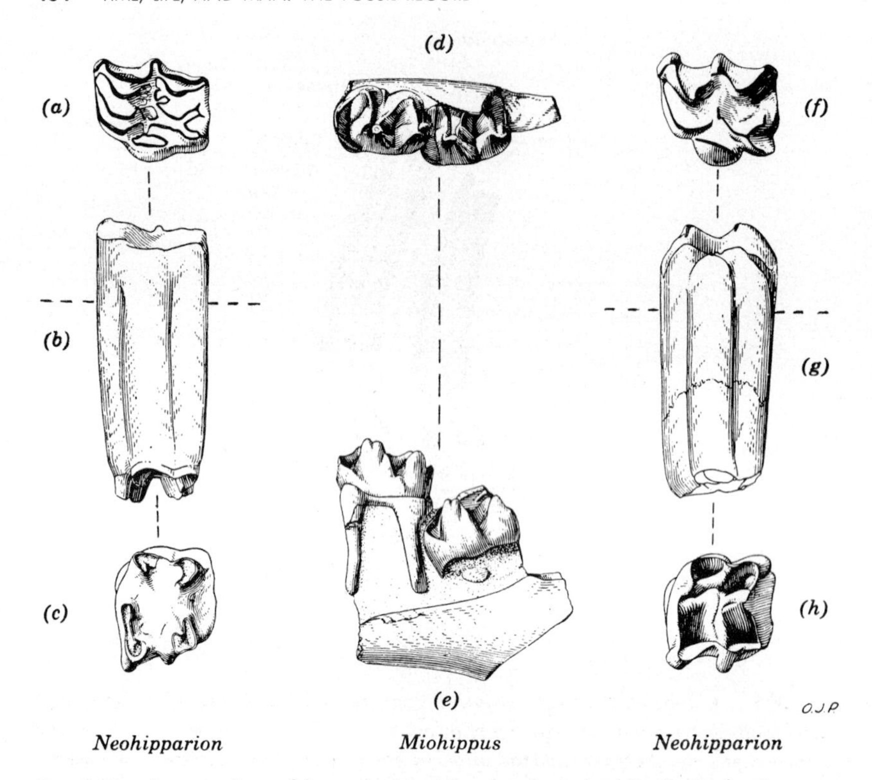

Fig. 267. A comparison of hypsodont and brachyodont cheek teeth: (a, b, c) *crown, side, and ventral views of a slightly worn* Neohipparion *hypsodont upper molar* (M^1) *in which the roots have fused to the basal dentinal plate;* (f, g, h) *the same views of an unworn upper molar* (M^2) *in the same animal showing that the roots are not fused in that animal when the tooth is erupting;* (d, e) *crown and side views of the first and second lower molars* (M_1, M_2) *of the brachyodont* Miohippus *showing that roots are fused before the enamel has worn through to expose the dentine. Only the roots extend into the jawbone in brachyodont teeth, but, as indicated by the horizontal spaced lines, much of the crown is embedded in the jaw in hypsodont teeth.*

hypsodont (high-crowned). In brachyodont teeth only the roots occur in the bony sockets of the jaw; the crowns are fully exposed. On the other hand, in hypsodont teeth not only the roots but part or much of the crown may be embedded in the jaw bones. Teeth are composed of ivorylike dentine covered with a coat of shiny hard enamel. Most hypsodont teeth and some brachyodont teeth are coated with a tartar substance called cement. As the teeth wear through the enamel cusps (high points) and lophs (crests), exposing the dentine, the more resistant enamel in contrast to the softer dentine and cement stands out in relief to form an excellent grinding surface (Figs. 266, 267).

The primitive pattern and basic plan from which all later horse teeth were formed is that of the early Eocene eohippus. This pattern was retained in all of the horses with brachyodont teeth. Even in these earliest horses most of the cusps, so prominent in the unworn teeth of later horses, are quite distinct. In the upper teeth there is one external crest (ectoloph) and two oblique inner crests (protoloph and metaloph). The lower teeth have four prominent cusps with incipient crescentic lophids connecting them. Names for cuspids and lophids in the lower teeth terminate in the letters *id* to differentiate them from those in the upper teeth (Fig. 265).

Transformation from a brachyodont to a hypsodont pattern, already referred to in the genus *Parahippus*, is not so complicated as it appears. In the upper teeth the protocone and hypocone mark the linguad (toward the tongue) terminations of the protoloph and metaloph, but near the middle of the lophs are the small cusps called protoconule and metaconule. On the posterior (back) border of the metaloph is a cusp named hypostyle that appeared for the first time in *Epihippus* and *Miohippus*. Extensions as crochets and plications from the metaconule to the protoconule, from the hypostyle to the metaconule, and from the hypostyle to the end of the ectoloph formed two centrally located lakes, or fossettes, termed prefossette and postfossette. This evolution initiated in *Parahippus* was completed in most species of *Merychippus*. In the lower teeth the main structural changes in these genera is in the anterior extension of the entoconid, together with the development and expansion anteroposteriorly of the metaconid and metastylid cusps. In this manner, and with the filling of the fossettes and inflections with cement to strengthen them, the molars and premolars become hypsodont without weakening their structure.

32

Lemurs, Tarsiers, Monkeys, and Apes

New World monkey

Tarsier

The fascinating story of the lower primates is frequently overlooked by beginning students in their anxiety to learn more about the immediate ancestry of man. Yet it is through these groups that man's early lineage evolved. It is a long history involving something like 70 million years after they become faintly distinguishable from their Cretaceous insectivore ancestors, the first truly placental mammals known.

The widespread, rather uniform warm temperate-to-tropical environ-

Fig. 268. Living species of the genus Tupaia *are distributed in India, China, the Philippines, and the countries of southeastern Asia; about the size of a red squirrel.*

ments of the Paleocene and Eocene in the Northern Hemisphere were well suited to the adaptive radiation of these small mammals. The angiosperms, with their flowers and fruits, offered an abundant food supply to these little creatures so well equipped to take advantage of it. It probably was a simple matter in the process of evolution to change partly or totally from an insect diet to the consumption of tender flowers and ripe fruits. Mutations toward better sight, together with an increase in the size of the brain, were perhaps the most important modifications for natural selection in these earliest primates. With these progressive changes, the front feet became more dexterous in functioning as hands. Primitive insectivorelike primates, with these adaptions and somewhat resembling the living tree "shrews," *Tupaia*, evidently were the first representatives of the order Primates (Fig. 268).

Order: Primates.
Infraorders: Lemuriformes; tupaiids and lemurs.
Lorisiformes; lorises and bush babies.
Tarsiiformes; tarsiers.
Anthropoidea; New World monkeys, Old World monkeys, apes, and man.

Tupaiids and Lemurs. The most primitive of the living lemurids are the tree "shrews." The name shrew in reference to these mammals is misleading, since they are not closely related to the true shrews. Unfortunately, the fossil record of the tupaiids is essentially unknown. One cranium of an old individual with heavily worn teeth, from the Oligocene of Mongolia displays many features similar to those in the insectivores, and other characters are suggestive of the true lemurs. There are several genera and species of the living *Tupaia* distributed in India, China, the Philippines, and in the countries of southeastern Asia, where they feed on insects and fruit. These little squirrellike animals with long narrow faces have probably changed relatively little since the early Cenozoic. In many anatomical characters they are like the true lemurid primates and for this reason have been placed in the order Primates. Nevertheless, some authorities insist that they should be included in the Insectivora, as they were in earlier classifications. Irrespective of the answer to that taxonomic problem, the evidence from tupaiids strongly supports the interpretation that primates arose from more primitive early Cenozoic or possibly late Cretaceous placental mammals.

Lemurs (Lat., ghosts) are well represented in the fossil record. It is fortunate that so many occur in the Paleocene and Eocene rocks of both North America and Europe. Later, they are missing from the record until we get to the Pleistocene and living forms that were isolated on the island of Madagascar off the east coast of Africa. The oldest known lemurid jaw fragment is from the middle Paleocene of North America. This specimen as well as those of members of other families still reflect insectivore relationships in their teeth and jaws. In the late Paleocene and throughout the Eocene these older lemurid families are specialized, especially in their teeth. Some Eocene species are as large as the American howler monkeys. The common lemurid genus in Europe is *Adapis*, and in North America the species of *Notharctus* are most frequently encountered. A fine skeleton of these little arboreal creatures was found several years ago in the Bridger Basin of Wyoming and is now on display in the American Museum of Natural History in New York, but the largest and most complete collection is in the United States National Museum in Washington, D.C. The only living lemurs are on Madagascar. The evidence, such as it is, suggests that they must have reached that island sometime during the Paleocene or Eocene when their nearest relatives were common in the Northern Hemisphere. Both the living and the Pleistocene lemurs on the island have evolved into varied specializations. *Archaeolemur* parallels the monkeys so closely in several features that for a while some mammalogists thought the genus was referable to the monkeys. One of the most astonishing lemurs is the living aye-aye (pronounced i-i). It is about the size of a domestic cat and lives in bamboo forests where it feeds on grubs. With its chisellike

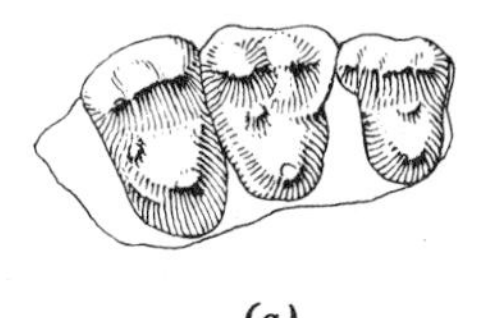

(a)

Fig. 269. Pronothodectes, *the oldest lemurid known. Three upper molars* (a) *and the fragmentary lower jaw* (b) *were found in the middle Paleocene in North America; × 4 natural size. After Simpson.*

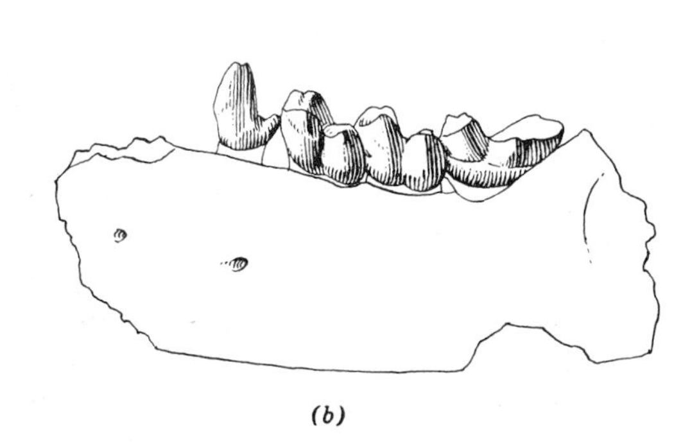

(b)

pairs of incisors it cuts into the woody stems, then extracts the choice morsel by means of its elongate flexible central finger. Some living lemurs are no larger than a mouse, but extinct kinds of the genus *Megaladapis* had skulls over a foot in length. These bulky creatures are said to have been living in Recent time, for they were described by natives to the French explorer De Flacourt when he visited the island in 1658. What a spectacular fauna that must have been, with the giant ground bird *Aepyornis* and these ponderous lemurs foraging about.

Lorises and Bush Babies. The lorises, sometimes called slow lorises because of their slow, deliberate movements through the trees at night, may be found living in Equatorial Africa, India, and the East Indies. Their large eyes, little evidence of external ears, furry coat, and stubby tail make them an object of curiosity wherever they may be seen. The fossil record is represented by one tooth from the Pliocene of India.

Bush babies, confined to Africa, are also nocturnal. They have cute little faces and naked ears. The ears double up and fold back when the animal is frightened. They are capable of making long leaps from branch to branch, where they cling with their flat padded toes. Their tails are long and bushy. Some excellent fossils have been found in the Miocene of Kenya in eastern Africa by field parties working under the direction of L. S. B. Leakey, director of the Coryndon Museum,

Nairobi. No other fossils of these interesting little creatures are known.

Tarsiers. In the forests of the Philippines and the East Indian archipelago are small nocturnal, arboreal primates called spectral tarsiers. The tarsier may be recognized by its big eyes, large erect ears, and long pencillike tail with its terminal brush of hairs. While foraging about at night these little owl-eyed creatures leap from branch to branch much in the manner of tree frogs. The fingers are equipped with disklike terminal pads by which the animal clings to the branches when it lands. Anatomically, these little fellows show so many similarities to the anthropoids that strong arguments have been presented to place one of their early Cenozoic forms in an ancestral position to the higher primates.

Tarsiers were even more abundant in the Paleocene and Eocene formations of North America and Europe than the lemurs. Not less than four genera were already in existence in the middle Paleocene, in which they were associated with other small primates of uncertain relationships. A more advanced and specialized form was found in the Oligocene of South Dakota. Tarsiers were plentiful in the Eocene. One of the best skulls found in the early Eocene of Wyoming was *Tetonius homunculus*. European genera, such as *Necrolemur* and *Microchoerus* (not lemur- and piglike animals, as earlier interpreted) showed characters in their skulls and teeth more like those expected in ancestors of the Old World monkeys. Part of the lower jaw of *Amphipithecus*, another primate, from the late Eocene of Burma, could also be a large tarsier.

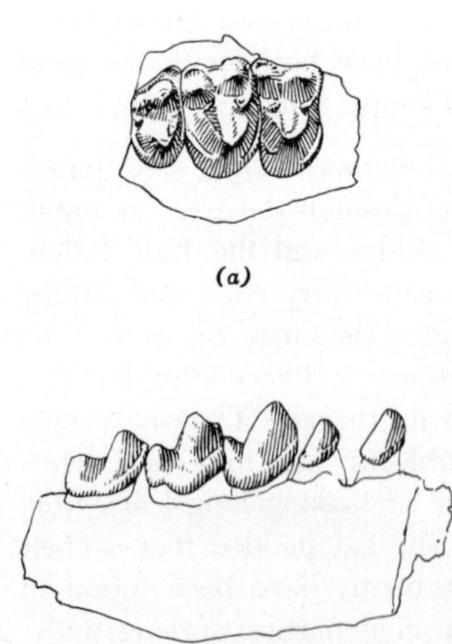

Fig. 270. Palaechthon, one of the oldest known genera of tarsiers. Three upper molars (a) and the lower jaw fragment (b) were found in the middle Paleocene in North America; × 4 natural size. After Simpson.

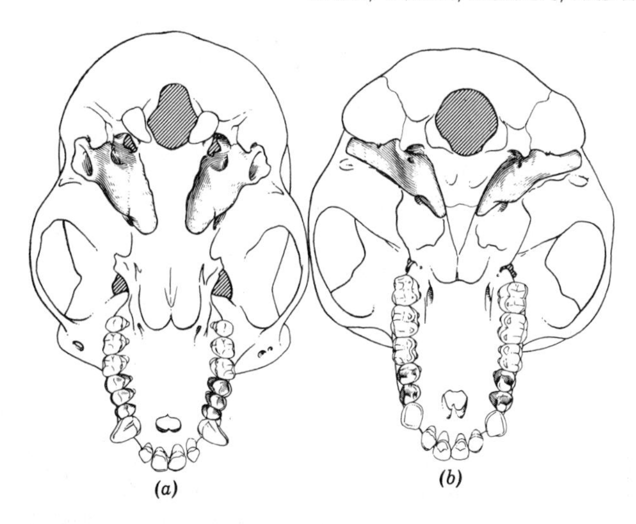

Fig. 271. Skulls of (a) *a New World monkey,* Ateles, *and* (b) *an Old World monkey,* Pygathrix, *showing difference in the numbers of premolars and in the construction of the ear bones, or auditory bullae; one half natural size.*

The Paleocene and Eocene record of lemurs and tarsiers in the Northern Hemisphere represents many families and numerous genera. It is interesting to note that there is no evidence of monkeys or apes at that time, though in the Oligocene and in later Epochs monkeys and apes were the dominate primates of Eurasia, Africa, and South America.

New World Monkeys. The relationships of the New World monkeys to those in the Old World is a perplexing problem. Superficially, they are much alike in their features and habits. Both groups, as well as the apes and man, have the orbits closed behind a bony partition extending from the zygomatic arch across to the side of the brain case. Yet there are other characters that indicate a long and distinct history for each group (Fig. 271).

NEW WORLD MONKEYS
Platyrrhines

1. Three premolars; each lower premolar with a single root.
2. Ear bone without long bony opening to the outside.
3. Nostrils, widely separated, tend to open directly to the front *(platyrrhine).*
4. Tail prehensile in many genera.

OLD WORLD MONKEYS
Catarrhines

1. Two premolars; each lower premolar with two roots.
2. Ear bone with long bony opening to the outside.
3. Nostrils, close together, tend to open downward *(catarrhine).*
4. Tail never prehensile.

Today the New World, or ceboid, monkeys range from tropical forests in Mexico deep into the forests of Brazil. Naturalists exploring in the American tropics retain permanent visions of *Ateles,* the agile, long-limbed, prehensile-tailed spider monkey, racing through the trees; the presence of the caupuchin, or white-faced, monkey, *Cebus,* feeding on wild figs near camp; or the occasional weird call of *Alouatta* the howler monkey. The tiny marmoset, *Callithrix,* and the larger squirrel monkey, *Saimiri,* are also frequently encountered. The only American genus with nocturnal habits is *Aotus* the goggle-eyed owl monkey.

The New World monkeys appear for the first time in the South American fossil record in the late Oligocene and early Miocene of Argentina. These are already typical ceboid monkeys, and they offer no clue to their ancestral relationships. It seems clear that they are not descended from the Old World monkeys but more likely that they arose from either lemurids or tarsiers of the early Cenozoic of North America. The early and perhaps rapid evolution of the New World monkeys may have taken place in a more favorable southern latitude between the present Isthmus of Panama and southern Mexico. There they may have been confined until chance dispersal carried them across the marine barrier between the two continents. Perhaps they were transported on a mangrove thicket, with its intertangled roots, sometimes dislodged by torrential streams and swept for miles out to the sea where it may have broken up on a distant shore. Islands such as Madagascar and those in the West Indies must have received their terrestrial and arboreal mammalian faunas in this way.

The first New World monkey remains, found in a Tertiary formation by the Argentine collector Carlos Ameghino, was part of a lower jaw named *Homunculus patagonicus.* After Florentino Ameghino described it in 1891, a distorted skull was discovered in the same badlands of Argentina. These specimens occurred with a large mammalian fauna that must have ranged through southern South America at that time. These Miocene monkeys attracted world-wide attention, partly because of the hominid interpretation that the author attributed to them. Actually they have been found to bear relationships to the howler monkeys.

Our information on New World monkeys was augmented in 1945 and 1946 when three specimens were discovered in badland exposures of the upper Magdalena Valley in Colombia. One of the lower jaws was referable to the genus *Homunculus* but belonged to a larger species than the one described by Ameghino. Another tiny pair of jaws named *Neosaimiri* was found to be related to the little squirrel monkeys. The third specimen was scattered over the surface where much of the skeleton had eroded out of the sandy claystone. This one was called *Cebu-*

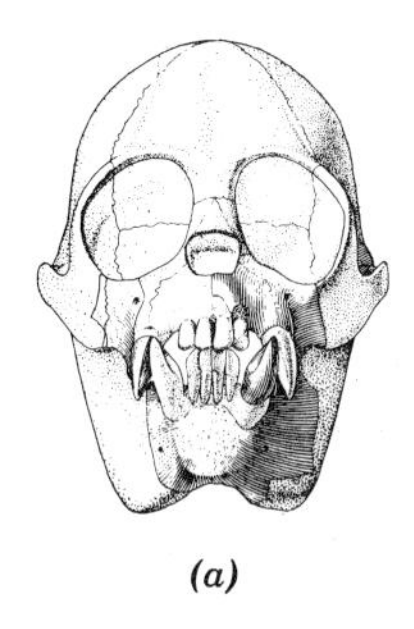

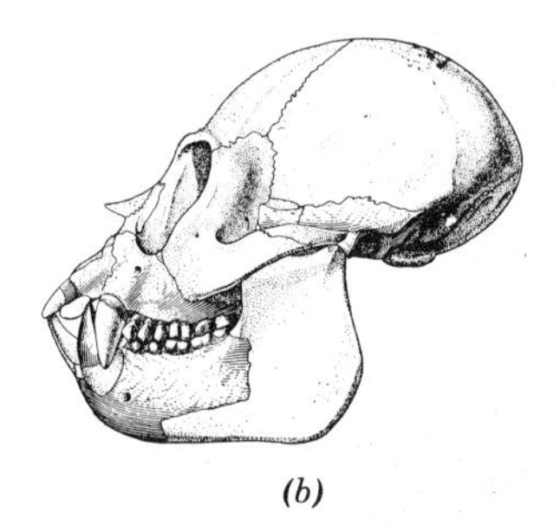

Fig. 272. A restored skull of a New World monkey, Cebupithecia sarmientoi, *from the Miocene of Colombia:* (a) *front view;* (b) *side view; one half natural size. After Stirton and Savage.*

pithecia, since it was related to the rather rare sakis and uakaris monkeys of South America. These late Miocene genera were all too advanced to throw any light on the origin of the monkeys of the Western Hemisphere.

More recently, Ernest A. Williams and Karl P. Koopman have described an extraordinary lower jaw of a New World monkey from a Kitchen midden on the island of Jamaica. They named the genus *Xenothrix* and found that it was quite distinct from all of the known genera. Its race had long since lost the last molar, as in the tiny marmosets, but otherwise looked nothing like those little creatures. How it reached Jamaica and where it came from remains a mystery.

Several fossils of monkeys and marmosets have been recorded from Pleistocene caves in Brazil, but they are so much like the living animals little of an ancestral nature can be determined about them.

Old World Monkeys. The Old World monkeys and apes seem to have descended as two quite distinct phylogenetic groups from late Eocene tarsier ancestors. The only Oligocene fossils are two lower jaws from the Fâyum beds of Egypt, where they were contemporaries of the earliest known mastodonts and other peculiar mammals. They were named *Apidium phiomensis* Osborn and *Moeripithecus markgrafi* Schlosser. Though they are advanced well beyond the Eocene tarsiers, they are not referable to Cercopithecidae, the Old World monkey family, nor do they belong with the apes. Evidently there was considerable diversification among the earliest monkeys and monkeylike creatures even in early Oligocene time. True catarrhines occur in the Pliocene and Pleistocene of Europe, Asia, and Africa. An excellent skull of *Mesopithecus* was found over a hundred years ago in a so-called *Hipparian* fauna at Pikirmi, Greece, and some 50 years later a skull of *Dolicho-*

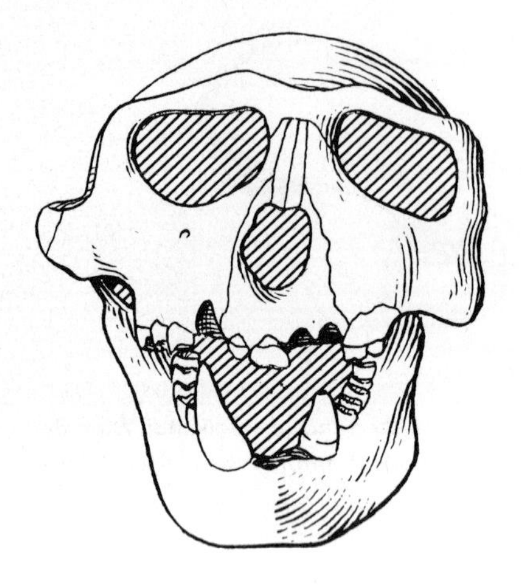

Fig. 273. The skull of a Pliocene Old World monkey, Mesopithecus pentelicus, *found near Pikirmi, Greece, in 1838 by a Bavarian soldier serving under the King of Greece;* $^2/_3$ *natural size.*

pithecus was discovered in rocks of slightly later age in southern France. Species such as the common African plains and bush monkeys and the macacs or rhesus monkeys occur in the Pliocene of India.

Another puzzling group of catarrhines are known primarily from the fossil remains of *Oreopithecus bamboli* Gervais from Pliocene rocks on Mount Bamboli, Italy. These small primates have features in the teeth resembling those in man. The canines are not enlarged, nor do their crowns extend higher than those in the premolars; the upper molars are squarish in outline and have four principle conical cusps; and the cusps are not connected by transverse crests as in the other Old World monkeys. On the other hand, there is an oblique crest on each of the molars that may reflect affinities with the earlier tarsiers, if the characters have not evolved independently.

Among the most interesting of the Old World monkeys are the baboons. Though they are common in Africa today, they appeared in India as early as the beginning of the Pliocene. In South Africa baboons move about in big familylike groups; some of them enter caves at night for shelter and protection, leaving the leader, a huge ferocious male, on guard near the entrance. Evidently they had these same habits during the Pleistocene, for their skulls and jaws are the fossils most frequently encountered in the many cave deposits in which the remains of australopithecine men and many other mammals occur. It has been suggested that these early men may have killed the baboons for food, since in some places clusters of the skulls are found in the cave breccias.

The Apes. No group of primates, other than man, has attracted so much interest as the apes. The well-known living forms include the gibbons, the orangutans, the chimpanzees, and the powerful gorillas. The fossil record shows clearly that there were many divergent lineages of ape and apelike creatures during the Tertiary. Ever since the appearance of Darwin's books on *Origin of Species* and *The Descent of Man* there has been rebellion among innumerable people at the thought of civilized man having descended from one of these anthropoids. Yet there is excellent evidence that man and the apes, as well as the other primates, constitute a group of ordinal magnitude in our scheme of classification of the animal kingdom and that the living apes and man were derived from a common ancestral form probably as far back as the Oligocene. There is no evidence whatsoever to conclude that the evolution in the higher primates did not follow the same basic principles as those so clearly displayed in the evolution of the horses or in other groups of organisms. There is a wealth of literature on the subject, supported by fossils, readily available to those who wish to take the time to look into the matter.

PARAPITHECUS. In the early Oligocene sedimentary rocks not far from the Egyptian pyramids were two pairs of primate lower jaws of even greater interest than the monkeylike *Apidium* and *Moeripithecus* with which they were associated. One of these specimens, called *Parapithecus fraasi* Schlosser, has been placed in a separate family, the Parapithecidae. This animal was smaller than most of the later monkeys but was unique in having several primitive dental features of the apes, a configuration of the mandible, and premolars more like those of the tarsiers. Indeed, it was not far from an intermediate position between the two groups. Yet it had the dental formula of two incisors, a canine, two premolars, and three molars so characteristic of the Old World monkeys, apes, and man. This specimen alone has offered evidence that the apes may have evolved directly from Eocene tarsiers without early catarrhine monkeys as intermediates in that chain of events.

GIBBONS. The other pair of lower jaws, also from the Fâyum beds of Egypt, was distinctly larger than that called *Parapithecus*. This specimen was named *Propliopithecus haeckeli* Schlosser. The teeth in the jaws bore a resemblance to those in the living gibbons, but they were much smaller and simpler in their patterns. Some authorities, however, believe that they were catarrhine monkeys. Tertiary gibbons are represented by *Limnopithecus* in Kenya, east central Africa, and by *Pliopithecus* from the Pliocene of Europe. Characters in the limb bones seem to show tendencies toward convergent characters with the baboons. Perhaps the limnopithecines spent more of their time on the ground than the living gibbons which are more arboreal in their habits than any of the existing apes.

Fig. 274. A white-faced gibbon. Hylobates lar, $^{1}/_{8}$ *natural size. From photograph, by permission of Philadelphia Zoological Garden.*

Gibbons, the smallest of the apes, inhabit the forests of southeastern Asia and the Malay archipelago. The largest animals are about 3 feet tall, and though they are almost exclusively arboreal when they descend to the ground they habitually walk upright. Though they live in warm tropical forests, they resemble infants dressed in furry garments for a trip out into the snow. Their small agile bodies with their long legs, even longer arms, and no visible tails attest to the prowess of these apes as arboreal acrobats. C. R. Carpenter has recorded gibbons swinging through the trees along their selected avenues of travel for perhaps a distance of 1000 yards at a pace equal to that of a fast-running man. Gibbons feed on buds, leaves, flowers, fruits, birds' eggs, young birds, insects, and a limited number of ants, and forage about in family groups emitting their calls, or sounds, for different occasions. The males, constantly on the alert to protect the family, flash into action at the first

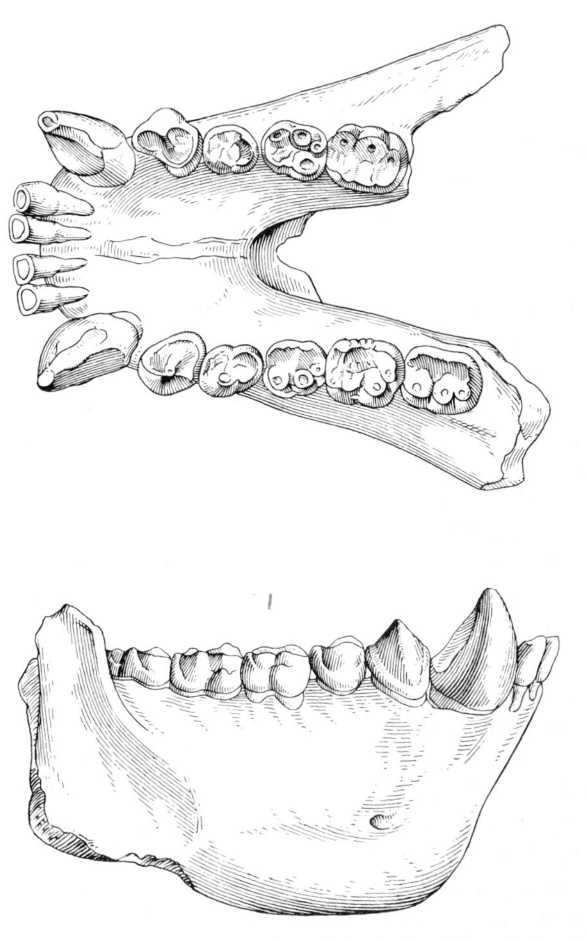

Fig. 275. Two views of mandible with the back part broken away: Dryopithecus fontani *from a Pliocene* Hipparion *fauna in France; ⅔ natural size. After Gregory and Hellman.*

distress call. The family group sleeps together in a tree near the center of their territory. They do not build nests (Fig. 274).

DRYOPITHECINE APES. The heritage of the larger apes, including the orangutans, chimpanzees and gorillas, appears to have stemmed from Miocene and Pliocene apes in central Africa, southern Asia, or Europe. Since the first fossil specimen was announced by the French paleontologist Édouard Lartet (1801–1871) in 1856, seven more genera and numerous species have added considerably to our knowledge of these most interesting apes. The genus most frequently encountered with the remains of *Hipparion* horses, giant giraffes, Old World antelopes, primitive hyenas, and many other mammals peculiar to Europe and southern Asia in Pliocene time was *Dryopithecus* (oak + ape). A character useful in recognizing the genus is the *Dryopithecus*, or y-patterns, of the lower molars. Critical studies on every detail of the teeth, jaws, and the two known limb bones have shown that some of these creatures were more like the orangutans, others showed characters in common with the chimpanzees or with gorillas, and usually there were combinations of characters of two or all three of the living apes. Furthermore, the attachment and insertion areas for muscles of the limb bones have indicated that *Dryopithecus* was not so well equipped to get about in the trees as the modern apes; but of even greater interest are the presence of many characters with a strong resemblance to those in early man. On the whole, though, they seem a bit closer to the large apes but have retained hominid characters that demonstrate a basic heritage not far removed (Fig. 275).

More recently the discoveries of L. S. B: Leakey, A. T. Hopwood, D. G. MacInnes, and others have given us a clear insight into the Miocene anthropoids from Kenya near Lake Victoria in east central Africa. The wealth of materials and the relationships indicated in the two genera and four species tend to support the conclusion that central Africa was the main center of evolution and radiation of the Anthropoidea in Miocene time. From there they could have spread to the north and northeast into Europe and southern Asia, where *Dryopithecus* and related genera were dominant in the Pliocene.

The most interesting of the African genera is *Proconsul.* A nearly complete cranium and mandibles of *Proconsul africanus* Hopwood, discovered on Rusinga Island, has given us the first skull of the subfamily Dryopithecinae. In addition to the numerous teeth and jaws, there are a few parts of the body skeleton. We can now say that these creatures were lighter in build than the living apes, and, as in the genus *Dryopithecus,* apparently they were not well adapted to life in the trees. On the contrary, they must have spent most of their time on the ground.

Fig. 276. Proconsul africanus, *based on restoration in the British Museum (Natural History).*

The careful investigations of Sir Wilfred E. Le Gross Clark and Dr. Leakey have shown that *Proconsul* lacked many of the specializations seen in the teeth and jaws of modern apes, and in these features they were much closer to early man. The rather squarish outline of the molars, the smaller canines, and the absence of a simian shelf at the back of the symphysis where the two lower jaws are fused together were some of the manlike characters observed. Unquestionably the species of *Proconsul* were very close to the lineage of man. How close will be determined by future discoveries (Fig. 276).

ORANGUTANS. We have seen that the species of *Dryopithecus* and related genera indicate relationships to the modern apes. They possess many characters seen in the orangutan. A fossil much more like the orangutans was found in the late Pliocene of India, and numerous remains of orangutans have been discovered in the Pleistocene of Java, Sumatra, and elsewhere in southeastern Asia.

The most interesting specimens of the fossil orangutans were three molar teeth purchased by G. H. R. von Koenigswald in Chinese drug-

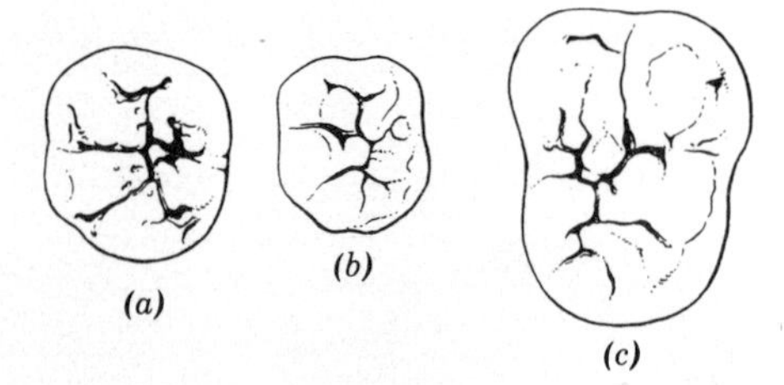

Fig. 277. Comparative sketches of the molars of (a) *Peking man, left* M_1, (b) *modern man, left* M_2, *and* (c) Gigantopithecus blacki, *right* M_3. *After Weidenreich.*

stores in Hong Kong, where they would soon have been pulverized with other fossil teeth and bones and sold for medicine. Evidently they had come from caves somewhere in Kiangsi Province in China. Von Koenigswald named them *Giganthopithecus blacki* and considered them as giant extinct apes. Later Franz Weidenreich thought these teeth belonged to an early man, said to have been the largest giant that ever lived. The molars were twice the dimensions of those in modern man. However, the consensus now favors the earlier opinion that the teeth belonged to an ape probably most closely related to the orangutans. Recently a lower jaw was found by Chinese paleontologists (Fig. 277c).

Orangutans are living today in swampy forests near the coasts or on interior lowlands in Borneo and Sumatra. The largest males are about 4 feet, 4 inches tall. The females are smaller. The long beards and heavy coat of long hair in the males give them a shaggy appearance. The legs are relatively short, but the arms are long, nearly reaching the ground when the animal is in an upright position. Orangutans are primarily fruit eaters but do eat leaves, buds, and bark. Such large arboreal animals must necessarily move about slowly in the trees, where they make shelters of boughs and leaves. Ernest Hooten reports that little is known of their social life but that the male leads a solitary existence much of the time (Fig. 278).

CHIMPANZEES. The chimpanzees are also descendants of a Dryopithecinae stock of the Miocene and Pliocene. Their characters are also reflected in some of the Miocene species of *Proconsul.* The immediate ancestry of the chimps is not known.

The living chimpanzees are widely distributed in equatorial Africa. Some of the males are nearly 5 feet in height, and the females are slightly smaller. Since they are gentle, lively, and intelligent, we frequently see them on theatrical programs, simulating humans in various amusing acts. In the wilds they are forest inhabitants; they spend much of their time on the ground, but they are more arboreal in habits than the gorillas. They forage about as a family or in small parties of several families, according to Henry Nissen. The older females may be seen walking along with an arm around a younger one. They eat fruit, blossoms, leaves, and even stems of desired plants. At night chimpanzees climb into the trees where they construct nests by bending and intertwining the branches (Fig. 279).

Fig. 278. Adult male orangutan, Pongo pygmaeus; *$^{1}/_{12}$ natural size. From photograph, by permission of Philadelphia Zoological Garden.*

Fig. 279. A chimpanzee, Pan troglodytes; *1/12 natural size. From photograph, by permission of Philadelphia Zoological Garden.*

GORILLAS. As in the orangutans and chimpanzees, the gorillas seem to have arisen from the Dryopithecinae. Gorillalike features are seen in species of the genera *Proconsul, Sivapithecus,* and *Dryopithecus,* but specializations in the dentitions of living gorillas are only faintly indicated or not present in species of the extinct genera. If late Pliocene fossils could be found somewhere in central Africa, perhaps we could learn much more about their direct line of descent (Fig. 280).

Gorillas are the largest and most powerful of the living apes. For the most part they inhabit the same areas in equatorial Africa as the chimpanzees. There are two kinds, the mountain and the lowland

Fig. 280. Nearly full grown adult mountain gorilla, Gorilla gorilla; $^1/_{12}$ natural size. From photograph, by permission of Philadelphia Zoological Garden.

gorillas. The largest males are about 5 feet, 6 inches tall. They are ferocious looking brutes and not so readily adaptable to captivity as the chimps. Gorillas feed on fruits and other plant materials to their liking. They live in family parties and move about on all four feet on the ground. In the evening they construct their nests on the ground or in bushes. At times the females and young may sleep in bushes or trees while the male sleeps on the ground below, perhaps as a guard for the family.

33 Prehistoric Men

Neanderthal man

Even when Linnaeus set up his system of nomenclature in 1758 he classified man in the order of Primates. Indeed, there was no other logical place for Linnaeus to put him after the anatomy of most of the major groups of mammals was understood. Further, it was gradually revealed by numerous anatomists that man's greatest similarity among the Primates was to the great apes. This classification has withstood a more critical review than that in any other group of organisms, but it has been strengthened by an ever-increasing wealth of fossil materials, notwithstanding the hoaxes that have been deliberately planned by pranksters. In Chapter 32 we reviewed the Tertiary record of the other Primates and outlined the relationships of the different groups and their most probable phylogenetic lineages. The fossil record of all the Primates has not been so complete as one would like, but as exploration parties have continued to add new discoveries minor discrepancies in the classification have been corrected; nevertheless, the broader concept has remained unaltered for the past 75 years or more.

Prehistoric men had developed brains superior to those in the other mammals with which they were associated. Usually they were too alert to be trapped in bogs or in other places in which so many mammals lost their lives. This, apparently, accounts for their relative rarity in the fossil record. Some of them inhabited forests, and when they died they were seldom in places in which their bones were likely to be quickly buried in sediments to become preserved as fossils. Scattered remains have been found in stream and flood-plain deposits, especially in partly open country where warm summer climates prevailed. Other races in the middle latitudes tended to occupy caves or grottoes for protection and shelter, and most of their fossil record has been derived from burials and cave accumulations.

Man is terrestrial and walks about in an upright posture, whereas most of the apes are better adapted to life in the trees, though some of them, like the chimpanzees and gorillas, now spend most of their time on the ground. Many of the characters that differentiate the families Hominidae and Pongidae tend to reflect their habits.

HOMINIDAE
Man

1. Face shortened, great expansion of brain.
2. Rows of cheek teeth tend to diverge posteriorly.
3. Canines short or tending toward that end.
4. With distinct bipedal posture.
5. Legs longer than arms.
6. Big toe not opposable.

PONGIDAE
Apes

1. Face longer, brain not so greatly expanded.
2. Rows of cheek teeth parallel.
3. Canines enlarged.
4. Usually with branchiating posture.
5. Legs shorter than arms.
6. Big toes opposable.

There are many more specialized features in both families.

Genera of Hominidae. The generic and specific classification of the fossil remains of man have been and still are a problem of considerable disagreement. During the past 75 years there has been a tendency on the part of most authorities to assign new generic and specific names rather liberally. The advisability of this practice is now being seriously questioned. Though it is advantageous to seek uniformity insofar as possible in the recognition of genera and species, in extinct groups this is almost impossible to attain. Sometimes it even confuses an understanding of their relationships to make the attempt.

Perhaps all fossil remains of men, including the most primitive australopithecines of South Africa, represent a taxonomic category of no greater magnitude than that of the genus *Felis* in the cat family. Possibly it is even less, but this we do not know, and as yet there is no accurate method of making a comparative evaluation of the two genera. In any given family, though, it is less difficult to recognize genera of about the same magnitude than to compare their rank with genera in another family. Most paleontologists and anthropologists recognize three groups with rather closely related species in the family Hominidae, though some even question the family rank of the Hominidae. Recognizable groups are the australopithecines, the pithecanthropines, and the hominines, though the last two groups appear to be closely related phyletically. As a matter of convenience, then, Sir Wilfred Le Gros Clark (1853) recognized two subfamilies *Australopithecinae*, and *Homininae.*

Taungs Child (*Australopithecus africanus* Dart). The attention of anthropologists and paleontologists was turned with startling suddenness to South Africa in 1924. For many years the cave breccias of the Transvaal had been worked for their lime content. The fossilized remains of many kinds of mammals and other animals had been found there in considerable numbers. The skulls and jaws of a small baboon were abundant. One day in December 1924 a skull somewhat larger and obviously different from the baboons turned up. This was dispatched to Professor Raymond A. Dart at The University of The Witwatersrand. The Professor studied it carefully, though much of it was still concealed by the hard stony matrix, and in 6 weeks he issued a preliminary notice describing it as an apelike child of about 6 years. The name he gave it was *Australopithecus africanus.* Most anthropologists and paleontologists tended to ridicule Dart's determination, as they are sometimes prone to do, but the eminent paleontologist Robert Broom also studied the specimen carefully and maintained that Dart was correct. Finally, in 1929 when the mandible was freed from the cranium and the teeth were cleaned Dart and Broom were satisfied that their interpretation was correct. The foramen magnum, or opening in the base of the skull where the spinal cord connects to the brain, was

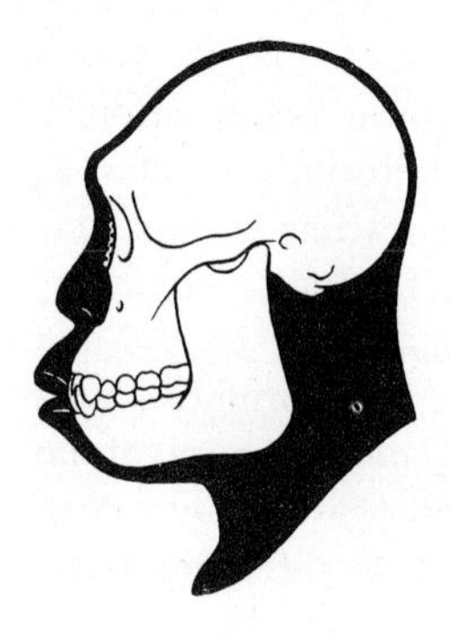

Fig. 281. Outline of head of Sterkfontein man, Australopithecus africanus; $^1/_6$ *natural size. Stippled parts restored.*

too far forward for an ape, and the teeth were like those in a human child, though larger. These and many other features clearly demonstrated its relationships to early man.

Sterkfontein Man (*Australopithecus africanus* Dart). When Field Marshal Jan C. Smuts, Prime Minister of the Union of South Africa, appointed Robert Broom Curator of Vertebrate Paleontology and Physical Anthropology at the Transvaal Museum the search for early man in South Africa started in earnest. Broom felt that the caves and fissures would yield more remains of early man. At Sterkfontein he urged G. W. Barlow, the manager in charge of quarrying the caves for lime, to be on the lookout for human remains. Not long after, Barlow found a brain cast, and Broom hastened to the site. Careful search in the blasted rocks resulted in the discovery of most of the skull of an adult. This was named *Plesianthropus transvaalensis,* but now is thought to belong to the same species as the Taungs Child. Operations in the quarries were watched carefully for 3 years. Skull fragments, parts of jaws, and pieces of limb bones of the Sterkfontein man continued to appear and with them the fossils of numerous mammals. More fine specimens were found after World War II. In 1947 a blast exposed an adult female skull which Broom referred to in his enthusiasm as "the most valuable specimen ever recovered." Mrs. Ples (*Plesianthropus*), as the skull was nicknamed, created a sensation everywhere. The possibilities in South Africa seemed almost limitless. What a far cry it was from the dubious days of Dart's Taungs Child of the late 20's (Fig. 281).

Kromdraai Man (*Paranthropus robustus* Broom). To retrace our story a bit, in June 1938, when Broom visited Sterkfontein, Barlow showed him a well-preserved palate with one molar still in its socket. It turned out that this specimen with fresh breaks on it was found just 2 miles away by Gert Terblanch, a schoolboy. Broom, on visiting the farm, learned that the boy had located the skull in an erosional remnant and had proceeded to extract it with a claw hammer. Pieces were still lying about, all of which Broom meticulously retrieved. Later at the school he located Gert among the other boys and found that he had four more of those "beautiful" teeth in his trouser pockets. He then learned that the boy had hidden part of the specimen elsewhere for safekeeping. To obtain permission from the principal for Gert to lead him to the cache, Broom lectured to the teacher and children about fossil bones and cave deposits, employing his skill in freehand artistry on the blackboard to illustrate the strange creatures he so clearly visualized. The cache yielded an excellent jaw and other pieces but no associated mammals. This larger and more powerful skull was named *Paranthropus robustus* Broom (Fig. 282).

The ever-enthusiastic Broom turned his attention to the Swartzkrans quarry in 1948, where numerous specimens with massive teeth were found. *Paranthropus crassidens* was named with part of a lower jaw as the type. As the work continued a fine skull appeared, as well as a pair of enormous lower jaws even larger than the giant *Meganthropus* from Java. The Swartzkrans specimens, more recently, have been considered as synonymous *P. robustus.* Though the remains of the South African early men are alike in many basic features, the detailed comparisons made by John Talbot Robinson has revealed some conspicuous differences between *Australopithecus* and *Paranthropus.* Among these characters in *Paranthropus* are the prominent insertion areas for the powerful masseter muscles, a saggital crest sharply elevated anteriorly, massive jaws and teeth, different ventral outlines of the symphysial region, and heavy supraorbital ridges. *Australopithecus* has been found at Taungs, Sterkfontein, and Makapan. *Paranthropus* has been taken at Kromdraai and Swartzkrans. The mammalian faunas from these sites are now considered to be of slightly different ages.

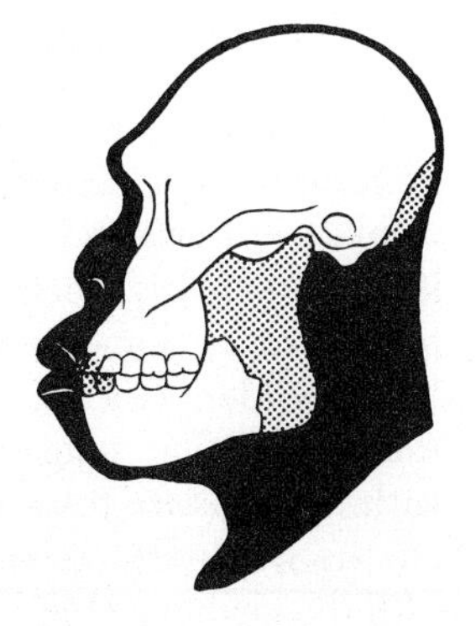

Fig. 282. Outline of head of Kromdraai man, Australopithecus robustus; *$^1/_6$ natural size. Stippled parts restored.*

Relationships of Australopithecinae. The large collections now assembled from the South African cave and fissure deposits gives us a rather clear picture of the australopithecines. They were short in stature, probably not exceeding 5 feet. The predominance of their characters fit in more closely with man than with the apes. The forward position of the occipital condyles, where the skull attaches to the vertebral column, as well as the construction of the body skeleton, especially the pelvis, attest to their manlike bipedal posture. Their most primitive attribute is displayed in their small brain capacity, which is only slightly greater than in the living apes and much less than in the ape men of Java, but the ratio of brain weight to body size in the Australopithecinae is large. The high and wide ascending ramus at the back part of the lower jaw is primitive, and the tooth-bearing anterior part of the jaw is relatively short.

Among the other conspicuous features are the concave face with its protruding premaxillaries and lower jaws and the elevated braincase. This gives them a different posture of the head and neck from that in the early Homininae (Figs. 281, 282, 284, 285). Since the skull is well balanced on the spinal column, the attachment areas for neck muscles at the back of the head are reduced. The last lower premolar, P_4, is two-rooted and larger than P_3. Though the teeth are large, they are basically much more like those in man than in the apes and resemble those in the ape men of Java. It has been suggested that *Australopithecus* and *Paranthropus* had different food habits.

Djetis Giant Man (*Meganthropus palaeojavonicus* von Koenigswald). While G. H. R. von Koenigswald was working over his precious early-man fossils from Java and the remains of their contemporary mammals, one of his collectors sent him part of an enormous human jaw. This was in Java in 1941 just before the Japanese opened with thunderous fury the war in the Pacific. The specimen, representing one of the oldest known men from the island, came from the Djetis fauna of north central Java where it was exposed along the bank of the Tjemoro River.

This amazing fossil was part of a right mandible with most of the symphysis, two bicuspid single-rooted premolars, and the first molar in place. It was so astonishingly large that von Koenigswald carried it around in his pocket for days, frequently handling it to accustom himself to its immense proportions. The jaw was 1 inch thick near its lower border, twice as thick as in the largest gorilla. In front there was no chin nor suggestion of a simian shelf behind the symphysis but with a diagastric fossa in that area as in some species of *Dryopithecus*. On the whole, the teeth were like those in man but of a size in keeping with the jawbone (Fig. 283).

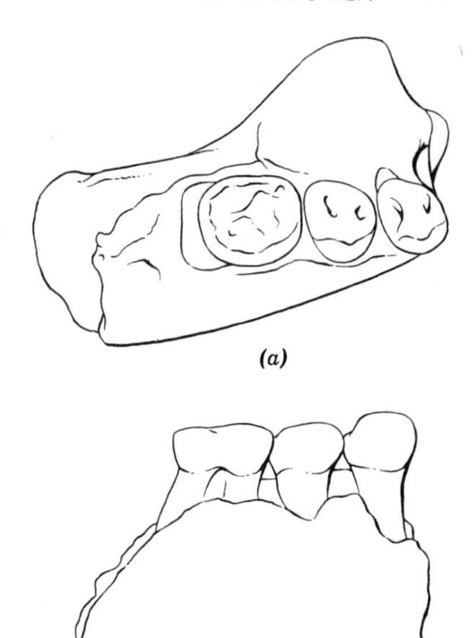

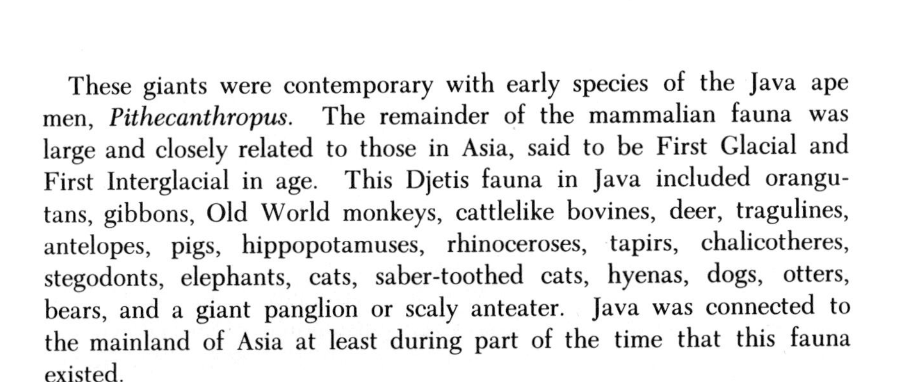

Fig. 283. Jaw fragment of Djetis giant man, Meganthropus palaeo-javonicus: (a) *from above;* (b) *from the side;* $^2/_3$ *natural size.*

These giants were contemporary with early species of the Java ape men, *Pithecanthropus.* The remainder of the mammalian fauna was large and closely related to those in Asia, said to be First Glacial and First Interglacial in age. This Djetis fauna in Java included orangutans, gibbons, Old World monkeys, cattlelike bovines, deer, tragulines, antelopes, pigs, hippopotamuses, rhinoceroses, tapirs, chalicotheres, stegodonts, elephants, cats, saber-toothed cats, hyenas, dogs, otters, bears, and a giant panglion or scaly anteater. Java was connected to the mainland of Asia at least during part of the time that this fauna existed.

Ape Men of Java (*Pithecanthropus erectus* Dubois; *P. modjokertensis* von Koenigswald; and *P. dubius* von Koenigswald). Toward the close of the nineteenth century when arguments were heated and furious in Europe over the relationships of the Neanderthal man, Eugene Dubois, a young Dutch surgeon, conceived the idea that early man should be looked for in warmer climates where the great apes were still living. Securing an appointment with the Dutch Army, he was sent to Suma-

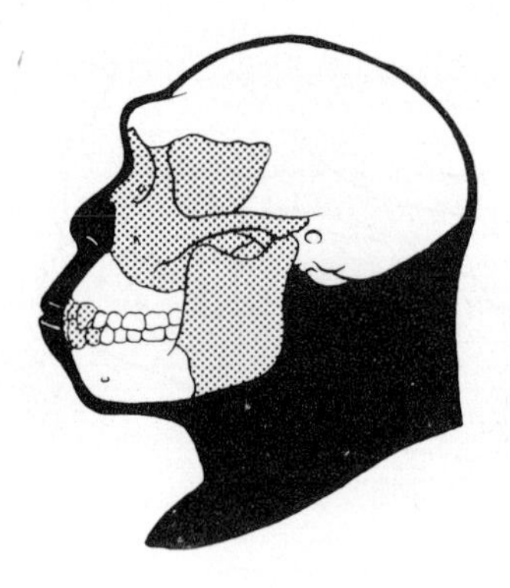

Fig. 284. Outline of head of ape man of Java, Pithecanthropus erectus; *1/6 natural size. Stippled parts restored.*

tra in 1887. In his spare time he searched for remains of early man but found only mammalian fossils on the island. Then word came to him of a fossil human skull found near Wadjak in Java. Though he found another skull at Wadjak, it was not so old as he had hoped to find. He then concentrated his efforts toward the center of the island. Eventually in the Poetjang volcanic sediments along the banks of the Solo River near the village of Trinil he found five specimens that were to excite even more controversy than the Neanderthal remains.

One was a low primitive skull cap with prominent eyebrow ridges; the others were three teeth (two molars and one premolar) and a femur so straight that the creature must have walked in an upright position. The two molars proved to belong to an orangutan, but the premolar belonged to a primitive man. This early man was what Dubois had visualized. He announced his discovery to the world in 1894 when he named his "missing link" *Pithecanthropus erectus.* Scientists were skeptical, and society for the most part was not receptive. Though Dubois was right, he eventually became so discouraged and insulted that he locked his controversial fossils in a strong box and refused to show them to anyone for 30 years (Fig. 284).

Though the discovery of Peking man strongly confirmed Dubois' opinion on *Pithecanthropus*, it was the outstanding discoveries of the young paleontologist G. H. R. von Koenigswald that settled the matter once and for all. In 1930 he went to Java as a member of the Netherlands East Indies Geological Survey with an objective firmly in mind of finding more of *Pithecanthropus.* He worked steadfastly for 11 years with additional funds from the Carnegie Institution of Washington, and by 1942 when he was taken prisoner of war by the Japanese he had found another skull cap even better than the one located by Dubois—a lower jaw and other parts and a large Trinil mammalian fauna. These animals had lived in a cool rainy climate which has been correlated with the Second Glacial Age of Asia by Hallam L. Movius of Harvard and Helmut DeTerra of India.

Von Koenigswald's efforts also resulted in the discovery of the Modjokert child skull and most of the cranium and lower jaws of an adult in the underlying strata, where they were associated with the giant *Meganthropus palaeojavonicus.* These he named *Pithecanthropus modjokertensis* (syn., *Pithecanthropus robustus* Weidenreich). These oldest known men from Java with their receding chins and low brain cases shared many characters with Peking man to which they were closely related. A character not seen in other men but present in the apes was a space, or diastem, between the upper canine and the outer incisor. As in *Meganthropus,* Peking men and all the later more progressive men, the second lower premolars were single rooted, and P_3 was larger than P_4. The brain capacity, on the other hand, was somewhat greater than in the australopithecines from South Africa. To another rather dubious jaw from the same beds he applied the name *Pithecanthropus dubius.* Thus von Koenigswald's discoveries not only confirmed the finds of Eugene Dubois but showed that *Pithecanthropus* ranged from the First to the Second Glacial Age. His work on the stratigraphy and mammalian faunas made these conclusions possible.

Peking Man [*Pithecanthropus pekinensis* (Black)]. The clue to one of the most outstanding discoveries of early man were two teeth found by Otto Zdansky, a Swedish paleontologist working on the Chinese Geological Survey. He uncovered the teeth in an exposure on Dragonbone Hill near the village of Chou-k'ou-tien only 37 miles southwest of Peking, China. Davidson Black, a young physician and professor of anatomy at the Peking Union Medical College who had come to China in quest of early man, became intensely interested and persuaded the Rockefeller Foundation to provide funds for extensive excavations into the hill of Ordovician limestones which had been honeycombed by water action; the limy materials had been carried away in solution. During part of the Ice Age these caverns had been occupied by man but subsequently had been partly filled with sediments and cave breccias which had become flinty hard. The excavators worked for 7 months before another molar turned up among the various kinds of fossil mammals. Black had the keen understanding to recognize the tooth as belonging to an early man, which he named *Sinanthropus pekinensis.*

The difficulties and dangers encountered by the parties working there in the field are now history as well as most interesting. Notwithstanding the hardships and danger, they continued to recover the precious jaws, teeth, and a few bones and stone implements of Peking man. In

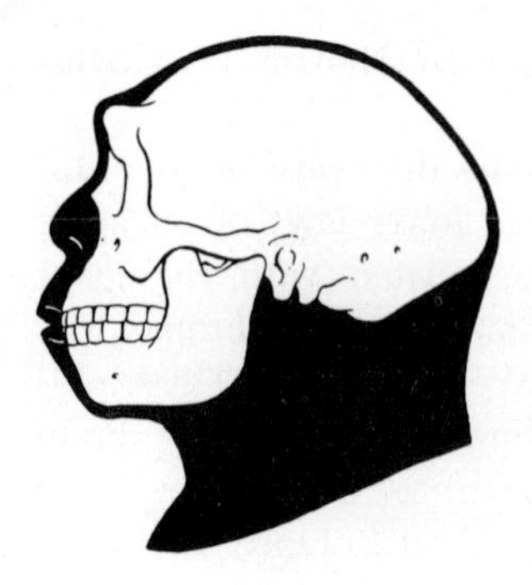

Fig. 285. Outline of head of Peking man, Pithecanthropus pekinensis; 1/6 *natural size.*

November 1929 W. C. Pei's assistants lowered him through another opening they had made. On the floor of the second cavern he came upon an almost complete skull of Peking man near the skull of a rhinoceros (Fig. 285).

In 1934 Davidson Black died suddenly from a heart attack. His work was continued the next year by Franz Weidenreich. Materials of Peking man belonging to individuals of all ages continued to come in from the excavations. Twelve more skulls and major parts of other skulls were recovered. There were remains of approximately forty individuals, most of which were children. The relationship of Peking man to *Pithecanthropus erectus* Dubois was soon apparent. Though the former took shelter in caves, at least part of the time, and used fire, the early men of Java probably lived in the open. Both species have been assigned to the genus *Pithecanthropus*.

The Peking and Java men display features that were much alike, and their cultures, as indicated by their stone tools, are similar but different from those used by early men in Europe. Careful and detailed study has revealed that the men from Java are somewhat more primitive, but in some characters the remains from Dragonbone Hill are less advanced. The skulls of *P. pekinensis* are slightly larger and the brain case more spacious; there is a slight elevation or bump on the frontal bone not apparent in *P. erectus*. The Java man has large frontal sinuses which are very small in Peking man, and *P. erectus* has a smooth surface on the palate which in *P. pekinensis* is rough. A diastem, or space, between the canine and outer incisor seen in an adult specimen referred to *P. modjokertensis* from Java does not occur in *P. pekinensis*. In addition, the increase in size in the lower molars from M_1 to M_3 in the Java men was not found in the people from Dragonbone Hill. For these reasons it has been concluded that these early men from China and Java belong to the same genus but to different species. Since the generic name *Pithecanthropus* has priority, Peking man has been called *Pithecanthropus pekinensis* (Black).

The correlations of Movius have shown that *P. pekinensis* lived during the Second Interglacial Age and that *P. erectus* slightly preceded it in the Second Glacial. The compositions of the mammalian faunas associated with these early men were different, but by the stage of evolution in some of the mammals and by correlation with faunas in India in which both the Chou-k'ou-tien and Trinil faunas had species in common the relative times of existence of the faunas could be established.

In November 1941 the rumors that Japanese troops would take Peking seemed to be approaching reality. Consequently, the director of the Chinese Geological Survey finally convinced the Americans that they should carry the fossils to safety under escort of a Marine detachment which was leaving on the American liner President Harrison. The colonel packed the precious specimens in his footlocker, but his Marine detachment was captured and the remains of Peking man disappeared, never to be seen again. There have been different stories as to what happened to those valuable fossils, but the specimens seem to have vanished completely.

Ternifine Man (*Pithecanthropus mauritanicus* [Arambourg]). One of the most recent discoveries of early men was made near the village of Palikas, Ternifine, Algeria. The three mandibles were described in 1954 by M. Camille Arambourg, Musée nationale de l'Histoire naturelle, Paris. These lower jaws, which show such remarkable resemblances to those of Peking man and the ape men of Java, were associated with a middle Pleistocene mammalian fauna and bifacial stone implements of early Palaeolithic Age. Their discovery has directed attention to a wide distribution of pithecanthropine men in the Old World.

Heidelberg Man (*Homo heidelbergensis* Schoetensack). After 20 years of patient, almost daily inspections of a huge sand pit near the village of Mauer, just 6 miles southeast of Heidelberg, Germany, Otto Schoetensack was rewarded with a fine pair of human lower jaws. The jaws were found in 1907 by one of the quarry workmen. Four teeth were broken off on the left side, but aside from that the specimen was remarkably complete. Though numerous fossil mammal teeth and bones had been found, this was the first remains of man from the pits. The specimen was a timely discovery, since it offered scientists more information in their controversy over the finds made by Dubois in Java. In addition, it extended the record of man in Europe back to a warm interlude before the Second Interglacial, or Mindel-Riss, Age. No artifacts were found (Fig. 286).

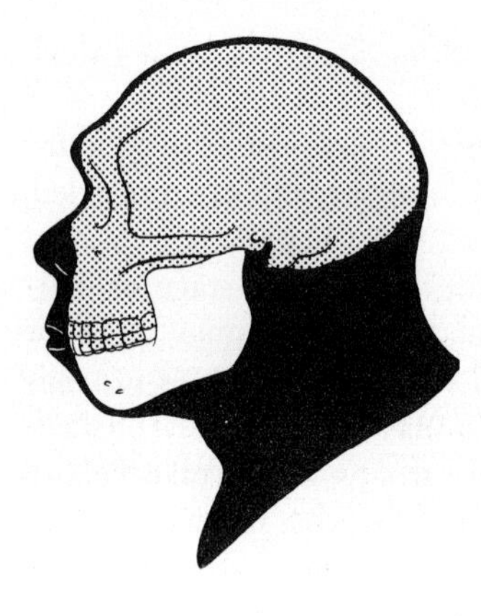

Fig. 286. Outline of head of Heidelberg man, Homo heidelbergensis; $^1/_6$ *natural size. Stippled parts restored.*

It is a bulky mandible but not so massive as some of those from South Africa and Java. The ascending ramus is wide, as seen from the side, and the sigmoid notch between the articulating condyle and the coronoid process is shallow. An apelike feature is shown in the receding chin, but there is no indication of an apelike simian shelf. Manlike characters are displayed in the tightly appressed teeth in the horseshoelike arc (parabolic) from one last molar around to the other. The canine is only slightly larger than the incisor next to it, and the last molar is slightly smaller than the molar in front of it. In fact the teeth are no larger than those in some living men. Nevertheless, it is an early man and is thought by some authorities to be near the ancestry of the Neanderthal men. It is still referred to as *Homo heidelbergensis.* If the teeth had not been found in their alveoli, the jawbone would surely have been considered as representative of a distinct genus of early man. In fact another generic name has been proposed for it.

Neanderthal Men. (*Homo neanderthalensis* King). The first skull of these most interesting people was found at a site known as Forbes Quarry on the north face of the Rock of Gibraltar. Lieutenant Flint of the British Army located the skull in 1848. It is said to be the skull of a female neanderthaloid, but at the time it was discovered it was evidently considered just another human skull, for its true relationships were not detected until the beginning of the twentieth century. Attention was first directed to *Homo neanderthalensis* in 1856 when Johann Karl Fuhlrott, a German scientist, rescued the scattered parts of a burial that had been dug into by workmen in a grotto of Devonian limestone in Neanderthal Valley near Dusseldorf, Germany. The skull cap, both femora, a humerus, lower arm bones, and other parts were retrieved. The description of these fossils by Herman Schaafhausen as representing a primitive race of men appeared in 1858, a year before Darwin's book *Origin of Species.* This initiated controversies on the authenticity and relationships of early man which lasted for more than half a century. Baron Cuvier, an advocate of special creation, even went so far as to say that fossil man did not exist. But as new finds continued to turn up throughout central and southern Europe and even in other parts of the Old World, the evidence became overwhelming. There was no doubt in the minds of serious-thinking people who informed themselves on the factual evidence that a species of man, *Homo neanderthalensis,* did exist during the last Interglacial and the first advance of the ice sheets in the Fourth or Würmian Glacial Age. Most of the remains were taken from cave burials. Their industry has been called the Mousterian (Fig. 287).

Two kinds of Neanderthal men have been recognized. Ernest A. Hooten of Harvard has described these conservative and progressive types in considerable detail (1947). Briefly, the conservative ones have flattened brain cases, a higher attachment of the neck muscles on the back of the skull, a marked projection of the jaws, and almost no chin. On the other hand, the progressive kinds have laterally compressed and higher brain cases, a lower attachment for the neck muscles at the back of the skull, jaws not so prominently projected, and moderately developed chins.

One of the finest skeletons, a conservative type, is the old man from La Chapelle-aux-Saints. His skeleton was uncovered in a small cave in France. Two skulls of adult males with fragmentary skeletons taken from a cave near Spy, Belgium, also have added much to our knowledge of the Neanderthal men. One is a progressive kind and the

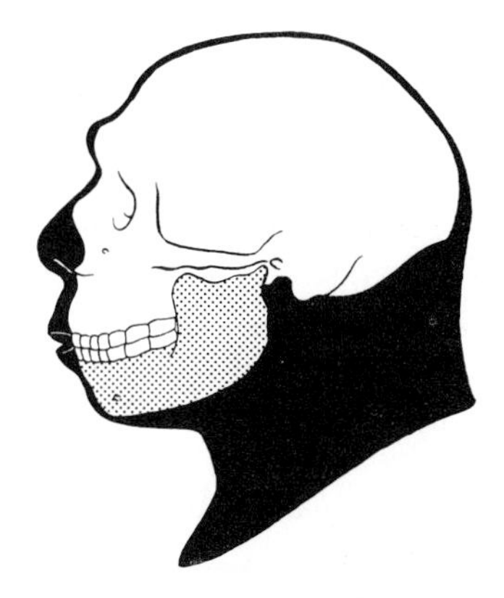

Fig. 287. Outline of head of Neanderthal man, Homo neanderthalensis; $^{1}/_{6}$ *natural size.*

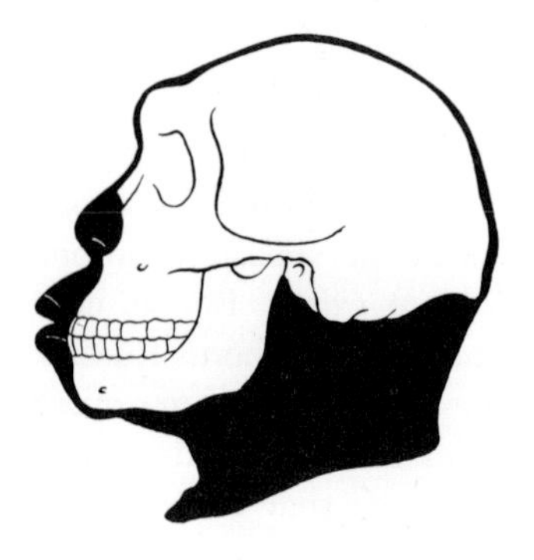

Fig. 288. Outline of head of Rhodesian man, Homo rhodesiensis; $^1/_6$ *natural size. Stippled parts restored.*

other is conservative. Associated with them are remains of the woolly mammoth, woolly rhinoceros, cave bear, cave hyena, and Mousterian artifacts. This discovery and others like it indicate that these hardy people lived in western Europe in extremely cold winters and ventured from their cave shelters to kill animal life for food.

Other skeletons, many skulls, and other parts have been taken from caves, rock shelters, quarries, and stream deposits in France and in other countries, but by far the best collection of these slightly stooped, beastly appearing men came from Mount Carmel in Palestine not far from Jerusalem. The excavations there were done in 1931 and 1932 by Dorothy Garrod, Theodore D. McCown, and Hallam L. Movius. The remains were taken from two adjoining caves called Skuhl and Tabun. Something like ten progressive individuals were encountered in the Skuhl cave, and the skeleton of an adult female as well as other remains of conservative types were found in Tabun. The progressive kinds have excited considerable interest. Here seemed to be excellent intermediates between the neanderthaloids and *Homo sapiens,* the species of modern man. It is still argued whether they offer evidence of a transitional type between the two species or are hybrids as a result of interbreeding when the two species came together.

Another interesting skull was found in a cave in Rhodesia at a place called Broken Hill. This skull shows some marked similarities to the Neanderthal men but differs in other features. Some authorities assign it to a distinct species *Homo rhodesiensis* (Fig. 288).

Ralph von Koenigswald and others found parts of seventy-seven skull caps and other parts of human remains near Ngandong, Java. They were associated with a mammalian fauna evidently of the Third Interglacial Age. These neanderthaloid creatures have been called *Homo soloensis* by W. F. F. Oppenoorth. Evidence from the broken skulls suggests that Solo man was cannibalistic. It has been suggested also that he may have been near to the ancestry of the New Guinea and Australian aborigines. Neanderthal man, Rhodesian man, and Solo man apparently are so closely related that some authorities have considered referring to them as one species.

Cro-Magñon Men (*Homo sapiens* Linnaeus). In the second advance of ice over western Europe in the Fourth or Würmian Glacial Age the Neanderthal men were replaced by the Cro-Magñon men of Aurignacian culture. Cro-Magñon occupied many of the caves and overhanging rock shelters that their predecessors had used. The first discovery of these people to attract attention was at Les Eyzies, France, in a rock shelter known as Cro-Magñon. In the year 1868 five fragmentary skeletons of four adults and one child were removed from the cave. Other discoveries throughout central and southern Europe followed. Remains of Cro-Magñons were more complete and more abundant in the grottoes than those of the neanderthaloids. The Grimaldi caves on the Riviera have yielded several skeletons which include those of a mother and her son huddled closely together in the famous Grotto des Enfants. At Solutre, France, it has been said that remains were found even before 1868. All in all something like seven skeletons were found at the site. Perhaps the most important of all the discoveries was at Predmost in Moravia, Poland. There a cemetery was found with twenty skeletons, most of them complete. These people had constructed sepulchers with the jaws and shoulder blades of mammoths and had carved designs in the ivory and bone. A fascinating cave at Brunquel in southern France thought to represent a Magdalenian industry yielded two Cro-Magñon skulls and the remains of a woolly mammoth, a woolly rhinoceros, a cave hyena, and about thirty horses. Evidently the horses were hunted for their meat (Fig. 289).

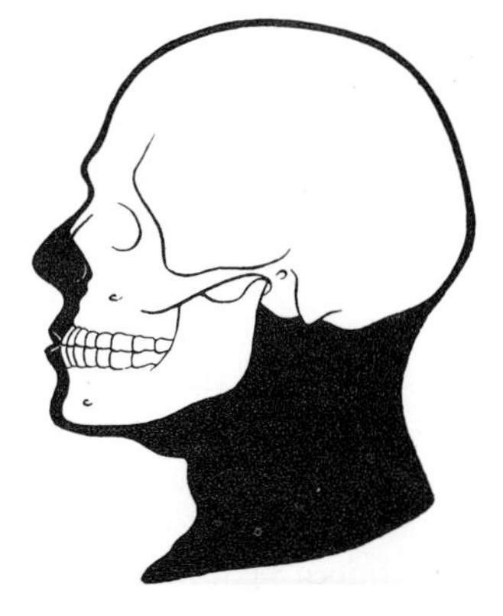

Fig. 289. Outline of head of Cro-Magñon man, Homo sapiens; *× 1/6 natural size.*

Fig. 290. Reindeer and part of landscape, from Kesslerloch, Switzerland, carved on fragment of deer antler by Cro-Magñon man. After Heim.

Cro-Magñons were skilled in artistry, as the polychrome paintings on the cavern walls, lithographs, and etchings on ivory and bone clearly demonstrate. Not only do these show the kinds of animals living in the area at that time but the ways in which these men hunted and killed them. Standing on the average slightly less than 6 feet in height, the Cro-Magñon was more erect than the Neanderthal. These and many other characters have demonstrated that Cro-Magñon belonged to the same species as modern man, *Homo sapiens*. The movements of the different races of men from Cro-Magñon on to recorded history were extremely complicated and are still not fully understood, but we can be relatively certain that the Cro-Magñon folks played an important role in our ancestry.

Antiquity of Neanthropic Men. There is some evidence suggesting that the ancestors of neanthropic, or highly developed, men were present in the Old World as far back as the Second Interglacial Age. One of the specimens is part of a cranium found in 1936 in a terrace of the lower Thames Valley at Swanscombe, England. The specimen consists of the occipital and parietal parts that cover the posterior and upper parts of the brain case. A careful search was made, but no other bones of Swanscombe man were recovered. This is extremely unfortunate because it is associated with early Paleolithic artifacts. Though the bone is rather thick and rounded, the shape of the cranium is more suggestive of modern man than of the neanderthaloids or of other more primitive men. Even so, we still cannot be certain that the bones of the face and the lower jaws may not have had features like those of more primitive men. Nevertheless, it is quite possible that the Swanscombe man is an early representative in the *Homo sapiens* lineage, or of a man much more closely related to the ancestry of modern man than is typical *Homo neanderthalensis*.

Another even better skull of the Second Interglacial Age is that of Steinheim man. This generalized neanderthaloid is comparable to the Swanscombe man and may have been near the ancestry of Neanderthal men on the one hand and Cro-Magñon on the other.

Man in America. The evolution of man took place in the Old World; he was a late arrival in the Western Hemisphere. The exact time of his crossing the Bering route from Siberia into North America has not been determined, but in his summary of early man in America E. H. Sellards states that man "had become widely distributed in North America more than 10,000 years ago" and "that he had reached the extreme southern tip of South America more than 8,000 years ago."

The Folsom culture, with its characteristic fluted projectile points, was widely spread over North America, especially in the interior, or Great Plains province. No skeletons or bones of Folsom man have been found, but the direct association of a projectile point lodged in a vertebra with charred bone have demonstrated that he hunted extinct bisons. The first site was found at Folsom, New Mexico, by a Negro cowboy. There nineteen Folsom points and evidence of about thirty extinct bisons, *Bison antiquus*, were uncovered. Another excellent camp site was located on the Lindenmeyer property in Colorado, near the Wyoming state line, and numerous localities have been found

Fig. 291. Bison antiquus *and the Folsom point; not drawn to scale.*

from Texas to Alaska. Radiocarbon tests have shown that the Folsom man existed close to 10,000 years ago.

Stratigraphically below the Folsom culture level near Clovis, New Mexico, is the Llano culture. Evidence of this culture also occurs in the panhandle of Texas at the Miami locality. Well-worked but non-fluted projectile points have been found with bones of the Columbian mammoth and other mammalian remains. Five other localities of the Llana culture all the way from Arizona to Nebraska are on record. Unfortunately, as with Folsom man, no bones of Llano man have been recovered. This culture may be 2000 to 3000 years older, but we do not know that these were the first men to have reached America. Indeed, there is some evidence that man may have spread into North America over 20,000 years ago, but this statement needs the substantiation of additional discoveries.

Innumerable Postfolsom cultural sites have been made in the United States, Mexico, and South America, but only a few have been accurately dated by the carbon 14 method.

The Piltdown Hoax. There have been repeated attempts by pranksters to confuse or make it possible to ridicule paleontologists by perpetrating hoaxes. They may bury fossils in formations in which years of experience in collecting have taught that fossils of that kind should not occur; or they may plant together certain combinations of specimens never encountered in undisturbed rocks. Nevertheless, it is a credit to paleontologists and anthropologists and their methods that sooner or later these fakes are disclosed.

In 1908, while Charles Dawson, an English lawyer, was passing along a country road in Sussex near Piltdown, he came upon some brown flints not previously known in that district. These were found in gravel that had been used in repairing the road and evidently had been taken from a shallow pit on a nearby farm. Consequently, Dawson urged the workmen to be on the alert for fossil bones. A few days later one of the workmen handed him part of the parietal bone of a human skull. When in 1911 he recovered part of the frontal bone evidently from the same skull he sought the advice and co-operation of Sir Arthur Smith–Woodward of the British Museum.

By sieving the gravel at the site and by carefully exploring the area they recovered more of the cranium and a broken lower jaw. These specimens were described a year later as *Eoanthropus dawsoni*. Eventually a number of mammalian remains, as well as some eolithic and Palaeolithic bone and stone artifacts, were also recovered from the locality.

From the beginning many anthropologists and paleontologists questioned the authenticity of the finds, especially the reference of the apelike lower jaw to the cranial parts that bore such a marked resemblance to modern man. Both points of view were argued at length. The situation was made even more disturbing by the unusual combination of associated mammalian teeth and bones, most of which had never been found in England. It was even observed that the worn surfaces on the teeth were quite abnormal. These and other observations tended to raise skepticism in the minds of many authorities.

Eventually, in 1950, the bones were put to radioactive and fluorine tests by Kenneth Page Oakley, C. R. Hoskins, and others, in which all of the evidence was critically reviewed. The conclusions were most enlightening. Most of the bones, the teeth, and even the artifacts had been treated with iron sulphate and chromium stain or Vandyke brown paint. The mammoth, mastodont, and rhinoceros tooth fragments had high fluorine content, indicating an older age, whereas the hippopotamus, deer, beaver, horse, and human cranial bones were low in fluorine. The lower jaw and the lower canine with no fluorine were those of a Recent orangutang. This was perhaps the most carefully planned hoax in the history of paleontology and anthropology.

Origin of Australian Aborigines. The Australian "blackfellows" were thought at first to be a primitive kind of man perhaps closely related to the neanderthaloids. But later studies have shown that they belong with other modern men in the species *Homo sapiens*. When the cultures of the numerous tribes were known it was found they used many kinds of stone tools quite like the different palaeoliths and mesoliths of early European cultures.

There have been different suggestions as to the origin, the time, and the number of invasions by these people of the Australian region. The meager evidence bearing on these problems now suggests that there was more than one infiltration, and that the dingo, or warrigal, did not accompany the first populations. Perhaps the extinct Tasmanian tribes were remnants of one of the first invasions. There is no good evidence of the time of the first arrival, but it may have been in the late Pleistocene or the Subrecent, over 6500 years ago.

Comparisons of the two Wadjak crania from Java, found in 1889, and the Keilor cranium, discovered near Melbourne in 1940, suggest that the Tasmanian and Australian aborigines came from the East Indies before the Negritos, the Papuans, and the Melanesians spread into the Indo-Pacific islands. Franz Weidenreich believed that the Wadjak and Keilor individuals belonged to the same human race, though he recognized certain differences in the crania which suggested that the Wadjak cranium might be slightly more primitive.

Bibliography

METHODS IN PALEONTOLOGY

Camp, C. L., and G. D. Hanna, *Methods in Paleontology,* University of California Press, 153 pp., 58 figs., 1937.

Glaessner, M. F., *Principles of Micropaleontology,* Wiley, New York, Chapter 3, pp. 33–51, 1947.

Jones, D. J., *Introduction to Microfossils,* Harper, New York, Chapter 2, pp. 7–18, 1956.

Norem, W. L., Separation of Spores and Pollen from Siliceous Rocks, *J. Paleont.,* **27,** pp. 881–883, 1953.

Ridgway, J. L., *Scientific Illustration,* Stanford University Press, Stanford, 173 pp., 23 figs., 22 pls., frontispiece, 1938.

Schenck, H. G., and B. C. Adams, Operations of Commercial Micropaleontologic Laboratories, *J. Paleont.,* **17,** pp. 554–583, 13 figs., 1 pl., 1943.

Trelease, S. F., *The Scientific Paper. How to Prepare It. How to Write It,* Williams and Wilkins, Baltimore, 2nd ed., 163 pp., figs. in text, 1951.

Walton, J., *An Introduction to the Study of Fossil Plants,* Adam and Charles Black, London, 201 pp., 138 figs., 1953.

ZOOGEOGRAPHY, SYSTEMATICS, EVOLUTION, AND NOMENCLATURE

Buchsbaum, R., *Animals Without Backbones, an Introduction to the Invertebrates,* University of Chicago Press., 2nd. ed., 405 pp., figs. and pls. in text, 1948.

Darlington, P. J., Geographic Distribution of Cold-Blooded Vertebrates, *Quart. Rev. Biol.,* **23;** Part 1, pp. 1–26; Part 2, pp. 105–123; 5 figs., 8 tables, 1948.

Darwin, C., *On the Origin of Species by Means of Natural Selection, or the Preservation of Favoured Races in the Struggle for Life,* John Murry, London, XI+ 502 pp., 1859.

Davies, A. M., *Tertiary Faunas,* Thomas Murby, London: Vol. I., 406 pp., 565 figs., 1935; Vol. II., 252 pp., 28 figs., frontispiece, 1934.

Dobzhansky, T., *Genetics and the Origin of Species,* Columbia University Press., New York, 3rd. ed., 360 pp., 1951.

———, *Evolution, Genetics, and Man,* Wiley, New York, 398 pp., figs. in text, 1955.

Ekman, S., *Zoogeography of the Sea* (translated from the Swedish by Elizabeth Palmer), Sidgwick and Jackson, London, 417 pp., 121 figs., 1953.

Goldschmidt, R. B., *Understanding Heredity,* Wiley, New York, 228 pp., 49 figs., 1952.

Hesse, R., W. C. Allee, and K. P. Schmitt, *Ecological Animal Geography,* Wiley, New York, 2nd. ed., 715 pp., 142 figs., 1951.

Huxley, J. S., *Evolution the Modern Synthesis,* Harper, New York, 645 pp., 1943.

Jepsen, G. L., E. Mayr and G. G. Simpson (editors), *Genetics, Paleontology and Evolution,* Princeton University Press., Princeton, 474 pp., figs. in text, 1949.

Ladd, H. S. (chairman), Report of the Committee on a Treatise on Marine Ecology and Paleontology, 1948–1949, *Nat. Research Council, Div. Geol. Geog.*, Washington, D. C., 121 pp., 8 figs., 1949.

Linnaeus, C., *Systema naturae per regna tria naturae, secundum classes, ordines, genera, species cum characteribus, differentiis, synonymis, locis, Editio decima, reformata, Tomus I. Laurentii Salvii, Holmiae,* 824 pp., 1758.

Matthew, W. D., Climate and Evolution, 2nd. ed., Special Publ., *N. Y. Acad. Sci.*, **1,** XI+ 223 pp., 33 figs., 5 tables, 1939.

Mayer, E., E. G. Linsley, and R. L. Usinger, *Methods and Principles of Systematic Zoology,* McGraw-Hill Book Co., New York, 328 pp., 45 figs., 1953.

Simpson, G. G., *Tempo and Mode in Evolution,* Columbia University Press, New York, 237 pp., 36 figs., 1944.

———, The Principles of Classification and the Classification of Mammals, *Bull. Am. Museum Nat. Hist.*, **85,** 350 pp., 1945.

———, *The Meaning of Evolution,* Yale University Press, New Haven, 364 pp., 38 figs., 1950.

———, Evolution and Geography, *Condon Lectures,* Oregon State System of Higher Education, 64 pp., 30 figs., 1953.

Stebbins, G. L., Variation and Evolution in Plants, Columbia University Press, New York, 643 pp., 55 figs., 9 tables, 1950.

Storer, T. I., *General Zoology,* McGraw-Hill, New York, 798 pp., 553 figs., 1943.

Wallace, A. R., *The Geographical Distribution of Animals,* Harper, New York: Vol. I, 503 pp., 13 pls., maps; Vol. II, 607 pp., 7 pls., 1876.

Woodward, S. P., *A Manual of the Mollusca; or Rudimentary Treatise of Recent and Fossil Shells,* John Weale, London, 486 pp., 272 figs., 24 pls., 1851–1856.

STRATIGRAPHY, GEOCHRONOLOGY, CORRELATION, AND GEOLOGIC TIME

Arkell, W. J., *The Jurassic System in Great Britain,* Clarendon Press, Oxford, 681 pp., 97 figs., 41 pls., 1933.

Dunbar, C. O., and J. Rogers, *Principles of Stratigraphy,* Wiley, New York, 356 pp., 123 figs., 21 tables, 1957.

Gilluly, J., Distribution of Mountain Building in Geologic Time, *Bull. Geol. Soc. Am.*, **60,** pp. 561–590, 1949.

Holmes, A., *The Age of the Earth,* Thomas Nelson and Sons, New York, 263 pp., 1937.

Hutton, J., *Theory of the Earth, with Proofs and Illustrations,* Cadell, Junior and Davies, London, and William Creech, Edinburgh: Vol. I, 620 pp.; Vol. II, 567 pp., 1795.

Jeletzky, J. A., Paleontology, Basis of Practical Geochronology, *Bull. Am. Assoc. Petrol. Geologists,* **40,** pp. 679–706, 1956.

Kleinpell, R. M., *Miocene Stratigraphy of California,* Am. Assoc. Petrol. Geologists, 450 pp., 22 pls., 13 figs., 18 tables, frontispiece, 1938.

Knopf, A., et. al., The Age of the Earth, *Nat. Research Council,* **80,** 487 pp., 1931.

Krumbein, W. C., and L. L. Sloss, *Stratigraphy and Sedimentation,* W. H. Freeman, San Francisco, 497 pp., figs. in text, 1951.

Libby, W. F., *Radiocarbon Dating,* University of Chicago Press, 2nd. ed., 175 pp., 11 figs., 1955.

Lyell, C., *Principles of Geology,* J. Murray, London, illustrations in text, with plates, wood-cuts, 1st., ed.: Vol. 1, 511 pp., 1830; Vol. 2, 330 pp., 1832; Vol. 3, 398 pp., with table of fossil shells, by G. P. Deshayes, 1833.

Oakley, K. P., The Fluorine Dating Method, in *Amer. Yearbook of Phys. Anthro.*, Vol. 5, pp. 9–44, New York, The Viking fund, 1949.

Oppel, C. A., *Die Juraformation Englands, Frankreichs und des Südwestlichen Deutschlands,* 857 pp., 1 table, 1 map, Stuttgart, 1856–1858.

Schenck, H. G., and S. W. Müller, Stratigraphic Terminology, *Bull. Geol. Soc. Am.*, **52,** pp. 1419–1426, 1941.

Zeuner, F. E., *Dating the Past: An Introduction to Geochronology,* Methuen, London, 3rd. rev. ed., 495 pp., 103 figs., 24 pls., 1952.

HISTORY OF PALEONTOLOGY

Adams, F. D., *The Birth and Development of the Geological Sciences,* Williams and Wilkins, Baltimore, 506 pp., 78 figs., 1938.

Brown, I. A., An Outline of the History of Paleontology in Australia, *Proc. Linn. Soc. N.S.W.*, **71,** 1946.

Fenton, C. L., and M. A. Fenton, *Giants of Geology,* Doubleday, New York, pp. 333, figs. in text, 1952.

Geikie, A., *The Founders of Geology,* 2nd., ed., Macmillan, London, 486 pp., 1905.

Merrill, G. P., Contributions to the History of American Geology, U. S. Government Printing Office, Washington, D. C., 733 pp., 141 figs., 37 pls., 1906.

Numerous authors, Geological Society of America, *Geology, 1888–1938,* Fiftieth Anniversary Volume, published by the Society, 578 pp., 1941.

Osborn, H. F., *Cope: Master Naturalist,* the life and letters of Edward Drinker Cope with a bibliography of his writings classified by subject, Princeton University Press, 740 pp., 30 figs., 1931.

Sahni, M. R., A Century of Palaeontology, Palaeobotany and Prehistory in India and Adjacent Countries, *T. Palaeont. Soc. India,* **1,** No. 1, pp. 7–51, 13 figs., 1956.

Schuchert, C., and C. M. LeVene, *O. C. Marsh, Pioneer in Paleontology,* Yale University Press, New Haven, Oxford University Press, London, 541 pp., 33 figs., 30 pls., 1940.

Simpson, G. G., The Beginning of Vertebrate Paleontology in North America, *Proc. Am. Phil. Soc.*, **86,** No. 1, pp. 130–188, 23 figs., 1942.

Wendt, H. W., *In Search of Adam,* Houghton Mifflin, Boston, The Riverside Press, Cambridge (translated from the German by James Cleugh), 540 pp., 48 pls., figs. in text, 1956.

Zittel, K. A. von, *History of Geology and Paleontology to the End of the Nineteenth Century,* Walter Scott, London, Charles Scribner's Sons, New York (translated by M. M. Ogilvie-Gordon), 562 pp., 13 pls., 1901.

PRE-CAMBRIAN

Brown, J. S., and A. E. J. Engel, Revision of Grenville Stratigraphy and Structure in the Balmot-Edwards District, Northwest Adirondacks, New York, *Bull. Geol. Soc. Am.*, **67,** pp. 1599–1622, 4 figs., 5 pls., 1956.

Cooke, H. J., The Canadian Shield, in *Geology and Economic Minerals of Canada, Geol. Survey Can., Econ. Geol. Ser.*, **1,** pp. 11–97, 1947.

David, T. W. E., and W. R. Browne, *The Geology of the Commonwealth of Australia,* Edward Arnold, London, Vol. 1, pp. 3–99, 31 figs., 1950.

Dey, A. K., Problems of Archaean Geology in India, *Quart. J. Geol. Mining Met. Soc. India,* **24,** pp. 39–42, 1952.

Du Toit, A. L., *The Geology of South Africa,* Hafner, New York, 3rd. ed., pp. 27–225, 34 figs., 12 pls., 1954.

Gignoux, M., *Stratigraphic Geology,* W. H. Freeman, San Francisco, English translation of 4th French edition, 1950, Chapter 1, pp. 28–45, 4 figs., 1955.

Tyler, S. A., and E. S. Barghoorn, Occurrence of Structurally Preserved Plants in Pre-Cambrian Rocks of the Canadian Shield, *Science,* **119,** pp. 606–608, 4 figs., 1954.

CAMBRIAN. PLIOCENE

Arnold, C. A., *An Introduction to Paleobotany,* McGraw-Hill, New York, 433 pp., 187 figs., 1947.

Bohlin, B., The Affinities of the Graptolites, *Bull. Geol. Inst. Univ. Uppsala,* **34,** pp. 107–113, 6 figs., 1950.

Brink, A. S., Speculations on Some Advanced Mammalian Characteristics in the Higher Mammal-like Reptiles, *Palaeo. Africana,* ann., Bernard Price Institute, Vol. 4, pp. 77–96, 5 figs., 1957.

Broom, R., The Mammallike Reptiles of South Africa and the Origin of Mammals, H. F. and G. Witherby, London, 376 pp., 111 figs., 1932.

Colbert, E. H., Siwalik mammals in the American Museum of Natural History, Trans. Am. Phil. Soc. (new ser.), **26,** 401 pp., 198 figs., 1 map, 1935.

David, T. W. E., and W. R. Browne, *The Geology of the Commonwealth of Australia,* Vol. 1, Edward Arnold, London, pp. 100–747, 178 figs., 1950.

Decker, C. E., Place of Graptolites in Animal Kingdom, *Bull. Am. Assoc. Petrol. Geologists,* **40,** pp. 1699–1704, 1 pl., 1956.

Dunbar, C. O., *Historical Geology,* Wiley, New York, 567 pp., 350 figs., 1949.

Durham, J. W., Cenozoic Marine Climates of the Pacific Coast, *Bull. Geol. Soc. Am.,* **61,** pp. 1243–1264, 3 figs., 1950.

DuToit, A. L., *The Geology of South Africa,* Hafner, New York, 3rd. ed., pp. 226–611, 39 figs., 29 pls., 1954.

Eardley, J. A., *Structural Geology of North America,* Harper, New York, 624 pp., 343 figs., 16 pls., 1951.

Gignoux, M., *Stratigraphic Geology,* W. H. Freeman, San Francisco, English translation of 4th French edition, 1950, pp. xvi–682, 155 figs., 1955.

Grabau, A. W., and H. W. Shimer, *North American Index Fossils, Invertebrates,* A. G. Seiler, New York: Vol. I, 853 pp., 1210 figs., 1909; Vol. II, 909 pp., 1937 figs., 1910.

Hamilton, E. L., Sunken Islands of the Mid-Pacific Mountains, *Geol. Soc. Am. Mem.,* **64,** 97 pp., 12 figs., 13 pls., 1956.

Hussey, R. C., *Historical Geology, the Geologic History of North America,* McGraw-Hill, New York, 465 pp., 322 figs., 1947.

Jenks, W. F., Handbook of South American Geology and Explanation of the Geologic Map of South America, *Geol. Soc. Am. Mem.* **65,** 378 pp., with pls., 1956.

Jones, D. J., *Introduction to Microfossils,* Harper, New York, 406 pp., figs. in text, 1956.

Kay, M., North American Geosynclines, *Geol. Soc. Am. Mem.* **48,** 143 pp., 20 figs., 1951.

Kozlowski, R., Les graptolithes et nous groupes d'animaux du tremadoc de La Rolonge, *Palaeontologia Polonica,* Tome III, 253 pp., 66 figs., 42 pls., 1948.

Kühne, W. G., *The Liassic Therapsid Oligokyphus,* British Museum (Natural History), 149 pp., 66 figs., 12 pls., 1956.

Laseron, C. F., *Ancient Australia,* Angus and Robertson, Sydney, 210 pp., 15 figs., 1954.

Moore, R. C., Introduction to Historical Geology, McGraw-Hill, New York, 582 pp., 364 figs., 1949.

——— (editor), Treatise on Invertebrate Paleontology, *Geol. Soc. Am.* and University of Kansas Press: D. Protista 3, 195 pp., 92 figs., 1954; E. Archaeocyatha and Porifera, 122 pp., 89 figs., 1955; F. Coelenterata, 498 pp., 357 figs., 1956; G. Bryozoa, 253 pp., 175 figs., 1953; L. Mollusca 4, 490 pp., 558 figs., 1957; P. Arthropoda 2, 181 pp., 123 figs., 1955. V. Graptolithina, 101 pp., 72 figs., 1955.

Moore, R. C., C. G. Lallicker, and A. G. Fischer, *Invertebrate Fossils,* McGraw-Hill, New York, 766 pp., figs. in text, 1952.

Neaverson, E., *Stratigraphical Paleontology,* Clarendon press, Oxford, 2nd. ed., 806 pp., 90 figs., 18 pls., 1955.

Osborn, H. F., *The Age of Mammals in Europe, Asia and North America,* Macmillan, New York, 635 pp., 220 figs., 1910.

Peyer, B., Über Zähne von Haramyden, von Triconodonten und von warscheinlich synapsiden Reptilien aus dem Rhät von Hallau Kt. Schaffenhausen, Schweiz, *Schweizerischen Paläont., Abh.,* Schweizerischen Natur. Gesellshaft, Zool. Ser. No. 148, 72 pp., 7 figs., 12 tafeln, 1956.

Piveteau, J., *Traité de paléontologie,* Masson, Paris: Tome I, Protistes, Spongiaires, Coelentérés, Bryozoaires, 782 pp., 142 figs., 10 pls., 1952; Tome II, Brachiopodes, Chétognathes, Annélides, Géphyriens, Mollusques, 790 pp., 102 figs., 24 pls., 1952; Tome III, Onychophores, Arthropodes, Échinodermes, Stomocordés, 1063 pp., 1275 figs., 17 pls., 1953; Tome V, Amphibiens, Reptiles, Oiseaux, 1113 pp., 979 figs., 7 pls., 1955; Tome VII, Primates, Paleontologie Humaine, 670 pp., 639 figs., 8 pls., 1957.

Romer, A. S., *Vertebrate Paleontology,* University of Chicago Press, 2nd. ed., 687 pp., 377 figs. 1945.

Schuchert, C., *Atlas of Paleogeographic Maps,* Wiley, New York, 84 charts, 1955.

Scott, W. B., A History of Land Mammals in the Western Hemisphere, Macmillan, New York, 786 pp., 420 figs., 1937.

Seward, A. C., *Plant Life Through the Ages,* Cambridge University Press, 2nd. ed., 603 pp., 139 figs., 1933.

Shimer, H. W., and R. R. Shrock, *Index Fossils of North America,* Wiley, New York, 837 pp., 303 pls., 1944.

Shrock, R. R., and W. H. Twenhofel, *Principles of Invertebrate Paleontology,* McGraw-Hill, New York, 816 pp., figs. in text, 1953.

Simpson, G. G., Holarctic Mammalian Faunas and Continental Relationships During the Cenozoic, *Bull. Geol. Soc. Am.,* **58,** pp. 613–688, 6 figs., 1947.

———, The Beginning of the Age of Mammals in South America, *Bull. Am. Museum Nat. Hist.,* **91,** 232 pp., 80 figs., 19 pls., 1948.

Smith, J. P., Climatic relations of the Tertiary and Quarternary faunas of the California region, *Proc. Calif. Acad. Sci.,* **9,** 4th ser., pp. 123–173, 1 pl., 1919.

Stirton, R. A., Principles in Correlation and Their Application to Later Cenozoic Holarctic Continental Mammalian Faunas, *Proc. Intern. Geol. Congress,* 18th Session, London, Part XI, pp. 74–84, 1 fig., 1951.

Taliaferro, N. L., Geology of the San Francisco Bay Counties, *Geol. Guidebook of the San Francisco Bay Counties, Bull.* **154,** *Calif. Div. Mines,* pp. 117–150, 8 figs., 1951.

Teichert, C., Stratigraphy of Western Australia, *Bull. Am. Assoc. Petrol. Geologists,* **31,** pp. 1–70, 29 figs., 1947.

Termier, H., and G. Termier, *Histoire géologique de la biosphère,* Masson, Paris, 721 pp., 105 figs., 1952.

Walton, J., An Introduction to the Study of Fossil Plants, Adam and Charles Black, London, 201 pp., 138 figs., 1953.

Watson, D. M. S., On the Skeleton of a Bauriamorph Reptile, *Proc. Zool. Soc.,* London, Part 3, pp. 1163–1205, 27 figs., 1931.

Weber, M., *Die Säugetiere*, Band II, Verlag Gustav Fischer, Jena, 898 pp., 573 figs., 1928.

Weeks, L. G., Paleography of South America, *Bull. Geol. Soc. Am.*, **59,** pp. 249–282, 16 pls., 1 fig., 1948.

Whitehouse, F. W., Early Cambrian Echinoderms Similar to the Larval Stages of Recent Forms, Part 4, *Mem. Queensland Museum*, **12,** Part 1, 64 pp., 1941.

Wilmarth, M. G., The Geologic Time Classification of the United States Geological Survey Compared with other Classifications, Accompanied by the Original Definitions of Era, Period, and Epoch Terms, a Compilation, *U. S. Geol. Survey, Bull. No. 769*, 138 pp., 1 pl., 1925.

Zittel, K. A., *Handbuch der Paleontologie:* Palaeophytologie, München, Leipzig, 958 pp., 433 figs., 1890: Mollusca und Arthropoda, 893 pp., 1109 figs., 1885; Pisces, Amphibia, Reptilia, Aves, 900 pp., 719 figs., 1887–1890; Mammalia, 799 pp., 590 figs., 1891–1893.

PLEISTOCENE

Anderson, C., The Fossil Mammals of Australia, Presidential address, *Proc. Linnean Soc. N. S. Wales*, **58,** pp. IX–XXV, 1933.

Azzaroli, A., The Deer of the Weybourn Crag and Forest Bed of Norfolk, *Bull. Brit. Museum* (Natural History) *Geol.*, **2,** No. 1, 96 pp., 50 figs., 1953.

Cooke, H. B. S., Mammals, Ape-Men and Stone Age Men in Southern Africa, Presidential address, 1951, *S. Afr. Archaeol. Bull.*, No. **26, 7,** pp. 59–69, 3 figs., 1952.

Emiliani, C., Pleistocene Temperatures, *J. Geol.*, **63,** pp. 538–578, 15 figs., 30 tables, 1955.

Ewing, M., and W. L. Donn, A theory of the Ice Ages, *Science*, **123,** No. 3207, pp. 1061–1066, 2 figs., 1956.

Flint, R. F., *Glacial and Pleistocene Geology*, Wiley, New York, 553 pp., figs. in text, 5 pls., 1957.

Hescheler, K., and E. Kuhn, Die Tierwelt der prähistorischen Siedelungen der Schweiz, in Otto Tschumi, *Urgeschichte der Schweiz*, Vol. I, pp. 121–368, figs. 38–146, 1949.

Lamberton, C., Contribution à la connaissance de la faune subfossile de Madagascar. *Mem. Acad. Malgache*, Notes IV–VIII, Lémuriens et Cryptoproctes, Part 27, 203 pp., 1939; Notes IX, Oreille osseuse des Lemurines, 132 pp., 10 pls., 1941.

Movius, H. L., Early Man and Pleistocene Stratigraphy in Southern and Eastern Asia, *Papers of the Peabody Museum Am. Arch. and Eth.*, Harvard University, Vol. 19, No. 3, 125 pp., 47 figs., 1944.

Osborn, H. F., *The Age of Mammals in Europe, Asia and North America*, Macmillan, New York, 635 pp., 220 figs., 1910.

Savage, D. E., Late Cenozoic Vertebrates of the San Francisco Bay Region, *University of Calif. Bull. Publ. Dept. Geol. Sci.*, Vol. 28, pp. 215–314, 51 figs., 1951.

Scott, W. B., *A History of Land Mammals in the Western Hemisphere*, Macmillan, New York, 786 pp., 420 figs., 1937.

Simpson, G. G., Possible Causes of Change in Climate and Their Limitations, *Proc. Linnean Soc.*, London, **152,** pp. 190–219, 1940.

Standing, H. F., On Recently Discovered Subfossil Primates from Madagascar, *Trans. Roy. Zool. Soc. London*, Vol. 18, Part 2, 216 pp., 52 figs., 18 pls., 1908.

Stock, C., Rancho La Brea. A Record of Pleistocene Life in California, *Los Angeles Museum Publ. No. 1*, 82 pp., 27 figs., 1930.

Stokes, W. L., Another Look at the Ice Age, *Science*, **122,** No. 3174, pp. 815–821, 1955.

Teilhard de Chardin, P., and C. C. Young, Fossil Mammals from the Late Cenozoic of Northern China, *Palaeont. Sinica*, Ser. C, Vol. 9, 88 pp., 10 pls., 1931.

Young, C. C., On the Artiodactyla from the *Sinanthropus* Site at Chou-k'ou-tien, *Paleont. Sinica,* Ser. C, Vol. 8, 158 pp., 29 pls., 1932.

———, On the Insectivora, Chiroptera, Rodentia and Primates other than *Sinanthropus* from Locality 1 at Chou-k'ou-tien, *Paleont. Sinica,* Ser. C, Vol. 8, 160 pp., 10 pls., 1934.

Zeuner, F. E., *The Pleistocene Period, Its Climate, Chronology and Faunal Succession,* printed for the Ray Society, Bernard Quaritch, London, 322 pp., 76 figs., 1945.

FORAMINIFERA

Cushman, J. A., *Foraminifera, Their Classification and Economic Use,* Cambridge, Harvard University Press, 4th ed., 605 pp., 85 pls., 1948.

Galloway, J. J., *A Manual of Foraminifera,* Principa Press, Bloomington, 483 pp., 42 pls., 1933.

Glaessner, M. F., *Principles of Micropaleontology,* Wiley, New York (reprinted with corrections), 296 pp., 64 figs., 14 pls., 1947.

Jones, D. J., *Introduction to Microfossils,* Harper, New York, 406 pp., figs. in text, 1956.

Kleinpell, R. M., Miocene Stratigraphy of California, Tulsa, *Am. Assoc. Petrol. Geologists,* 450 pp., 22 pls., 13 figs., 18 tables, frontispiece, 1938.

CEPHALOPODS AND STRUCTURES IN AMMONITES

Arkell, W. J., A classification of the Jurassic ammonites, *J. Paleont.,* Vol. 24, pp. 354–364, 2 figs., 1950.

Flower, R. H., and B. Kummel, A classification of the nautiloids, *J. Paleont.,* **24,** pp. 604–616, 1 fig. 1950.

Hyatt, A., Cephalopoda in Zittel-Eastman's *Textbook of Paleontology,* Vol. 1, pp. 502–604, 210 figs., Macmillan, New York, London, 1900.

Kummel, B., A Classification of the Triassic ammonoids, *J. Paleont.,* **26,** pp. 847–853, 3 figs., 1952.

Miller, A. K., Devonian Ammonoids of America, *Geol. Soc. Am., Spec. Paper* **14,** 262 pp., 41 figs., 39 pls., 1938.

Moore, R. C. (editor), Treatise on Invertebrate Paleontology. Part I. Mollusca 4, Cephalopoda, Ammonoidea, *Geol. Soc. Am.,* Univ. Kansas Press, Part L, pp. L1–L490, 558 figs., 1957.

Wright, C. W., A Classification of the Cretaceous Ammonites, *J. Paleont.,* **26,** pp. 213–222, 2 figs., 1952.

HIGHLIGHTS IN THE EVOLUTION OF PLANTS

Arnold, C. A., *An Introduction to Paleobotany,* McGraw-Hill, New York, 433 pp., 187 figs., 1947.

Axelrod, D. I., A Theory of Angiosperm Evolution, *Evolution,* **6,** pp., 29–60, 14 figs., 1952.

Brown, R. W., Palmlike Plants from the Dolores Formation (Triassic) Southwestern Colorado. *Profess. Paper U. S. Geol. Survey,* **274-H,** pp., 205–209, 1 fig., 2 pls., 1956.

Chaney, R. W., Paleobotany, *Encyclopedia Britannica,* 1948.

———, A Revision of Fossil *Sequoia* and *Taxodium* in Western North America Based on the Recent Discovery of *Metasequoia. Trans. Am. Phil. Soc.,* **40,** pp. 171–262, 12 pls., 1951.

Darrah, W. C., *Textbook of Paleobotany,* D. Appleton-Century, 441 pp., 180 figs., 1939.

Elias, M. K., Tertiary Prairie Grasses and other Herbs from the High Plains, *Spec. Paper Geol. Soc. Am.*, **41,** 76 pp., 16 pls., 1942.

Emburger, L., *Les plantes fossiles dans leurs rapports avec les végétaux vivants,* Masson, Paris, 489 pp., 457 figs., 1944.

Erdtman, G., *An Introduction to Pollen Analysis: Angiosperms*, Cronica Botanica, Waltham, 539 pp., 261 figs., 1952.

Florin, R., Evolution in cordaites and conifers, *Acta Horti Bergiani,* Band 15, No. 11, pp. 285–388, 70 figs., 1 pl., 1951.

Gothan, W., and H. Weyland, *Lehrbuch der Paläobotanik,* Akademie-Verlag, Berlin, 535 pp., 419 Abb., 1954.

Lang, W. H., and I. C. Cookson, On A Flora Including Vascular Plants Associated with *Monograptus* in Rocks of Silurian Age from Victoria, Australia, *Phil. Trans. Roy. Soc. London, Ser. B*, **224,** pp. 421–449, 4 pls., 1935.

La Motte, R. S., Catalogue of the Cenozoic Plants of North America Through 1950, *Mem. Geol. Soc. Am.*, **51,** 381 pp., 1952.

LeClercq, S., Evidence of Vascular Plants in the Cambrian, *Evolution,* Vol. 10, pp. 109–114, 1956.

Plumstead, E. P., Bisexual fructifications borne on *Glossopteris* leaves from South Africa, *Palaeontographica,* Beiträge zur Naturgeschichte der Vorzeit, Stuttgart, Band 100, Abt. B., 25 pp., 5 figs., 14 pls., 1956.

———, On *Ottokaria,* the fructifications of *Gangamopteris, Trans. Proc. Geol. Soc. S. Africa,* **59,** pp. 211–236, 6 figs., 7 pls., 1956.

Seward, A. C., *Plant Life through the Ages,* Cambridge University Press, 2nd. ed., 601 pp., 139 figs., 1933.

Tyler, S. A., and E. S. Barghoorn, Occurrence of Structurally Preserved Plants in Pre-Cambrian Rocks of the Canadian Shield, *Science,* **119,** pp. 606–608, 4 figs., 1954.

Walton, J., *An Introduction to the Study of Fossil Plants,* Adam and Charles Black, London, 201 pp., 138 figs., 1953.

HIGHLIGHTS IN THE EVOLUTION OF VERTEBRATES

Colbert, E. H., *Evolution of the Vertebrates,* Wiley, New York, 479 pp., 122 figs., 1955.

Gregory, W. K., *Evolution Emerging,* Macmillan, New York: Vol. 1, 736 pp.; Vol. 2, 1013 pp., with figures and plates, 1951.

Kühne, W. G., *The Liassic Therapsid Oligokyphus,* British Museum (Natural History), 149 pp., 66 figs., 12 pls., 1956.

Patterson, B., Early Cretaceous Mammals and the Evolution of Mammalian Molar Teeth, *Fieldiana; Geol.*, Chicago Nat. Hist. Museum, **13,** 105 pp., 17 figs., 1956.

Peyer, B., *Geschichte der Tierwelt,* Büchergilde Gutenberg, Zürich, 288 pp., 184 abb., 1950.

Romer, A. S., *Vertebrate Paleontology,* University of Chicago Press, 2nd ed., 687 pp., 377 figs., 1945.

Watson, D. M. S., *Paleontology and Modern Biology,* Yale University Press, New Haven, 216 pp., 77 figs., 1951.

DINOSAURS

Brown, B., and E. M. Schlaikjer, The Structure and Relationships of *Protoceratops. Ann. N. Y. Acad. Sci.*, **40,** pp. 133–266, 33 figs., 13 pls., 1940.

———, A Study of the Troödont Dinosaurs with the Descriptions of a New Genus and Four New Species, *Bull. Am. Museum Nat. Hist.*, **82,** pp. 121–249, 12 pls., 1943.

Colbert, E. H., *The Dinosaur Book,* Handbook No. 14, American Museum Natural History, New York, 156 pp., figs. in text, 1945.

Huene, F. von, Die fossile Reptilordnung Saurischia, ihre Entwicklung und Geschichte, *Mono. Geol. Pal.,* 1st Ser. Vol. 4, 361 pp., 36 figs., 56 pls., 1932.

Lull, R. S., A Revision of the Ceratopsia or Horned Dinosaurs, *Mem. Peabody Museum Nat. Hist.,* **3,** Part 3, 135 pp., 42 figs., 17 pls., 1933.

Matthew, W. D., *Dinosaurs,* Handbook, No. 5, American Museum Natural History, pp. 162, 46 figs., 1915.

Osborn, H. F., Skeletal Adaptations of *Ornitholestes, Struthiomimus, Tyrannosaurus. Bull. Am. Museum Nat. Hist.,* **35,** pp. 733–771, 21 figs., 4 pls., 1917.

Parkinson, J., The Dinosaur in East Africa, H. F. and G. Witherby, London, 188 pp., 8 figs., 12 pls., frontispiece, 1930.

Piveteau, J., *Traitê de Palêontologie,* Masson, Paris, Tome V, Dinosauriéns, par A. F. de Lapparet et R. Lavocat, pp. 785–962, 156 figs., 1955.

Romer, A. S., *Vertebrate Paleontology,* University of Chicago Press, 2nd ed., Chapter 12, pp. 229–256, 598–600, figs., 187–205, and bibliography, pp. 643–645, 1945.

Swinton, W. E., *The Dinosaurs:* A Short Story of a Great Group of Extinct Reptiles, Thomas Murby, London, pp. 233, 20 figs., 25 pls., 1934.

BIRDS

Buick, T. L., *The Mystery of the Moa,* Thomas Avery and Sons, New Plymouth, New Zealand, 357 pp., 27 pls., 1931.

De Beer, G. R., *Archaeopteryx lithographica: A Study Based upon the British Museum Specimen,* British Museum (Natural History), 68 pp., 9 figs., 15 pls., 1954.

Gregory, J. T., The Jaws of the Cretaceous Toothed Birds, *Ichthyornis and Hesperornis, Condor,* **54,** pp. 73–88, 9 figs., 1952.

Heillman, G., *The Origin of Birds,* J. F. and G. Witherby, London, 208 pp., 142 figs., 2 pls., 1926.

Howard, H., Fossil Birds, with Special Reference to the Birds of Rancho La Brea, *Los Angeles Co. Museum Paleont. Publ. No. 10,* 40 pp., 21 figs., 1955.

———, Fossil Evidence of Avian Evolution, Ibis, Vol. 92, 21 pp., 5 charts, 1950.

Howard, H., A Gigantic "Toothed" Marine Bird from the Miocene of California, *Santa Barbara Museum Nat. Hist., Dept. Geol. Bull.* **1,** 23 pp., 8 figs., 1957.

Lambrecht, K., *Handbuch der Palaeornithologie,* Brontraeger, Berlin, 1024 pp., 209 figs., 4 pls., 1933.

Mcdowell, S., The Bony Palate of Birds, Part I, The Palaeognathae, Auk, Vol. 65, pp. 520–549, 6 figs., 1948.

Miller, A. H., An Avifauna from the Lower Miocene of South Dakota, *Univ. Calif. Publs. Geol. Sci.,* **27,** pp. 85–100, 8 figs., 1944.

Miller, L. H., Avian Remains from the Miocene of Lompoc, California, *Carnegie Inst. Wash. Publ.* **349,** pp. 107–117, 1 fig., 9 pls., 1925.

Miller, L. H., and H. Howard, The Flightless Bird *Mancalla, Carnegie Inst. Wash. Publ.* **584,** pp. 201–228, 6 pls., 1949.

Simpson, G. G., Fossil Penguins, *Bull. Am. Museum Nat. Hist.,* **87,** 99 pp., 33 figs., 1946.

Wetmore, A., A Check-List of the Fossil and Prehistoric Birds of North America and the West Indies, *Smithsonian Misc. Coll.,* **131,** No. 5, 105 pp., 1956.

EVOLUTION OF THE HORSES

Camp, C. L., and N. Smith, Phylogeny and Function of the Digital Ligaments of the Horse, *Mem. Univ. Calif.,* **13,** No. 2, pp. 69–124, 41 figs., 11 pls., 1942.

Edinger, T., Evolution of the Horse Brain, *Mem. Geol. Soc. Am.* **25,** 177 pp., 24 figs., 4 pls., 1948.

Matthew, W. D., The Evolution of the Horse, a Record and Its Interpretation, *Quart. Rev. Biol.,* **1,** pp. 139–185, 27 figs., 1926.

Osborn, H. F., Equidae of the Oligocene, Miocene and Pliocene of North America, Iconographic Type Revision, *Mem. Am. Museum Nat. Hist.,* **2,** 330 pp., 173 figs., 54 pls., 1918.

Simpson, G. G., *Horses,* Oxford University Press, New York, 247 pp., 34 figs., 32 pls., 1951.

Stirton, R. A., Phylogeny of North American Equidae, Univ. Calif. *Publs. Geol. Sci.,* **25,** pp. 165–198, 52 figs., 1940.

LEMURS, TARSIERS, MONKEYS AND APES

Gregory, W. K., Origin and Evolution of the Human Dentition, *J. Dental Research,* Vol. II, Part I, pp. 89–182, 42 figs., 4 pls., 1920; Part II, pp. 215–282, 33 figs., 3 pls., 1920; Part III, pp. 357–426, 75 figs., 6 pls., 1920; Part IV, pp. 607–717, 84 figs., 1920; Part V, pp. 87–228, 53 figs., 1 pl., 1921.

Hill, W. C. O., *Primates, Comparative Anatomy and Taxonomy,* Edinburgh University Press. Vol. I, Strepsirhini, 798 pp., 199 figs., 34 pls., 1953; Vol. II, Haplorhini: Tarsioidea, 347 pp., 49 figs., 14 pls., 1955; Vol. III, Pithecoidea: Platyrrhini, 354 pp. 102 figs., 27 pls., 1957.

Hooton, E., *Man's Poor Relations,* Doubleday, Doran, New York, 412 pp., 11 figs., pls. in text, 1942.

Le Gros Clark, W. E., *History of Primates, an Introduction to the Study of Man,* 3rd ed., British Museum (Natural History), London, 117 pp., 40 figs., 1953.

Le Gros Clark, W. E., and L. S. B. Leakey, *Fossil Mammals of Africa,* No. 1. The Miocene Hominoidea of East Africa, British Museum (Natural History), 117 pp., 28 figs., 9 pls., 1957.

Piveteau, J., *Traité de paléontologie.* Masson, Paris, Tome VII, Vers la forme humaine, le problème biologique de l'homme, les époques l'intelligence, 670 pp., 639 figs., 8 pls., 1957.

Simpson, G. G., Studies on the Earliest Primates, *Bull. Am. Museum Nat. Hist.,* **77,** pp. 185–212, 8 figs., 1940.

Stirton, R. A., Ceboid Monkeys from the Miocene of Colombia, *Univ. Calif. Publs. Geol. Sci.,* **28,** pp. 315–356, 2 figs., 8 pls., 1951.

Williams, E. E., and K. F. Koopman, West Indian Fossil Monkeys, *Am. Museum Novitates, No. 1546,* 16 pp., 4 figs., 1952.

PREHISTORIC MAN

Arambourg, G., Deuxième supplément à la notice sur les travaux scientifiques, résumé analytique des publications, II, paléontologie humaine, 1, L'Atlanthropus de Ternifine, Paris, pp. 11–13, figs. 7–14, 1955.

Buole, M., *Les Hommes Fossiles,* 4th ed., revised by H. V. Hallois, Paris, 587 pp., 294 figs., 1952.

Brodrick, A. H., *Early Man, a Story of Human Origins,* Hutchison's Scientific and Technical Publications, London, 288 pp., 21 figs., 1948.

Leakey, L. S. B., *Stone Age Africa,* Oxford University Press, London, 218 pp., 13 pls., 1936.

Le Gros Clark, W. E., *History of the Primates, an Introduction to the Study of Man,* 3rd ed., British Museum (Natural History), London, 117 pp., 40 figs., 1953.

Moore, R., *Man, Time, and Fossils,* Knopf, New York, 411 pp., figs. in text, 32 pls., 1953.

Robinson, J. T., The Dentition of the Australopithecinae, *Transvaal Museum Mem. No.* 9, 179 pp., 508 figs., 1956.

Romer, A. S., *Man and the Vertebrates,* 3rd ed., University of Chicago Press, 405 pp., figs. and pls., 1941.

Sellards, E. H., *Early Man in America,* University of Texas Press, Austin, 211 pp., 47 figs., and pls., 1952.

Von Koenigswald, G. H. R., Begegnungen mit dem Vormenschen, Eugen Diederichs Verlag, 230 pp., 41 Abb., 20 Tafel, 1955.

Weiner, J. S., *The Piltdown Forgery,* Oxford University Press, London, New York, 214 pp., 9 figs., 1955.

CORRELATION CHARTS FOR NORTH AMERICA

Cooke, C. W. (chairman of committee), Correlation of the Cenozoic Formations of the Atlantic and the Caribbean Regions. *Bull. Geol. Soc. Am.,* **54,** pp. 1713–1722, pl. 1, 1943.

Cooper, G. A. (chairman of committee), Correlation of the Devonian Sedimentary Formations of North America, *Bull. Geol. Soc. Am.,* **53,** pp. 1729–1794, 1 fig., 1 pl., 1942.

Howell, B. F. (chairman of committee), Correlation of Cambrian Formations of North America, *Bull. Geol. Soc. Am.,* **55,** pp. 993–1004, 1 pl., 1944.

Imlay, R. W., and J. B. Reeside, Jr., Correlation of Cretaceous Formations of Greenland and Alaska, *Bull. Geol. Soc. Am.,* **65,** pp. 223–246, 2 figs., 1 pl., 1954.

Moore, R. C. (chairman of committee), Correlation of Pennsylvanian Formations in North America, *Bull. Geol. Soc. Am.,* **55,** pp. 657–706, 1 pl., 1944.

Reeside, J. B. (chairman of committee), Correlation of the Triassic Formations of North America Exclusive of Canada, *Bull. Geol. Soc. Am.,* **68,** pp. 1451–1514, 1 pl., 1957.

Stephenson, L. W. (chairman of committee), Correlation of Outcropping Cretaceous Formations of Atlantic and Gulf Coastal Plain and Trans-Pecos Texas, *Bull. Geol. Soc. Am.,* **53,** pp. 435–448, 1 pl., 1942.

Swartz, C. K. (chairman of committee), Correlation of Silurian Formations of North America, *Bull. Geol. Soc. Am.,* **53,** pp. 533–538, 1 pl., 1942.

Twenhofel, W. H. (chairman of committee), Correlation of Ordovician Formations of North America, *Bull. Geol. Soc. Am.,* **65,** pp. 247–298, 2 figs., 1 pl., 1954.

Weaver, C. E. (chairman of committee), Correlation of Marine Cenozoic Formations of Western North America, *Bull. Geol. Soc. Am.,* **55,** pp. 569–598, 1 pl., 1944.

Wood, H. E., 2nd. (chairman of committee), Nomenclature and Correlation of the North American Continental Tertiary, *Bull. Geol. Soc. Am.,* **52,** 48 pp., 1 pl., 1941.

Citations to Bibliographies

FOSSIL PLANTS AND ANIMALS

Nickles, J. M., et al., Bibliographies of North American Geology, *U. S. Geol. Survey Bull.* Covering the years since 1732; published since 1886.

Nickles, J. M., M. Siegrist, et al., *Bibliographies and Indices of Geology Exclusive of North America*, published by Geol. Soc. Am. since 1934, 21 vols.

FOSSIL PLANTS IN NORTH AMERICA

Lesquereux, L., Catalogue of the Cretaceous and Tertiary Plants of North America with Reference to the Descriptions, *U. S. Geol., Geog. Survey Terr., Ann. Rept.* 1876, pp. 487–520, 1878.

Knowlton, F. H., A Catalogue of the Cretaceous and Tertiary Plants of North America. *U. S. Geol. Survey Bull., No. 152*, 247 pp., 1898.

———, Catalogue of the Mesozoic and Cenozoic Plants of North America, *U. S. Geol. Survey Bull., No. 696*, 815 pp., correlation chart, 1919.

La Motte, R. S., Supplement to Catalogue of Mesozoic and Cenozoic Plants of North America, 1919–1937, *U. S. Geol. Survey Bull., No. 924*, 330 pp., 1944.

———, Catalogue of the Cenozoic Plants of North America through 1950, *Mem. Geol. Soc. Am.*, **51**, 381 pp., 1952.

FOSSIL VERTEBRATES

Camp, Charles L., and V. L. VanderHoof, Bibliography of Fossil Vertebrates, 1928–1933, *Spec. Paper Geol. Soc. Am.*, **27**, 503 pp., 1940.

Camp, Charles L., D. N. Taylor, and S. W. Welles, Bibliography of Fossil Vertebrates, 1934–1938, *Spec. Paper Geol. Soc. Am.*, **42**, 663 pp., 1942.

Camp, Charles L., S. P. Welles, and M. Green, Bibliography of Fossil Vertebrates, 1939–1943, *Mem. Geol. Soc. Am.*, **37**, 371 pp., 1949.

———, Bibliography of Fossil Vertebrates, 1944–1948, *Mem. Geol. Soc. Am.*, **57**, 465 pp., 1953.

Hay, Oliver P., Bibliography and catalogue of the fossil Vertebrata of North America, *U. S. Geol. Survey Bull., No. 179*, 868 pp., 1902.

———, Second Bibliography and Catalogue of the Fossil Vertebrata of North America, *Carnegie Inst. Washington Publ. No. 390*, Vol. 1, 916 pp., 1929.

———, Second Bibliography and Catalogue of the Fossil Vertebrata of North America, *Carnegie Inst. Washington Publ. No. 390*, Vol. 2, 1074 pp., 1930.

Romer, Alfred S., Vertebrate Paleontology, 2nd ed., University of Chicago Press, pp. 628–661, 1945.

Index